INTERNET GIS

INTERNET GIS
Distributed Geographic Information
Services for the Internet and
Wireless Networks

Zhong-Ren Peng

Ming-Hsiang Tsou

WILEY

JOHN WILEY & SONS, INC.

Library of Congress Cataloging-in-Publication Data:

Peng, Zhong-Ren.
    Internet GIS: distributed geographic information services for the internet and wireless networks / Zhong-Ren Peng, Ming-Hsiang Tsou
        p. cm.
    Includes bibliographical references (p.).
    ISBN 0-471-35923-8 (cloth)
    1. Geographic information systems. 2. Wireless communication systems. 3. Internet. I. Tsou, Ming-Hsiang. II. Title.
G70.212.P47 2003
025.06′91—dc21

2002027427

To Ching, Lucinda, and Naomi

—Zhong-Ren Peng

To Chun-Yi (Eva)

—Ming-Hsiang Tsou

CONTENTS

12 INTERNET GIS APPLICATIONS IN INTELLIGENT TRANSPORTATION SYSTEMS

FOREWORD

New technologies have always been a key driving force in geographic information science, as they have in science and society generally. On the one hand, technologies may allow us to do things more quickly and more cheaply or to see things we have not been able to see before: "GIS technology is to geographical analysis what the microscope, the telescope and computers have been to other sciences" (Abler, 1987, p. 332). On the other hand, we recognize, despite the obvious importance of tools, that the primary drivers of science should be human curiosity and the need to solve problems of significance to society and that "to someone with a hammer everything eventually begins to look like a nail."

This book is about a suite of new technologies that have emerged since the popularization of the Internet in the early 1990s and is variously aimed at making Geographic Information Systems (GIS) more mobile, powerful, and flexible and better able to share and communicate geographic knowledge. They include search engines, digital libraries, open and interoperable systems, personal digital assistants, wireless communication, new modes of interaction, and many more. Early GIS was impeded by the lack of good algorithms and powerful computers to analyze data and by the difficulties associated with digitizing, which were such that 80% or more of a project's resources were often consumed in the task of converting paper records to digital databases. Today, a new source of frustration has emerged. Despite all of the on-line digital data now available and the vastly increased power of GIS, it is still common to spend 80% or more of a project's resources on searching, discovering, assessing, retrieving, and reformatting data. Internet GIS is about removing that very significant impediment to effective GIS use.

But the potential of Internet GIS goes much further than that essentially data-centric view would imply. Good data are essential, but the other components of a GIS project are essential also and may be more valuable. Human resources are an often-overlooked part of the project, and the ability of humans to direct and control GIS projects depends on our ability as a society to educate and train and on the complexity of the user interfaces that provide access to systems. Software, algorithms, and methods are also vital and if assessed per bit will almost certainly have greater value than data bits. Methods and tools are generic and likely to be of interest to a much larger collection of users than a given item of data. Hence effective sharing of methods and tools has potentially much higher return than sharing of data—yet historically our interest in the GIS community has been almost entirely on the sharing of data. Many of the new technologies discussed in this book are about sharing of methods and tools, including distributed geographic information services and component-based software architectures. The book represents a welcome turn from a data-centric to a service-centric perspective in Internet GIS.

When Galileo turned his telescope to the heavens, there were no organizations devoted to international standards, and there was almost certainly no thought in his mind that the science of astronomy would be aided if standards could be established regarding telescope construction or the sharing of information derived from telescopes. Today, it seems that no new technology can emerge, and particularly no new information technology, without extensive standards-making activities. Wireless GIS, for example, requires coordination between numerous components, each of which must function together for wireless GIS to be possible. Standards are now developed and promulgated well in advance of the release of new technologies and clearly play an important part in the development of national and multinational industrial strategies (compare the European success with the Global Systems for Mobile Communications (GSM) standard to the complex situation obtained in the United States). Thus much of this book is about standards and their role in making Internet GIS possible and in promoting new applications.

In writing this book, Zhong-Ren Peng and Ming-Hsiang Tsou's objectives are to introduce the fundamentals of Internet GIS technologies, to provide an applicable framework for Internet GIS applications, to illustrate available software packages, and to demonstrate real-world applications. It is a welcome addition to a literature that already includes books by Plewe (1997) and Green and Bossomaier (2001) and increasing coverage of Internet GIS in standard GIS textbooks. This new book is by far the most comprehensive treatment of a topic that grows increasingly complex every year. It begins to ask the kinds of questions that academics should be asking in this technology-dominated field: What are the pros and cons of each new technology and each new set of standards; what new kinds of scientific insights will be possible when these technologies are fully developed and available and what important societal problems will be addressable; and what will be the broader impacts of these technologies on society? At the same time it provides a very useful guide and reference to help the reader in navigating the field and to help the user get the best out of what currently exists and what is likely to exist in the next few years.

REFERENCES

Abler, R. F. (1987). The National Science Foundation National Center for Geographic Information and Analysis. *International Journal of Geographic Information Systems,* 1(4), pp. 303–326.

Green, D. G., and Bossomaier, T. (2001). *Online GIS and Spatial Metadata.* New York: Taylor and Francis.

Plewe, B. (1997). *GIS Online: Information Retrieval, Mapping, and the Internet.* Santa Fe, New Mexico: OnWord Press.

MICHAEL F. GOODCHILD

University of California
Santa Barbara, California

PREFACE

During the writing of this book, we have been pondering its title. Several names have been considered, including "Distributed GIS," "Internet GIS and Mobile GIS," "Web GIS," "Network-Based GIServices," and so on. The numerous choices of an appropriate book title essentially reflect the wide range of technologies and applications Internet GIS covers. The name "Distributed GIS" reflects the software architecture, that is, Internet GIS software and data are distributed via different computers located in different physical locations across the Internet. The name "Internet GIS and Mobile GIS" reflects the two physical elements of the Internet networks, the wired Internet and the wireless mobile networks. It indicates that the user can access Internet GIS functions and data through the wired Internet and wireless devices such as Personal Digital Assistants (PDAs) and mobile phones. The name "Web GIS" focuses on the use of the most common application on the Internet, the World Wide Web, as the common user interface to the Internet GIS programs. The name "Network-Based GIServices" focuses on the concept of geographic information services. That is, the main function of Internet GIS is to provide a certain service to consumers, such as travel information services, land use information services, and environmental information services.

The indecisiveness of the book title also reflects the current development state of Internet GIS. This is similar to the early development of Geographic Information Systems (GIS), which were represented in various names, such as spatial databases, land information systems, and automatic mapping facilities. Maybe the best title of this book should be determined in 10 years, when the GIS community will conclude a new vocabulary for Internet GIS.

Regardless of how irresolute this may be, a book cannot be without a name. We finally chose *Internet GIS* as the book title, reflecting our view that Internet GIS is centered on the use of Internet technology and is accessed through the use of the Internet, wired and wireless. Internet GIS is quite often used in the literature and is easy to remember. We also use the subtitle *Distributed Geographic Information Services for the Internet and Wireless Networks* to reflect the nature of distributed-computing systems and focus on the service-oriented concept.

What then is Internet GIS? This book defines Internet GIS as network-based geographic information services that utilize both wired and wireless Internet to access geographic information, spatial analytical tools, and GIS Web services.

Internet GIS, which focuses on the Internet technology and utilizes a distributed architecture framework, symbolizes an invisible revolution of GIS— from closed, centralized GIS to open, distributed GIService. It not only signifies the advancement of technologies but also implies a paradigm shift, from collecting data and finishing a GIS project on a desktop to considering universal accessibility to remotely distributed data and analysis functions. Take a simple example of finding a hotel close to a particular national park. In the traditional GIS environment, one needs to find data for the national park system, road networks, and separate data for hotel locations; one also needs to purchase a complete GIS software program that is capable of conducting a network analysis with the shortest path-finding function, even though the user only needs to conduct this particular analysis and nothing else. Furthermore, the next time the user wants to plan other trips, the old data may not be valid as road network and hotel information may have already changed. He or she has to obtain the data again. In an Internet GIS environment, however, the user needs only to search for Web sites that provide this kind of services. He or she would care less about where the data are and what other functions the software program provides as long as the Web site does the job that he or she wants. What the user needs is the ability to access to the service, including the data and analysis tools. He or she pays for the services at the time the services are used. Since the Internet services link the data with the data sources, the data are always up to date. Internet GIS would allow users to be more productive in applying Internet GIS technologies in actual applications rather than spending valuable time in operating the software and collecting the data.

From the service provider's point of view, the adoption of Internet GIS can also foster paradigm shift in GIS program design and development and encourage the specialization of data providers and service providers. The traditional desktop GIS model encouraged the development of proprietary and one-size-fits-all GIS software programs. Each GIS vendor strives to provide the most comprehensive GIS software to meet the diverse user needs, despite the fact that only a small portion of the functions provided are actually used by most users. On the other hand, some special functions geographic information scientists need may not be available due to the small amount of users. By adopting the Internet GIS design concept, the GIS program developers would focus on standards-based components with specialized functions that can interface with other components from other service providers. This would foster the emergence of many small Internet GIS service providers that can offer their unique services.

From the data provider point of view, the adoption of Internet GIS provides an efficient means of advertising, publishing, and distributing data through the Internet to a wide variety of users. The distributed nature of Internet GIS allows data providers to keep the data directly at the data sources so that the data are always up to date. At the same time it allows any Internet GIS service provider to remotely access the database over the Internet without downloading.

These shifts in thinking in the access and use of geographic information will produce many more services with brand-new functions and would allow GIS users, GIS professionals, and the general public many opportunities to explore, enjoy, and appreciate the spatial world in which we live. As Thomas S. Kuhn indicated in his famous book, *The Structure of Scientific Revolutions* (1962, Chicago: The University of Chicago Press, page 122), "The scientist who embraces a new paradigm is like the man wearing inverting lenses. Confronting the same constellation of objects as before and knowing that he does so, he nevertheless finds them transformed through and through in many of their details." Internet GIS is like the inverting lenses for geographers, urban planners, and GIS professionals to transform traditional GIS projects and solutions to new information services and strategies.

The goal of this book is to introduce the concepts, theories, technologies, and applications of Internet GIS, to illustrate available software packages, to provide an applicable framework for Internet GIS applications, and to demonstrate real-world applications. This book is intended to provide a systematic introduction of Internet GIS technologies from a vendor-neutral perspective. It covers the fundamental theories of Internet GIS, state-of-the-art Internet GIS technologies, various on-line applications, as well as the emerging future trends. This book also includes many Internet GIS showcase boxes to illustrate real-life applications and research projects of Internet GIS. The purpose of Internet GIS showcases is to emphasize that the real value of Internet GIS lies not in the technologies but in the actual services provided by the GIS professionals and applications produced by users. Information content providers and users are the key to the success of Internet GIS. The almost ubiquitous Internet and the realized value of geographic information services on the Internet will attract more people to participate and ensure the sustainable development of Internet GIS to cope with rapidly changing needs and technologies in the long run.

The Internet GIS theories, technologies, and applications are multidisciplinary and complex. We have tried our best in this book to cover many different aspects of Internet GIS, including basic conceptual models of Internet GIS and its underlying distributed-component frameworks and various GIS architectures and working mechanisms of different Internet GIS models. The book is organized as follows.

Chapter 1 is an introduction to the concept of Internet GIS and distributed GIServices. It introduces the basic concept of the Internet, the Web, and their impact on GIS as well as the brief history of Internet GIS.

Chapters 2 and 3 present the technological foundations of Internet GIS. Chapter 2 introduces the computer networking basics, the hardware part, on which Internet GIS runs. It starts with two network communication models, then moves to the local-area networks, wide-area networks, and the Internet. Chapter 3 covers the software architecture on which Internet GIS relies. It discusses two important frameworks underpinning Internet GIS: the client/ server computing model and the distributed-component frameworks.

Chapters 4 and 5 cover the evolution of the Internet GIS programs, from Web mapping to distributed GIServices. Chapter 4 talks about the brief history of Web mapping, starting from embedding static map images in an HTML document to making static HTML-based maps on the Web with limited map-rendering functions to more interactive Web mapping based on Java applets and ActiveX controls. Chapter 5 discusses the architecture, components, and implementations of distributed geographic information services, a more advanced topic of Internet GIS technology.

Chapter 6 presents three standard efforts that will have fundamental impacts on the development of Internet GIS, that is, the standard and specification effort by the Open GIS Consortium (OGC), the standard effort by the International Standards Organization (ISO), particularly the ISO reference model and geospatial data model, and the development of geospatial metadata standards.

Chapter 7 introduces the OGC-led new standard derived from the Extensible Markup Language (XML), the Geography Markup Language (GML), to encode and transport geospatial data over the Internet. This is an important standard to achieve interoperability of data access over the heterogeneous Internet environment. GML is winning wide support from a variety of GIS vendors and is expected to play an important role in the future Internet GIS development.

Chapter 8 introduces four commonly used commercial Internet GIS programs, including ArcIMS from ESRI, GeoMedia WebMap Professional from Intergraph, MapXtreme from MapInfo, and MapGuide from Autodesk. This chapter presents and compares the architecture and major components of each program.

Chapter 9 introduces the framework and applications of mobile GIS, that is, the access to and use of GIS through mobile and wireless devices such as personal digital devices and mobile phones.

Chapter 10 discusses the quality of service and security issues, including what affects performance of Internet GIS and how to enhance Internet GIS performance, what are security concerns, and how to improve security when setting up Internet GIS sites.

Chapters 11–13 are specific applications of Internet GIS. Chapter 11 focuses on data discovery, access, sharing, and dissemination of geospatial data. Chapter 12 focuses on Internet GIS applications in transportation, particularly in intelligent transportation systems. Chapter 13 presents several examples of how Internet GIS has been used in planning and resource management.

The last chapter, Chapter 14, makes a few conjectures about the possible impact of Internet GIS on the GIS industry, geographers, and the general public. We also speculate on some visions of the development of future internet GIS.

As you can see, this is big book that covers a lot of material. Depending on your background, you may be more comfortable with some chapters than with others. One suggestion is to focus only on what really interests you. For example, if you are a software developer and programmer, Chapters 2–6 can help you to understand the programming environment, tools, and specifications that are associated with the development of Internet GIS. If you are a GIS project manager, Chapter 8 introduces currently available software packages, and Chapters 10–13 can give you an overall view of different types of Internet GIS applications. If you are a GIS educator or student, Chapters 3–7 cover the basic concept of Internet Map Server (IMS) and the GIServices mechanisms. In general, this book could be used as a reference book for GIS professionals, as an advanced GIS textbook for GIS educators and graduate students, or as a strategy book for GIS managers.

One unique characteristic about Internet GIS is the rapid change of its technologies. Keep in mind that Internet GIS technologies change VERY, VERY QUICKLY. New tools and new standards of the Internet technologies are released every month. It seemed impossible for us to keep up with every state-of-the-art technology when we wrote this book. With this book we tried to provide a comprehensive and rigorous discussion of the theory, architecture, and implementation technology, with some examples of Internet GIS applications. But as information technology evolves, new architectures and technologies will emerge. Therefore, the best strategy to keep up with the new development of Internet GIS is to create your own "information networks" and keep your communication channel open. For example, access GIS-related Web sites frequently, browse related Internet computing and GIS magazines, attend GIS conferences, and talk to your GIS colleagues and staff. The Web resources given at the end of each chapter may be a good resource for you to update related information and learn more about Internet GIS. We have also created a companion Web site to update you on new developments in the field of Internet GIS: http://map.sdsu.edu/gisbook.

ACKNOWLEDGMENTS

This book was initiated by John Wiley editor Philip Manor in early 1999, who talked Zhong-Ren into writing it. With the retirement of Phil, editor Jim Harper saw the book to completion. Jim has been very instrumental in enriching the contents of the book and has been very patient with us. We owe him a great deal. Much of the contents of the book have been taught in the advanced GIS class offered at the Department of Urban Planning, University of Wisconsin-Milwaukee. The students have made numerous suggestions. Some of the contents were also presented at the workshop "Internet GIS: State of the Art" sponsored by the Urban and Regional Information Systems Association (URISA) and co-taught by Professor Zhong-Ren Peng, Professor Joseph Ferreira, and Mr. Thomas H. Grayson. Both Joe and Tom have made many comments and suggestions on the workshop materials, part of which were incorporated into this book.

This book has benefited from many colleagues who have reviewed the early drafts and book plans, including Professor Kenneth J. Deuker of Portland State University and Professors William Huxhold, Nancy Frank, Sammis White, and Robert Greenstreet of the University of Wisconsin-Milwaukee. Professor Lyna Wiggins of Rutgers University and Professor Richard Klosterman of the University of Akron have also make many direct and indirect comments about this work. Their comments, suggestions, and encouragement have been invaluable in the writing of this book. We are especially grateful to two anonymous reviewers, whose comments have improved the book considerably.

Parts of this book are derived from Tsou's Ph.D. dissertation at University of Colorado at Boulder. We would like to take this opportunity to thank Tsou's Ph.D advisor, Dr. Barbara P. Buttenfield, who offered her valuable comments and guidance throughout the dissertation process. She helped Tsou with academic, financial, and spiritual support, especially in the last year of Tsou's Ph.D. study, when Tsou's first child was born with medical problems. Many thanks also go to Dr. Michael F. Goodchild at the University of California at Santa Barbara, whose comments were very important in revising and improving the dissertation. Thanks go to Dr. Clayton Lewis from the Department of Computer Science at the University of Colorado, who helped Tsou with revising the research focus and building a more feasible research framework. Thanks also go to Dr. Gary L. Gaile and Dr. Rene F. Reitsma, who gave their valuable comments, support, and advice from the geographer's and spatial scientist's perspectives. And to all friends and colleagues in Boulder and San Diego, thank you for your support and help through the writing of this book.

During the writing process, we had substantial assistance from many graduate students at the University of Wisconsin-Milwaukee. Mr. Jason Valerius helped proofread the whole document. Ms. Jacqueline Eastwood helped with the application cases in Chapter 13. Mr. Ruihong Huang helped with Chapter 12 and some of the "key concepts" in the book. Ms. Chuanrong Zhang helped with Chapter 7. Ms. Qi Xu and Mr. Young S. Lee painstakingly drew all the graphics. Mr. Xinyang Zhang helped with some showcase applications and many other tedious tasks, and Ms. Yanlin Weng helped with the indexing.

Many thanks go to the friends and organizations who worked directly with us on some contents in the book, including Mike Koutnik and James Westman from ESRI; Ignacio Guerrero, Roger Harwell, and Neil Crisp at Integraph; Douglas Gordon from MapInfo; and Dan Ahern and Stephen Billias from Autodesk. In addition, we would like to thank those who allowed us to use their materials in this book. Although every effort has been made to obtain copyright permission, some owners of copyrighted materials could not be reached.

Finally, we owe special thanks to our families. Special thanks go to Peng's wife, Ching, who has given her unconditional love and support and who has endured so much during the writing of this book; to Peng's two lovely daughters, Lucinda and Naomi, whose love and smiles always bring happiness and joy; and to Peng's parents, who taught Peng much about dedication, care, and life. Tsou would also like to thank his parents in Taiwan, who have always been there throughout all these years, encouraging Tsou to explore the world. Special thanks also goes to Tsou's wife, Chun-Yi, who did everything she could to offer her support at all times. Her love and care have been essential to the success of Tsou's Ph.D. study and the writing of this book. Thanks also goes to Tsou's lovely daughter, Shu-An, who demonstrates the meanings of patience and love.

INTERNET GIS

CHAPTER 1

GIS, INTERNET GIS, AND DISTRIBUTED GISERVICES

The enlightened possess understanding. So profound they cannot be understood.
Because they cannot be understood, I can only describe their appearance.
—Lao Tse (*Tao Te Ching,* Chapter 15)

Internet GIS is an exciting research and application direction in Geographic Information Systems (GIS), and represents an important advancement over the traditional desktop GIS. It has been widely accepted in governmental agencies and educational institutions and among geospatial data producers and users, GIS vendors, and GIS professionals. The Internet-based GIS software has been developed to fill various needs and demands from plain-vanilla-style mapping functions to advanced GIS applications with Cherry-Garcia-type user interfaces and highly interactive functionality.

As a GIS professional or a GIS manager, you may be overwhelmed with different choices of Internet GIS software programs, different applications, and different terminologies. Which Internet GIS programs should we choose—server-side approach or client-side approach? Should I use thin client or thick client? Which server platform can carry our needs and tasks? Which client platform will be used by our customers? Should I pick up Hypertext Markup Language (HTML) client or Java client? What about ActiveX controls? What is a distributed-component model? How does the client talk with the server? How good is the system performance? How do I improve the performance of my Internet GIS site? How should I handle security issues? What is mobile GIS? For what applications should I use mobile GIS? Is the Internet GIS the same as the Web GIS?

1

Getting these questions answered is not an easy task. To choose, run, and develop an efficient and effective Internet GIS site, you need to know the answers to these questions. A successful site must have the right Internet GIS platforms, architecture, vendor supports, and software programs. The server systems must be dependable, reliable, scalable, secure, and long lasting. To identify and ride the *right* Internet GIS wave, you need to have a comprehensive plan for the whole system and to avoid ad hoc solutions and legacy technologies that might be outdated within two to three years.

It is a very challenging task to understand the whole Internet GIS technology. Of course, you could get some answers from individual Internet GIS vendors, but they are likely to be more enthusiastic about selling their own products than providing neutral answers. Ultimately, you are on your own to decide what Internet GIS programs you want to buy and what future trend you want to ride. Ideally, you should become knowledgeable about the fundamentals of Internet GIS technology and applications, including recent developments and future trends. At the minimum, you need to be able to ask the right questions when you talk with the vendors.

This book is intended to provide some of these answers from a vendor-neutral perspective. It covers the fundamental theories of Internet GIS, state-of-the-art Internet GIS technologies, various on-line applications, as well as the emerging future trends. It will first introduce the basic architecture of Internet GIS and its building blocks, including the mechanisms of networking, the client/server model, and the client and server components. It will then present some examples of Internet GIS applications and industry standards and specifications and conclude with some discussion of future trends.

1.1 IMPACTS OF THE INTERNET ON GIS

The increasing popularity of the Internet, from on-line surfing to e-commerce to interactive chatting, has made the Internet an integral part of our society. The nearly ubiquitous access to the Internet and interactive content of the World Wide Web (WWW) have made them a powerful means for people to access, exchange, and process information. Many applications in journalism, sciences, publishing, and other fields have been changed by and adapted for use on the Internet (Plewe, 1997). Likewise, the Internet has changed how GIS data and processing are accessed, shared, and manipulated.

The Internet is a modern information relay system that connects hundreds of thousands of telecommunication networks and creates an "internetworking" framework. Internet GIS is a research and application area that utilizes the Internet and other internetworking systems (including wireless communications and intranets) to facilitate the access, processing, and dissemination of geographic information and spatial analysis knowledge.

The development of information relay systems can be traced back thousands of years. The earliest information relay systems were developed in Egypt and in China by using human beings on horseback and relay stations situated on major roads. Maps, military documents, and letters were delivered by postmen or messengers who traveled between villages, cities, and countries. Another type of early information relay systems was created by using pigeons (Figure 1.1). Some historical documents indicate that pigeons were used to deliver the outcomes of the Olympic Games in ancient Greece, around 776 B.C. The Romans, around the fourth century, also used homing pigeons as the messengers. Two thousand years later, the electric modern information relay systems were invented by Samuel Finley Breese Morse, the electric telegraph in 1838 (Holzmann and Pehrson, 1994).

Figure 1.1 Pigeon post, woodcut from 1481. (*Sources:* G. J. Holzmann and B. Pehrson, *The Early History of Data Network,* Los Alamitos, California: IEEE Computer Society Press, 1994, p. 7; Fabre, Maurice, *A History of Communications* [Translated from the original French edition], New York: Hawthorne Books, 1963. Figure reprinted with permission from IEEE Computer Science Press.)

Key Concepts

The **Internet** *is a modern information relay system that connects hundreds of thousands of telecommunication networks and creates an* **"internetworking"** *framework. Internet GIS utilizes both wired and wireless network systems.*

One hundred and twenty years after Morse created the telegraph, the U.S. Department of Defense (DOD) developed the progenitor of the modern Internet in the 1970s: a self-adjustable, decentralized networking system called ARPANET. The original goal of the ARPANET project was to provide a reliable telecommunications network that could survive a nuclear war (http://www.cybergeography.org/atlas/historical.html). In 1983, the ARPANET project adopted the Transmission Control Protocol/Internet Protocol (TCP/IP) as the standardized protocol for communications across interconnected networks, between computers with diverse hardware, and between various operating systems. The dramatic success of the Internet and the popular adoption of TCP/IP pushed the development of telecommunication into a new age. Along with the rapid development of the Internet, many applications and programs have been developed, such as Newsgroup, Gopher, Bulletin Board System (BBS), Telnet, and so on. The GIS community also began to utilize the Internet to develop Internet GIS and other Internet-based applications.

The Internet is affecting GIS in three major areas: GIS data access, spatial information dissemination, and GIS modeling/processing. The Internet provides GIS users easy access to acquire GIS data from different data providers. GIS data warehouse/clearinghouse and digital libraries are two common forms of Internet data access systems. The U.S. Geospatial Data Clearinghouse Activities under the Federal Geographic Data Committee (FGDC) in the United States has been working to build a distributed archive of information for universal access (http://www.fgdc.gov). The Alexandria Digital Library (ADL) project is constructing an indexed repository of spatial information from diverse collections to make it available to the public over the Internet (Chen et al., 1977; Frew et al., 1998; Goodchild and Proctor, 1997).

The Internet also enables the dissemination of GIS analysis results and spatial information to a much wider audience than does traditional GIS. The general public can now directly access spatial information and explore spatial patterns and relationships from their Web browsers at home or public libraries. People can even conduct an on-line, free search and query analysis for spatial objects without purchasing any expensive GIS software.

Furthermore, the Internet is becoming a means to conduct GIS processing. It enhances the accessibility and reusability of GIS analysis tools by dynamically downloading or uploading GIS processing components. In the future, GIS users could work on GIS data interactively by using their Web browsers without installing GIS software on their local machines.

Access to and transfer of GIS data over the Internet are the first steps toward true Internet GIS. On-line data access allows GIS users who have stand-alone GIS software installed on their local machine to access and transmit GIS data across the Internet. This method of use of the Internet is efficient for data access, but the user's ability to view and analyze the data is limited by his or her desktop GIS software. The ability to access to GIS analysis functions and to conduct GIS analysis anywhere over the Internet is the next important step. The major framework of on-line GIS processing is still under development and will be available very soon to provide a true Internet GIS or distributed GIService.

The next section will examine issues and emerging techniques in the evolution of GIS in more detail. It will first briefly review the history of the GIS technology, from centralized GISystems to distributed GIServices. It will then discuss the definition and functions of Internet/distributed GIS.

1.2   GIS TECHNOLOGY: FROM CENTRALIZED GISYSTEMS TO DISTRIBUTED GISERVICES

The development of GIS has been highly influenced by the progress of Information Technology (IT). In fact, the development of GIS technology has closely mirrored the development of computer technologies. It evolved from mainframe GIS to desktop GIS to distributed GIS, which includes Internet GIS and mobile GIS (Figure 1.2). Mainframe GIS refers to GIS programs hosted on a mainframe with terminal access. Desktop GIS refers to either stand-alone programs with no information exchange between computers or networked programs in which desktop GIS programs share data, applications, and other resources within Local-Area Networks (LANs). Distributed GIS refers to GIS programs working on the Internet (Internet GIS) or wireless network environments (mobile GIS). Distributed GIS was made possible by the recent development of the Internet and wireless data communication technologies. These changes include (1) the rapid expansion of low-cost bandwidth on the Internet and (2) a new generation of Web-enabled desktop computers and mobile devices.

Mainframe GIS adopted the monolithic computing model; that is, all programs were in the same mainframe computers. User access to GIS data and analysis functions on the mainframe server were through dumb terminals over LANs.

Desktop GIS relies on GIS programs on the desktop computers. It has two categories: one is the stand-alone desktop GIS and the other is the LAN-based desktop GIS. The stand-alone desktop GIS has all the GIS functions, user interface, and data in one stand-alone computer. There is no data communication between one and the other. The LAN-based desktop GIS usually adopts the two-tier client/server model. GIS programs on the desktop computers (clients) communicate with servers inside a departmental LAN. GIS

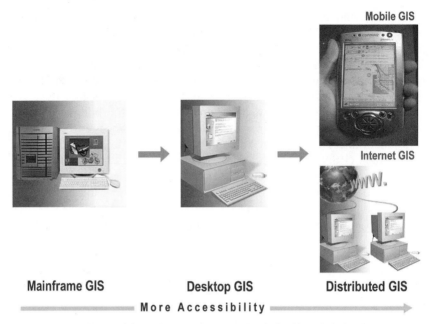

Figure 1.2   A development path of distributed GIS.

programs have to be installed on every desktop computer. Users have to be able to access desktop GIS programs in order to use it. This greatly limits the number of users who can access GIS programs.

Key Concepts

> *The development of GIS technology has evolved from mainframe GIS to desktop GIS to distributed GIS, which includes wired Internet GIS and wireless mobile GIS. The mainframe GIS and the desktop GIS is also traditionally called GISystems, while the distributed GIS is referred to as distributed GIServices.*

Distributed GIS represents a dramatic departure from the traditional two-tier client/server model. Rather than relying on desktop GIS programs, distributed GIS, when fully implemented, does not necessarily require the user to install GIS programs on the user's desktop. It relies on the Internet and wireless networks for data and processing communication. Users can access the GIS analysis tools and data from anywhere with Internet access or wireless data service coverage. The client could be a desktop computer, a laptop computer, a Personal Digital Assistant (PDA), or a mobile phone. There are two categories of distributed GIS: Internet GIS and mobile GIS. The major difference between them is that *Internet GIS* works on the wired Internet, while *mobile*

GIS works through the wireless telecommunication networks. The other difference is that the client for Internet GIS is usually a desktop computer, while the client for mobile GIS may be a laptop computer, a PDA, or a mobile phone. Because of the client difference, there are also major differences in functionality and applications. (We will discuss it in detail in Chapter 9.) Table 1.1 lists the main differences among these four eras of GIS development based on the computing architectural models used, the main components of client and server, and the networks. (Some of the terms may be new to you but will be discussed in the following chapters.)

The mainframe GIS and desktop GIS are traditionally referred to as GISystems, and distributed GIS is referred to as GIServices or distributed GIServices. The term *services* here refers to component services; that is, components with certain functions can be downloaded and reassembled together to build larger, more comprehensive services to perform certain tasks. GISystems provide several capabilities to handle georeferenced data, including data input, storage, retrieval, management, manipulation, analysis, and output (Aronoff, 1989). Due to the popular use of the Internet and the dramatic progress of telecommunications technology, the paradigm of GIS is shifting into a new direction—distributed GIS or GIServices. The new architecture of GIServices is platform independent and application independent. It can provide flexible and distributed GIServices on the Internet without the constraints of computer hardware and operating systems. Figure 1.3 shows three types of GIS architecture from the GIServices perspective.

Traditional GISystems are closed, centralized systems that incorporate interfaces, programs (or logic), and data. Each system is platform dependent and application dependent. Migrating traditional GISystems into different operating systems or platforms is difficult. Different GIS applications may require different GIS packages and architecture design. Every element is embedded inside traditional GISystems and cannot be separated from the rest of the architecture. Traditional GISystems include mainframe GIS and stand-alone desktop GIS.

Client/server GISystems or current desktop GIS are based on generic client/server architecture in network design (Tsou and Buttenfield, 1998). The client-side components are separated from server-side components (databases and programs). Client/server architecture allows distributed clients to access a server remotely by using distributed computing techniques such as Remote Procedure Calls (RPCs) or database connectivity techniques such as Open Database Connectivity (ODBC). The client-side components are usually platform dependent. Each client component can access only one specified server at one time. Different geographic information servers come with different client/server connection frameworks, which cannot be shared.

Distributed GIServices are built upon a more advanced networking scheme. The most significant difference is the adoption of distributed-component technology, which can connect to and interact with multiple and heterogeneous systems and platforms and without the constraints of traditional client/server

TABLE 1.1 Distributed GIS and Desktop GIS Comparisons

Application Characteristics	Mainframe GIS	Desktop GIS	Distributed GIS	
			Internet GIS	Mobile GIS
Architectural models	Monolithic	Ethernet era client/server (two-tier)	Web client/server (three-tier or *n*-tier)	Wireless client/server (three-tier or *n*-tier)
Client	Dumb terminals	Desktop computers	Web client	Wireless devices
Client interface	—	Fat Graphic User Interface (GUI) clients	Web browser, JavaBeans, ActiveX controls	Minibrowser, Wireless Application Protocol (WAP)
Networks	Local area networks	LANs or Wide Area Networks (WANs)	The Internet	Wireless networks and the Internet
Server	Mainframe	Application servers and data servers	Web servers, application server, GIS server, and data servers	Gateway server, Web server, and GIS servers
Number of accessible servers	One	One or a limited few	Thousands or more	Thousands or more

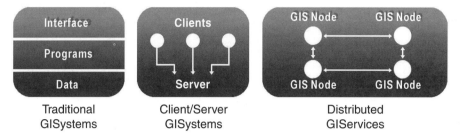

Figure 1.3   Three types of GIS architecture.

relationships (Montgomery, 1997). Under a distributed GIServices architecture, there is no difference between a client and a server. Every GIS node embeds GIS programs and geodata and can become a client or a server based on the task at hand. A client is defined as the requester of a service in a network. A server provides a service. A distributed GIServices architecture permits dynamic combinations and linkages of geodata objects and GIS programs via networking. In fact, the architecture of distributed GIServices is very similar to the "peer-to-peer" (P2P) computing, while the architecture allows Personal Computers (PCs) or workstations to communicate directly with one another with or without any help from a server (Roberts-Witt, 2001). The only differences between P2P computing and distributed GIServices architecture is that while P2P can only allow one-to-one or one-to-many communications, a truly distributed GIServices can allow many-to-many communications among computers at the same time.

Key Concepts

> **Traditional GISystems** *are closed, centralized systems.* **Client/server** *architecture allows distributed clients to access a server remotely by using distributed computing techniques.* **Distributed GIServices** *can connect to and interact with multiple and heterogeneous systems and servers at the same time and without the constraints of traditional client/server relationships.*

In the GIS community, many research projects have provided Internet GIServices and its applications. Most early popular Internet GIServices were using the Web browser via HTML format and Common Gateway Interface (CGI) programs. Examples include the Xerox Map Viewer (Putz, 1994) and GRASSLinks (Huse, 1995). Research projects, such as the Alexandria Digital Library Project (Frew et al., 1998; Buttenfield and Goodchild, 1996), adopted advanced Java technologies to explore more comprehensive services for online spatial queries, map browsing, and metadata indexing. Many projects and organizations also focus on the issue of Internet GIS standardization, addressing OpenGIS specifications (Buehler and McKee, 1996, 1998), International Standards Organization Technical Committee ISO/TC211 (Ostensen, 1995), component-oriented GIS (Li and Zhang, 1997), and Virtual Data Sets (Vckovski, 1998).

From an information service perspective, both GISystems and GIServices are value-adding processors that add meaning value to data. The main goal of information services, to provide users with information in the right form, requires selection and abstraction (Shuey, 1989). In the GIS community, many research projects in academia and industry focus on the need to provide GIS to the public and to researchers (Buttenfield, 1997; Li, 1996; Plewe, 1997; Zhang and Lin, 1996). For example, the recent development of digital libraries provides library services to dispersed populations (Goodchild, 1997), and the prototype of on-line GIS courses provides a virtual GIS classroom for distance learning (Buttenfield and Tsou, 1999). In general, the long-term goal of GIServices is to facilitate the synergy of the GIS community by sharing geographical information, spatial analysis methods, and users' experiences and knowledge. On-line, distributed GIServices will encourage multidisciplinary cooperation between the GIS community and other communities, including library information science, computer science, telecommunications, education, civil engineering, and so on.

With the progress of computer networking technologies, distributed GIServices can provide broader capabilities and functions compared to traditional GISystems: "Information services include tools for data management, browsing, access, cleaning, processing, interpretation, presentation, and exchange" (Buttenfield, 1998, p. 161). GIServices will broaden the usage of geographic information into a wide range of on-line geospatial applications and services, including digital libraries [National Science Foundation (NSF), 1994], digital governments (NSF, 1998), digital earth (Goodchild, 2000), on-line mapping (Kraak and Brown, 2001; Peterson, 1997), data clearinghouses (Peng and Nebert, 1997), real-time spatial decision support tools (Craig, 1998), dynamic hydrological modeling (Huang and Worboys, 2001), distance learning modules (Buttenfield and Tsou, 1999), and so on.

1.3 WHAT IS DISTRIBUTED GIS?

Like many other new fields, there is no general agreement on the term to describe GIS programs based on the Internet. Several names are used, such as Internet GIS (Peng 1999; Peng and Beimborn, 1998), GIS on-line, Distributed Geographic Information (DGI) (Plewe, 1997), and Web-based GIS, or simply Web GIS. These different terms are similar but sometimes have different meanings. They all seem to mean GIS data access and processing on the Internet. But, as discussed below, the Internet and the Web are not the same, and on-line can refer to either the Internet and/or the WWW. Therefore, Internet GIS may not be the same as Web-based GIS.

An Internet is any network composed of multiple, geographically dispersed networks connected through communication devices and a common set of communication protocols (Hall, 1994). The WWW is a networking application supporting a HyperText Transfer Protocol (HTTP) that runs on top of the Internet. It is a means of accessing information over the Internet. There are many other applications that run on the top of the Internet but are not part of the Web, such as e-mail, File Transfer Protocol (FTP), and Telnet (Shan and Earle, 1998). In other words, the Internet is an infrastructure that hosts many applications (usually based on the client/server model), including the Web, as well as more advanced client/server applications that are yet to emerge.

Key Concepts

The **Internet** *is any network composed of multiple, geographically dispersed networks connected through communication devices and a common set of communication protocols.*

The **World Wide Web** *is a networking application supporting a HTTP that runs on top of the Internet.*

The Internet is not synonymous with the World Wide Web. The Internet refers to the network infrastructure, while the Web refers to one of many applications that run on top of the Internet. Therefore, **Internet GIS is not the same as Web-based GIS.**

Therefore, *Internet GIS* is not necessarily synonymous with *Web-based GIS.* Internet GIS refers to the use of the Internet as a means to exchange data, perform GIS analysis, and present results, whereas Web-based GIS refers to the use of the WWW as a primary means. Both Internet GIS and Web-based GIS use the client/server computing model. The Web-based GIS uses the Web as a client, but Internet GIS may not necessarily use the Web as a sole client; it can use other clients. Although the Web is a major part of the Internet and the most important application running on top of the Internet for the time being and the current Internet GIS programs center on the use of the Web, the term Internet GIS has a broader and more enduring meaning than Web-based GIS. It leaves room to include other and new applications on the Internet.

Plewe (1997) uses the term distributed geographic information to refer to the use of Internet technologies to distribute geographic information in a variety of forms, including maps, images, data sets, analysis operations, and reports. It is similar to the term Internet GIS.

Since the scope of Internet GIS technologies is very broad and diversified, this book will use several different terms to describe the nature of Internet GIS, including distributed GIS, on-line GIS, and distributed GIServices. Some of them are interchangeable but may imply different aspects of Internet GIS. The following is the explanation of terminology used in this book.

- *GIS* is the abbreviation for geographic information systems. It is useful to view GIS as a research domain or an abstract concept. In this book, *GISystems* will be used to indicate the system perspective of GIS, which focuses on software/hardware implementation and operations. *GIServices* will be used to illustrate the service perspective of GIS, that is, delivering geographic information and processing tools to users over the Internet: "GIServices are a rapidly growing form of electronic commerce" (Longley et al., 2001, p. 19).
- *Internet GIS* is the framework of network-based GIS that utilizes the Internet to access remote geographic information and geoprocessing tools. It is used interchangeably with GIServices and *on-line* GIS in this book.
- *Distributed GIS* represents a broader framework including both Internet GIS and mobile GIS. This term emphasizes the software characteristics of Internet GIS and mobile GIS, which are distributed and dynamic.
- *Distributed GIServices* focus on the on-line processes of information services and task-oriented Internet GIS applications.

This book adopts the term distributed GIS to cover a wide range of GIS applications distributed in a networked environment, wired or wireless, including wire-based Internet GIS and wireless GIS or mobile GIS. Since intranets are essentially a subset of the Internet that use the same IP and other services, but with limited access, intranet GIS is a subset of Internet GIS.

What, then, is distributed GIS? Distributed GIS is defined as a network-centric (wired or wireless) GIS tool that uses the Internet or a wireless network as a primary means of providing access to distributed data and other information, disseminating spatial information and conducting GIS analysis. Distributed GIS allows a variety of client devices to access geospatial data and processing tools in servers anywhere and at any time. The client devices could be desktop computers, laptop computers, PDAs, or cellular phones. The servers could be distributed in multiple locations.

In addition to many of the functions of traditional stand-alone GIS software packages, distributed GIS has additional functions that take advantage of the Internet and its associated protocols, such as HTTP, WAP, FTP, Common Object Request Broker Architecture (CORBA) protocols, and ODBC libraries. Different types of GIS data and functionality could reside on different servers on the Internet and could be assembled and integrated locally on demand. The next section will illustrate the unique characteristics of distributed GIS.

1.3.1 Distributed GIS Is an Integrated Client/Server Computing System

Distributed GIS applies the dynamic client/server concept in performing GIS analysis tasks. The client can request data and analysis tools from the server.

The server either performs the job itself and sends the results back to the client through the network or sends the data and analysis tools to the client for processing. The connections between the client and server are established according to a communication protocol, mainly TCP/IP. Depending on the amount of processes performed on the client side, the client could be "thick" or "thin." If most of the processing is performed at the server side, and the client is merely used to request user input and present output, it is a thin client. But if most processing is performed at the client side, it is a thick client. In addition, the concepts of client and server are relative; any computer could be both a server if it provides services to other computers and a client if it requests services from other computers.

Key Concepts

Thin client: *There is little or no logic processing at the client side; most of the processing is performed at the server side. The client is merely used as a user interface for the user to input requests and to view processing output from the server. The client computer does not need to be very powerful, but the server needs a very powerful computer.*

Thick client: *Most logic processing is performed at the client side. The server usually sends data to the clients. The powerful client-side computer is needed to handle client-side processing.*

1.3.2 Distributed GIS Is a Web-Based Interactive System

While the traditional desktop GIS relies on a GUI for users to interact with GIS programs, the distributed GIS relies on the WWW and its add-ons (at least currently or in the near future) to provide interactivity between the user and the distributed GIS programs. In addition to the interactivity provided by HTML, Extensible Hypertext Markup Language (XHTML), or WAP, distributed GIS can also handle vector-based GIS data. It enables users to manipulate GIS data and maps interactively over the Internet or wireless networks. Users can perform GIS functions such as map rendering, spatial queries, and spatial analysis using a Web browser or other Internet-based client programs.

1.3.3 Distributed GIS Is a Distributed and Dynamic System

Distributed GIS takes advantage of the Internet as a giant distributed system so that the GIS data and analysis tools can reside in different computers (or servers) on the Internet. Users can access those data and application programs on demand from anywhere across the Internet or wireless networks. Geospatial data are usually distributed across different departments within an organization and among organizations, either on intranets or on the Internet. There are also an increasing number of databases available over the Internet that are provided by public agencies or private data providers. Distributed GIS can take advantage of these distributed data systems and can potentially query and extract these distributed databases in situ rather than simply downloading the data directly into the end users' local machines to combine with local data.

In addition to distributed databases, geospatial analysis tools can also be distributed across the Internet. Distributed GIS clients should be able to search, download (with or without fees), and assemble these analysis tools on demand. The clients have control over the functions required for a particular task. Therefore, different clients may choose different client applications, and the same user may choose different client applications from different vendors at different times based on the data needed and the tasks the users will be performing.

Because distributed GIS is a distributed system, databases and application programs reside on computers that serve them. This distributed system keeps data and application programs current. In other words, distributed GIS is dynamically linked to the data sources. This dynamic nature allows distributed GIS to be more capable in linking with real-time information in real-time connection, such as real-time satellite images, traffic movements, and emergency response information.

1.3.4 Distributed GIS Is Cross-Platform and Interoperable

Distributed GIS can be accessible across platforms regardless of what operating system the user is running. Distributed GIS is not limited to any one kind of machine or operating system. As long as one has access to the Internet or the wireless communication services, everyone could access and use distributed GIS from anywhere (although we are not there yet) provided that distributed GIS providers offer platform-neutral or cross-platform and interoperable GIS tools. The clients in distributed GIS tend to be able to run in a variety of computing environments and platforms, including different desktop and laptop computers with different operating systems, different PDAs, and cellular phones.

The challenge of distributed GIS is the ability to access many forms of GIS data and functions in the heterogeneous environment. To be able to access and share remote GIS data and functions, Internet GIS programs requires high interoperability (Bishr Yaser, 1996). The Open Geodata Interoperability Specification and Geography Markup Language (GML) by the OpenGIS Consortium (OGC) are attempting to lay the ground rules for GIS interoperability (http://www.opengis.org).

In short, distributed GIS is a special type of GIS tool that uses Internet and wireless networks as a major means to access and transmit distributed data and analysis tools, to conduct spatial analysis, and to create multimedia and multidevice GIS presentations. At its essence, distributed GIS is object oriented, distributed, and interoperable. Ideally, the end user does not necessarily need to have GIS data and software installed in his or her local computer, because all data and analysis modules can be available on network servers that can be requested by and delivered to the end user on demand. Local GIS users on different operating systems are able to access remote data and analysis tools as if they are stored locally.

1.4 WHY DO WE NEED DISTRIBUTED GIS?

1.4.1 Uniqueness of Geographic Information on the Internet

Geographic information is one of the most complicated information types stored in computer systems. Due to the uniqueness of geographic information, distributed GIServices require a different solution from other types of information services, such as financial information services or medical information services. This section will discuss the unique characteristics of on-line geographic information, with an emphasis on how geographic information is represented and disseminated across the networks.

First of all, the contents of geographic information vary in different resolutions, scales, times, and domains. Thus, it is a challenge to integrate heterogeneous data formats or set up a standardized data transfer procedure for distributing geographic information across the networks. For example, a series of raster-based remotely sensed images with 40 meters resolution will require different protocols and transferring procedures compared to vector-based Digital Line Graphs (DLGs) with double-precision accuracy. Current GIS software solutions have difficulty providing interoperable geospatial data sets and automatic data conversion/sharing tasks (Buehler and McKee, 1998). Geographic information scientists and GIS professionals, with appropriate knowledge to deal with geographic information and spatial phenomena, need to formalize the different characteristics of geographic information and to help software engineers design comprehensive GIServices architecture. Together, the GIS industry may be able to provide more reasonable and feasible frameworks for on-line GIServices applications.

Another unique characteristic of geographic information is the power of GIS operation/overlay, which can process geographic information and generate new layers of information. For example, a road map will become more valuable for tourists if the data layer can be overlaid with points of interest (e.g., hotels, gas stations, parks, restaurants). Another example is the overlay of a population change map with available housing units to predict the potential needs for housing. These examples indicate that the value of geographic information will increase dramatically by providing GIS users with the capability of GIS operations and overlay procedures. However, current Internet GIS programs mainly focus on the display of geographic information without providing many comprehensive Internet GIS operation tools. One of the major problems is the lack of appropriate mechanisms for exchanging or uploading GIS operations to servers. The current software architecture cannot provide GIS users with distributed GIS operations and modeling procedures (OGC, 1998). From a geographic information scientist and GIS professional's perspective, the study of Internet GIS should emphasize spatial analysis, modeling, and distributed GIS operations. The concepts of interoperable GIS programs, models, and analysis procedures need to be emphasized, with the participation of geographers, during the design process of distributed GIServices architecture. Although the idea of program interoperability has been around in computer science for a few decades, the development of GIS software rarely focuses on the actual implementation of interoperable GIS programs. It is clear for GIS users that the design of distributed GIServices needs to provide a balance between data interoperability and program interoperability.

Finally, in order to achieve both data interoperability and program interoperability, the GIS community needs to revise the metadata scheme for geographic information and emphasize the operational meaning of metadata. Traditional GIS research only uses descriptive metadata for tracking data lineage or facilitating the correct use of data (Gardels, 1996). On the other hand, the metadata research in computer science emphasizes machine-readable metadata for storing, searching, and integrating software components (Orfali et al., 1996). The research of distributed GIServices should adopt both ideas and design an integrated metadata scheme for geospatial data and software components. The integrated metadata scheme is one of the key points for the successful deployment of distributed GIServices architecture.

1.4.2 Why Do We Need Distributed GIS?

Traditional GISystems have difficulty in delivering on-line, distributed GIServices and providing flexible, friendly GIS solutions for users. Along with the progress of computer software engineering and the increasing volume of available geospatial data sets, traditional GISystems with legacy database engines are being superseded by distributed GIServices because they cannot communicate with other programs or access heterogeneous data via networking. Different GISystems have unique functions and data formats, which cannot be shared. The computer programs inside traditional GISystems are fixed and difficult to customize for network-oriented, distributed GIS tasks. Many users have problems in designing their own GIS solutions due to the unfriendly, complicated programming environment and modeling tools. The GIS industry cannot adopt the state-of-the-art technologies into legacy GISystems because of the lack of software compatibility and networking capability. Thus, the architecture of legacy GISystems has limited the power of GIS operations due to the lack of interoperability, reusability, and flexibility.

What GIS users need today is a distributed, Internet-based GIServices architecture that will provide a flexible and dynamic scheme for Internet geographic information services. The following sections describe the reasons for adopting a distributed GIS architecture for Internet geographic information services and the major problems in building a distributed GIServices environment. These discussions are organized from three different viewpoints: the management perspective, the user perspective, and the implementation perspective.

1.4.2.1 Management Perspective From the management perspective, there are two main reasons for Internet-based geographic information services. The first reason is the globalization of geographic information access and distribution. Currently, federal agencies face ever-greater demand and expectations from the user to make information available to the public and meet research needs via effective and efficient methods. Traditionally, geographic information has been distributed via paper maps or off-line disks or tapes, which are costly and difficult to update: "We must put in place a global data and information system that makes environmental data, past and current, available to all who need it, in a form that they can use" (Eddy, 1993, p. 6). In order to build such a global information system/service, the GIS community should provide on-line geographic information services on the Internet accessible to the GIS users around the world. A global geographic information service will facilitate a large scope of geographic research in the scientific community as well as assist the general GIS users to gain information about the community and the environment and other location-based services.

A second reason for Internet geographic information services relates to the decentralization of geographic information management and update. Along with the progress of data gathering techniques such as Global Positioning Systems (GPSs), remote sensing, and satellite images, more and more GIS applications and projects deal with huge databases. Huge and bulky GIS databases cause serious data management problems for maintaining, updating, and exchanging geographic information. Federal agencies are looking for new ways to more widely and effectively disseminate data, primarily via the Internet (Jones, 1997). Internet GIS under a distributed architecture provides one possible solution. One advantage of the Internet GIS is that data sets may be more appropriately maintained in its source site rather than a centralized location. For example, the certification and quality control of specific data sets will be granted only from specialized agencies, such as demographic information from the U.S. Census Bureau or the topographic map data from the U.S. Geological Survey (USGS). Another advantage is increased reliability, where failure at one site will not mean failure of the entire geographic information service (Worboys, 1995). In general, establishing open and distributed GIServices will improve the efficiency of GIS database management and reduce the cost of GIS database maintenance.

1.4.2.2 User Perspective From the user's perspective, there are three main reasons for Internet-based distributed GIServices. The first reason is the need of distributed GIS processing to cope with increasing size and variety of geospatial data sets, which impede GIS processing. Large files are time consuming to download and convert, and processing may not always be possible on smaller workstations. With expected increases in data volume and variety, traditional GISystems will be less able to handle increasingly complex geospatial data sets in a single, centralized architecture. One possible solution is to establish a dynamic, distributed processing arrangement whereby one can send encapsulated GIS processing components to a large data clearinghouse. Data would be processed dynamically at the server and results encapsulated within the processing component to be returned to the client. Distributed processing capability will facilitate the usage of distributed geospatial data sets and energize GIS processing without the constraints caused by running on local machines.

The second reason for Internet-based distributed GIServices is the need for customizable GIS modules for software package specialization. Most GIS software platforms have acuity for specific processing tasks. For example, some but not all packages can handle dynamic segmentation (breaking up linear features on the basis of a particular attribute); others are adept at merging field data with vector features; still others provide excellent address matching as a primary function. The complexity of modeling tasks undertaken by most GIS analysts increasingly demands a working knowledge of several GIS packages. In a truly distributed geographic processing environment, GIS analysts can federate GIS processing commands to the most appropriate GIS package available on the distributed network in order to conquer the complexity of spatial modeling. Also, in traditional GIS software, 90% of users utilize less than 10% of an application's features. These users must nonetheless pay for the full monolithic software suite, as opposed to licensing only those modules they require. The remaining 10% of advanced users requiring more complex features are dependent upon version update cycles that dictate when new features become available. By using the distributed-component technology, individual software modules may be updated independently. Distributed GIServices architecture will provide more flexible services for GIS users, where users can combine individual components based on their needs, plugging selected modules together. They will not be constrained to a single GIS package or software vendor. The pricing of GIS software licenses should also become more flexible and lower for individual GIS users.

The third main reason for distributed GIServices is the demand for location-based information from the general public due to the popularity of the Internet and mobile devices. Traditional GISystems can only be accessed from inside the office by GIS professionals, but distributed GIServices opens the door to the general public. The provision of GIS data and processing functions on the Internet and through the wireless networks offers the great opportunity for any Internet and wireless users to use distributed GIServices, a technology that is previously inaccessible. With the popularity of GPS, in-car navigations, and wireless access to Internet services, distributed GIServices have the real potential to bring GIS to the masses.

1.4.2.3 Implementation Perspective From the implementation perspective, the first problem in developing Internet-based, distributed GIServices is the lack of a high-level architecture that can support logical construction methods. Most current Internet geographic information services and research projects adopt a quick, ad hoc, technology-centered approach to provide a temporary solution for open and distributed GIS. Once the technology changes, every component in the old system is abandoned and a whole new system has to be designed and implemented. Without an appropriate architecture, distributed GIServices could not be achieved due to the short-term life cycle and rapid change in technology. A dynamic, upgradable architecture will facilitate the development of open and distributed GIS from a short-term strategy to a sustainable development strategy.

The second implementation problem is that current development of open architectures mainly focuses on data interoperability issues. However, GIS is both data oriented and process oriented. The GIS community needs to focus on GIS processing, and on the interactions between GIServices. This book will introduce a high-level GIServices framework that focuses on the dynamic integration of distributed GIS processing.

Three operational issues must be addressed to implement a dynamic GIServices architecture in distributed-network environments. The first issue is the definition of client/server relationships among distributed GIS components and geospatial data objects. In distributed-network environments, the major obstacle is the integration and the interactions among heterogeneous software (GIS components) and databases (geospatial data objects). A key issue for the integration is the development of modular, independent GIS components along with the comprehensive definitions of interactions and relationships between components.

The second issue is the formalization of comprehensive metadata descriptions and GIS functionality. Metadata provide a mechanism for objects and processes to describe themselves, to communicate, and thus to interoperate. In distributed-network environments, users can copy or download data objects and programs from one machine to another. Data sets and GIS operators become more dynamic, movable, and interoperable on the Internet. By defining the behaviors and requirements for geospatial data objects and GIS operators, a comprehensive metadata scheme will facilitate the effective and correct use of data sets and GIS components.

The third consideration is the problem of information overload in distributed-network environments. Distributed-network environments enlarge the scope and variety of available data. In distributed-computing environments, users may wish to fuse heterogeneous data models in different GIS software. The two aspects (large data files and incompatible data models) will inhibit the implementation of distributed GIServices. Some research projects in the GIS community have addressed this data compatibility issue, by means of the Virtual Data Set (Vckovski, 1998) and the Open Geodata Model (OGM) (Buehler and Mckee, 1996). However, another type of information overload is the complexity of GIS operations and modeling. Distributed-network environments enable users to access hundreds of different GIS programs and models on-line. Most users may not have adequate knowledge to bridge different models and programs together for their own GIS tasks. Thus, GIS users need some help in integrating heterogeneous GIS programs and models besides the data compatibility.

Key Concepts

A dynamic, upgradable architecture will facilitate the development of open and distributed GIS from short-term projects to a sustainable development strategy. Data sets and GIS operators will be dynamic, movable, and interoperable on the Internet.

1.5   BASIC COMPONENTS OF DISTRIBUTED GIS

Distributed GIS adopts the three-tier or *n*-tier client/server computing architectural model. It typically has the client component, a Web server and application server, one or many GIS servers, and data servers. Mobile GIS has additional components. Therefore, the wired Internet GIS and the wireless mobile GIS are discussed separately.

1.5.1   Basic Components of Internet GIS

Internet GIS generally has four major components: the client, Web server with application server, map server, and data server (Figure 1.4). The client serves as user interface for users to interact with the Internet GIS programs. The Web server receives client requests, serves static Web pages, and invokes application servers. The application server manages server transactions, securities, and load balance. The map server processes client requests and generates results. The data server serves geospatial and nonspatial data and provides data access and management through a Structured Query Language (SQL).

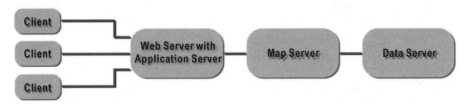

Figure 1.4   Basic components in Internet GIS.

1.5.1.1 The Client The client is a place for users to interact with spatial objects and analysis functions in Internet GIS. It is also a place for Internet GIS programs to present outputs to the users. While traditional desktop GIS uses graphic user interface to construct the client, Internet GIS usually relies on the Web and Web add-ons as its client. A typical Web interface with HTML and forms is a simple client of Internet GIS. But this simple HTML-based client has very limited user interactivity. It is particularly inadequate for users to interface with maps and spatial objects. Users cannot draw a box or a circle and cannot select spatial objects directly from maps.

To increase user interactivity and help users to interact directly with spatial objects on a map, alternative clients that use Web add-ons have been developed. These interactive clients include dynamic HTML and client-side applications such as plug-ins or help programs, Java applets or Java beans, and ActiveX controls.

Dynamic HTML uses client-side scripting like JavaScript or VBScript to make the plain HTML dynamic. For example, when a mouse moves over a spatial feature like a line or a polygon on a map, the color of that spatial feature could change or a text box could be popped up to show the attributes of that spatial feature.

Browser plug-in clients are software executables that run on the browser to extend the capabilities of Web browsers. Plug-in clients for distributed GIS are developed to provide the user interactivity with geospatial data and map images so that the user can view the maps and select features and make queries directly on the map. Plug-ins can support both vector data and raster images. The plug-in client can communicate with vector geospatial data on the server and fetch data on demand. The geospatial data arrive to the plug-in as a data stream. Map rendering and spatial analysis functions are built in the plug-in to allow the user to manipulate spatial features directly. Therefore, some geospatial processes can be conducted at the client side.

Java applets are another type of client for displaying geospatial information and conducting spatial query and analysis. Java applets reside at the Web server and are downloaded from the server and executed on the client at runtime. They allow the user to interact directly with the spatial features on the map. Map-rendering and data processing functions are usually built into the Java applet so that the user can render maps, make queries, and do other processing at the client side if the feature data are streamed to the Java applet. While Java applets are browser add-ons, Java beans are independent components that can be downloaded from the server and run at the client machine.

Finally, the client could also be created using ActiveX controls. ActiveX controls are general component ware that can plug into any application that supports Microsoft's Object Linking and Embedding (OLE) standard. Like plug-ins and Java applets within Web pages, clients in distributed GIS implemented in ActiveX controls have map rendering and data processing functions built in. Data are streamed asynchronously to and displayed by the ActiveX controls. Users rely on built-in functions and user interfaces to manipulate the feature data and/or maps over the Web browser. Furthermore, ActiveX controls allow local data to be combined with data retrieved from remote sources.

These different kinds of clients have their advantages and disadvantages as clients with distributed GIS. A dynamic HTML (DHTML) client is generally an interactive display of maps and analysis outputs, with spatial processing in the server. Except for map rendering, processing in DHTML at the client side is quite limited. Plug-ins are platform and browser dependent and work only with certain browsers and thus have problems in interoperability. Java applets and ActiveX controls as distributed GIS clients are more versatile. Java applets have the advantage of being platform neutral and more secure, while ActiveX control has an edge in performance but suffers from less rigorous security and platform dependence. All of these client types have been implemented in some distributed GIS programs. The detailed descriptions of these Internet technologies will be introduced in Chapter 4.

1.5.1.2 *Web Server and Application Server*   The second component in Internet GIS is comprised of the Web server and application server. The Web server is also called the HTTP server; its major function is to respond to requests from Web browsers via HTTP. There are several ways for the Web server to respond to client requests: (1) by sending existing HTML document or ready-made map images to the client, (2) by sending Java applets or ActiveX controls to the Web client, and (3) by passing requests to other programs and invoking other programs such as a CGI that could process the queries.

When the Web server passes client requests to other programs, it requests services from application servers. An application server can be a glue program or middleware that connects the Web server and server-side applications such as a map server. An application server acts as a translator or connector between the Web server and the map server.

The major functions of an application server include establishing, maintaining, and terminating the connection between the Web server and the map server; interpreting client requests and passing them to the map server; managing the concurrent requests and balancing loads among map servers and data servers; and managing the state, transaction, and security.

The application server could include the CGI model or CGI extensions such Netscape's Netscape Server Application Program Interface (NSAPI), Microsoft's Internet Server Application Program Interface (ISAPI) and Active Server Pages (ASP), Apple's WebObjects, Javasoft's servlets, Allaire's ColdFusion, and many others.

1.5.1.3 *Map Server*   A *map server* is a major workhorse component that fulfills spatial queries, conducts spatial analysis, and generates and delivers maps to the client based on the user's request. The map server is also called

a spatial server in ArcIMS by Environmental Systems Research Institute (ESRI). The map server provides specific traditional GIS functions or services, including query filtering and data extraction, geocode service, spatial analysis service, map-making services, and so on. These services could be located in different servers as individual components.

The output of the map server can be in one of two forms: (1) filtered feature data that are sent to the client program for user manipulating and (2) a simple map image in a graphic format [e.g., Graphics Interchange Format (GIF) or Joint Photographic Experts Group (JPEG)], or a graphic element map that is composed of discernible map elements with predefined colors, styles, legends, and so on. Detailed descriptions of different types of map servers will be given in Chapters 4 and 5.

1.5.1.4 Data Server A data server serves data, spatial and nonspatial, in a relational or nonrelational database structure. A client application such as a Web client or a map server gains access to the database through the SQL. Therefore, a database server is often referred to as a SQL server. Although SQL is an international standard language, the implementation by different vendors results in different versions of SQL for different databases. Therefore, database middleware is often used to access different databases. There are three major database middleware: ODBC, Java Database Connectivity (JDBC), and Object Linking and Embedding Database (OLE DB) ActiveX Data Object (ADO). Through SQL, ODBC, or JDBC drive, the client application can query, retrieve, and even modify database records in the database server. Detailed descriptions of different types of data servers will be provided in Chapter 5.

1.5.2 Basic Components of Mobile GIS

Mobile GIS refers to the use of GIS through mobile and wireless devices such as laptop computers, PDAs, and cellular phones. The architecture of mobile GIS is very similar to the wireline-based Internet GIS. It has three major components: the client, the server (the combination of Web server, map server, and data server), and the network service providers. The major differences are the client devices and the wireless communication service providers. The server functions and structures are similar to those in Internet GIS.

Mobile GIS relies on wireless communication networks for data exchanges and hand-held devices for display and data input. The communication environment—the wireless network—is more constrained than the wired Internet connections. The hand-held device is also different from the desktop or laptop computers. It has limited display device and data input devices. Detailed descriptions of mobile GIS and its networks will be given in Chapter 9.

1.5.2.1 Mobile Device Clients While the client for the Internet GIS is a PC, the client for mobile GIS includes a full range of client devices, from mainstream laptop computers with all the computing power of a PC, to palm-sized PDAs, to cellular telephones with smaller displays, simpler input devices, and limited processing capability.

All hand-held wireless devices have limited processing capability [e.g., less powerful Central Processing Units (CPUs), less memory in terms of Read-Only Memory (ROM) and Random-Access Memory (RAM)], less supply of power, smaller screen size, and limited input devices. This presents a different and more constrained computing environment compared to desktop computers.

Because of the limited screen size and the lack of mouse, the user interface of a wireless handset is fundamentally different from that of a desktop computer. It has to use different input devices such as a phone keypad, handwriting recognition, and voice input. Since the display of maps requires a larger screen than the display of text, this poses another big challenge to the mobile GIS design. This requires us to rethink the design of the mobile GIS. For example, graphics and maps have to be highly simplified in order for them to fit in a small hand-held screen due to low bandwidth and the small screen; in addition, most business logics have to be processed in the server side due to limited processing power on the client side.

1.5.2.2 Wireless Communication Networks

Wireless communication networks have much lower bandwidth than the wired Internet networks. Most existing wireless networks only allow data transmission speed of between 9.6 to 128 kbps (kilobytes per second), which represents a fraction of the transmission speed of between 10 Mbps (megabytes per second) to 1 Gbps (gigabytes per second) for wired networks. Furthermore, wireless networks are much less reliable and stable than the wired Internet networks. The wireless networks also have more latency. That is, it takes much longer for a request to make a round trip from the client to the server and back. This poses big challenges to the construction of mobile GIS since geospatial data are usually bulky and require high bandwidth.

1.5.2.3 Gateway Services

The gateway server is a piece of middleware that links the wireless device and the Web server. It offers the wireless devices access to Web servers and other servers. The presence of a gateway is a major difference between the wireline Internet GIS and the mobile GIS. The functions of a gateway include translating the user requests from the wireless devices to HTTP requests for the Web server and repurposing outputs into different formats to fit the requirements of the mobile devices as well as other protocol conversion functions.

1.5.2.4 Internet GIS Servers

The gateway servers link directly with Internet GIS servers via wired Internet connection. The Internet GIS server includes a Web server, application server, map server, and data servers. These servers are system components that make data and applications available to client devices. They are not much different from the wired Internet GIS servers. But since the mobile GIS uses different protocols, the servers have to support a variety of protocols and Application Programming Interfaces (APIs). For example, the Web server has to support both HTML and WAP contents. The Web server will send out either WAP or HTML contents depending on which type of browser the requests come from (WAP microbrowser or Web browser).

In summary, mobile GIS is very similar to the thin-client applications of the Internet GIS. All user requests are processed at the server. The difference is the packaging and presentation of the processed results. The Internet GIS uses HTML to package the results that are sent to the Web browser, while the mobile GIS uses Wireless Markup Language (WML) and WAP to package the results of applications on the server. Most contents and process results on the existing Web and map servers can be repackaged and sent to wireless devices via the gateways.

1.6 APPLICATIONS OF DISTRIBUTED GIS

Distributed GIS has a wide range of applications. These applications can be categorized in four general groups: from data sharing and disseminations, to simple geospatial data search and queries, to online data processing, and to location-based services.

1.6.1 Data Sharing

Distributed GIS provides an ideal mechanism for data sharing and exchanges over the Internet. In addition to simple raw data exchanges over the Internet through FTP, distributed GIS can directly search and use remote data as if they are local. For example, suppose you wanted to obtain some data about the Mississippi River basin. In the old days, you had to call around to locate the data. Once you found it, you then waited for the data to be shipped in. Data from different sources may have been in different spatial reference systems and required comparisons for accuracy and consistency and conversion to matching formats.

Now consider how distributed GIS could help. Using distributed GIS, data providers could put data on the Web in one of three ways:

1. Put the raw data in the original format along with the metadata in their own Web site.
2. Join a data clearinghouse network or a GIS data portal and list your data using the same standard or protocol of that data clearinghouse network or GIS data portal. The data clearinghouse includes the U.S. National Geospatial Data Clearinghouse (NGDC) managed by the

FGDC (1995), and the GIS portal includes the Geography Network managed by ESRI (http://www.geographynetwork.com).

3. Put data in a standard format like GML and make it available on the Web.

On the user side, the user could search for the data across the Web using a data search engine, read the metadata, and download the data. In some cases, the user could also directly load the data on the fly as if the data are locally stored. The distributed GIS program could also make conversion of the spatial reference systems on the fly if the data are in different spatial reference systems.

1.6.2 Geospatial Information Disseminations

The largest current application of distributed GIS is geospatial information dissemination. Distributed GIS makes it easier to disseminate information such as land use plans, zoning information, environmental information, and traffic information. Such systems can improve public access to this information. It can also foster information sharing and exchange among different departments within an agency or across agencies. For example, environmental protection agencies can publish their maps of environmentally sensitive areas on a Web page. This information can then be used by transportation planners in the development of transportation plans or by land use planners to make land use plans. This open data-sharing system will greatly reduce the barriers between departments within an agency and among different agencies.

The primary current means to disseminate geospatial information from desktop GIS is to print out paper maps, which are static and passive. Users can only passively observe maps prepared by GIS specialists. With distributed GIS, however, geospatial information can be disseminated over the Internet. Furthermore, users can choose the information they would like to see and query. For example, many local governments have published their land use and zoning information on the Internet using Web mapping. Users can look, search, and query land use, zoning, and other information about a particular parcel. This would free up city planners' and/or zoning administrators' time to answer telephone queries. It also saves developers and other users the time and effort of driving to city hall to get updated information. Another example is real-time traffic information and weather-related road conditions. Distributed GIS could greatly improve customer services and reduce the burden on toll-free telephone road information services. With Internet GIS, road condition information can be easily updated and maintained. A weather-related road condition can be linked with weather information and changed in real time.

1.6.3 Online Data Processing

Analysis tools in traditional GISystems can only be accessed by GIS professionals. The vast majority of geospatial information users cannot access these

Internet GIS Showcase: Geospatial Data Clearinghouse

The FGDC has coordinated an effort creating a geospatial data clearinghouse. The clearinghouse provides a search mechanism for users to search for available geospatial data that are located in different servers. It provides the opportunity for data providers to make their data available on the Internet. But the FGDC's clearinghouse does not necessarily include data for download on-line. It simply provides a metadata search. Once the data have been identified, the user could contact the data source to request the actual data.

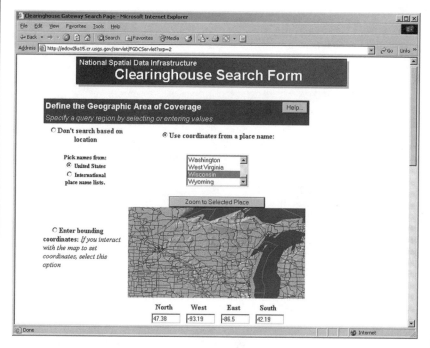

Graphic image reprinted with permission from FGDC.

Source: http://www.fgdc.gov/clearinghouse/clearinghouse.html.

Internet GIS Showcase: San Diego MSCP Habitat Monitoring

The accompanying picture indicates an Internet map service for remote sensing data and habitat monitoring in San Diego, California (Mission Trail Regional Park). The research was conducted by the Department of Geography at San Diego State University in support of San Diego's Multiple Species Conservation Program (MSCP). The MSCP was developed by the City and County of San Diego in support of the Natural Community Conservation Plan (NCCP) for five southern California counties. This information can help local governments and private landholders to establish management policies to better manage habitat reserves.

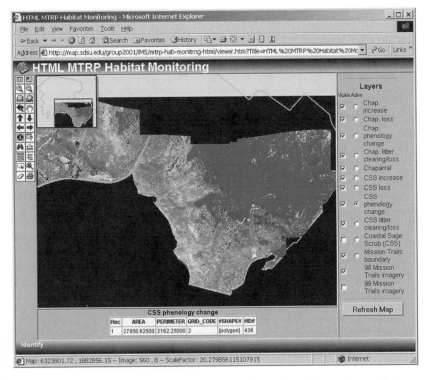

Graphic image reprinted with permission from San Diego State University.

Source: http://map.sdsu.edu/arc.

tools. Distributed GIS makes these GIS analysis tools available on the Web to anyone who has Internet access. Therefore, data processing could be done on line. Furthermore, the user-friendly Web browser interface makes the GIS analysis tools easier to use. So distributed GIS increases the potential for GIS technology to reach the masses.

Furthermore, the component-based distributed GIS architecture allows the development of interoperable geospatial analysis tools as individual and sellable geoprocessing components. This would foster more software component innovations.

1.6.4 Location-Based Services

Another important application of distributed GIS is in Location-Based Services (LBSs) and field work. LBSs refer to a system that offers real-time information about a location and its surrounding area. This information allows users to determine the location of a destination and the best route to get there. LBS systems and mobile devices have emerged as an important means of conducting business in a variety of fields.

The use of hand-held devices has increased dramatically due to demand for information, real-time information, and location-specific information. These real-time, location-based information needs are difficult to meet with traditional desktop GIS, but distributed GIS can meet these needs easily. Mobile GIS can offer mobile users location-based services, such as real-time traffic information and the locations of nearby businesses and attractions.

Traditionally, field workers have to take a laptop computer with preloaded data when going into the field. There is no direct connection with office data systems, and field workers have no way to know whether the data have been updated since last downloaded into the laptop. The lack of contact with the office also means that field data updated by the field workers cannot be made immediately available in the office database. This time delay is not acceptable for time-critical applications such as utility repairs and facility security. Furthermore, managing and maintaining the synchronization of data and software between the field computer and office computers can be costly and complex. With distributed GIS, specifically mobile GIS coupled with GPS, there will be a direct connection between the field worker's mobile devices and the database in the server. Any update in the office database will be instantly transmitted to the field workers, and vice versa.

1.6.5 Distributed GIS and Intelligent Transportation Systems

Distributed GIS can be an important tool in the development of Intelligent Transportation Systems (ITSs). For example, it can be used to disseminate real-time travel information and to make real-time trip plans.

Interactive distributed GIS is a perfect presentation tool of real-time travel information once it is linked with real-time traffic information. Several trans-

Mobile GIS Showcase: Maps on a Pocket PC

The accompanying picture shows a pocket PC (Compaq iPAQ H3670 with 64 MB memory) loaded with ESRI's ArcPad 6.0 software. The information displayed in this small device is the map of downtown San Diego with aerial photos, topographic maps, freeways, and trolley lines. The menu on the screen also indicates that this pocket PC can access geographic information or add a new data layer via an Internet map server or ESRI's Geography Network services.

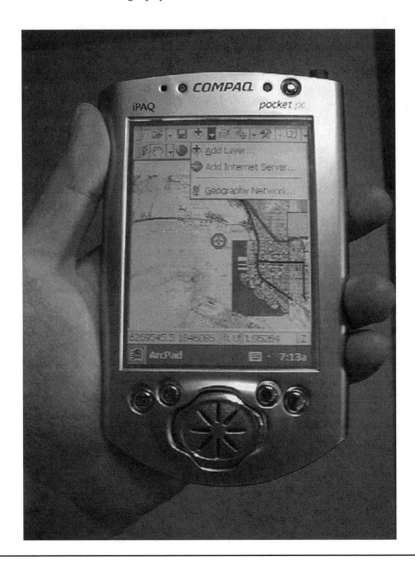

portation agencies have created real-time traffic maps on the Internet. For example, under the ITS model deployment initiative, the Washington Department of Transportation (DOT) created TrafficView (http://www.wsdot.wa.gov/PugetSoundTraffic) to present real-time traffic information on the Web. Users can view the traffic flows on the entire freeway system in the city. They can also zoom into a smaller area for a more detailed view. Users can quickly find the travel speed along a specific link on a freeway. The road link is also connected with real-time video snapshot so that the user can see the traffic flow.

The graphic presentation of Internet GIS makes it an ideal tool for travelers to plan their trips on-line. By linking Internet GIS with real-time traffic information, a traveler can plan a trip on the Internet based on his or her origin and destinations as well as real-time traffic conditions. For example, Seattle travelers can plan their travel itinerary according to real-time traffic conditions on the Puget Sound Traffic Web page (http://www.wsdot.wa.gov/PugetSoundTraffic). Similarly, a transit user can also plan his or her transit trip on the Internet based on the real-time transit schedule and bus locations by linking with an Automatic Vehicle Location System.

Distributed GIS also offers an excellent channel for public involvement in the planning and community decision-making process. Governmental agencies can publish information on neighborhood planning or other related issues on the Internet. The public can interact with the plan and offer their input directly from the Web. Rather than going to town hall meetings, the public can directly access the information right from their homes. Hopefully, this will give the public more input in the planning process.

1.7 WHERE DO I START?

Now you have a basic idea of what distributed GIS is and what its potential applications are, what's next? The remainder of the book introduces you to the fundamentals of distributed GIS theories and technologies, as well as examples of applications. Chapters 2–3 cover the technological background of computer networking hardware and software. Chapters 4–5 cover the evolution and theories of Internet mapping and distributed GIS. Chapters 6–7 cover some standards of distributed GIS development. Chapter 8 introduces four major commercial Internet GIS software programs. Chapter 9 presents the concept and architecture of mobile GIS. Chapter 10 discusses the issue of quality of services. Chapters 11–13 present some examples of Internet GIS applications. The final chapter concludes the book with a discussion of possible future impact of Internet GIS.

Depending on your interests and background, you could follow the sequence of the chapters to read the whole book (this is our recommendation). You could also focus on specific chapters.

With this brief introduction to a big topic, let's start with the exploration journey!

Internet GIS Showcase: Waukesha Transit Trip Planner

The picture below is a snapshot of a transit trip planning system designed by the University of Wisconsin-Milwaukee for the City of Waukesha, Wisconsin. This is an Internet GIS-based system that allows users to plan a trip based on a specified trip origin, destination, and desired travel time. The user can directly identify the origin and destination from the map, enter a street address, or select landmarks from a pull-down menu. Transit Trip Planner was developed using ESRI's MapObject along with a custom-built path-finding algorithm and is served by the MapObject Internet map server.

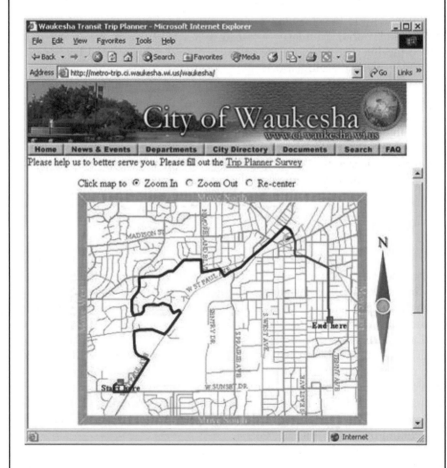

Source: http://metro-trip.ci.waukesha.wi.us/waukesha.

WEB RESOURCES

Descriptions	URL
FGDC	http://www.fgdc.gov
Alexandria Digital Library	http://www.alexandria.ucsb.edu
History of the Internet	http://www.cybergeography.org/atlas/historical.html
Internet GIS course	http://map.sdsu.edu/geog596
Puget Sound Traffic	http://www.wsdot.wa.gov/PugetSoundTraffic/

REFERENCES

Aronoff, S. (1989). *Geographic Information Systems: A Management Perspective.* Ottawa, Canada: WDL Publications.

Bishr Yaser, M. Sc. (1996). A Mechanism for Object Identification and Transfer in a Heterogeneous Distributed GIS. In *Proceedings of the 7th International Symposium on Spatial Data Handling,* August 12–16, 1996, Delft, The Netherlands, International Geographical Unikon, pp. A.1–A.13.

Buehler, K., and McKee, L. (Eds.) (1998). *The OpenGIS® Guide: Introduction to Interoperable Geoprocessing and the OpenGIS Specification,* 3rd ed. Wayland, Massachusetts: Open GIS Consortium.

Buehler, K., and McKee, L. (Eds.) (1996), *The OpenGIS® Guide: Introduction to Interoperable Geoprocessing,* Wayland, Massachusetts: Open GIS Consortium, Inc.

Buttenfield, B. P. (1997). The Future of the Spatial Data Infrastructure: Delivering Geospatial Data. *GeoInfo Systems,* June 1997, pp. 18–21.

Buttenfield, B. P. (1998). Looking Forward: Geographic Information Services and Libraries in the Future. *Cartography and Geographic Information Systems,* 25(3), pp. 161–171.

Buttenfield, B. P., and Tsou, M. H. (1999). Distributing an Internet-Based GIS to Remote College Classrooms. In *Proceedings of ESRI International User Conference, San Diego, CA:* ESRI, CD-ROM. URL: http://greenwich.colorado.edu/babs/esri/P634.htm.

Chen, H., Smith, T. R., Larsgaard, M. L., Hill, L. L., and Ramsey, M. (1997). A Geographic Knowledge Representation System for Multimedia Geospatial Retrieval and Analysis. *International Journal on Digital Libraries,* 1(2), pp. 132–152.

Craig, W. J. (1998). The Internet Aids Community Participation in the Planning Process. *Computer, Environment and Urban Systems,* 22(4), pp. 393–404.

Eddy, J. A. (1993). Environmental Research: What We Must Do. In M. F. Goodchild, B. O. Parks, and L. T. Steyaert (Eds.), *Environmental Modeling with GIS.* New York: Oxford University Press, pp. 3–7.

Federal Geographic Data Committee (FGDC). (1995). *Content Standards for Digital Geospatial Metadata Workbook,* Version 1.0. Reston, Virginia: FGDC/USGS.

Frew, J., Freitas, N., Hill, L., Lovette, K., Nideffer, R., and Zheng, Q. (1998). The Alexandria Digital Library System Architecture. In J. Strobel and C. Best (Eds.),

Proceedings of the Earth Observation and Geo-Spatial Web and Internet Workshop '98 (Salzburger Geographische Materialien, Vol. 27). Salsburg: Instituts für Geographie der Universität Salzburg. URL: http://www.sbg.ac.at/geo/eogeo/authors/frew/frew.htm, May 11, 2000.

Gardels, K. (1996). The Open GIS Approach to Distributed Geodata and Geoprocessing. In *Proceedings of the Third International Conference on Integrating GIS and Environmental Modeling,* Santa Fe, New Mexico: National Center for Geographic Information and Analysis, CD-ROM.

Goodchild, M. F. (1997). Towards a Geography of Geographic Information in a Digital World. *Computers, Environment and Urban Systems,* 21(6), pp. 377–391.

Goodchild, M. F. (2000). Communicating Geographic Information in a Digital Age. *Annals of the Association of American Geographers,* June 2000, 90(2), pp. 344–355.

Goodchild, M. F., and Proctor, J. (1997). Scale in a Digital Geographic World. *Geographical and Environmental Modeling,* 1(1), pp. 5–23.

Hall, C. L. (1994). *Technical Foundations of Client/Server Systems,* New York: Wiley.

Holzmann, G. J., and Pehrson, B. (1994). *The Early History of Data Networks.* Los Alamitos, California: IEEE Computer Society Press.

Huang, B., and Worboys, M. F. (2001). Dynamic Modelling and Visualization on the Internet. *Transactions in GIS,* 5(2), pp. 131–139.

Huse, S. M. (1995). GRASSLinks: A New Model for Spatial Information Access in Environmental Planning. Unpublished Ph.D. dissertation, University of California at Berkeley, Berkeley, California.

Jones, J. (1997). Federal GIS Projects Decentralize. *GIS World,* 10(8), pp. 46–51.

Kraak, M.-J, and Brown, A. (2001). *Web Cartography.* London: Taylor & Francis.

Li, B. (1996). Issues in Designing Distributed Geographic Information Systems. In *Proceedings of GIS/LIS'96,* November 19–21, 1996, Denver, Colorado. Bethesda, Maryland: American Society for Photogrammetry and Remote Sensing, pp. 1275–1284.

Li, B., and Zhang, L. (1997). A Model of Component-Oriented GIS. In *Proceedings of GIS/LIS'97,* October 28–30, 1997, Cincinnati, Ohio. Bethesda, Maryland: American Society for Photogrammetry and Remote Sensing, pp. 523–528.

Longley, P. A., Goodchild, M. F., Maguire, D. J., and Rhind, D.W. (2001) *Geographic Information Systems and Science,* New York: Wiley.

Montgomery, J. (1997). Distributing Components. *BYTE,* April 1997, 22(4), pp. 93–98.

Open GIS Consortium (OGC). (1998). *The OpenGIS Abstract Specification,* Version 3. Wayland, Massachusetts: Open GIS Consortium. URL: http://www.opengis.org/techno/specs.htm, May 11, 2000.

Orfali, R., Harkey, D., and Edwards, J. (1996). *The Essential Distributed Objects Survival Guide.* New York: Wiley.

Ostensen, O. (1995). Mapping the Future of Geomatics. *ISO Bulletin,* December 1995, pp. 13–15.

Peng, Z.-R. (1999). An Assessment Framework of the Development Strategies of Internet GIS. *Environment and Planning B: Planning and Design,* 26(1), pp. 117–132.

Peng, Z.-R., and Beimborn, E., (1998). Internet GIS: Applications in Transportation. *Transportation Research* (*TR*) *News,* March/April 1998, No. 195, pp. 22–26

Peng, Z.-R., and Nebert, D. D. (1997). An Internet-Based GIS Data Access System. *Journal of Urban and Regional Information Systems,* 9(1), pp. 20–30.

Peterson, M. (1997). Cartography and the Internet: Introduction and Research Agenda. *Cartographic Perspectives,* 26, pp. 3–12.

Plewe, B. (1997). *GIS Online: Information Retrieval, Mapping, and the Internet.* Santa Fe, New Mexico: OnWord Press.

Putz, S. (1994). Interactive Information Services Using World Wide Web Hypertext. In *Proceedings of the First International Conference on the World-Wide Web,* May 25–27, 1994, Geneva, Switzerland. Bethesda, Maryland: American Society for Photogrammetry and Remote Sensing.

Roberts-Witt, S. (2001). Peer Pressure. *PC Magazine Internet Business,* June 26, pp. 8–16.

Shan, Y.-P., and Earle, R. H., (1998). *Enterprise Computing with Objects: From Client/Server Environments to the Internet.* Reading, Massachusetts: Addison Wesley Longman.

Shuey, R. (1989). Data Engineering and Information Systems. In A. Gupta (Ed.), *Integration of Information Systems: Bridging Heterogeneous Databases.* New York: IEEE Press, pp. 11–23.

Tsou, M.-H., & Buttenfield, B. P. (1998). Client/Server Components and Metadata Objects for Distributed Geographic Information Services. In *Proceedings of GIS/LIS' 98,* November 10–12, 1998, Fort Worth, Texas. Bethesda, Maryland: American Society for Photogrammetry and Remote Sensing, pp. 590–599.

Vckovski, A. (1998). *Interoperable and Distributed Processing in GIS.* London: Taylor & Francis.

Worboys, M. F. (1995). *GIS: A Computing Perspective.* London: Taylor & Francis.

Zhang, L., and Lin, H., (1996), A Client/Server Approach to 3D Modeling Support System for Coast Change Study. In *Proceedings of GIS/LIS'96,* November 19–21, 1996, Denver, Colorado. Bethesda, Maryland: American Society for Photogrammetry and Remote Sensing, pp. 1265–1274.

CHAPTER 2

NETWORKING FUNDAMENTALS OF INTERNET GIS

The network is the computer.
—Sun Microsystems

Internet GIS, as the name suggests, relies on the wired or wireless Internet to communicate. This is in contrast with desktop GIS, which only utilizes the LAN to exchange information. But what are exactly the Internet and the LANs? Why can we not simply expand the desktop GIS from the LAN to the Internet? Why are the current Internet GIS applications slower than the desktop GIS? To answer these questions, it is necessary to understand the concept of computer networking, the information communication process between computers, and the differences between the LAN and the Internet.

The purpose of this chapter is to provide you with the fundamentals of network computing underneath the Internet GIS and help you understand the major differences in networking between desktop GIS in the LAN environments and the Internet GIS applications in the WAN and Internet environments. The chapter starts with the discussion of an Internet GIS application to illustrate the different networks the application needs to use followed by the network communication model and protocols—the Open Systems Interconnection (OSI) model and TCP/IP. The chapter further discusses information exchanges in the network environment, starting with the information flow between two computers, to that within a LAN and the WAN and ending with the information flows on the Internet. The chapter also introduces some future developments of the Next Generation of the Internet (NGI) and Internet 2.

2.1 INTRODUCTION TO NETWORK ENVIRONMENTS

To start with, we will use an Internet GIS application scenario to illustrate the roles of different network environments on the Internet GIS application. The scenario will also illustrate some of the basic concepts of computer networking.

2.1.1 Internet GIS Application Scenario

A general GIS user, Mike, plans a trip from Boulder, Colorado, to Utah's Arches National Park. He needs to acquire map information and make a hotel reservation for two nights in the city of Moab, Utah. Based on the scenario, Mike will need the following geographic information:

- Colorado/Utah state highway road map,
- Arches National Park trails map,
- Moab city road map, and
- hotel/motel locations and reservation systems in Moab, Utah.

In this scenario, there are three major actors (Figure 2.1). Mike, a general GIS user who has no formal training in GIS, wants to plan a trip. Tina, a map designer, designs appropriate symbols and layouts for several national park maps, including the Arches National Park trails, and puts them into a GIS server for on-line access. There may be other map designers for highway roads, city roads, and hotel locations. Since the map designers' activities are similar, this scenario will only identify Tina as a representative map designer. Kevin, the hotel reservation manager, will provide the hotel reservation information (e.g., price, room availability) and accept the reservation from Mike.

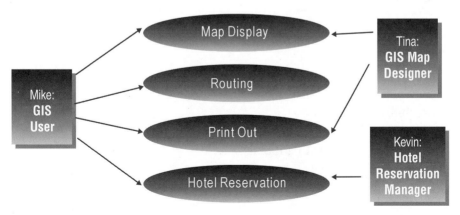

Figure 2.1 Travel plan scenario.

2.1.2   Traditional GISystems Solution

Being a novice GIS user, Mike does not have desktop GIS software. He needs to purchase a CD-ROM called Travel Plan U.S.A. in order to design his trip plan. The CD-ROM includes detailed roads for all 50 states and hotel information as well as trip-planning software. However, the data sets in the CD-ROM do not include trail maps of the Arches National Park. Mike has to purchase another CD-ROM called U.S. National Parks. Since the two CD-ROMs use different types of databases and map formats, there is no way for Mike to integrate the trail maps with the road maps without buying another map format conversion program. Therefore, Mike prints out two different sets of travel maps: one for highways/hotel/roads and one for national park trails. However, the trail maps use many color-based symbols and the printouts look terrible in Mike's black-and-white laser printer. Mike decides to use the color laser printer in the main office to print out the trail map again. The next task for Mike is to make a hotel reservation. Mike uses the CD-ROMs to retrieve 10 hotels located in Moab, Utah, and their phone numbers. He makes 10 phone calls, one to each hotel, to compare room availability and prices. Finally, he makes the reservation after spending roughly 20 hours to complete his travel plan.

2.1.3   Internet GIServices Solution

Using Internet GIServices, Mike will take entirely different actions. In addition, the three players need to collaborate with Internet GIS software and to create telecommunication channels for accessing different geographic information. Their actions are described in the following paragraphs.

MIKE'S ACTIONS   Mike's first action is to log on to Internet Service Providers (ISPs) and open a GIS component called Map Display, possibly from within a Web browser. In the menu, Mike checks out the Extended Component: Travel Plan option. The Map Display component loads the Travel Plan component. Mike describes his travel itinerary to the Travel Plan component with the following information:

Start point: Boulder, Colorado.
End point: Moab, Utah.
Required map layers:
1. highway roads,
2. Arches National Park trails,
3. Moab city roads, and
4. hotel locations.

One minute later, Mike sees four maps being overlaid and displayed on the Map Display window. Mike's machine accesses the remote data sets via the

Internet and has the match index for their qualification in this case (Figure 2.2). He selects *Generate the shortest route* function and the component generates the shortest road path from Boulder, Colorado, to Moab, Utah, and displays the route on his screen. Then, Mike picks a few hotels in Moab and opens the On-line Reservation function from the Software Agents menu. He describes the possible prices and date for his trip. The hotel reservation (virtual) agent retrieves the information Mike requests and sends the results back in 10 seconds. Mike compares prices and availability. Finally, Mike selects one hotel and makes an on-line reservation for his two-day vacation. Twenty seconds later, Mike gets a confirmed message from the on-line hotel reservation system.

After Mike finishes the hotel reservation, he prints out the travel maps via the printer server located in his office. Mike spends 10 minutes or less to finish his travel plan.

TINA'S ACTIONS   Tina updates and maintains the databases of the National Park maps on the www.national-park.gov GIS server. In order to provide

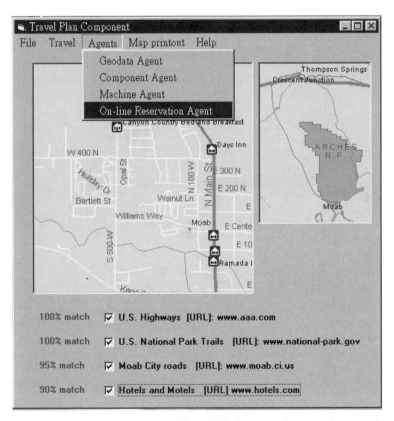

Figure 2.2 Travel Plan component. (*Note:* This interface is a mockup window, designed with Microsoft Visual Basic.)

smart map display and printout functions, Tina designs appropriate legends and predefined symbols for different types of media, including Cathode-Ray Tube (CRT) screens, black-and-white printers, and color printers. Tina used her GIS workstation to finish the design and then served these final maps on the Web server and the GIS server.

KEVIN'S ACTIONS   Kevin updates hotel information (e.g., pricing, room availability, reservation list) by accessing the on-line hotel reservation system via the telephone modem from his hotel. The on-line hotel reservation system can accept the reservation from Mike by collaborating with the hotel reservation agent.

2.1.4 Network Environments

To establish the telecommunication channels among Mike, Tina, and Kevin, the network environments should provide three kinds of elements: links, nodes, and clouds (Figure 2.3):

- *Links* are the physical media connecting computers, such as network wires or cables.
- *Nodes* are the computers connected by links.
- *Clouds* are the icons representing any types of networks.

This scenario is established dynamically on nine GIS nodes or computers:

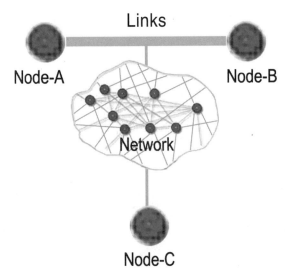

Figure 2.3 Three elements of network environments: links, nodes, and clouds.

1. Mike's workstation,
2. Mike's office printer server,
3. the www.aaa.com GIS node,
4. the www.national-park.gov data warehouse server,
5. Tina's workstation,
6. the www.moab.ci.us GIS node,
7. the www.hotels.com GIS node,
8. Utah-hotel-reservation.com (on-line Utah hotel reservation system), and
9. Kevin's PC.

Key Concepts

A PC by itself without connecting to other computers is called a **stand-alone** *computer. Two or more computers and other devices connected together are called a* **network,** *and the concept of connected computers sharing resources is called* **networking.**

There are four types of network environments utilized in this scenario:

1. a direct link between two computers,
2. LANs,
3. WANs, and
4. the Internet.

The first type of network environment is the direct link between two computers. In this scenario, Mike's computer connects to his office printer server directly in order to print out his travel map. This type of network environment is called a *point-to-point* connection (Figure 2.4). There are other types of direct links where more than two nodes may share a single link, called a *multiple-access* connection. Usually, direct links (point to point or multiple access) can only cover small areas and nodes. One exception is the satellite communication link, which can be applied in a wider area and across a large numbers of nodes (Peterson and Davie, 1996).

The second type of network environment is the Local-Area Network (LAN). In this scenario, Tina uses the LAN to move the finalized maps from her workstation to the data warehouse server located in the National Park office (Figure 2.5). Other cartographers in the office can use the LAN to communicate with each other, share or transfer data and maps, and publish their own maps to the data warehouse server. The LAN enables file exchanging, file serving, and message broadcasting in a small geographical area.

The third type of network environment is a Wide-Area Network (WAN). In this scenario, Kevin needs to use his telephone line and modem to connect his PC to the hotel reservation server (hotel-reserve.com) (Figure 2.6). The

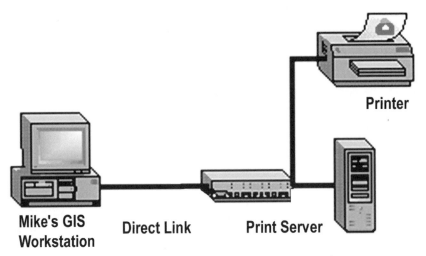

Printer

Mike's GIS Workstation   **Direct Link**   **Print Server**

Figure 2.4   Direct-link (point-to-point) network.

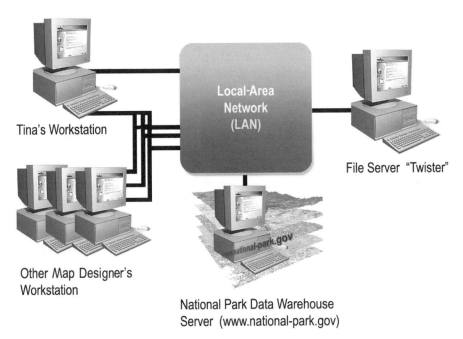

Tina's Workstation

Local-Area Network (LAN)

File Server "Twister"

Other Map Designer's Workstation

National Park Data Warehouse Server (www.national-park.gov)

Figure 2.5   Local-Area Networks (LANs).

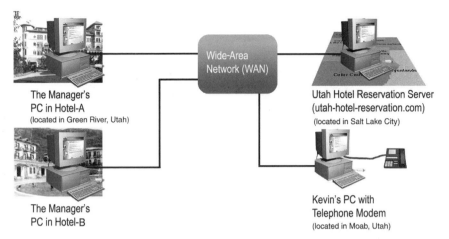

The Manager's
PC in Hotel-A
(located in Green River, Utah)

Utah Hotel Reservation Server
(utah-hotel-reservation.com)
(located in Salt Lake City)

The Manager's
PC in Hotel-B

Kevin's PC with
Telephone Modem
(located in Moab, Utah)

Figure 2.6   Wide-Area Networks (WANs).

WAN is designed to enable the interconnection among LANs and individual machines via telephone lines, cables, or other network media for geographically dispersed users (Cisco on-line documents, 1999). There are many different types of WANs available to provide different users with different types of services, including cable modems, Asymmetric Digital Subscriber Line (ADSL), Integrated Services Digial Network (ISDN), and so on. (We'll discuss them in detail later in this chapter.)

The final type of network environment is the Internet. In this scenario, the Internet connects multiple networks together and allows different users to access or to distribute GIS data sets across different machines. The Internet connects Mike's workstation to the National Park data warehouse server in a LAN environment and to the Utah hotel reservation system in a WAN environment (Figure 2.7). The Internet also provides the internetworking capabilities and protocols to integrate multiple applications and to allow different users to access these applications in different platforms. For example, Mike can access the on-line hotel reservation system from his Mac or PC and Kevin can receive the reservation requests and confirm the reservation from his PC or Unix workstation.

The Internet relies on a common set of rules for communications between computers and networks. The set of rules or standards is called *protocols.* Internet protocols are used to guide the communications between two or more computers and other communication devices so that requests from one computer can be understood by the other. Standard protocols are also needed to allow hardware and software from various vendors to interoperate with each other. Different kinds of network environments may use different protocols and physical media.

The following sections will discuss these four types of network environments in detail, from direct point-to-point connection between two network

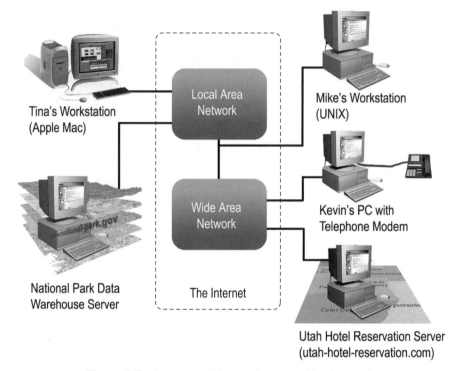

Figure 2.7 Internetworking environment (the Internet).

devices to LANs, WANs, and the Internet. It will also introduce how the data sets are transferred between machines via a LAN, WAN, or Internet environment. We will introduce two fundamental network communication frameworks—the OSI model—and the TCP/IP model, and the most popular and important protocol for the Internet applications—TCP/IP.

2.2 NETWORK COMMUNICATION MODELS

Information exchange among different network devices such as computers, printers, and servers in the network environment is a complex process. It involves different devices made by different manufactures and different software programs produced by different software vendors. This is similar to personal communications among a group of people from different countries and speaking different languages. Communication is impossible without following the same rule or speaking the same language. To guide network communications, a set of models and communication protocols have been developed. We will discuss two common network communication models and one universally used protocol: the OSI reference model, the TCP/IP model, and the TCP/IP.

2.2.1 OSI Model

The OSI reference model is an international standard to guide computer networking. The OSI model is a set of specifications that describes network architecture for connecting dissimilar devices and exchanging information in a network environment. It provides a standard to guide how network hardware and software components work together to make network communications possible. The OSI model was first introduced in 1978 and subsequently revised in 1984 by the ISO [see Cisco Web site (http://www.cisco.com/univercd/cc/td/doc/cisintwk/intsolns/index.htm) and Microsoft, 1999].

Information communications and device coordination between computers are very complex tasks. The OSI model divided these complex tasks into smaller and more manageable subtasks. Each subtask is responsible for certain functions and is modeled as an OSI layer. The whole network communication process within a computer is divided into seven layers (Figure 2.8). Each OSI

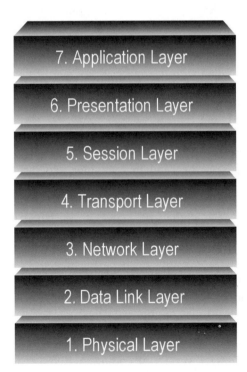

Figure 2.8 Seven-layer OSI model. (The first letter of each layer corresponds to the beginning letter of each word in *All People Seem To Need Data Processing.*)

layer has well-defined networking functions and protocols and covers different network hardware equipments.

In general, the layers at the top in the OSI model are responsible for communicating with applications, making sure that the data bits from the applications get to where they are supposed to go, and determining what to do if they do not. The layers at the bottom specify the tasks and functions in the network's physical media (or hardware), including how to put data bits from the top layers onto the network adapter cards and wires as electrical pulses. The higher level layers have more complex tasks than the lower level layers. Each layer in the OSI model has certain functions and performs specific services. The actual application program, such as a GIS software program, is a higher level application on the top of the application layer and falls outside the scope of the OSI model. The following sections describe the purposes and functions of each layer.

2.2.1.1 Application Layer   The application layer is the top layer of the OSI model that provides network access to user application software such as word processor, spreadsheet, or GIS software. The application layer covers such applications as file transfers (e.g., Telnet, FTP), data access software, e-mail, and even network management software (Wookcock, 1999). The application layer provides for higher level applications with network-related functions such as database and file access, printer-sharing access, and directory services. For example, if a GIS program from one computer wants to access data from another computer, such as a data server, the GIS program has to first communicate with the application layer of the OSI model through the directory services to identify if that data server is available and accessible on the network. The application layer handles such issues as general network access, flow control, and error recovery.

2.2.1.2 Presentation Layer   The presentation layer is concerned with data representation issues in the sending computer and the receiving computer. Different computers may code the same information differently. For example, the small letter "a" is coded as decimal 97 in the ASCII (American Standard Code for Information Interchange) character set, but it is coded as decimal 129 in the EBCDIC (Extended Binary Coded Decimal Interchange Code) character set that is used in the IBM mainframe (Woodcock, 1999). So the presentation layer is responsible for ensuring that the sending machine and the receiving computer use the same data representation or understand each other's representation. In other words, the presentation layer determines the data format for data exchanges between devices to ensure that information sent from the application layer of one computer system will be readable by the application layer of another system (Cisco Web site). It acts like a data translator by providing a variety of coding and conversion functions on the data sent down from the application layer. At the sending computer, the presentation layer converts data from a format into a commonly recognized, intermediary format. At the receiving computer, the presentation layer converts the intermediary format into another format required by that computer's application layer (Microsoft, 1999). The functions of this layer are usually implemented as part of an operating system.

2.2.1.3 Session Layer   A session refers to a connection between two network devices, similar to a connection in a phone conversation. The session layer establishes, manages, and terminates a session between two applications on different computers, known as Application Entities (AEs) (Woodcock, 1999). To start the communication process between two AEs, the session layer first establishes a connection or session, taking care of security measures (e.g., validating a password) if necessary. Once the session has been established, the session layer then manages and synchronizes the data flow, determining which side transmits, when, for how long, and so on. Finally, it is the responsibility of the session layer to terminate the connection once the information exchange is complete.

2.2.1.4 Transport Layer   While the session layer is responsible for managing a connection between two applications, the transport layer is responsible for ensuring that data packets are delivered to the other computer error free. That is, the functions of the transport layer are to make sure that data packets get to where they are supposed to and to check for and repair transmission errors. It uses flow control to manage and coordinate the speed and the amount of data transmission between the sending device and the receiving device so that the sending device does not send more data than the receiving device can handle.

At the sending end, the transport service repackages messages by breaking down long information into several packets and collecting small pieces together in one package (information before this layer has been transmitted in its original-sized chunk) (Woodcock, 1999). This allows the packets to be transmitted efficiently over the network. When the packets get to the receiving machine, the transport layer unpacks the messages, reassembles the packets into their original forms, and acknowledges receipt to the sending machine. The TCP is a typical transport layer implementation. The TCP is the protocol in the TCP/IP suite that provides reliable transmission of data.

2.2.1.5 Network Layer   The network layer is responsible for network addressing and packet routing. It defines an address like an IP address (e.g., 129.89.72.108) and translates logical addresses and names [like a Uniform Resource Locator (URL)] into physical addresses such as a Media Access Control (MAC) address that is literally "burned in" or hardwired to its network adapter card (see the next session) (Woodcock, 1999). It also determines the routing from the source to the destination computer, that is, the path the data will be taken based on network conditions, priority of service, and other factors. This is done by transferring a packet to a router, which is a piece of

Internet GIS Showcase: Dublin, Ohio, Web-Based City Information System

Dublin, Ohio's, GIS allows city planners, surveyors, utility engineers, and financial analysts—as well as ordinary citizens on the Internet—to view the city via data-rich maps, photographs, and other diagrams. Property lines and orthophotos, plotted together on a common coordinate system, form the foundation of the city's GIS, and today they remain its most popular features. Many types of data have been collected on the site. A small sample includes street paving and maintenance dates for the public works division, police and fire incident reports for the police department, road centerlines and edges, zoning and ward boundaries, and city and county limits.

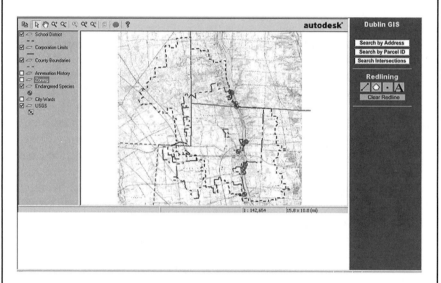

As many as 200 employees throughout the city government view map data from Dublin's GIS via Autodesk MapGuide, which is built onto the city's in-house, Web-based intranet. A free tool, MapGuide Viewer, allows any employee to use the system's pan-and-zoom functionality. Users can query the GIS by typing in street names, addresses, or homeowner names. They can specify the layers they want to see and then magnify areas of interest. Similar features are available on the public site, though fewer information layers are provided and residents' names have been removed.

Sources: http://www.autodeskgovernment.com/News/successstory1/Dublin.PDF and http://www.dublin.oh.us/business/gis/intro/index.html.

hardware that acts as a transfer point and maintains routing tables indicating where things are (e.g., LANs, hosts) and how to get frames to them (Microsoft, 1999).

The frame size, known as the Maximum Transmission Unit (MTU), varies on different kinds of networks. Therefore, another function of the network layer is to allow the router to break the packet into smaller units if the original packet size is bigger than what the router can handle (Woodcock, 1999). The network layer at the receiving end will reassemble the packet into its original form. Also, the network layer manages traffic problems on the network, such as packet switching, routing, and network congestion control (Microsoft, 1999). The IP in the TCP/IP suite and the X.25 protocol used on packet-switching networks serve this function.

2.2.1.6 Data Link Layer The data link layer provides services to reliably transmit data from the network layer to the physical network link. It involves two basic functions: assemble packets from higher level into data frames (the data part), and establish links and provide reliable transmission of data frames between network nodes (the link part). That is, the data link layer takes the packets from the network layer and assembles them into data frames. It then sends data frames to the physical layer to ensure the transit of these data frames is error free from one computer to another through the physical layer.

A data frame is an information unit at the data link layer with an organized, logical structure in which data can be placed (Microsoft, 1999). It is similar to a packet, which is an information unit at the network layer and above. Sometimes packet and data frame are used interchangably (Cisco web site). An example of a simple data frame is shown in Figure 2.9. A data frame has three parts: the header, the data, and the trailer.

The header information includes destination ID, send ID, and control information. The destination ID represents the address of the destination or recipient computer, while the send ID represents the address of the source computer. The control information represents frame type, routing, and segmentation information, which instructs network components how to pass the data along and how to assemble the data frames together in the receiving computer (Microsoft, 1999).

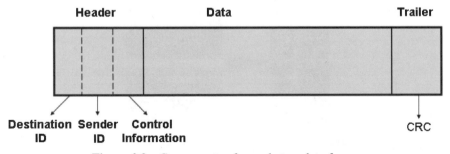

Figure 2.9 Components of a packet or data frame.

The data are the original information itself. The data part can be of various sizes, depending on the network. The data section on most networks varies from 512 bytes (0.5K) to 4K. Any data that are larger than 4K in size must be broken into smaller units in order to be transmitted over the network.

The trailer usually contains an error-checking component called a Cyclical Redundancy Check (CRC) that represents information about error checking, correction, and verification to ensure that the data arrives intact.

The concepts of headers, trailers, and data are relative. A header, trailer, and data from an upper layer become data at the lower level. For example, an information unit at the network layer includes data and headers and/or trailers at the network layer. When this information passes to the data link layer, the whole information unit from the network layer becomes data. New headers and/or trailers are added at the data link layer. In other words, the data portion of an information unit at a given OSI layer potentially can contain headers, trailers, and data from all the higher layers. The lower layer does not distinguish header, trailer, and data from the upper layers. This is known as encapsulation. The size of individual data frame differs from network to network.

The header and trailer information as a whole are also called control information or addressing information. The control information in the sending machine is intended to communicate with peer layers in the receiving machine. This control information at each layer consists of specific requests and instructions that can only be understood by its peer OSI layer in the receiving machine. For instance, header information added at the network layer in the sending computer will only be read by the network layer in the receiving computer.

The Institute of Electrical and Electronics Engineers (IEEE) has subdivided the data link layer into two sub layers: Logical Link Control (LLC) and MAC, as shown in Figure 2.10.

The LLC sublayer creates, terminates, and manages the data link communication between network nodes. It is also responsible for sequencing frames and acknowledging the receipt of frames as well as controlling for frame traffic (Woodcock, 1999).

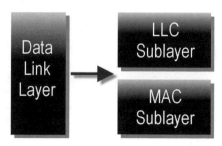

Figure 2.10   Data link layer contains two sublayers.

The MAC sublayer manages access to the physical network. It communicates directly with the network adapter card and is responsible for delivering error-free data frames between two computers on the network (Woodcock, 1999). The IEEE MAC specification defines MAC addresses, which enable multiple devices to uniquely identify one another at the data link layer.

2.2.1.7 *Physical Layer*
The physical layer transmits the unstructured raw bits stream (0s and 1s) over a physical medium such as the network cable. Therefore, the physical layer is concerned with the specifications of physical components of the network, such as cabling, adapter cards, connectors and pins, and electrical signals. There are numerous specifications that are used to define these physical elements, such as specifications for the functions of pins in a connector and how different cables are connected with the network adapter cards (Woodcock, 1999). These specifications are defined to activate, maintain, and deactivate the physical link between network devices by specifying characteristics such as voltage levels, timing of voltage changes, physical data rates, maximum transmission distances, and physical connectors (Microsoft, 1999). LAN and WAN have different physical layer implementations.

2.2.1.8 *Relationships among OSI Model Layers*
Each layer in the OSI model has its own standard rules or protocols to define its functions. Each layer can only communicate with its adjacent layer, that is, the layers immediately above and below it. Adjacent layers communicate with each other through an interface. The interface in a layer defines the services it offers to the layer above it and how it can access the services below it (Woodcock, 1999).

Logic or virtual communication is also allowed between the peer layers on the sending machine and the receiving computer (Microsoft, 1999). For example, the network layer protocol on the receiving machine communicates only with a network layer protocol at the sending machine. This communication is not direct and has to be carried out by actual communication between adjacent layers in one computer through the network to the other computer.

2.2.2 TCP/IP Reference Model

The OSI reference model is the commonly used model to describe computer networking. But it is not the only model. The other commonly used ones include the TCP/IP reference model and IBM's network architecture model SNA (Systems Network Architecture). The TCP/IP reference model is commonly used for network communication in the Internet environment, while the SNA is usually used in IBM products: mainframes, terminals, and printers (Wookcock, 1999). For our discussion of Internet GIS, we are more concerned with the TCP/IP reference model and its relationship with the OSI model.

The development of the TCP/IP reference model is closely related to the development of the ARPANET and the Defense Data Network supported by the DOD in the 1970s and 1980s. The TCP/IP model and its companion protocol suite are part of the military standard protocols issued by the DOD to support the development of the ARPANET. The original goal of the ARPANET was to provide a military communication network to meet the requirements of availability, survivability, security, network interoperability, and surge traffic control ability (Stallings, 1997). The TCP/IP model is widely used today in the Internet environment and is often referred to as the *Internet reference model* (Wookcock, 1999).

Key Concepts

The **TCP/IP model** *is a combination of two protocols: Transmission Control Protocol and Internet Protocol. Similar to our real-world mailing systems, TCP provides the mechanisms for transferring information packages across networks and IP is the mailing address system for the Internet machines.*

The design of TCP/IP did not follow the original concepts proposed in the OSI model—it did not provide more flexible adoptions for different applications and environments. One big difference is that the TCP/IP model does not have a clearly defined communication session like the OSI model does. Instead, it focuses on the transfer and routing of information among different networks. This is because, although establishing and maintaining a session is very important in the traditional mainframe/terminals and the constrained client/server applications, its importance becomes secondary in the Internet environments, where transferring and routing information among different applications and networks are more critical. In short, the TCP/IP model deals with information exchanges in the internetworking environment (Woodcock, 1999). Thus, the TCP/IP reference model defines a four-layered structure as shown in Figure 2.11.

Notice there is no equivalent layer to the TCP/IP internetworking layer in the OSI model. There is also no equivalent OSI physical layer in the TCP/IP model. The TCP/IP model does not include the physical layer within its model structure; rather it relies on other international standards to define the physical media environments.

2.2.2.1 Application Layer   The functions of the application layer in the TCP/IP model are very similar to the application layer and the presentation layer in the OSI model, that is, to provide higher level applications with network access and services as in the application layer in the OSI model and to ensure the use of standard or interoperable data representations, as in the presentation layer in the OSI model (Wookcock, 1999). Protocols used in this

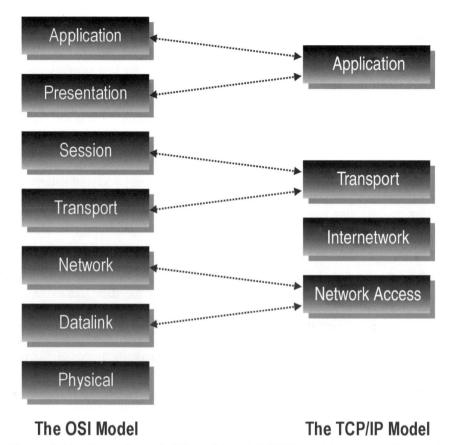

The OSI Model   The TCP/IP Model

Figure 2.11   Comparison of OSI model and TCP/IP model. (*Source:* Woodcock, 1999, p. 79, Fig. 4.4. Figure reprinted with permission from Microsoft Press.)

layer include FTP, Simple Mail Transfer Protocol (SMTP), and Telnet protocol.

2.2.2.2 Transport Layer   The TCP/IP transport layer is similar to the transport layer in the OSI model and has some functions of the session layer as well. It is responsible for reliable data transmission from a sending machine to the receiving machine, including flow control, error checking and recovery, and such functions as in the transport layer in the OSI model.

In addition, the TCP/IP transport layer is also responsible for helping communication computers to establish a *connection,* or *virtual circuit* (Woodcock, 1999), which is similar to but not exactly equivalent with the *session* in the OSI session layer. While data transfer over the session in the OSI session layer refers to a connection transfer—that is, the communication computers must be physically connected at the same time and data transfers follow the fixed path—data transfer over the TCP/IP transport layer is *connectionless*. That is, not all data transmission in the TCP/IP model requires an actual physical connection between two computers. The data packets (usually called *datagrams* in the TCP/IP model and include data packets with header and trailer information attached) can be routed in different paths from the origin to the destination based on the network traffic at the time. TCP is the typical protocol used in this layer.

2.2.2.3 Internetwork Layer The TCP/IP internetwork layer has no equivalent layer in the OSI model. It is responsible for routing packets or datagrams from the sending machine to the receiving machine across networks. This routing process may require forward datagrams to hop through intermediary networks or network segments through routers and gateways (Woodcock, 1999). IP is the typical protocol used in this layer.

2.2.2.4 Network Access Layer The network access layer, also called link layer, is responsible for managing and transmitting the data frames to the physical networks, including creating and addressing the data frames. It has some of the functions of the network and data link layer in the OSI model, including transmitting and delivering data frames to the hardwired network adapter cards. Some national and international standards are used for the network access layer, such as X.25, IEEE 802.3 (Ethernet), 802.4 (token passing), and 802.5 (Token Ring). The Network Driver Interface Specification (NDIS) driver in Windows NT is another implementation example.

In general, the OSI model was designed for applications in LANs, the SNA was designed for the mainframe/terminal environment, and the TCP/IP was designed for the Internet environment. However, they are not mutually exclusive. Both the OSI model and TCP/IP model can run on the same LAN, which can also communicate with the SNA-based mainframe network. It all depends on what protocols are used to implement the models.

2.3 PROTOCOLS

Network communication models such as OSI and TCP/IP describe the conceptual structure of network communication in a series of self-contained layers. The models are simply frameworks that describe the functions in each layer. However, there needs to be software in each layer to make those functions work. Protocols are rules and procedures embedded in the software in each layer for communicating and interacting among different layers in the same computer and between computers in the network. The models do not define protocols at each layer; rather, they simply standardize the services and interfaces the protocol should provide. So the relationship between the model and the protocol is "pretty simple: The model describes what needs to be done, and the protocol makes it—whatever the 'it' is—happen" (Wood-cock, 1999, p. 81). •

The layered structure of the model simplifies the job of protocols. The protocols need only concentrate on the services and functions in a particular layer as well as the communication with the peer in the receiving computer and protocols in the layer immediately above it and below it. The layer at which a protocol works describes its function (Microsoft, 1999). A protocol does not need to worry about other functions in other layers and can safely assume those services are taken care of by other protocols (Woodcock, 1999).

There are many protocols involved in the whole process of information communication from one layer to the next; each has its own purposes and functions. Some protocols work at one OSI or TCP/IP layer; some work at several layers. But there is no single protocol that provides all the services in the data communication process, from packaging, sending, and receiving data packets across the network. There is, however, a set of protocols, each functioning at a different layer, that together handle all the tasks in the data communication process. That set of protocols working together as one unit is called a *protocol stack* (Woodcock, 1999; Microsoft, 1999).

Key Concepts

Protocols *are rules and procedures defined by software for communicating and interacting among different applications in the same computer and between computers in the network.*

A **protocol stack,** *or* **protocol suite,** *is a combination of multiple protocols that work together to enable communication on a network.*

A *protocol stack,* often also known as *protocol suite,* is a complete set of protocols, usually one for each layer, that work together to enable communication on a network (Woodcock, 1999), such as the Novell Netware stack, Apple AppleTalk stack, and TCP/IP stack. Protocol stacks are independent from each other, but they are not mutually exclusive. One network can support more than one stack of protocols. For instance, Both Novell Netware stack and TCP/IP stack can run on the same network.

As the popularity of the Internet has increased, so has the importance of the TCP/IP stack. For Internet GIS, TCP/IP is particularly important. We are now turning to the introduction of TCP/IP.

TCP/IP is an industry standard suite of protocols for accessing resources and exchanging information in a heterogeneous Internet environment. It provides information routing and guarantees information delivery over the network.

One of the unique and most important characteristics of TCP/IP is its support of different network typologies and environments, or interoperability among many different types of computers in the network environment. This interoperability is the prime reason for its popularity in the Internet environment. Almost all networks support TCP/IP now (Microsoft, 1999).

2.3.1 Transmission Control Protocol

TCP provides services to the transport layer in both the OSI model and the TCP/IP model. The design of TCP focuses on packaging/unpackaging information for efficient transmission and provides a connection-oriented, reliable transport service. TCP is a connection-based protocol. That is, the data transmission relies on the actual connection between the sending and receiving computers, and every data transmission involves error checking. TCP is also a full-duplex (i.e., to send and receive data simultaneously) protocol and includes a flow control mechanism for data streams.

TCP provides a framework to package data into small sets of packets, called *segments*. It also indicates how to transfer these segments between sender machines (source node) and receiver machines (destination). Figure 2.12 illustrates the working process of TCP between senders and receivers. First, the sender machine packages the original data into hundreds of segments and issues each segment a unique number. Then the sender machine sends out segment 1 to the receiver. When the receiver gets segment 1, it sends back an acknowledgement message to the sender. So the sender will not need to send segment 1 again. Next, the sender begins to send out segment 2. This is an iterative process until all segments have been sent to the receiver. Then, the receiver can combine these segments together and reassemble the original data on the receiver's machine. TCP provides this kind of transmission control protocol and specifies possible network change situations, such as how to recover the transmission if the network is broken down during the transmission. This kind of transmission method is also called "three-way handshake algorithm" (Peterson and Davie, 1996).

2.3.2 Internet Protocol

IP is created to provide services at the network layer in the OSI model and the internetwork layer in the TCP/IP model. Its primary function is to route

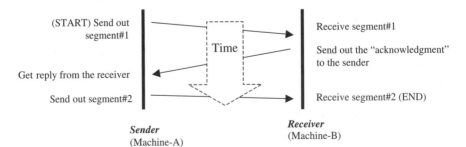

Figure 2.12 Segment transfer mechanism in TCP.

the data segments or datagrams from the source to the destination. It does this by assigning a unique address (called an IP address) for computers (such as sender machines or receiver machines) and attaching the sending address and receiving addresses with each data segment. It then checks the sending and receiving addresses of each data segment and forwards the data segments to the destination based on a *routing table*. In addition, if a data segment or datagram must be broken down into smaller units, IP must perform the task of fragmenting and reassembling the data segment or datagram (Woodcock, 1999). Unlike TCP, IP is a connectionless protocol, which means that the datagram can be routed along different paths to get to the same destination and there is no constant error checking to verify whether the datagram has arrived at the destination. The verification falls to the responsibility of the application.

An IP address is a 32-bit integer and has four levels to make up a complete address name, such as 128.130.170.14. Each level can use 0–255 values. So, the total maximum number of machines that can be used in the Internet is 4,294,967,296, which at present time is still adequate to accommodate a great number of sites. But with the massive growth of the Internet, IP numbers are facing a serious challenge in the future. The IP address implies the topology for the network connection and a LAN may define a "mask" to filter some unnecessary signals from outside networks by using a "bridge" or "router" (network devices to be discussed later in this chapter) as a gateway for LANs.

More recently, a new type of IP is under development, called IPv6, which is the major protocol for the next generation of the Internet. The goal of the next-generation IP (IPv6) is to provide a scalable framework for future Internet applications. IPv6 provides an 128-bit address space that is capable of adding 3.4×1038 nodes. An example IPv6 address would be

34DC:1133:4343:ABC2:0044:1153:3A34:0012

Besides the much bigger capacity of potential nodes, IPv6 also provides many new functions, such as auto-configuration, security support, real-time services, and mobile hosts.

Since IP addresses are difficult to remember, people who use the Internet would like to have an easier way to assign their servers and workstations, similar to "real-world" mailing addresses. Therefore, Paul Mockapetris designed the Domain Name System (DNS) in 1984 at the Stanford Research Institute's Network Information Center (SRI-NIC). The DNS is a distributed database used by TCP/IP applications to map between "hostnames" and "IP addresses" and to provide electronic mail-routing information. The DNS has a tree structure (Figure 2.13) and is more flexible and memorable than IP addresses.

Each LAN (university department, campus, or department within a company) maintains its own database of information and runs a server program for which other systems across the Internet can query its DNS. For example, San Diego State University has two servers for its DNS (130.191.1.1 and

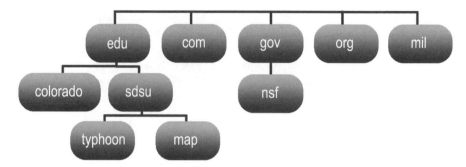

Figure 2.13   Domain Name System (map.sdsu.edu).

130.191.200.1). The level of hierarchy is unlimited and the total number in a domain name address could be 256 characters.

A hostname is usually assigned to an IP address, such as WWW.distributedgis.org. WWW (World Wide Web) in the first part of the hostname is purely arbitrary and conventional. The server does not use this data to determine the type of service to provide. Rather, the type of service (e.g., HTTP, FTP, Telnet) is determined by the port on which the connection was established. Therefore, it is just as valid to make the hostname service .dgis.com or public.distributedgis.org. The last part of the hostname, .org, is referred to as a top-level domain, which defines the type of organization within the United States or the country of origin (e.g., .ca = Canada, .uk = United Kingdom). Lastly, the middle part of the hostname plus the top-level domain name (distributedgis.org) is the unique domain name that is used to imply the origin of the Web site.

The IP address is used to identify unique host machines. But the IP port number is used to identify applications running on the host machine. A port is a logical channel or an entry point to an application running on a host machine. By using multiple IP port numbers, a single machine can provide multiple Internet applications, such as FTP, Telnet, e-mail, and Web server, at the same time. Note that this refers to a logical port, not a physical port, such as the serial or parallel port on the computer. A port number is represented by a 16-bit integer, ranging from 1 to 65535. Ports 1–1023 are reserved. The purpose of the port is to route requests to the proper applications that service them. For example, Web servers typically "listen" for connections on port 80, and 21 is a reserved port number for FTP. The combination of IP address and port number becomes a socket address. If for some reason you want to specify another port (e.g., 81) for the Web server, the user has to specify the whole socket address: www.distributedgis.org:81. A socket address is used to establish a connection between any two computers, such as a client computer and a server (Orfali et al., 1999).

The combination of TCP and IP supports routing and worldwide information access among many different types of computers and networks. It guarantees reliable information delivery to destinations over the worldwide network. Some protocols are written at the higher level application layer specifically for the TCP/IP stack, including SMTP, FTP, and Simple Network Management Protocol (SNMP).

2.3.3 Network Setting

From an Internet GIS application perspective, the most important task is to create a successful network connection for your Internet GIS servers. The following figure is one example of how to configure the TCP/IP setting under a Windows 2000 server. The first thing to do is to make sure that the TCP/IP suite has been installed in your machine. Then you need to get a unique IP address assigned by your network administrator. Some network systems utilize a dynamic IP mechanism. But a dynamic IP address may not be appropriate for a permanent Web server or Internet map server. The next step is to know the kind of subnet mask in your local network and the gateway IP. Usually, it could be 255.255.255.0 or 255.255.0.0. The gateway IP is usually very similar to your own IP address except the last number. The final step is to find out the DNS server located in your LAN (Figure 2.14).

We have built a foundation of networking models and protocols, so we now turn to how the information communicates from one computer to the other, starting from information exchanges between two computers, moving to the LAN, and ending with the WAN and the Internet.

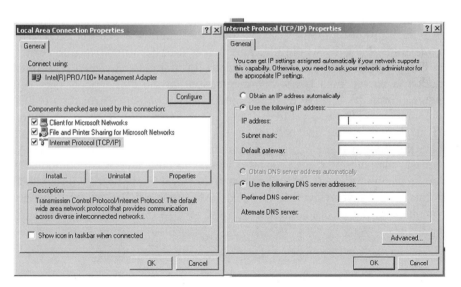

Figure 2.14 Network setting in Windows 2000 server.

2.4 INFORMATION EXCHANGE PROCESS BETWEEN TWO COMPUTERS

Suppose there are only two computers in a simple network that is connected with a cable: computers A and B. This is the simplest network, or a simple peer-to-peer network. (A P2P network means computers in the network are peers and have the same functions.) Assume a GIS program in computer A requests spatial data from computer B. We now see how this can be done.

Key Concepts

A P2P network means computers in the network have similar configu-rations and network functions. There is no client/server hierarchy in the network. Each machine can become a client or server based on assigned tasks.

As we all know, GIS data tend to exist as very large files. Transporting large files through the network at one time slows down the network. This is because large amounts of data sent as one chunk ties up the network and makes other kinds of communication through the network cable difficult or even impossible. The second reason is that when there are errors in the data transmission, the whole data file has to be retransmitted (Microsoft, 1999). This is a very inefficient way for information exchange over the network.

Therefore, the large chunk of data has to be broken into smaller and more manageable pieces. These small pieces are called packets or data frames. By breaking down large data chunk into smaller packets, the data transmission over the network cable can be done more quickly and efficiently. If there is error in transmitting some packets, only those affected packets need to be retransmitted, not the whole file. This makes it easier to recover the error.

The previous example can be used to illustrate step by step how infor-mation is passed in network communications in the OSI model. The com-munication in the TCP/IP model is similar. First, computer A makes a request to computer B. The request goes from the top application layer to the bottom physical layer, across the network to computer B, where the request travels from the bottom layer to the top application layer. Based on computer A's request, computer B starts to send data files to computer A. In computer B, the information flows from the highest layer (layer 7, application layer) down to the lowest layer (layer 1, physical layer) sequentially. At the receiving computer A, the information flows in the reverse order, from the lowest to the highest layer:

1. First, the GIS program in computer A sends a data request to the application layer in computer A.

2. The application layer in computer A first identifies if network resources connecting with computer B are available. It then invokes database access software to send the data request to computer B.

3. The data request is then moved down to the presentation layer, where original data request information is transformed into ASCII format. In addition, a header is added to indicate the original and destination information as well as other control information.

4. The data request packet with header information is moved down to the session layer, where a session or connection is to be established with computer B once the data packet is received by the session layer in computer B. More header and trailer information is added.

5. The original block of data gets broken into the actual packets at the transport layer, if necessary. In addition, sequence information is added at the transport layer that will guide the receiving computer B in reassembling the data from packets.

6. Addressing and routing information is added at the network layer and is used for delivering the data packets to the right destination.

7. The data packet then moves down to the data link layer, where the data packets are reformatted into data frames and the logical addresses are transformed into physical addresses. Now the data frames are prepared to move to the physical layer.

8. At the physical layer, the entire data frames are placed onto the network medium (e.g., network card, cable) and are sent across the medium. There is a receiver's address at the header of the packets that indicates that the packets are intended for computer B.

9. The network adapter card in computer B examines the receiver's address on all data frames sent on its connecting cable. If the destination address of the data frame matches the address of the network adapter card, the adapter card will intercept the data frame, which will then enter through the cable into the network card.

10. The physical layer in computer B receives the packet unit and passes it to the data link layer. The data link layer in computer B then reads the control information contained in the header added by the data link layer in computer A. The header is then removed, and the remainder of the information unit is passed to the network layer. Each layer performs the same actions: A software utility at each layer reads the corresponding addressing information on the data frame, strips it away, and passes the data frame up to the next highest layer. Therefore, the header becomes smaller and smaller as the data frame is passed from the lower layer to the upper layer.

11. When the packet gets to the session layer, the session layer software protocol reads the header information from its peer layer in computer A and establishes a connection or session for data communication.

12. When the packet finally gets to the application layer, all control information has been stripped away. What remains is the packet containing the original data request, in exactly the same form in which it was transmitted by the application in computer A. Finally, the application layer of system B passes the data to the recipient application program (a data server or a file server).

13. The application program (e.g., a data server) reads the data request from computer B and starts to send the data file to its application layer. The application layer in computer B then adds any control information (header) to specify the destination of the information and informs the application layer in computer A about what the information is and how it should be handled. The resulting information unit (a header and the data) is passed to the presentation layer, which adds its own header containing control information intended for the presentation layer in computer A. The information unit grows in size as each layer adds its own header (and in some cases a trailer) that contains control information to be used by its peer layer in computer A.

14. The whole spatial data are broken down into actual packets in the transport layer for efficient transmission. The data packets are then passed to the lower level layers in computer B to the physical layer and on to the application layer in computer A in the reverse order.

As you can see, the data transmission is created starting from the application layer of the OSI model and descends through all seven layers in the sending machine. At each layer, additional information relevant to that layer is added to the data. That is, additional header information is added to be read by the corresponding layer in the receiving machine (Figure 2.15). Except for the lowest physical layer in the OSI model, no layer can pass information directly to its counterpart on another computer. The information on the sending computer must be passed through all of the lower layers to the network cable and moves up to the receiving computer to arrive at the same level of the sending computer where the information originated (Figure 2.15).

2.5 INFORMATION COMMUNICATIONS IN LANs

The communication process in the simple two-computer network can be expanded to more computers in the LAN and WAN. The basic information flow from the sending machine to the receiving machine is the same. But it gets more complicated as more machines are added to the network, because the sending machine needs to know where the recipient machine is in the network and how to send the packets there. As the distance of the network grows, there is also an issue of signal degradation over distance. We are now turning to these issues. We start with the LAN and progress to the WAN.

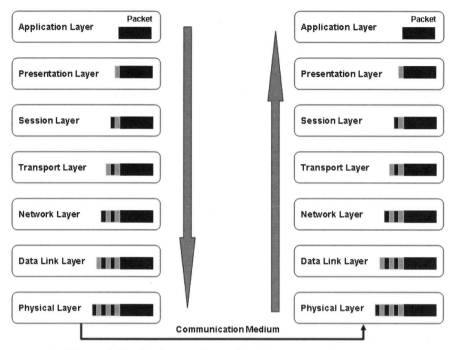

Figure 2.15 Packet-creating and packet-transporting process.

Networks have different sizes and cover different geographic areas. A network that consists of a small number of computers within a limited area is known as a LAN. LANs link computers and other devices within a single building or among several adjacent buildings over distances from a few feet to 6 miles (Microsoft, 1999).

As the LAN expands its geographic limits into a city, it becomes a Metropolitan-Area Network (MAN). Different offices of the same firm networked within the same city limits would form a MAN. MANs extend up to about 10 miles. As the geographic scope of the network spans into different cities or different states, the LAN grows into a WAN. The Internet is based on a WAN.

Besides the geographic areas covered, the major difference between the LAN and the WAN is the network topology and network communication protocols. A WAN is not simply an expansion of the LAN by laying out more cables and adding more computers; rather it involves different communication devices and software protocols.

2.5.1 LAN Architectures

A network's architecture defines the overall structure (topology) and its associated components, including hardware, system software, and/or protocols.

As more computers are added to form a LAN, we cannot simply link these computers together using network cables. Instead, we have to construct the LAN based on some architecture models. There are many types of network architectures for the LAN. This chapter introduces three types that are the most commonly used and the most relevant to Internet GIS. They include Ethernet, Token Ring, and wireless LANs. These are lower level protocols that deal with the two lower layers of the OSI model and the lowest layer of the TCP/IP model. For each type of LAN architecture, we will focus on how each architecture works and its relative performance.

2.5.1.1 Ethernet   Ethernet is currently the most popular baseband-type architecture for LANs. It typically uses a bus topology and relies on CSMA/CD (Carrier Sense with Multiple Access and Collision Detection) as a method to access and regulate traffic on a segment of a LAN. The typical transmission speed of Ethernet is 10–100 Mbps.

2.5.1.1.1 Network Topology of Ethernet

Key Concepts

> **Ethernet** *is typically constructed using a linear bus, star, or star bus topology. It uses CSMA/CD as an access method to regulate traffic on a segment of a LAN. The* **topology** *of a network refers to the way the computers are connected together using a cable.*

Ethernet is typically constructed using a linear bus, star, or star bus topology. The topology of a network refers to the way the computers are connected together using a cable. A *linear bus* is the simplest and most common method of connecting computers together in the Ethernet environment. All computers are connected with each other via a single cable that is also called a trunk line (or backbone or segment), as shown in Figure 2.16. All the computers (nodes) on a bus network are *passive* participants, always listening in on the line and checking for the destination addresses attached in the messages (Woodcock, 1999). Any nodes can send messages to the network, but only one computer at a time to avoid contention. The message is broadcast to the entire network, but only the node that the message is sent to can receive it. At the end of the cable there must be a device called a *terminator* to absorb signals and to prevent them from bouncing back to the network.

The advantage of this linear bus topology is its simplicity. It is very easy to add another node in the network. But more nodes in the network may affect the performance since the wait time is longer for any computer to send a message. In addition, when a cable is broken at one segment, the whole network will no longer function, which is called the network being "down."

In the *star* topology, computers are connected through a hub, as shown in Figure 2.17. Messages or data packets are sent from the sending machine

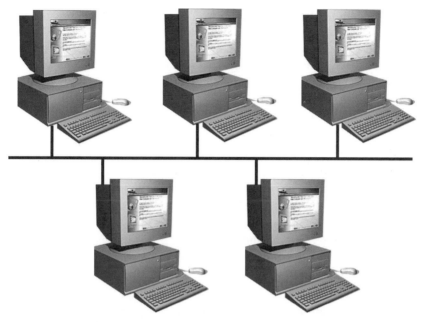

Figure 2.16   LAN with bus topology network.

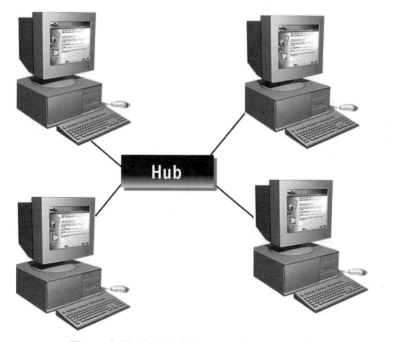

Figure 2.17   LAN with star topology network.

through the hub to all other computers on the network. The hub is the central component in a star topology. There are *active hubs* and *passive hubs.* Active hubs are just like repeaters (see Section 2.4.2) that regenerate and retransmit the data packets from the sending machine. Passive hubs act as connection points and do not amplify or regenerate the signal. While active hubs require electrical power to run, passive hubs do not. The function of the star-based network does not rely on individual computers. That is, when one computer or the cable connecting with that computer in the network fails to function, the problem is isolated to this one particular computer or cable while the rest of the network can function normally. The star topology also allows for placing LAN diagnostic equipment in the hub. But if the hub fails to function properly, the network will be down.

The *star bus* topology is naturally a combination of the bus and star typologies. In the star bus topology, two or more hubs are connected together in a linear bus trunk line, as shown in Figure 2.18. In this star bus topology, one failed computer will not affect the function of the network but a failed hub would bring the whole network down.

2.5.1.1.2 Ethernet Access Method In the Ethernet networks all machines connected to the network have equal access to the network, but only one machine can use it at a time. So there are some conflicts among the computers. To avoid this conflict, the Ethernet networks use the CSMA/CD as its access method, which is sometimes referred to as a contentious networking method (Microsoft, 1999). CSMD/CD works at the MAC sublayer of the data link layer in the OSI model. Its major functions are to allow computers to gain

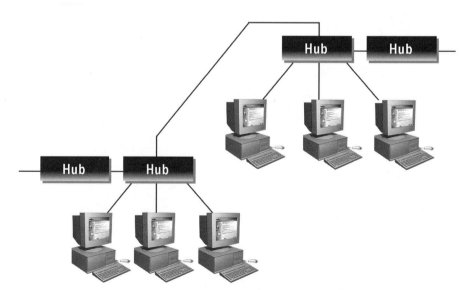

Figure 2.18 LAN with a star bus topology network.

access to the network when they have packets to transmit and to ensure two computers do not transmit packets at the same time.

The CSMA/CD works mainly in an Ethernet environment such as the linear bus or star bus network. It works like this: When a computer or other devices such as a printer are ready to transmit something, it first "listens" to its segment of the network to see if anyone else is using the network. If so, it will remain in the listen mode until it detects the network is free. It then transmits the data packets onto the network cable. The amount of data in each Ethernet-unique frame ranges from 64 to 1518 bytes in size (Woodcock, 1999). If the data segment to be transmitted is larger than 1518 bytes, it has to be broken down into smaller packets before transmission. On the receiving end, the receiving computer captures the packets and reassembles them into the data's original format. Once the data packets have been received, the network becomes available for other nodes to transmit data.

Sometimes, two computers may detect that the network is free and thus send data packets onto the network at the same time. This will result in what is called a data collision. When a collision occurs, the sending computer is able to detect it and then send out a jamming pulse to tell other computers to wait. The sending machines will then wait for a random time interval before sending out the data packets to the network again.

CSMA/CD works best when the number of nodes is small or the load of network traffic is light. But when the number of nodes and the amount of data exchange on the network segment increases, the performance of the CSMA/CD will be negatively impacted.

2.5.1.2 Ethernet Performance and High-Speed Ethernet

Ethernet's performance is determined by the Ethernet protocol. That is, no matter what transmission media it uses, whether copper wire or fiber, the transmission speed is the same, usually 10 Mbps (megabits per second). It can increase up to 100 Mbps for some high-speed Ethernet protocols. Under the moderate load of network traffic, Ethernet's performance is pretty good. But under heavy load of network usage, as the number of computers in the network increases and the network communication frequencies increase, the Ethernet network's bandwidth will reduce and its performance becomes unpredictable. Even though each transmission still commands the 10-Mbps Ethernet maximum transmission rate, other computers have to wait longer to obtain transmission rights.

To improve the Ethernet network's performance, we could use a switched Ethernet network. The difference between the switched Ethernet and the normal Ethernet network is that the normal Ethernet network uses passive hubs while the switched Ethernet networks employ switching hubs. The switching hubs can allow several pairs of senders and receivers to establish separate communication channels at the same time with the same 10-Mbps speed. Therefore, more nodes can send messages onto the network at the same time, which reduces the network contention among computers in the Ethernet trunk line and increases the bandwidth.

Another tactic to improve Ethernet performance is to make the Ethernet network segment smaller by dividing a crowded segment trunk line into two or more less-populated segments, which are then connected using either a bridge or a router (Microsoft, 1999). In this arrangement there are fewer computers in each segment and therefore fewer data collisions. Segment division works especially well for high-bandwidth applications such as Internet GIS, image process, and video streaming.

In addition to Ethernet switching and segment division, there are other ways to increase Ethernet transmission speed, most notably the 100BaseVG (Voice Grade) AnyLAN Ethernet and 100BaseX Ethernet (also referred as fast Ethernet), with transmission speed increased to 100 Mbps.

Though Ethernet networks vary, there are several features that are common to all Ethernet networks. First, they all use broadcast transmission to deliver signals to the entire network at the same time. Second, all Ethernet networks adopt the same CSMA/CD access method. Third, all Ethernet networks are defined in the IEEE 802.3 Ethernet specification. Lastly, most Ethernet networks are baseband networks (information travels on one single channel) rather than broadband networks (information can be multiplexed or travel on many different frequencies or channels within the band concurrently).

2.5.1.3 Gigabit Ethernet

The dramatic growth of network traffic is pushing the IT industry to look to higher speed network technologies that can offer a greater bandwidth. Among the new technologies, Gigabit Ethernet and its next generation 10 Gigabit Ethernet, with data rates aimed at 1 and 10 Gbps, seem to be the solution to congested backbones.

Gigabit Ethernet is an extension of the 10- and 100-Mbps IEEE 802.3 Ethernet standards. It has a raw data bandwidth of 1000 Mbps and supports new full-duplex operating modes. Gigabit Ethernet uses a combination of two proven network technologies—the standard adopts both the original IEEE 802.3 Ethernet specification and the ANSI X3T11 fiber channel specification for the physical interface. In other words, Gigabit Ethernet employs Ethernet protocols to manage frame transfer and media access on layer 2 and fiber channel optics, connectors, and cables on layer 1 (Gigabit Ethernet Technology, http://www.10gea.org)

Initially, Gigabit Ethernet will operate over optical fiber, but it can also be applied over twisted-pair cabling and coaxial cable due to the expected advances in silicon technology and digital signal processing. The initial applications of Gigabit Ethernet were for campuses or buildings requiring greater bandwidth between routers, switches, hubs and repeaters, and servers. That is, this technology first arrived in corporate or campus backbones, either to increase total bandwidth or to replace multiple 100-Mbps networking links (Lo, 1997).

2.5.1.3.1 Strengths of the Gigabit Ethernet

The first and foremost strength of Gigabit Ethernet is its relative low cost in terms of performance when compared to other technologies such as fast Ethernet and Asynchronous

Transfer Mode (ATM). The cost of ownership includes not only the purchase price of equipment but also the cost of training, maintenance, and trouble-shooting. These "follow-up costs" will be far lower than other technologies because the installed base of users is already familiar with Ethernet technology, maintenance, and troubleshooting tools.

The second strength of Gigabit Ethernet is its support of multiple standards and vendor interoperability. The members of Gigabit Ethernet Alliance are committed to developing this backbone technology to be able to support existing applications, Network Operating Systems (NOSs), network management platforms and applications, and network protocols such as IP, Internetwork Package Exchange (IPX), and AppleTalk.

Finally, easy migration is another attraction of Gigabit Ethernet. For example, a fast Ethernet backbone's existing switches can be interconnected by the addition of Gigabit Ethernet links and switches to form a high-capacity backbone network. This backbone can support a large number of switched and shared segments, all without displacing any desktop hardware or software.

2.5.1.3.2 Limitations of the Gigabit Ethernet   The first limitation of Gigabit Ethernet technology is its use of fiber. The initial Gigabit Ethernet standard calls for Ethernet frames over a fiber channel physical layer. The cost of installing fiber-optic cabling, if not already in place, will raise the overall price of a Gigabit Ethernet installation. In addition, the gigabit switches and network interface cards are also more expensive than the 10-Mbps and fast Ethernet counterparts due to the cost of fiber-optic transceivers.

Another limitation of Gigabit Ethernet is its segment length limitations, which are common in all Ethernet topologies. With multimode fiber, Gigabit Ethernet has a segment limitation of 500 m, compared to 2 km for both fast Ethernet and Fiber-Distributed Data Interface (FDDI). With Unshielded Twisted Pair (UTP) category 5 cabling, Gigabit Ethernet backbones likely will be limited to 25 m, compared to 100 m under both fast Ethernet and FDDI. This distance limitation may slow down the adoption of Gigabit Ethernet as a backbone technology.

Lastly, Gigabit Ethernet may not be an ideal candidate for voice and video delivery, even though it has a bandwidth of 1000 Mbps. Since Ethernet frames vary in size from 64 bytes to about 1500 bytes, the traffic streams are much grainier and less predictable. This is a problem because smooth, consistent traffic flow is very important for real-time, delay-sensitive packet transmission such as voice and video.

2.5.1.4 10 Gigabit Ethernet   The 10 Gigabit Ethernet adopts the IEEE 802.3 Ethernet MAC protocol, the IEEE 802.3 Ethernet frame format, and the minimum and maximum IEEE 802.3 frame size. The new 10 Gigabit Ethernet continues the natural evolution of Ethernet in speed and distance. Compared to Gigabit Ethernet, the major differences of 10 Gigabit Ethernet is that it is a full-duplex only and fiber-only technology, and it does not need the CSMA/CD protocol that defines slower, half-duplex Ethernet.

In conclusion, Gigabit Ethernet with the features of easy, straightforward migration, scalability, low cost of ownership, and support of multiple applications is a good choice for high-speed backbone networking. However, as Internet 2 will adopt the ATM technology to connect the gigapops, ATM may become the dominant backbone technology for most universities. Once network administrators have put in their money and efforts to purchase the expensive ATM and train their personnel, it would be impossible for them to give up on ATM (unless it proves to be a failure) and go back for Gigabit Ethernet. (We will discuss ATM in Section 2.6.2.3.)

2.5.1.5 Token Ring Distinct from Ethernet networks, Token Ring typically uses a Token Ring network topology and relies on token passing as a method to access and regulate traffic on a segment of a LAN. The typical transmission speed of Token Ring is 4 and 16 Mbps.

2.5.1.5.1 Token Ring Network Topology A typical Token Ring network uses a ring topology. The ring topology connects computers on a single circle of cable (Figure 2.19). Unlike a linear bus topology, the ring topology does not have terminated ends or terminators. The signals or data packets travel around the loop in one direction and pass through each computer or node on

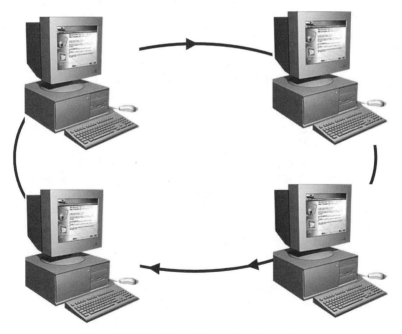

Figure 2.19 Token Ring network topology.

the ring. Each computer acts as a repeater to boost the signal and send it on to the next computer in line. Therefore, if one computer fails to function, the entire ring network will be down. This is not an ideal fault-tolerant situation since any computer or cable can fail at any time.

Therefore, the actual implementation of the Token Ring topology is usually a star shape, as shown in Figure 2.20. All computers are connected with the central concentrator, or hub, called the Multistation Access Unit (MSAU). The MSAU acts like a logical ring. That is, each computer is connected with a port in the MSAU and the data frames pass from port to port similar to a ring. Each MSAU has a finite number of connection ports. For example, an IBM MSAU has 10 connection ports. It can connect up to eight computers. An MSAU can also connect with other MSAUs. Each ring can have as many as 33 MSAUs or hubs (Woodcock, 1999).

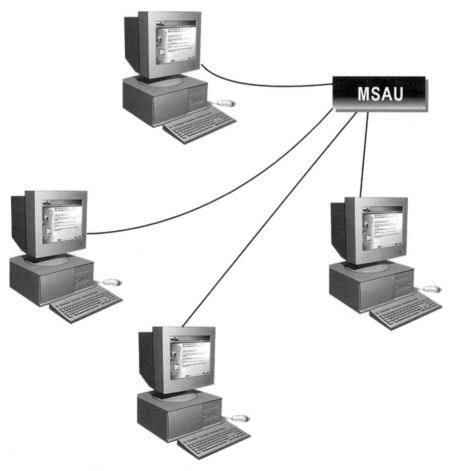

Figure 2.20   Token Ring network implementation.

MSAUs were designed to watch for failed network adapter cards or off-line stations. MSAUs will bypass any bad connections or failed computers to preserve the ring. Therefore, a faulty computer or connection will not affect the rest of the Token Ring network.

2.5.1.5.2 Token Ring Access Method: Token Passing   While Ethernet uses a contentious networking method, Token Ring adopts a noncontentious networking method—token passing. That is, any computer has the right to transmit data to the network but only one computer can do that at a time. A token is thus used to determine who gets this right. Whoever gets the token (there is only one token in the Token Ring network at a given time) gets the right to transmit the data; once the token is taken, other computers in the network cannot transmit any data and have to wait for the token to be released. Unlike Ethernet's CSMD/CD method, there is no contention in token passing.

A token is a predetermined piece of information or data frame that is generated by computers in the Token Ring network whenever there is a computer on-line. When a token is generated, it is continuously moving from one computer node to the next around the ring in one direction. When a computer wants to transmit data, it has to grab the token when the token passes through this computer. If the token is free, the computer will grab the token and mark it as busy. It then loads the data onto the network. If the data frame is too large, it has to be broken down into smaller frames. The data frame then travels around the ring until it reaches the destination computer whose address matches the destination ID in the frame.

The receiving computer unloads the data into its receive buffer and marks the data frame in the frame status field to indicate that the information has been received. It then releases the frame to continue traveling around the ring to the sending machine. The sending machine checks the frame header information to make sure the data have been successfully received. It then releases the token for other computers to use. Whoever gets the token can only transmit one single data frame before being required to pass the token to the next node.

2.5.1.5.3 Token Ring Performance   There are two speed levels for a Token Ring network: 4 and 16 Mbps. Token Ring network adapter cards are available in both 4- and 16-Mbps models. The 16-Mbps cards accommodate larger frame lengths than the 4-Mbps cards, so they can transmit the same amount of data with fewer transmissions. But in order for the 16-Mbps card to function at full capacity, the Token Ring network has to be run on the 16-Mbps card as well. The speed of the Token Ring network can be doubled by implementing a "full-duplex" version of the Token Ring, which allows computers on the ring to send and receive data simultaneously. Recently, the speed of Token Ring has improved to a maximum of 100 Mbps or even 1 Gbps (Woodcock, 1999).

Token Ring networks have better performance than Ethernet networks in a very heavy network traffic situation because the token-passing scheme has better predictability than the Ethernet's CSMA/CD scheme. But in the moderate-network-traffic load, the performance difference is small and does not justify the higher cost of Token Ring. On the other hand, the price of Token Ring has dropped significantly over the last few years. The price advantage of Ethernet is eroding.

Token Ring has a much smaller user base than the Ethernet. Therefore, while waiting to migrate to higher speed local data communications, much of the current activities are focused on faster versions of the Ethernet rather than faster versions of Token Ring.

Furthermore, a WAN technology, the 155-Mbps ATM (to be discussed in the WAN section), has found its way to the desktop LAN environment. Therefore, it is difficult to foresee what will be the ultimate winner as the fast network technology in the LAN. Nevertheless, the transmission speed will become faster and more predictable, laying a strong foundation for Internet GIS.

2.5.2 Components of LANs

In a LAN environment, computers are directly connected with cables in a segment. Different network segments are connected with bridges and routers. If different segments use different protocol stacks like AppleTalk or SNA, gateways are also needed to link these heterogeneous network segments. Some basic hardware components of LANs include repeaters, bridges, routers, and gateways.

Repeaters, bridges, routers, and gateways are network-connecting devices that are frequently used to construct a LAN. Repeaters and bridges forward data packets to different network segments within a network. Routers forward data packets between networks, while gateways are used to forward data packets between networks with different protocols, data-formatting structures, and network architectures. They each work at different layers of the OSI model. Repeaters function in the physical layer, bridges work at the data link layer, routers work at the network layer, and gateways usually use the application layer. A device that works at the higher layers is more intelligent and has more functions.

Repeaters and bridges are commonly used in smaller LANs. Routers and gateways are usually used in larger LANs and in the WAN. Data packets are usually handled first by repeaters and bridges and then by routers and, if necessary, gateways (Microsoft, 1999).

2.5.3 Wireless LANs

With the proliferation of wireless devices—laptops, PDAs and pocket PCs— and with the increasing need for mobility, users find it very convenient to use the wireless network to communicate with their office network. For example, employees in the meeting room often have a laptop and may need to check in the network for data, files, schedules, and e-mails. A wireless network would enable their laptops to communicate with the server and with desktop computers.

A Wireless LAN (WLAN) refers to the use of wireless communication technologies such as infrared and radio waves to connect PCs, laptops, printers, and hand-held devices with the wired network within a short distance. A WLAN is usually used to extend rather than replace wired LAN, at least for now. For devices to communicate with each other through wireless media, a wireless network adapter card with a transceiver must be installed in each device or computer. The transceivers act as access points through which signals are broadcast and received from each other.

It is also possible for two mobile devices, such as two laptop computers, to communicate with each other directly. In this case they can establish a temporary P2P or *ad-hoc network* (Woodcock, 1999). One device could be a nonmobile station on a wired LAN, the other device could be a mobile device such as a laptop or a PDA. They can form a more or less permanent wireless network. The mobile device can gain access to resources in the wired LAN. Data can then be exchanged between the wireless devices and the wired LAN.

There are several techniques for data communication in the LAN, including infrared, radio, microwave, and laser transmissions. However, the most common ones are infrared and radio. In addition, there are also some new technologies, such short-range bluetooth and 3G mobile communication.

2.5.3.1 IEEE 802.11 Wireless LANs
IEEE 802.11 specifies the physical and MAC layers for operation of WLANs and addresses the Direct-Sequence Spread Spectrum (DSSS) and Frequency-Hopping Spread Spectrum (FHSS) access methods for the radio medium. The standard provides for data rates of 1 and 2 MbPs—the later being optional (Pandya, 2000).

- 802.11 has three extensions: 802.11a, 802.11b and 802.11g (http://www.webopedia.com/TERM/8/802_11.html).
- 802.11a is an extension to 802.11 that provides up to 54 Mpbs in the 5 GHz band (also referred to as Wi-Fi5). It uses an orthogonal frequency division multiplexing encoding scheme rather than FHSS or DSSS.
- 802.11b (also referred as 801.11 High Rate or Wi-Fi) is an extension to 802.11 that provides 11 Mbps transmission (with a fallback to 5.5, 2 and 1 Mbps) in the 2.4 GHz band. 802.11b uses only DSSS, and it allows wireless functionality comparable to Ethernet. This specification has been implemented by many laptop wireless networking cards.
- 802.11g applies to wireless LANs and provides 20+ Mbps in the 2.4-GHz band.

2.5.3.2 HIPERLAN: High-Performance European Radio LAN
The starting objectives for the European wireless LAN specification were to provide performance (throughput, security) equivalent to a wireline LAN such as Ethernet and to provide some support for isochronous services such as

video and image. Some additional requirements included seamless roaming, a range exceeding 50 m, and low power consumption. The HIPERLAN standard operates at 23.529 Mbps and supports multihop routing, time-sensitive traffic, and power-conserving methods. The specification was developed by the European Telecommunications Standards Institute (ETSI) and is limited to the specification of the physical and data link layers, the latter being divided into Channel Access Control and MAC sublayers (Pandya, 2000).

2.5.3.3 Bluetooth Bluetooth is a promising new technology for small-area wireless communication. Bluetooth utilizes Radio Frequency (RF) channels to create a short-range, point-to-multi-point voice and data transfer communications among bluetooth-enabled devices. For example, you can insert a bluetooth network card into your notebook and your notebook can automatically be connected to the Internet by accessing a bluetooth station nearby your notebook. Bluetooth is becoming the wireless standard for connecting local computer devices and peripherals, such as printers, pocket PC, and network devices. The bluetooth technology enables multiple devices to work together under the same wireless communication standard. The link range of bluetooth devices is 10–100 m depending on the setting of relay stations. The bandwidth of bluetooth technology could provide up to 1-Mbps communication channels. One advantage of bluetooth is its omnidirectional links, which are better than single-directional links by the Infra-red Data Association (IrDA) standard. Bluetooth also can provide security and encryption functions for data protection. In general, bluetooth will be more and more popular in the next few years and may become the major wireless technology to support mobile GIS (www.mobileblietooth.com).

2.6 INFORMATION COMMUNICATIONS IN WANs

Though we have compared the speed of Ethernet and Token Ring architectures in LANs, the difference is negligible when compared to the speed of WANs. WANs currently operate at substantially slower speeds than LANs (Dixon, 1996). How much slower? About 5–100 times slower. For example, the speed for standard Ethernet is 10 Mbps, but the speed for T-1 lines is only 1.544 Mbps, and the speed of a 56K modem is only 56 kbps. Since the Internet GIS is a WAN application, the speed of WANs is much more important than that of LANs. In other words, the bigger challenge for Internet GIS in terms of performance is the speed limitation of WANs. (We will discuss Internet GIS performance in Chapter 9.) But why is there a huge difference in transmission speed between LANs and WANs? Why cannot we simply expand the LAN into a wider area? This section discusses different approaches and technologies in information communications in WANs.

There are generally two types of WANs: circuit switched and packet switched. Circuit switching is similar to the common phone system. When two communication parties initiate a communication link by phone or by modem through a central switching authority, a physical link or session is established. This physical link or session is maintained for as long as the two parties communicate (Dixon, 1996). Circuit switching can be used to transmit analog voice and video as well as digital data.

Packet switching, on the other hand, is a digital communication system. Unlike a central switching system that is employed in circuit switching, packet switching spreads switching responsibility across the entire network. Transmissions are broken down into small pieces or packets. These data packets will find their own way to the destination in the huge interconnected network. When part of the network is down, they can always find other ways to get to the destination (Dixon, 1996). Packet switching is a fast and efficient method to transmit data over WANs.

It should be noted that the difference between circuit-switched and packet-switched WANs is not that circuit switching is for analog transmission while packet switching is for digital transmission. Circuit switching can be used to transmit digital data as well. The difference is the way they package and route data. Transmissions between the sender and receiver using circuit switching operate in a point-to-point fashion. That is, there has to be a physical link established between the sender and receiver for as long as the communication is still on. All data or voice communications go through the same link from the sending point to the receiving point. As the number of points increases, the required point-to-point lines will increase exponentially (Microsoft, 1999; Woodcock, 1999). For example, a five-point communication requires establishing 10 point-to-point lines, as shown in Figure 2.21. A six-point communication would require 15 lines and so on (Dixon, 1996). Packet switching, on the other hand, does not rely on point-to-point communication. Instead, it follows a point-to-many-point communication pattern. That is, data transmitted from one point can follow different routes to get to the destination based on the traffic conditions on the network.

Using an analogy, circuit switching is like loading everything onto a train and sending it along a dedicated train track, while packet switching is like

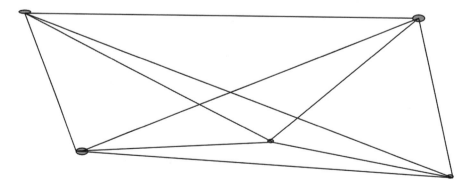

Figure 2.21 Point-to-point communication.

using a number of different trucks (Microsoft, 1999). Each truck driver can decide on the best route to the destination. If something happens to one truck, it is easier and cheaper to fix it and reload than if a train runs off the track.

2.6.1 Circuit-Switched WANs

Circuit-switched technologies are just like the telephone system in that a physical connection must be established between the sender and the receiver and maintained throughout the communication process. Circuit-switched WANs include analog connectivity technologies such as dial-up analog lines and digital connectivity technologies such as Switched 56, the T-carrier service, and ISDN.

2.6.1.1 Analog Connectivity The full phone network is an analog network that can be used for computer communication. This worldwide network is sometimes called the Public Switched Telephone Network (PSTN). Since data can be transferred over the voice-grade dial-up telephone line through modems, the PSTN can be thought of as one large WAN link.

2.6.1.1.1 Dial-Up Lines Dial-up lines are currently the common communication means for individuals to connect with the network from homes and even small businesses. Users connect with the networks through modems over the telephone line. However, computers cannot communicate with each other directly over the telephone line because computers communicate with a digital signal and a telephone line can only carry analog signals (sound) (Microsoft, 1999). A digital signal has only the two binary values of 0 or 1, but the analog signal is a smooth curve that can represent an infinite range of values (as shown in Figure 2.22).

In order for computers to send and receive information over telephone lines, modems are needed to convert or modulate the digital signals to analog signals at the sending end (e.g., a server) and transmit the analog signals onto the telephone lines. At the receiving end (e.g., individual user's machine), the modem will convert or demodulate the analog signals to digital signals for the computer to understand. When modems first make a connection, they have to establish a handshaking process. That is, they have to make sure they have the same standard and agree on the appropriate transmission speed. The transmission speed is the lowest of the two modem speeds. For example, if the server's modem speed is 56 kbps (kilobits per second) while the remote user's modem speed is 28.8 kbps, the actual communication speed would be 28.8 kbps.

The transmission rate is sometimes referred to as the *baud rate*. But the baud rate is not necessarily equal to the number of bits transmitted per second. Baud refers to the speed of the oscillation of the sound wave or the number of analog signal changes in a transmission per second. But by compression, each modulation of sound wave (or signal change) can actually encode more

Internet GIS Showcase: Batchawana Bay Interactive Tourism Guide

The Batchawana Bay Interactive Tourism Guide is an effort by the Community Development Corporation of Sault Ste. Marie & Area and its partners to promote tourism and recreation resources in rural Northern Ontario. It sees the region as underutilized for tourism relative to similar areas in the United States and attributes this lack of visitor traffic to travelers' ignorance about the area's geography, recreation opportunities, services, accommodations, and amenities.

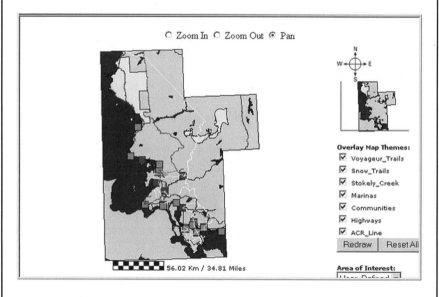

By delivering customizable maps over the Internet, potential visitors can research eating, sleepng, and shopping locations in this unique area. In addition, they can have locations mapped out for them before they visit.

This Web site marks the beginning of an initiative to promote outdoor recreaton across all of Northern Ontario. Batchawana Bay including the surrounding area was chosen as a pilot site to determine the effectiveness of integrating GISs and Internet technology for the purpose of rural tourism promotion. More than a simple advertising medium, the interactive guide contains useful information. The guide uses ArcView and MapObjects Internet Map Server (IMS) to allow visitors to build and customize maps that highlight activities, trails, and points of interest. The trails database is searchable by length, surface type, and suitable purposes (e.g., biking, hiking, driving).

Source: http://www.explorernorthernontario.com.

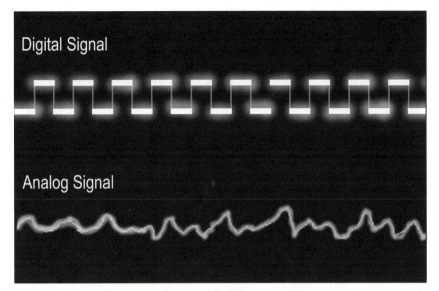

Figure 2.22 Digital signals versus analog signals.

than one bit of data. For example, a modem that modulates at 28,800 baud can actually send data at the speed of 115,200 bps (Microsoft, 1999). Therefore, bits per second should be used in describing transmission speed of dial-up lines.

There are two types of modems: asynchronous and synchronous. Asynchronous, or async, is widely used at home to connect home computers with the Internet over the telephone line. A synchronous modem is widely used to connect the computer to the digital line. It is faster and more efficient than the async modem, but it is too complex and too costly for home use.

Asynchronous modems transmit data in continuous bit strings. Each character—letter, number, or symbol—is turned into a string of bits. Each bit string is marked with start and stop information for identification purpose. The receiving computer uses the start and stop bit markers to check for the timing of the next block of data and to make sure the received data match what was sent. About 25% of the data traffic in async communication is spent on data traffic control and coordination.

Synchronous modems, on the other hand, transmit data in blocks known as frames. Communications between two modems are in a timed, controlled (synchronized) fashion. There is no need for start and stop bits. Transmission automatically stops at the end of the frame and starts at the beginning of the next frame. Therefore, the overhead of error control and coordination is much smaller than with the async communication. This results in a faster communication with synchronous modems than with the async modems. Synchronous communication is a more efficient way to send data but is more costly than async.

The speed of the dial-up service limits the transmission of large data and the downloading of client-side programs. Therefore, it is more suitable for server-side Internet GIS processing; for example, the server processes user requests and returns the results as a static image to the end user. It is barely enough to support more bandwidth-hungry client-side Internet GIS processing that involves the transmission of large data files and client-side applications such as Java applets.

2.6.1.2 Digital Connectivity Dial-up analog lines were designed for transmitting voice. Since the human ear is very versatile, a little noise will not prevent us from understanding what the people on the other end said. But the noise in the analog line poses a bigger problem for transmitting digital data. To reduce the errors in transmitting data on the WAN, Digital Data Service (DDS) lines could be used. Digital lines can transmit data 99% error free and thus are more reliable.

DDS uses digital communication; it thus does not require modems to modulate and demodulate signals. But it requires a modemlike device called a Data Service Unit/Channel Server Unit (DSU/CSU) to convert the standard digital signals generated by the computer into the type of digital signals (bipolar) transmitted in the synchronous communication environment (Microsoft, 1999). Another role of the DSU/CSU is to protect the DDS service provider's network. All forms of digital communication need DSU/CSU.

Besides DDS, there are other different forms of digital line, including ISDN and T-carrier facility.

2.6.1.2.1 Integrated Services Digital Network ISDN is a circuit-switched inter-LAN digital connectivity specification that accommodates the digital transmission of voice, data, and image over the same line. The goal of ISDN is to connect homes and businesses over copper telephone wires by converting existing telephone circuits from analog to digital.

ISDN divides its available bandwidth into two types of communication channels: a "B" channel for carrying voice, data, and images and a "D" channel for carrying only signaling control information, no data. Two B channels can move data at 64 kbps each with a total combined transmission rate of 128 kbps. Higher transmission rates or throughput can be achieved if both end stations can also support data compression (Dixon, 1996).

With a maximum 1.536 Mbps, ISDN presents a significant improvement over the dial-up analog connection. With a moderate data transmission, Internet GIS can run on ISDN with a reasonable response time. But Internet GIS still cannot achieve the same speed of GIS processing as is possible on the LAN environment.

2.6.1.2.2 Cable Modem Cable modems utilize existing TV cables as the media to transfer digital signals between users and create a high-speed WAN. Cable modems are deemed an important next step for cable companies in their efforts to add new services and, more importantly, to generate a significant amount of income.

According to cable modem manufacturers, cable modems have a downstream date rate that ranges from 6 to 30 Mbps and an upstream transmission rate from 515 kbps to 10 Mbps. The transmit speed of a cable modem is much faster than for the traditional telephone modem or ISDN.

Cable modems do have their limitations. The most critical limitation is the incapability to carry a two-way transmission in some CATV systems. In this case the subscribers have to tie up a phone line to connect to the Internet for the upstream transmission, which causes speed decreases, line occupation, and phone bill increases. Another limitation is that users may run into a bottleneck as more people get on the Internet or watch TV.

2.6.1.2.3 Asymmetric Digital Subscriber Line   ADSL is a technology being developed by telecommunication companies in the hope of clearing the bad name created by ISDN and to compete with CATV companies offering Internet service. Similar to ISDN, a pair of ADSL modems are connected to each end of a twisted-pair copper phone line, creating three information channels—a high-speed downstream channel, a medium-speed upstream channel, and a Plain Old Telephone Service (POTS) channel. The speeds of the downstream channel range from 1.5 to 9 Mbps, while the upstream channel ranges from 16 kbps to 1 Mbps. Speeds vary depending on the loop length. POTS is generally restricted to about 52 kbps (http://www.webopedia.com/TERM/P/POTS.html).

One limitation with ADSL is that it will not work on any line with a loading coil in place. Estimates are that 15–20% of the U.S. local loops have loading coils. This will create problems as telecommunication companies try to connect ADSL modems to phone lines.

2.6.1.3 T-Carrier Service   Like other circuit-switched WANs, T-1 is a point-to-point digital transmission technology that is the most widely used high-speed WAN connection. T-1 uses two-wire pairs (one pair to send and the other to receive) to transmit digital voice, video signals, and data at a speed of 1.544 Mbps (Microsoft, 1999).

T-1 uses technology called multiplexing, or muxing, developed by Bell Labs. Multiplexing collects several signals from different sources into a component called a multiplexer and sends them into one single physical line for transmission. At the receiving end, the data are demultiplexed back into the original form (Dixon, 1996).

Since T-1 is quite expensive, some opt to subscribe to only one or more T-1 channels in 64-kbps increments called fractional T-1 (Microsoft, 1999). On the other end of the T-carrier service spectrum is the T-3 and fractional

T-3, which provide voice and data-grade service from 6 to 45 Mbps. T-3 could have the same or even higher speed than the LAN. One T-3 line can be used to replace several T-1 lines. Copper wire can accommodate the speed of T-1, but fiber-optic or microwave is required to support T-3 or T-4.

T-1 is adequate to support some Internet GIS applications so long as the data communication between the client and server is moderate. T-3 is certainly enough to support most Internet GIS applications. Running Internet GIS on T-3 would be equivalent to running desktop GIS programs on the LAN environment. But T-3 has not yet been widely implemented, and wider use is uncertain given the development of packet-switched communication technologies.

2.6.2 Packet-Switched WANs

While circuit switching is a point-to-point technology that requires a dedicated pathway between sender and receiver, packet switching does not require a dedicated pathway; it breaks up the data at the sending end and then sends packets over many different possible paths to get to their destination.

In general, packet switching works like this. The original data package is broken down into smaller packets and each packet has its own destination and other control information at its header. Each packet is switched separately and is directed by switches and exchanges along the best route available between the source and the destination at the time of travel. Each packet may follow a totally different route to get to the same destination. The computer at the receiving end receives all packets and reassembles them. If there is an error in a particular packet, only that packet needs to be retransmitted (Microsoft, 1999).

Packet-switching networks are fast, efficient, and reliable and thus are ideal to be used for Internet GIS applications. There are several technologies for packet-switching networks, including X.25, frame relay, ATM, FDDI, Synchronous Optical Network (SONET), and Switched Multimegabit Data Service (SMDS).

2.6.2.1 X.25 X.25 is a set of protocols used specifically in a packet-switching network. It was originally designed to use telephone lines to transmit data. Because the telephone line is a very unreliable medium, particularly for data transmission, X.25 incorporates extensive error checking to ensure the reliability of the data transmission. It is thus very reliable. But the error checking takes away much possible bandwidth and therefore limits the top speed of any X.25 networks to about 56 kbps (Dixon, 1996; Woodcock, 1999). With a transmission rate of 56 kbps, X.25 is not really adequate for Internet GIS applications.

2.6.2.2 Frame Relay Frame relay is an enhancement to X.25, but it is mainly concerned with fiber-optic media. Frame relay defines a Public Data

Network (PDN) interface by dropping many error-checking and accounting functions that are not necessary in a reliable and secure fiber-optic circuit environment. Since the fiber-optic network is much more reliable and secure than the copper telephone line, it becomes unnecessary to check errors in each step of the packet forward process. By dropping much of the error-checking process in between, frame relay increases its transmission speed up to that of a T-1. Typically, the transmission rate of frame relay is in 56-kbps increments up to T-1 speed. This is faster than most other switching systems that perform the same basic packet-switching operations (Microsoft, 1999).

Frame relay is a point-to-point system that uses Permanent Virtual Circuits (PVCs). This means that there is no need to find a best route for transmission and all packet deliveries follow the same route. The frame relay networks require frame-relay-compatible routers and switching equipment to link LANs with WANs.

2.6.2.3 Asynchronous Transfer Mode
Both X.25 and frame relay deliver data packets in variable sizes over the network. This variable-sized packet is not the most efficient way to transmit. ATM uses a small, fixed packet unit (53 bytes) called a *cell* in network transmission. Each cell has 48 bytes of data and 5 bytes of header information. These small, consistent, and uniformly sized packets are transmitted and routed faster in the network (Woodcock, 1999). This is one of several reasons that ATM is so fast. ATM works in both LANs and WANs with a transmission speed of between 155 and 622 Mbps or more (Microsoft, 1999).

ATM supports simultaneous delivery of data, voice, and video over the same communication line. It supports different cable media as well, such as traditional coaxial cable, twisted pair, fiber-optic, emerging fiber channel, FDDI, and SONET. ATM is fast, reliable, and scalable and holds great promise for communication in WANs. However, ATM networks require special ATM-compatible hardware and exceptional bandwidth, which are expensive and not widely available. This is one reason that ATM has not yet been widely implemented.

2.6.2.4 Fiber-Distributed Data Interface
FDDI is another emerging technology that promises to deliver high-speed (100-Mbps) transmissions on a network. It adopts the token-passing network architecture and uses fiber-optic media. FDDI's transmission speed is much higher than existing Ethernet (10-Mbps) and Token Ring (4-Mbps) architecture (Microsoft, 1999).

FDDI uses a standard token-passing architecture but with three important improvements. First, the standard Token Ring network allows only one frame at a time to circulate in the ring. But FDDI allows more than one frame to circulate in the ring at the same time. Within a predetermined time slot when the computer has the token, the computer can transmit as many frames as it can produce. Therefore, there may be several frames circulating the ring at a time. This is one reason FDDI is much faster than the standard Token Ring network.

Second, while the standard Token Ring network uses a single cable in the ring, FDDI uses two cables in the ring. One ring is the primary ring and the other is the secondary ring. Frames in two rings travel in opposite directions. But packets normally travel in the primary ring. The secondary ring is used only when the primary ring has failed. In other words, the dual-ring system is used for redundancy and backup. This increases the system reliability. Each FDDI network can have a maximum of 500 computers and 100 km of cable (Microsoft, 1999).

Third, FDDI uses beaconing to isolate and identify failures. When a computer discovers a fault, it sends out a beacon to the network. Other computers also send out beacons. When a computer receives a beacon from its upstream neighbor, it stops. If the computer does not receive the beacon from its upstream neighbor, it successfully pinpoints the computer that has a fault, the one upstream from the computer that is still sending beacons. When every computer receives beacons, it assumes the problem has been fixed and normal operation will then resume.

Strictly speaking, FDDI is not a WAN technology because its maximum ring length is limited to 100 km (62 miles). It is often considered a fast backbone network to link LANs within a city or metropolitan area. It can also be used in LANs. It will be a great network for Internet GIS applications as well as for other high-bandwidth applications such as video, Computer-Aided Design (CAD), and real-time large-scale distributed scientific modeling and simulation.

2.6.2.5 Synchronous Optical Network

SONET is an American National Standards Institute (ANSI) standard to transport different information, including data, voice, and video, through optical cables. It has been incorporated with and is now almost equivalent with the Synchronous Digital Hierarchy (SDH) recommended by the International Telecommunications Union (ITU), a standards organization for international telecommunications. SONET is the name used in North America and Japan, while SDH is used in Europe.

SONET is a long-distance transport service on an optical network. But it can handle information coming from and going to nonoptical networks at the origins and destinations. It is similar to transporting goods from a manufacturer to a consumer using both trucks and trains. The goods are first transported from the manufacture using trucks to a train station. They are then repackaged in a container in a train and shipped on a train following rail tracks to a destination rail station, where they are transferred to a truck again to ship to the consumer. SONET first gets the data, voice, or video in the form of Synchronous Transfer Signals (STSs) from a nonfiber media like a T-1 line; it then converts them into Optical Carrier (OC) levels and transports them on the fiber-optical lines. When the data reach the destination of the fiber-optic line, the SONET converts the data again into STS; the data will then be transported on another nonmedia carrier to the final destination. This process is illustrated in Figure 2.23.

In Figure 2.23, information in the form of STS comes from different sources or tributaries and is converted into OC at the multiplexer station. It is then transported in a fiber-optic network. The fiber-optic network consists of fiber-optic cables, switches, routers, and repeaters. The OC information is then converted again at the SONET demultiplexer station into STS format and is transported into other tributaries to its final destinations.

The transmission speed of SONET is based on multiples of a single base signal rate of STS-1 (Synchronous Transport Signal level 1) or equivalent optical OC-1, which is 51.84 Mbps. For example, STS-3 is three times the rate of STS-1, which is 155.52 Mbps (3×51.84). The top rate STS-48 could reach the speed of 2.488 Gbps (gigabits per second). As you can see, this is incredibly fast for any application. But then the bottleneck will be the speed of different tributaries at the origins and destinations.

2.6.2.6 Switched Multimegabit Data Service SMDS is a broadband switching service offered by telecommunications carriers to provide businesses with a high-bandwidth service to connect their LANs at different locations (Microsoft, 1999; Woodcock, 1999). SMDS is a connectionless packet-switched service. That is, it does not require the connection to be always "on." But it is always available when needed. The bandwidth can also be shared by many users. Therefore, SMDS is less expensive than a dedicated leased line. SMDS simply passes the information from one LAN to the other without the complicated error checking, flow control, or format conversion. The speed of SMDS ranges from 1 to 45 Mbps.

Incoming STS Tributaries

Outgoing STS Tributaries

Optical SONET Transmission

SONET Multiplexer

SONET Demultiplexer

Figure 2.23 Illustration of SONET transport process. (Adapted from Woodcock, 1999, p. 214. Reprinted with permission from Microsoft Press.)

2.7 THE INTERNET AND FUTURE DEVELOPMENT

Key Concepts

The **Internet** *is an interconnected internetwork including* **backbone networks, regional networks, LANS, WANs, ISDN Lines, DSLs, cable modems,** *and so on.*

The Internet is an interconnected internetwork including LANs and WANs. Different LANs and WANs may have totally different network protocols and architectures. The Internet consists of vast amounts of communication lines, servers, gateways, routers, and computers. The structure of the Internet includes different network levels. The first level is the *backbones,* which are major communication lines across major cities. In the United States, the early development of a national backbone connects four major cities: San Francisco, Chicago, New York, and Washington, DC. The connection points near these four cities are called Network Access Points (NAPs) (Woodcock, 1999).

The second level of the Internet is the regional networks. In the United States, these regional networks include the northwest, midwest, west, east, southeast, northwest, and central California (Woodcock, 1999). The regional networks communicate with others through their connections with the *backbones.*

The third level of the Internet is the individual LANs and WANs within the regional networks, also called stub networks (Woodcock, 1999). These stub networks are connected with the regional networks and the backbone through routers, gateways, and other network devices.

For individual Internet users, the connection with the Internet is through modems and telephone lines, ISDN lines, or Digital Subscriber Lines (DSL). But they have to first dial up with the Internet server at one of the many ISPs such as AOL Time Warner, AT&T, MCI, or Microsoft Network (MSN). The ISPs provide a pipeline for individual Internet users to connect with the Internet. The server at the ISP connects with the Internet by connecting with the regional networks and the backbone. Once the user is connected with the ISP, he or she can explore all of the worldwide resources available on the Internet.

2.7.1 Connections with the Internet

Most current Internet users rely on existing phone lines to connect with the Internet. The user computers at home or office have to establish a connection or session with a remote server at an ISP. To do that, both machines have to agree on certain communication details such as the size of frames and error controls. These tasks are handled by IP protocols at the data link layer, particularly the Point-to-Point Protocol (PPP).

PPP is an Internet standard protocol that supports other multiple protocols such as TCP/IP, IPX, AppleTalk, and so on. It has two components: the Link Control Protocol (LCP) and the Network Control Protocol (NCP). LCP is used to initiate, negotiate, and end a computer-to-computer link, while the NCP is used to handle the details of data transmission.

Here is a brief description of the working process of PPP (Woodcock, 1999):

1. A user dials up from a computer modem to the modem on the server at the ISP.
2. LCP establishes a link between the user's PC and the server at the ISP and negotiates the communication details like the frame type and the packet size.
3. The NCP is then used to configure protocol-specific options to be used during the connection session, such as assigning dynamic IP addresses.
4. Once the session is established and the user's PC can directly talk with the server, the data communication process starts.
5. When the user terminates the connection, NCP dismantles the network layer connection.
6. The LCP terminates the connection gracefully.

Besides PPP, there are other IP protocols such as Serial Line Internet Protocol (SLIP) and Compressed Serial Line Internet Protocol (CSLIP). Those may be replaced by the PPP in the near future and hence are not discussed here.

2.7.2   Future Development of the Internet

The Internet is still growing very fast every day. Hundreds of thousands of Internet applications are under development every month. Millions of computers are joining the Internet every year. However, the growth of the Internet applications and computers is faster than the capacity of the current Internet infrastructure, which was originally developed in 1980s. Currently, the most challenging task for the Internet is to provide very high speed communication channels and very broad bandwidths for multimedia and high-end computing applications, such as Internet GIS, real-time video, and virtual reality. It is therefore essential to establish new technologies and infrastructures for the future Internet applications. There are many projects and research initiatives focusing on the future Internet technologies and applications. The Internet 2 and the NGI are two examples. The Internet 2 is a university-led project and the NGI is a federal government–supported initiative in the United States.

2.7.3 Internet 2

The goal of the Internet 2 is to facilitate research on advanced network technologies and applications. The role of the Internet 2 is not to replace the current Internet but to improve the current capabilities of the Internet by using high-speed communication technologies and creating new network backbone infrastructures.

Different from the current Internet, which was developed originally by the DOD, the Internet 2 is a university-led effort to develop advanced network technology and applications. In October 1996, thirty-four U.S. universities formed the Internet 2 Consortium and created the University Corporation for Advanced Internet Development (UCAID) to support the Internet 2. Currently more than 170 universities participate in the project and work closely with industry and the federal government (Fingerman, 1999)

The Internet 2 receives over $80 million in annual funding from UCAID's university members and corporate members (Internet 2 frequently asked questions, http://www.internet2.edu). The research of the Internet 2 focuses on several new network technologies, such as IPv6, multicasting, Quality-of-Service (QOS), and very high speed networks.

The Internet 2 also created a high-speed network infrastructure by connecting over 150 Internet 2 universities and institutes as "gigapops" that are connected to high-performance backbone networks. Gigapops are the high-bandwidth nodes or hub universities that can serve as the portal to various Internet 2 applications.

There are two kinds of technologies used in such high-speed backbones. The very high performance Backbone Network Service (vBNS) was developed and supported by the NSF and MCI (now part of WorldCom) and can provide 622 Mbps–2.4 Gbps bandwidth communication between two backbone nodes. Another network is the Abilene network supported by UCAID and its partners. Abilene is developed by Qwest, Nortel, and Cisco together and utilizes high-speed SONET facilities to provide connections between gigapops. The bandwidth of the Abilene network is between 2.4 Gbps and 9.6 Gbps. The two types of backbone infrastructures can be interconnected by vBNS+ (the next generation of vBNS) (see Abilene Project frequently asked questions: http://www.internet2.edu/abilene/html/faq-general.html).

Besides the creation of network infrastructure, the Internet 2 also focuses on network applications such as QOS and multicasting. QOS is a collection of technologies that allow users to receive predicable service levels and ensure that high-performance applications get enough bandwidth when they need it. Different types of Internet applications will get different kinds of service levels under the QOS control. For example, a real-time distance medical operation will have the highest level of services (high bandwidth) compared to the low-level, on-line chatting services (WorldCom white paper: Ipv6 Services Overview, June 2001: http://www.vbns.net/index.html). Multicasting focuses on the network technologies that can support large numbers of users to access broadcasting materials that require high-speed communications and broadcasting capabilities. Multicasting technology can be applied for distance learning and daily news.

Actually, one of the biggest potential applications for Internet 2 is Internet GIS. Currently, many Internet GIS packages and software are limited by bandwidth constraints and cannot provide interactive mapping and spatial analysis functions. Users may need to wait for as long as 10 minutes to download a Java viewer or plug-in viewer. Under the Internet 2 high-speed infrastructure and QOS control, many Internet GIS applications could be developed and operated with reasonable performance, such as real-time flood control and traffic control.

2.7.4 NGI and Large-Scale Networking (LSN)

A parallel but different project is the NGI sponsored by the U.S. federal government (http://www.ngi.gov). The NGI project was a three-year project started in 1997 with a $300 million commitment. The focus of NGI was the need for collaboration among federal government agencies and to connect existing high-speed network projects within the federal government, including the NSF, the U.S. National Aeronautics and Space Administration (NASA), National Institutes of Health, Environmental Protection Agency (EPA), and National Institute of Standards and Technology (NIST).

In 2001, the NGI was replaced by a new project, LSN, which also focuses on future network technologies. There are three research areas identified by the LSN (Workshop report: http://www.itrd.gov/iwg/pca/lsn/lsn-workshop-12mar01):

1. *Network technology research* will include fundamental research (agile optical transport networks and wireless technologies), network management and security, and network modeling/scaling.
2. *Network applications research* will focus on, for example, data-intensive computing, collaboration technology, visualization, and remote operation.
3. LSN will create several *testbeds* for these technologies and applications.

In general, "The goal of LSN R&D is to provide leadership in network communications through advances in high performance network components; technologies that enable wireless, optical, mobile, and wireline communications; large scale network engineering, management, and services; and systems software and program development environments for network-centric computing" (National Coordination Office for Computing, Information, and Communications, 1999, p. 2).

In conclusion, the development of the advanced Internet technology like the Internet 2, NGI, and LSN will no doubt improve network performance, which laid a strong foundation for the development of Internet GIS in the future.

This chapter introduced the theoretical and practical frameworks of the Internet, such as OSI model and TCP/IP, and related networking technologies, including LAN, WAN, and the Internet. These network technologies and protocols can provide the fundamental communication channels and network bandwidths for Internet GIS applications. The next chapter will begin to illustrate software components and programming tools of Internet GIS that build upon these networking technologies.

WEB RESOURCES

Descriptions	URL
Cisco on-line document	http://www.cisco.com/univercd/cc/td/doc/ cisintwk/intsolns/index.htm
10 Gigabit Ethernet Alliance	www.10gea.org
History of ARPANET	http://www.cybergeography.org/atlas/ historical.html)
Network communications (course note)	http://www.eli.sdsu.edu/courses/spring97/cs596/ notes/networks/networks.html
DNS	http://www.rad.com/networks/1995/dns/dns.htm
Internet 2	http://www.internet2.edu/
NGI	http://www.ngi.gov/
Abilene project	http://www.internet2.edu/abilene/html/faq-general.html

REFERENCES

Dixon, R. (1996). *Client/Server and Open Systems.* New York: Wiley.

Fingerman, S. (1999). Internet2 and Next Generation Internet: Two for the Future. *Information Outlook,* November. URL: http://www.findarticles.com/cf_dls/m0FWE/ 11_3/57785870/print.jhtml.

Lo, S. (1997). Inside Gigabit Ethernet. *Byte,* May. URL: http://www.byte.com/art/9705/ sec5/art2.htm.

Microsoft. (1999). *MCSE Training Kit Networking Essentials Plus,* 3rd Ed. Redmond, Washington: Microsoft Press.

National Coordination Office for Computing, Information, and Communications. (1999). *High Performance Computing and Communication FY-1999-FY2000 Implementation Plan.* URL: http://www.itrd.gov/pubs/imp99/ip99-00.pdf.

Orfali, R., Harkey, D., and Edwards, J. (1996). *The Essential Distributed Objects Survival Guide.* New York: Wiley.

Pandya, R. (2000). *Mobile and Personal Communication Systems and Services.* New York: IEEE Press.

Peterson, L. L., and Davie, B. S. (1996). *Computer Networks, a Systems Approach.* San Francisco, California: Morgan Kaufmann.

Stallings, W. (1997). *Data and Computer Communications.* Upper Saddle River, New Jersey: Prentice-Hall.

Woodcock, J. (1999). *Step Up to Networking.* Redmond, Washington: Microsoft Press.

CHAPTER 3

CLIENT/SERVER COMPUTING AND DISTRIBUTED-COMPONENT FRAMEWORK

- *Past: Mainframes = Sanity and Order;*
- *Present: Open Systems = Insanity & Chaos;*
- *Future: Closed & Open Systems = Peaceful Coexistence*
  —M. Malek (1995, p. xii)

3.1 INTRODUCTION TO DISTRIBUTED SYSTEMS AND CLIENT/SERVER COMPUTING

The implementation of Internet GIS requires not only network infrastructures to move geospatial information but also software architecture to provide interactive GIS functions and applications. In Chapter 2, we discussed the network infrastructures and hardware specifications for wired Internet GIS and wireless mobile GIS. These network technologies provide high-speed communication channels for publishing and accessing geographic information via networks. This chapter will focus on the software technologies that empower the capabilities of Internet GIS to allow distributed GIS users to access, download, and operate GIS applications remotely in real time. The first section will introduce the basic concepts of distributed systems from a software perspective and introduce a GIS user scenario. The user scenario will be used to demonstrate two types of client/server software architecture: traditional client/server systems and distributed-component frameworks. This chapter will also provide a comprehensive introduction to distributed-component technologies, which include Microsoft's Distributed-Component Object Model (DCOM)/.NET, Sun Microsystems's Java platform, and OMG's CORBA. Finally this chapter will discuss the advantages and disadvantages of these distributed software architecture and technologies and their implications to Internet GIS.

3.1.1 Site Selection Scenario

To illustrate the different uses of GIS software in a typical site selection process, we consider the following example. We will use this example to introduce different ways of conducting spatial analysis: using the traditional desktop GIS that implements the client/server architecture and the emerging Internet GIS approach that adopts the distributed-component systems.

SCENARIO DESCRIPTION   A GIS spatial analyst, Dick, wants to locate a new Wal-Mart store in Boulder. He needs to obtain related map information and perform a GIS overlay analysis for this task. The following criteria must be used to guide the Wal-Mart site selection:

1. The land use must be close or adjacent to an urban residential area.
2. The site must lie above the 500-year flood plain.
3. The site must be located within 200 meters of a major road.
4. The neighborhood of this site (within 1 mile) should have appropriate demographic characteristics: medium to high income (annual household income above $30,000), high consumption age group (median age below 40), higher population density (population density above 1000 per square mile), and a low crime rate (crime risk index not greater than 2).*
5. The new store will need 50,000 square feet in a compact shape.
6. The site fulfills the previous criteria (1–5) and has the lowest land value.

Based on this scenario, Dick will need the following data for his analysis:

- land use,
- flood zone,
- roads,
- census data and crime risk index, and
- land values and parcel records.

This scenario also requires four kinds of GIS task: spatial overlay, buffering, shape fitting, and data conversion (Figure 3. 1). Three human actors need

---

*The crime risk index scales from 1 (lowest) to 5 (highest). The crime risk index was generated by the Boulder County Policy Department based on the annual crime records for each census tract in Boulder County.

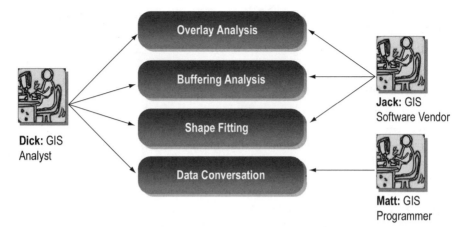

Figure 3.1 Wal-Mart site selection scenario.

to collaborate. Dick, a GIS analyst, wants to select a site for Wal-Mart. Jack is a GIS software vendor who sells GIS packages and provides a general technology solution for GIS users like Dick. Matt is a freelance software programmer who can customize software functions by writing stand-alone and distributed GIS components.

3.1.2 Traditional GISystems Solution

Dick goes to the Boulder County Planning Department and asks for the required data sets. The Planning Department has land use, flood zone, and census data sets. Dick finds out that land values and parcel records are stored in the Tax Assessor Office, information about roads is stored in the Colorado Department of Transportation (CODOT), and the crime risk index is stored in the Police Department. Dick goes to each department one by one and finally gets all the required data sets. Dick spends two weeks contacting people and requesting data from these departments. Dick also finds out that the Police Department does not have the crime risk index for the current year. The Police Department tells Dick that the updated index will be released in their annual reports six months later. So, the Police Department gives him the original crime report records of this year in a text-based format. Dick comes back to his office and uses his GIS software to display the data sets provided by the Planning Department, CODOT, and Tax Assessor office. He uses the overlay function and buffering function in his desktop GIS program and generates a map called procedure A, which includes the area above the 500-year flood plain, close to the urban residential area and within the 200 meters of major roads. The procedure A map includes four candidate polygons (Figure 3.2).

Next, Dick converts the crime data from thousands of text records into the crime risk index in a map format. Dick gets the equation for calculating the

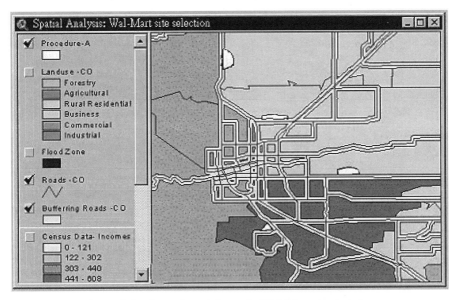

Figure 3.2 Procedure A layer in Wal-Mart site location.

crime risk index from the Police Department. Dick uses desktop GIS software to convert these text data by address matching, classifying, and overlaying the crime rate areas with the census tracts. The conversion task is very time consuming because Dick needs to reformat the crime rate records, import them into the GIS database, and write a program to calculate the index. Dick spends one week finishing the crime rate data conversion. Finally, Dick generates another map layer of the crime risk index that integrates the demographic data and crime rate in census tracts.

Dick's next step is to generate a 1-mile buffer from the center points of the four procedure A polygons and to further identify the demographic characteristics and crime risk index within these buffer zones. Dick compares the demographics statistically and chooses one polygon with the lowest crime rate, high population density, and highest incomes as the potential site for the location of the Wal-Mart site (Figure 3.3).

The final step is to determine footprints for the exact location of the Wal-Mart building that can fit inside the candidate land parcels. However, Dick finds out that in his GIS software there is no such rectangle shape-fitting function. He makes a phone call to Jack, the GIS software vendor, and asks if extension modules are available. A few hours later, Jack calls back and says that no such function is available right now, but the GIS software company is willing to develop this new function in the next version to be released next year. Dick decides to take an alternative approach. Dick uses the graphic tools to draw several squares inside the potential areas by hand (Figure 3.4).

Dick inserts these squares into another map layer and identifies them to the land value parcel in order to generate the cost of land for each potential

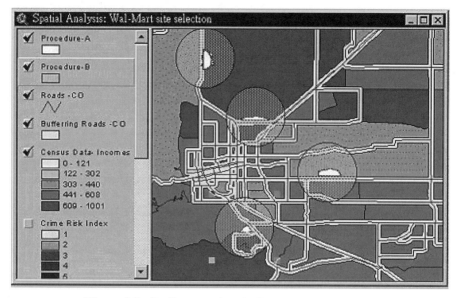

Figure 3.3 Buffer procedure in Wal-Mart site location.

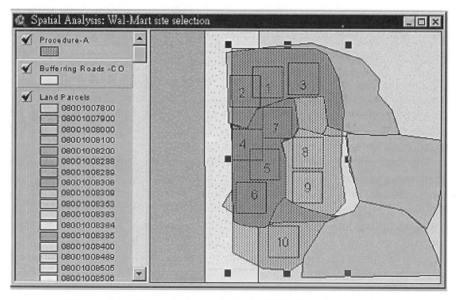

Figure 3.4 Shape-fitting analysis for Wal-Mart site selection.

site. Since Dick is manually doing the drawing and calculation, he can only test 10 possible squares in the candidate sites. Finally, Dick picks four sites with the lowest land value. The four sites will be considered as the possible locations for the new Wal-Mart store in Boulder. Dick spends about four weeks to finish this GIS project.

3.1.3 Internet GIS Solution

In order to complete the site selection using Internet GIS, three players must collaborate: Dick, Matt, and Jack.

DICK'S ACTIONS   On his local machine (a GIS node), Dick launches his GIS software. He uses the map display component and asks for the required Boulder maps (flood zone, land use, census data, roads, crime risk index, parcel records). These data sets are displayed on his computer screen 20 minutes later. The screen indicates that these data are retrieved from the GIS servers in the Boulder Planning Department, the Police Department, the Tax Assessor office, and the CODOT. The screen also displays a message that an automatic procedure for converting the crime risk index resides on the server of Boulder's Police Department. Five minutes later, a message from Boulder's Police Department server indicates that the conversion is complete and displays the crime risk index map on Dick's screen automatically.

Dick creates a GIS operation model provided by his GIS software to formalize his GIS modeling procedures (buffering and overlaying) for the Wal-Mart site selection. Dick sends out his GIS operation procedure from his GIS workstation. Twenty minutes later, the final result layer, procedure A, is sent back to Dick's workstation. Dick overlays four procedure A polygons with the crime risk index and census data and identifies a candidate area that meets the site selection criteria for the Wal-Mart site.

The next step is to find a rectangle shape-fitting function for the land parcel records within the candidate area. The local GIS workstation tells Dick that there is no such function on his workstation. However, there may be a similar component on other nodes on the Internet. Dick sends a request to search for a shape-fitting function via networks. A few seconds later, the search results indicate three similar GIS components available for downloading. Dick reviews the descriptions and functionality of the three GIS components and purchases one for $30 for one day's usage.

Dick downloads the new GIS component and plugs it into his GIS software. Then he uses this new component to perform the rectangle shape-fitting analysis. He gets 100 candidate sites and automatically generates a list of the cost of land parcels for 100 candidate sites. The new GIS component helps Dick to display the 10 sites that have the lowest land value. The total process of rectangle shape fitting takes less than 1 hour to finish. Dick spends a total of 6 hours to finish this GIS analysis of the Wal-Mart site selection.

JACK'S ACTIONS   Jack is the GIS software vendor who provides desktop GIS software for general use. His responsibility is to provide modularized GIS software modules with external interfaces that can accept new components or functions from third-party developers.

MATT'S ACTIONS   Matt is a freelance GIS software developer. He develops customizable GIS components for specific tasks, such as shape-fitting analysis, hydrological models, and three-dimensional visualization. He puts his GIS components on the Internet for downloading and charges a usage fee.

In the first scenario, Dick uses the traditional desktop GIS program that adopts the client/server architecture; while in the second scenario, Dick relies on Internet GIS programs that are constructed in distributed-computing architecture. As you can see, there is a dramatic difference between these two kinds of software architecture models. We will now discuss the client/server model, the distributed-component model, and the major differences between these different architecture models.

3.2   INTRODUCTION TO CLIENT/SERVER AND DISTRIBUTED-SYSTEMS ARCHITECTURE

Most current desktop GIS applications adopt the traditional client/server architecture. But what is the client/server architecture? Take any application, such as a GIS program, as an example. A typical application usually includes three essential elements: presentation, logic, and data (Figure 3.5). *Presentation* represents user interface, *logic* (sometimes called business logic) refers to processing, and *data* refers to the database or database management system. These three elements play their important roles in every application. Any GIS program would typically have these three elements. Typically, a GIS program's graphic user interface handles user's input from the keyboard or mouse; its logic element processes the user's requests (e.g., query, spatial overlay). The logic element has to rely on data to process the request; the query results can also be stored as data for future use (persistent data).

 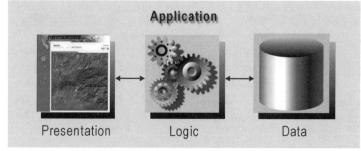

Application

Presentation   Logic   Data

Figure 3.5   Elements of a typical application.

Key Concepts

In a client/server computing model, the element that makes a request is called a **client,** *and the element that fulfills the request is called a* **server.**

The relationship between these three elements is that one element makes a request to the other element and the other element then fulfills the requests. This making and fulfilling of requests is called a client/server computing model (Shan and Earle, 1998). The element that makes a request is called a client, and the element that fulfills the request is called a server. The user typically uses the presentation element to make requests to the logic element; the logic element would perform some operation and may make additional requests to the data element to retrieve data in order to fulfill those user requests. For example, a user using a GUI in a GIS program sends a request to have a query operation; this request will be sent to the logic element, which will make a further request to the data element to search for the relevant data. The results are then returned to the user via the GUI.

These three elements can reside in the same machine or different machines. If they reside in the same mainframe computer, the application is called a mainframe application with centralized computing. The early GIS programs based on the mainframe computer are examples of a centralized computing application. If the three elements reside in the same personal computer, the application would be a stand-alone one. If the three elements reside in different machines and each machine is responsible for one or two specific functional elements, the application is called a client/server application using the client/server model. Many desktop GIS programs running on the LAN are examples of client/server applications. A Web application is another example of the client/server model implementation.

If the three elements reside in different computers that are connected through a network, the computer that the user uses to make a request is called a client, and the computer that processes and fulfills the request is called a server. Therefore, there is a difference between the physical concept of client/server and the logical concept of client/server.

In the physical client/server model, the client refers to the computer that interacts with the end user and the server refers to the computer that manages and stores data and/or other processing functions. The logical client/server model, on the other hand, defines client and server based on the service request. The requesting element is the client, and the elements that provide services and fulfill the request are the servers. In the physical model, a client computer and a server computer are clearly defined and fixed. But in the logical model, the roles of client and server are not fixed. A server can also request service from other elements of the system to help fulfill a request, in which case the server is also acting as a client. For example, a client can request a map from a Web server. The Web server, in turn, can request a map server to produce the map. The Web server acts as a server to the Web browser

Internet GIS Showcase: BuzzGB in London

Buzzlondon is a tourism Web site for London, England. This site allows tourists or residents to get maps, plan routes, book hotels on line, or find other local information. The site can also provide on-line hotel reservations, city maps, air photos, and routes for traveling to U.K. towns and cities. The routing service provides the best route between towns or postcodes, and it indicates the location of speed cameras.

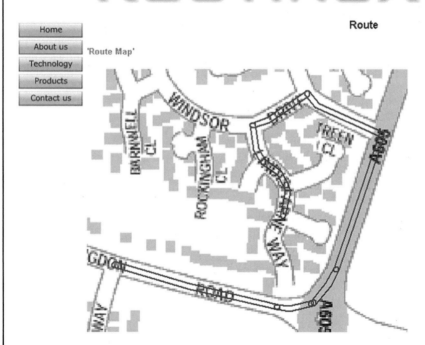

In addition to its up-to-date on-line maps, BuzzGB also provides photomaps that combine aerial photos with maps in a single image. Interactive icons on the maps provide click-through links to hotels, pubs, and other features, plus reviews of those places. This map-based system allows visitors to, for example, pick the hotel with the best view. For residents and U.K. visitors alike, this application also provides the locations and contact details for doctors, dentists, surgeons and hospitals, hotels, and other local information.

All the functions of BuzzGB can easily be integrated into personal Web sites or intranets to add routing and mapping, editable points of interest, or photomaps. BuzzGB is a GeoMedia WebMap application.

Source: http://www.buzzlondon.com.

(the client) and as a client to the map server. In fact, in a P2P network environment, any computer is both a client and a server.

Take a simple query operation in a desktop GIS program as an example, where the GUI (presentation element) and an application program (logic element) reside in the client computer (usually a PC), and the database resides in the server in a relational database management system like Oracle or SQL server. If the user makes a query request for data that meet certain criteria (e.g., all properties with assessed value between $100,000 and $150,000) from the client computer, the whole process of requesting and receiving information consists of the following six steps:

1. The user at the client computer makes a specific request by specifying a query criterion.
2. The request is translated into SQL through the GIS software at the client computer.
3. The SQL request is sent over the network to the server.
4. The database server carries out a search on the computer where the data exist.
5. The requested records are returned to the client.
6. The data are presented and displayed to the user at the client computer.

The logical client/server model also allows client and server elements to exist on distinct physical computers or on the same machine. The location of the three elements on the client computer or server computer is called *partitioning*. The point where the function of an element is split between a client computer and server computer is called a partitioning point. For example, if the partition point is between the presentation element and the logic element, the presentation is then located at the client computer, and the logic and data elements are located at the server. The following sections briefly describe the components of the logical client/server system—the client, the server, and the connection between them (the glue)—followed by a discussion of client /server partitioning.

3.2.1 The Client

Key Concepts

The **client** *usually refers to the presentation element or the front-end component of the client/server system. The* **server** *usually refers to the entity that provides shared resources such as databases or server-side applications to multiple clients.*

The client usually refers to the presentation element or the front-end component of the client/server system. The client is responsible to request the services of a server and to present the data and information in a user-recognized form through a user interface (Orfali et al., 1999). The client usually has the following functions:

- presents an interface for human interactions,
- formats requests for data or services from a server, and
- displays data or query results it receives from the server.

When the user makes a request for data, the client computer accepts query instructions, prepares them for the server, and then forwards the request over the network to the server. The server receives the request from the client and processes the request. It then locates the appropriate information and sends it across the network back to the client. The client then feeds the information to the interface that presents the information to the user.

There are different types of client, including PCs, mobile devices such as PDAs (e.g., Palm pilot, a pocket PC), or cellular phones. Generally, the clients can be classified into three categories: non-GUI clients, GUI clients, and Object-Oriented User Interface (OOUI) clients (Orfali et al., 1999). Non-GUI clients are a simple display interface with limited human interactions. These include cellular phones (except for smart phones), automatic teller machines, barcode readers, and fax machines.

GUI clients use graphic dialog, menu bars, scroll boxes, pull-down menus, and pop-up windows as user interfaces. The user usually first chooses an object (e.g., a file) and then clicks on the actions to be performed (e.g., open) on the chosen objects. This is called the object/action model (Orfali et al., 1999). User interfaces in Windows 3.X and most formed-based static Web pages are two examples of GUI clients. If the client is a Web browser, we sometimes call it a Web client/server or browser/server model. That is, the presentation element is the Web browser.

OOUI clients hold a collection of objects in the form of icons on the desktop, such as documents, folders, and printers. Users interact with these objects directly by double clicking and drag and drop rather than by selecting an object and then the action. The Apple Macintosh operation system and Windows 98 and beyond are examples of OOUI clients.

3.2.2 The Server

The server usually refers to the entity that provides shared resources such as databases or server-side applications to multiple clients (Shan and Earle, 1998). It is the back end of the client/server model to fulfill client requests. The server receives the requests from the clients, processes them, and sends the processed information back to the client. Because the server usually stores and manages data and processes requests, the server in a typical client/server environment should be more powerful than the clients. The server usually has the following functions (Orfali et al., 1999):

- *Wait for and respond to client requests.* The server spends most of its time waiting and receiving client requests. Once it receives the requests, it will respond by assigning a dedicated session to every client, by creating a dynamic pool of reusable sessions, or by providing a mix of the two.
- *Process client requests.* This is the main function of the server. A server has to respond to many simultaneous client requests, so it needs to have the multithread capability. That is, the server can fulfill many user requests simultaneously. Servers without multithread capability are only good to handle few requests and their performance is thus slow. Some existing desktop GIS and Internet GIS servers still do not have multithread capability.
- *Perform other service chores on the background.* Besides the main tasks, the server has to take care of other chores like backup, data downloading, load balancing, security checking, and task prioritizing.

There are many servers with very different functions; some are used to serve files (file servers), some are used to process data (data servers), and some are built to manage service transactions (transaction servers). From the viewpoint of distributed systems, the servers have evolved from stand-alone file servers to generic database servers, distributed database servers, and file servers to distributed-component object servers (Table 3.1).

STAND-ALONE FILE SERVERS (1982–PRESENT) A file server is a device that delivers files to everyone on a LAN. It allows everyone on the network to get to files in a central storage space, typically on one server. A file server directs movement of files and data on a multiuser communication network. Users can store information and access application software on the file server (Newton, 1996). For example, a desktop GIS program makes a request for a specific file over a network to the file server at the dedicated server machine. The user can map the file directory on the server as a drive so that it can use the file as if it is locally stored.

File servers also store and serve programs. Programs stored on the LAN file server are not actually run from the file server. Rather, file servers only store programs. When a client machine requests the use of a program, the file server sends the program out and loads it into the client machine's memory. The program is then run from that client machine. This is in contrast to mainframe and minicomputer programs, which are run on the host computer and are not transferred to the terminal.

From a network management perspective, this is a very primitive form of providing data and files to the end users. The file servers usually handle a

TABLE 3.1 The Major Development Stages of Distributed Systems/Services

Feature	Major Stages	Major Functions	Network Topology
Closed	Stand-alone file servers (UNIX NFS, Netware, Windows NT shared directory)	Files and disk space sharing	Many clients to one server with restricted access
	Generic database servers (Oracle, MS Access)	Query database and get results from servers	Many clients to one server with dynamic access
	Distributed database servers (Oracle) and distributed file servers (Windows 2000)	Query database or file sharing from an integrated server group	Many clients to one integrated server group, homogeneous servers
Open	Distributed-component object servers (CORBA, DCOM, Java)	Distributed-component object manipulation by sending requests	Many clients to many distributed servers, heterogeneous servers

huge amount of transactions, which typically becomes a significant bottleneck in a LAN. The system structure of file servers is fixed in both clients and servers. Different file servers have their own protocol and file formats, which may not be compatible with others (Schroeder, 1993).

GENERIC DATABASE SERVERS (1986–PRESENT)   A generic database server is a stand-alone computer that sends out data to users on a LAN. Unlike the file server, which sends the whole file to the user and lets the user manipulate the whole file, a database server does the picking, sending only the requested part of the database to the user's workstations. Typically, the client sends a SQL query to the data sever, which processes the SQL query and returns the query results to the client. Thus, a database server incurs less network traffic than a file server in a multiuser database system. It also provides better data integrity, since one computer handles all the record and file locking (Newton, 1996). This is a more efficient way to manage and use a central database than the file server. In addition, database servers are more flexible than file server systems, especially on the client side. Multiple users can easily establish new client-side applications to access the same database server. However, the server-side applications are fixed in most cases. It is impossible to access multiple databases at one time or integrate heterogeneous databases under a single server architecture.

DISTRIBUTED DATABASE SERVERS AND FILE SERVERS (1992–PRESENT)   "A distributed database server appears to a user as a single logical database, but is in fact a set of databases stored on multiple computers. The data on several computers can be simultaneously accessed and modified using a network" (Oracle, 1992, p. 21-2). Basically, the main functions and capabilities of distributed database servers mimic generic database servers, but the physical locations of databases are distributed across a network. Similar to the architecture of distributed databases, distributed file servers appear to a user as a single logical file server but physically are distributed in different places. However, distributed file servers are designed for file sharing instead of database access. Distributed file servers can provide users with a virtual integration of distributed file servers on a LAN. An example of this is the active services functionality in the architecture of Microsoft Windows 2000 (Seltzer, 1998). Both generic and distributed database/file server systems basically follow the traditional client/server architecture, which is restricted to specific internal communications and processing capabilities. There are several problems with the traditional client/server architecture for GIS requests and processes because it cannot provide rich transaction processing and rich data management or handle overly complex queries or operations. For example, if a traditional database server receives requests from 500 client-side applications at the same time, the server's operating system may hang. Without a transaction control function, traditional database architecture is not appropriate for complex GIS applications. In some cases, transaction processing monitors (TP monitors) or transaction servers have been used to assist major enterprise databases with their transaction services (Orfali et al., 1996).

DISTRIBUTED-COMPONENT OBJECT SERVERS (1995–PRESENT) Distributed-component object servers are advanced client/server systems that can handle complex transactions and requests from heterogeneous systems. Distributed-component technology adopts the concepts of Object-Oriented Modeling (OOM) and Distributed-Computing Environment (DCE). Currently, both academic and industrial studies of distributed systems are focusing on distributed components in open environments that can provide new capabilities for the next-generation client/server architecture (Montgomery, 1997). CORBA, developed by the Object Management Group (OMG), and DCOM, developed by Microsoft Corporation, are two examples of distributed-component frameworks (Orfali and Harkey, 1997). Comparing the distributed database/file servers, the main advantage of distributed-component object servers is the interoperability, reusability, and flexibility for cross-platform applications. A detailed description of distributed components will be addressed later in this chapter.

In addition, there are other servers, such as transaction servers, groupware servers, and Web servers. Transaction servers are created to manage user requests (or transactions) such as simultaneous SQL statements and to make efficient use of server resources. Groupware servers are used to manage shared information such as e-mail, bulletin boards, scheduling, and images. They have been used in the office environment but have no direct application for GIS programs. Web servers are used to serve static files and documents to the Web client.

3.2.3 The Glue

The client and the server are two independent components in the client/server system. In order for the client to communicate with the server, we need a connection component or translator to link them together. This intermediate connection component is called glue or middleware [*glue* is a term used by Shan and Earle (1998)]. The role of the glue is to allow the client and the server to talk in the same language and follow the same rules (protocols) so that when the client makes a request, the server understands what it wants. In other words, the role of the glue is to mask complex systems to give users the illusion of a single system—to make the client, server, and network components transparent to users. Transparency means using glues to hide the complexity of the server and network from the users and even the application programmers so that the client/server system is really a seamless, single-system illusion (Orfali et al., 1999).

The glue is also commonly called middleware. The term "middleware" generally refers to any software that goes in between two other technologies to allow them to work together. In the client/server system, middleware is software that helps clients and server to communicate and cooperate. Therefore, there are many forms of middleware. Two common forms of middleware include database middleware and more general module-to-module middleware. For example, networking software, such as TCP/IP, can be considered middleware, as can RPCs. "Glue" may be a preferred term since middleware can be ambiguous because of the broad spectrum of software that it covers (Shan and Earle, 1998).

There are a large number of different glues to link and utilize objects in the client and server or between objects among different servers. These glues include RPCs, Message-Oriented Middleware (MOM), Publish-and-Subscribe Glue (PUSH), DCOM, and CORBA. RPC makes subroutine or function calls to remote hosts as if they are local. It is often used for remote data access. DCOM is a standard that allows clients to transparently access and utilize remote objects. CORBA is another standard for object sharing in a heterogeneous network environment by OMG. All these glues deal with interoperability between clients and remote servers so that the client and server can communicate. They are important to the development of distributed GIS.

3.2.3.1 Remote Procedure Call   A RPC is a direct way to allow clients and servers to talk with each other. It works like this: A client process calls functions on the remote server and waits for the server to respond. The server then fulfills the requests from the client by calling the remote procedure and sends the results back to the client. The client receives the results and closes the connection with the server.

RPC is just like an ordinary procedure call to the local machine. The difference is that the procedure is called to the remote server. Parameters passed between the client and the server are similar to any ordinary procedure call. RPC is synchronous, meaning that the client initiating a call waits until it receives the results from the server. Therefore, a dedicated and logical connection between the client and server must be established in order to complete the communication process. If the client or the server is down or the connection is broken, no communication is possible. To prevent disruptions, procedures have to be developed to handle sudden failure of the client, the server, or the network.

3.2.3.2 *Message-Oriented Middleware*   While RPC requires the client and server to run at the same time, MOM does not. MOM is based on the message-queuing model. The client makes one or more requests that are put on a queue. The server will take those requests from the queue at a later time. The server then services the requests and puts the results back on the queue. The client takes the results from the queue afterward. As you can see, the MOM model does not require a real-time connection between the client and server.

MOM provides a very flexible and versatile means of communication between the client and server. Data can be transferred between client and server even if one or both are temporarily unavailable. Applications communicate with the server by using queues to temporarily store outgoing or incoming messages.

MOM can be used to create one-to-many or many-to-many relationships. It means that many clients can send requests to one server queue or many clients can send many requests that require multiple server responses. The server will later pick up the requests from the queue by priority or on a first-come, first-served basis.

An analogy can be made to compare the RPC and MOM. RPC is like making a telephone call without an answer machine (Orfali et al., 1999). Both parties have to be present in order to communicate. But MOM is like using e-mail or fax. Neither party is required to be present at the same time. The e-mail or fax sender can send them at any time and the recipients can simply read and respond to the e-mail or fax at a later time. This system has advantages and disadvantages. On the positive side, the sender (the clients) can send requests at any time, even when the network or server is down. This is good for mobile users that do not have an active connection with the server. On the negative side, the sender may become frustrated by delayed response times. This is especially true for systems that perform sequential operations; that is, the later operation depends on the early operation. On the server side, there is overhead associated with sorting through the pile of user requests and deciding who to respond to first (load balancing).

Therefore, RPC is generally good for applications that need immediate response from the server, when the next step is dependent on the server's response to the initial query. For example, a buffer operation is necessary before you can do a spatial overlay. In this case an RPC is needed. On the other hand, if the user wants to be notified when the GIS data are updated, a MOM approach can be used.

3.2.3.3 Publish-and-Subscribe Glue   PUSH is driven by events. The publisher (usually a server) is a producer of an event and the subscriber (usually a client) is a consumer of an event. When an event occurs, the publisher informs the subscriber and the subscriber (a client application) then performs some action based on the event. For example, in the business world, when a credit check is completed by a credit checker application, the other application either approves or rejects the loan. In the GIS world, when a server sends new information to a client, such as an updated street map, the client application may need to rerun the query based on the updated database.

The producers and consumers of events do not talk with each other; rather, they go through an event broker. An event broker establishes a communication channel between subscribers and publishers of events (Orfali et al., 1999). When something happens, the event broker tells the subscriber to perform some action.

RPC and MOMs are commonly used with LAN-based client/server systems. But they are being replaced by Object Request Brokers (ORBs), which are commonly used in object-oriented applications as a preferred intercomponent communication method over the Internet. ORBs provide all three functions of the RPC, MOM, and PUSH models: (1) RPC-like request–reply, (2) MOM-like asynchronous messaging, and (3) PUSH via events.

3.2.3.4 Service-Specific Glue   There are many service-specific glues, such as database service glues and client/server glues. Typical database server glues provide the client with access to the database server. They include SQL, ODBC, and JDBC. SQL is data access language to connect the client with access to a relational database management system. But ODBC is designed to access multiple and heterogeneous databases. ODBC runs only in a Windows environment. JDBC is a data access method for Java objects.

In addition, the client/server glues also include NOSs, which cover such services as security, directories, and distributed file systems. These services make it easy for clients to communicate with the servers.

3.2.4 Client/Server System Partition

A partition is commonly used in computer hardware to logically divide (or format) a hard disk so that you can have separate hard drives for different functions such as file management and multiple-user access. Partition is also used in software for different purposes such as database management or client/server operation. Each partition works independently but can communicate seamlessly with the other partitions. Each partition can work together in one or more computers.

The function of the partition in the client/server system is to divide an application into three basic functional elements—presentation, logic, and data–and to allocate them into different computers in terms of client and server computers. For example, the presentation element can be in one computer, the logic component in another computer, and the data in yet another computer. From the user's perspective, the location of these elements is in either a client computer or one or more server computers. The dividing points of these elements between the client and the server computer are called *partitioning points*.

The partitioning point could be in any of the three application elements—presentation, logic and data—or the boundaries between them. Based on the Gartner Group's terminology [cited in Shan and Earle (1998)], if the partitioning point is at the presentation element, it is called *distributed presentation;* if at the boundary between presentation and logicm, it is called *remote presentation;* if at the logic element, it is called *distributed function;* if at the boundary between the logic and data element, it is called *remote data access;* if at the data element, it is called *distributed database.*

(a) *Distributed presentation* is a very thin client configuration in which the data and logic as well as part of the client are located in the server. The client is responsible for only a portion of or the mirror image of the user interface. All data processes are located in the server (Fig. 3.6). An example is using X-Windows to access GIS programs like Arc/Info at a Unix workstation, where the user interface at the X-Windows is a reflection of that at the other Unix workstation. Another example is the ArcView IMS and MapObject IMS by ESRI, where maps on the Web are map images produced at the ArcView or MapObject map server. A simple Web browser that renders only HTML (i.e., no plug-ins and applets) is another example of a distributed presentation.

(b) *Remote presentation* partitions the entire presentation to the user interface at the client side, whereas the logic and data reside in the server (Fig.

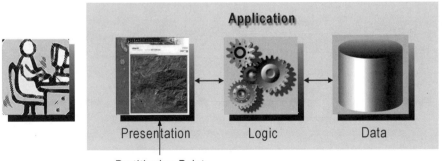

Figure 3.6 Partitioning point of distributed presentation.

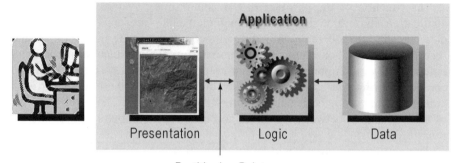

Partitioning Point

Figure 3.7   Partitioning point of remote presentation.

3.7). An example is the server-based Internet GIS using CGI, where the user interface (in the form of HTML) is at the Web client, and the user request is processed at the server via a CGI program. The query results are then presented at the Web client.

(c) *Distributed function* splits the logic element between the client and the server and puts the presentation on the client machine. It allows some functions to execute in the client computer while it sends other presumably more complex functions to the server for processing (Fig. 3.8). Some Internet GIS that use Java applets or ActiveX controls fit into this category. Some basic functions such as query, zoom, and pan can be performed in the client machine, while other functions like address matching and image analysis are performed in the server. Other examples include distributed transaction servers and RPCs used in the LAN-based desktop GIS programs.

(d) *Remote data access* puts presentation and application logic on a client that retrieves data from a remote database (Fig. 3.9). This is the architecture of a thick client, in which the client is responsible for all logic operations.

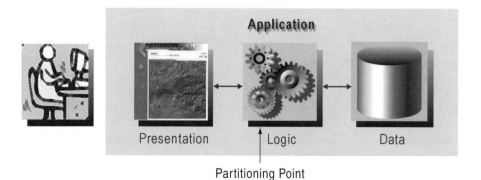

Partitioning Point

Figure 3.8   Partitioning point of distributed function.

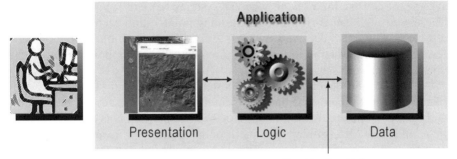

Figure 3.9 Partitioning point of remote data access.

For example, remote data access might use SQL APIs to make calls to a relational database. Most LAN-based desktop GIS programs that are capable of accessing relational database management systems at the data server using SQL query belong to this category. In the Internet GIS area, Intergraph's GeoMedia Enterprise is an example of remote data access.

(e) *Distributed database* splits the data management functions between the client and one or more servers, while it allocates the logic and presentation elements to the client (Fig. 3.10). IBM's Distributed Relational Database Architecture (DRDA) and Webcasting that leverages "push technology" are two examples (Shan and Earle, 1998). Any Internet GIS programs that are capable of streaming vector data to the client belong to this category, such as feature server inside ArcIMS.

As you may have noticed in the above list of topologies, each successive type shifted more functions from the server to the client, and so the client changes from very thin to very thick. Systems (*a*) and (*b*) are thin clients,

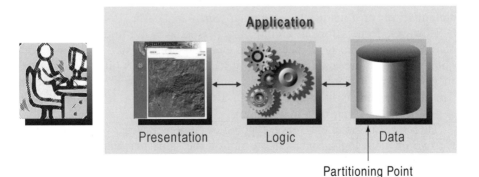

Figure 3.10 Partitioning point of distributed database.

system (*c*) is a medium client, whereas systems (*d*) and (*e*) are thick clients. These are rough categories. The partitioning point within each category could also vary. For example, even within a "distributed function," there are differences among systems in terms of how many logic components are in the client side and the server side.

An application developer can decide on the partitioning point for any applications depending on the hardware topology, protocol used, business needs, and nature of the application itself. However, some protocols are less flexible than others in determining the partitioning point. For example, SQL has clearly defined partition points (*d*) between logic and data. It does not leave any room for developers to redefine the partitioning point. On the other hand, RPC allow the developer much more flexibility to decide where the partition point (*c*) should fall in the spectrum.

3.2.5 Two-Tier, Three-Tier, and *n*-Tier Architectures

The partition of the client/server system can be further implemented in a tier structure. For example, if the partition point is at the boundary of presentation and logic components, this could be implemented as at least two tier structures: a two- a three-tier structure. For the two-tier structure, the presentation would be located at the client machine and the logic component and the data component could be located in the server machine. For the three-tier structure, the presentation would be located at the client machine, the logic component would be located at the server machine, and the data component could be located in the data server machine. Therefore, the partition and the tier structure are related but not the same. The tier structure is concerned with the locations of different partitioned components in the client/server systems in terms of hardware and software configurations.

From a hardware platform perspective, three-tier hardware architecture involves three classes of computers: The client is usually a PC, a PDA, or a cellular phone; the middle tier is usually a workstation server or a minicomputer; and the back-end server is usually a mainframe computer. Two-tier hardware architecture generally involves only the PC client and either a workstation server or a mainframe computer (Shan and Earle, 1998).

However, from a logical architecture perspective, the three tiers are presentation, logic, and data. These three elements could reside in one or more computers. The presentation is usually located in a PC, while the logic element could be located either in a server or in the PC on a client computer. The data element is usually located in a data server (e.g., a Windows 2000 server, a Unix server, or a mainframe).

Therefore, as the client/server typologies demonstrated, it is quite common for two or more logical elements to run on the same machine. For example, Figures 3.11*a* and *b* show a configuration that is physically two-tier but logically three-tier. In Figure 3.11*a*, both presentation and logic elements run on the PC, and the database runs on the server (Shan and Earle, 1998). This is

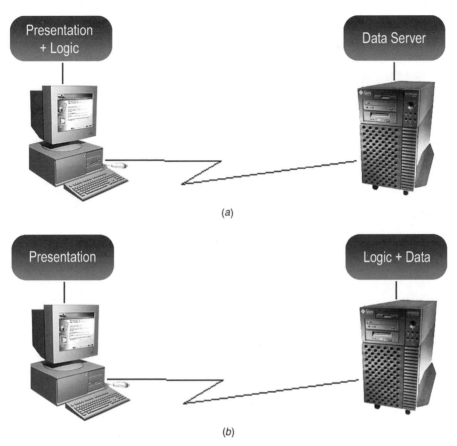

Figure 3.11 Logically three-tier and physically two-tier systems: (*a*) presentation and logic element run on PC and database runs on server; (*b*) presentation runs on PC and both logic and data runs on server.

typically the case for LAN-based desktop GIS, where user interface and GIS processing run on the PC whereas the database is stored in a server. In Figure 3.11*b*, presentation runs on the PC, and both logic and data run on the server. This is the case for a Web server and server-based Internet GIS, such as ArcView and MapObject IMS, as well as for mobile GIS applications.

An *n*-tier system expands the logic or data element into multiple computers or multiple components. For example, in Figure 3.12 the partition point is at the logic element, which can be run in multiple machines—the so-called distributed computing. In other words, the logic element can be broken down into specific functions that can run in multiple programs. In this case, it is a logically three-tier architecture but a physically four-or-more-tier system. Similarly, the database can be further partitioned and stored in multiple-database-management servers (Figure 3.13). But logically it is still a single

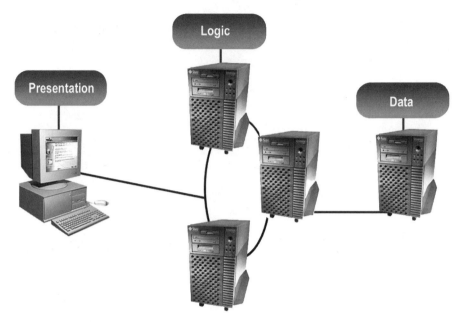

Figure 3.12   The *n*-tier logic components.

database. These multiple-database servers can be linked together by a data category server or a data access program like ODBC or JDBC. The request goes to the category server first, and the category server redirects the request to different database servers. (We will discuss data category servers in Chapter 5.)

3.2.6   Advantages and Disadvantages of Client/Server Architecture

The client/server architecture presents a dramatic departure from and a great improvement over the centralized computing model as in mainframe computers. Applications using the centralized computing model such as the first generation of GIS—mainframe GIS—put all data and processing in the centralized mainframe computer. Users gain access to the data and process through dumb terminals. Users send requests through terminals to the server that retrieve the information and send it back to the terminal for display. Dumb terminals as a presentation element are useful for inputting and displaying information, but the terminal is dumb because there is no computing power in the terminal, and every request has to go back to the mainframe through the network. In other words, no processing power can be shared between the central machine and the dumb terminal. There is also not much coordination between the client (the termina) and the server to determine what data to retrieve and send. Every single-user request has to go to the mainframe

Figure 3.13   The *n*-tier data structure.

computer. This causes a large increase in network traffic between the main-frame and the dumb terminals. Furthermore, as more users make requests at the same time, mainframes still have to deal with the requests one at a time. This leads to delays as users wait for the mainframe to "turn around" their requests.

With the desktop computers and the client/server architecture, users can use their own computers on their desktop to process the data and get instant results. There is no need to wait for the mainframe to respond. The network traffic with the client/server model is also reduced.

In addition, the client/server architecture model is more scalable than the mainframe model. If the data server is slower to respond due to heavy user demand or larger data volume, it is very easy to add another data server. Furthermore, more mirror servers can be added so that when the primary server fails, the backup server can function without affecting the client side of the application. To the user at the client machine, nothing is different.

Powerful client computers have also made it possible to support computation-intensive GUI. The GUIs are much more interactive and much easier for users to understand and to interact with. This is especially important for GIS users, because GIS users have to deal with graphics and maps all the time.

However, this migration from centralized computing to client/server computing has caused many problems, especially for the two-tier client/server model with presentation and processing logic locating in the client machine. As data interaction and logic processing moved to the client computer, the client machines have become "thick." As the applications become more and more complex, the demand for more powerful machines has dramatically increased, leading to frequent and expensive upgrades of personal computers.

The two-tier client/server computing lacks central control, which is costly to manage, support, and upgrade. When the time comes for a software upgrade, it only takes one upgrade for the mainframe computer, but it would take many more people and much more time to upgrade hundreds or thousands of personal computers. Similarly, training and technical support for all the client computers are very costly for corporations.

The use of department-level LANs can bring down the number of copies that need to be updated. But the size of the LANs must be limited, because, as discussed in Chapter 2, a large "thick-client" software may affect the performance of the LAN. The task of making simultaneous changes to hundreds of copies of client software is still a daunting one.

For client/server-based desktop GIS, there is also an accessibility issue. LAN-based desktop GIS can only be accessed by people working in the office. It would be very difficult, if not impossible, for people beyond the LAN to be able to gain access to it. This limits the number of potential GIS users. Hence, moving GIS programs from the desktop to the Internet could dramatically increase the base of GIS users. This leads to our discussion on the Web client/server architecture.

3.3 WEB CLIENT/SERVER ARCHITECTURE

The Web client/server architecture is a special case of the general client/server architecture, with the Web browser as the client. The presentation element is the Web browser or downloadable clients. The Web client/server architecture has the advantages of the client/server model but avoids some of its shortcomings. It has the flexibility of being a thin client or a thick client. In the thin-client case, the user interface components (a Web browser) run on the user machine but the logic and data elements usually remain on the server. In the case of the thick client, the user interface and the logic element are located at the client's computer in the forms of client-side applications such as plug-ins, ActiveX controls, Java applets, or even the downloadable front-end Internet applications. Only the data are located in the server. But in either case the software components are centrally controlled, which makes them easier to maintain, update, and support.

Software updates are only needed for the server. Once the software is updated at the server, it is available instantly to every client. For example, Java applets are initially stored in a Web server. They are automatically loaded into the browser on the user's machine as needed. If a newer version becomes available, that version is automatically loaded each time the users access the site. This would reduce the cost of ownership. It is more cost effective to distribute applications across the Internet because it eliminates the need to manually install components on every machine.

Another advantage of the Web client/server model is the wide accessibility. Anyone with a Web browser can gain access (with proper permission in some cases) to the computing resources and data anywhere at any time.

Furthermore, the Web client/server architecture is more scalable and secure than the traditional client/server model. Web applications are served from servers. If the number of users increases, more servers can be added. Since security is centralized, there is more control over how to implement it.

The Web client/server model has evolved from a static Web to a dynamic Web to distributed-component applications (Orfali et al., 1999; Shan and Earle, 1998). The static Web is used to publish static Web pages with no processing capability on the Web browser. The dynamic Web adds more interactivity and processing power to the Web page by using client-side applications like Java applets and plug-ins. The distributed-component applications do not necessarily use the Web browser as the front end; rather they create and disseminate their own front-end GUIs to the end user. The Web is simply used to download these front-end client GUIs. The front-end client then communicates directly with the other portion of the Web client/server application on the server (Orfali et al., 1999).

In the first-generation Web model, the Web client is only responsible for the presentation element and to display static HTML pages. No logic and data are available at the client. As the Web technology evolves, some logic and data elements can shift to reside in the Web browser. The Web browser has also progressed from static browsers to dynamic browsers that are accompanied with client-side applications such as plug-ins, Java applets, and ActiveX components to Web casting. The partition points thus move progressively from the presentation element to logic to data element.

3.3.1 Web Client: Web Browser with Client-Side Applications

The typical Web client is a Web browser with HTML. To make the Web user interface dynamic and more interactive, DHTML and some client-side applications are used to enhance the static HTML into a dynamic one. With dynamic HTML, the Web page becomes more interactive. For example, with a mouse click over a Web page users can move a graphic across a page, draw a line or a box on a graphic, or even change the attributes of a font. These actions can be done directly on the Web page without any plug-ins or Java applets.

Key Concepts

Plug-ins, Java applets, *and* **ActiveX controls** *are three common examples of client-side applications working together with the Web browser. Plug-ins are software executables that run on the Web browser. They are used to extend the browser capabilities to process data that are not supported by the HTML; Applets are executable Java code that are downloaded from the server and executed on the client at runtime; an ActiveX component is a piece of executable code that can run on Windows platforms.*

However, DHTML is limited only to animate the Web page. Its functions are very limited and it cannot be used to handle many sophisticated GIS functions. Plug-ins, Java applets, and ActiveX controls are three common examples of client-side applications working together with the Web browser. We will discuss these three client-side applications in detail in Chapter 4.

The partitioning of these client-side applications is at the logic level of a Web application. In addition, partitioning can also happen at the data level. For example, many Internet GIS and other applications have a data-streaming feature that streams data to the Web client, something we will discuss later, in Chapter 8. The server contains the complete data while the client receives a subset of the data that is of interest to the user and that is cached on the client side. The presentation (display and visualization) and logic components (e.g., map rendering) of the application also reside on the client side to directly process and display the data. This pushes the partitioning point into the data area.

As the partitioning point moves from the presentation to the data, more flexibility and interactivity are offered to the Web browser (the client). This is very useful for Internet GIS applications since the map interface requires higher levels of interactivity. On the other hand, more functions at the client side make the Web browser become a thicker client rather than a thin client.

3.3.2 Server: Web Server with Server-Side Applications

A Web server is a daemon on a server machine listening to the Internet traffic and serving HTML pages and other information. It relies on HTTP to communicate with the Web client. Therefore, it is often referred to as an HTTP server. The major role of the HTTP server is publishing Web pages, including HTML, text, images, and so on. An HTTP server is stateless, which means once it sends out the information to the client, it automatically breaks the connection with the client. It treats every request as a new request even if the request comes from the same client. This stateless nature does not work well to handle more sophisticated client requests. Therefore, additional application servers and data servers are needed to work together with the HTTP server. The role of the Web server has thus evolved from publishing static information to offering dynamic contents providing enterprise applications (Shan and Earle, 1998). The form of Web server has also changed from the stateless HTTP server to a combination of HTTP server and many applications and data servers.

Various mechanisms are introduced to support dynamic contents on the server side. The simplest one is to get the HTTP server to invoke CGI programs and servlets, in addition to simply providing static HTML documents. So a client query can be passed to the CGI to generate a response.

Another mechanism is to use server-side scripting and a back-end application server such as Microsoft ASP and ColdFusion to expand the functions of the Web server. A linkage with these server-side scripts and back-end application servers can be embedded inside the HTML page. For example, users can use Jscript and VBScript statements within an HTML page to invoke programs and query the database in Microsoft ASP. Java, JavaScript, and Java servlets can also be used along with the Web server. These server-side scripts can be invoked directly or from within an HTML page. Many Internet GIS programs uses Java servelets, ASP, and ColdFusion as server-side adds-on to serve dynamic and interactive queries.

3.3.3 Glue: HTTP

Key Concepts

HTTP *is the set of rules for exchanging files between the Web browser and the Web server. It functions as a message carrier.*

HTTP daemon *is a program that is designed to listen to HTTP requests from the Web browser.*

The Web relies on HTTP as the main glue to link the Web client and the Web servers. HTTP is the set of rules for exchanging files between the Web browser and the Web server (or more specifically the HTTP server). HTTP functions as a message carrier. It carries user requests from browsers to servers and takes the requested information (text, graphic images, sound, video, and other multimedia files) from servers back to browsers.

How does it do this?

The HTTP uses various commands or methods to communicate between the Web client and the Web server. In each Web server there is an HTTP daemon, a program that is designed to listen to HTTP requests from the Web browser. Once the HTTP daemon receives the request, it processes the request based on some standard methods. These methods include GET, HEAD, PUT, POST, REPLY, and so on. (http://www.w3.org/Protocols/HTTP/Methods.html). For example:

- GET is the most commonly used method in retrieving whatever data are identified by the URL, including running scripts. It is also used for searches.
- HEAD returns only HTTP headers and not the actual document body.
- PUT stores the data to the supplied URL in the Web server. The URL must already exist. POST and REPLY should be used for creating new documents.
- DELETE will ask the server to delete the information corresponding to the given URL.
- POST creates a new object or appends information linked to the specified object (like a discussion bulletin board) identified by a URL.
- LINK links an existing object to the specified object.
- UNLINK removes link information from an object.

Equipped with these HTTP commands or methods, it would be an easy task for the Web browser to communicate with the Web server. When the user enters file requests by either typing in a URL or clicking on a hypertext link, the browser builds an HTTP request (e.g., GET, HEAD, or PUT) and sends it to the Web server indicated by the URL. The HTTP daemon in the destination server machine receives the request and, based on the HTTP request methods, returns the requested file (if available) to the Web browser. Once the requested file is returned to the Web browser, the communication between the Web browser and the Web server is completed.

3.4 DISTRIBUTED-COMPONENT FRAMEWORKS

The WWW is one of the most exciting and pervasive applications on the Internet. However, Web applications that rely on the Web browser as user interface are only one means of distributed client/server computing on the Internet. As we discussed before, the Internet is much more than the Web. Client/server applications based on the Internet do not have to use the Web browser as the front end. They can have their own graphic user interface that is downloadable over the Internet from the server as long as they are developed based on the HTTP and TCP/IP standards. The Internet is used as an infrastructure to run those programs.

Some applications such as sophisticated GIS image processing can be difficult to handle on the Web browser but work much more easily with an image-processing client. One of the important functions of the Web is its search and downloading capability. The Web browser can be used to search and download HTML pages, plug-ins, and applets as well as executables unrelated to the Web browser. This function can be used to download the whole client portion of the application, Web-based or not. The client can be distributed to any machine with Internet access through the Web from the HTTP server. Once installed at the client machine, the client can run independently of the Web and communicate directly with the remote server over the Internet. This is the distributed-component model.

The original concept of distributed-components came from the development of distributed systems. Different from traditional client/server system models, distributed-component frameworks break up the client and server sides of an application into smart components that can interoperate across operating systems, networks, languages, applications, tools, and multivendor hardware. Examples of distributed components include roaming agents, rich data management, abstract and generalized interfaces, self-managing entities, and intelligent middleware (Orfali et al., 1996). The current commercial market provides three major infrastructures for distributed-component technology: CORBA, developed by OMG; DCOM and .NET, developed by Microsoft; and Java technology by Sun Microsystems and its subsidiaries, Sunsoft and Javasoft.

The original idea of distributed components came from object-oriented modeling technology, which has developed over the past 20 years (Orfali and Harkey, 1997). Recently, distributed components have become the most important trend in the development of software technology. The generic features of distributed components adopt concepts of object-oriented modeling, including encapsulation, polymorphism, inheritance, framework and classification, and object relationships (Rumbaugh et al., 1991; Taylor, 1992).

The most important contribution of object-oriented technology is to provide an efficient way to make software constructed by standard and reusable components (Taylor, 1992). Objects correspond to real-world entities such as cars or people. Each object encapsulates related procedures (methods) and data (variables). The method of encapsulation can prevent a program from being interfered with by other programs. Communication between objects depends on the calling of methods or functions for each object. Some methods can carry multiple meanings in a single form, which is called polymorphism. Polymorphism can simplify complex systems and improve the efficiency of the programming. Many objects can be organized and grouped as hierarchic classes. The classes of objects are similar to our real world. Different classes share different properties by using a mechanism called inheritance. Object-oriented modeling allows different parts of the software to be developed simultaneously, to be easily maintained and modified when necessary (Graham, 1994). It also improves the reliability of software and makes the information system more useful and flexible.

By adopting the object-oriented modeling technology, distributed components can handle rich and complex requests and prioritize the sequence of requests from the client side. For example, when a data component server is busy, the next distributed request can wait in a queue instead of being canceled. Another important feature is that distributed components provide more flexible access and application on both the client side and server side. A single system can play both a server's role and a client's role. For example, a Colorado local GIS site can access many federal database servers as a client. When other GIS projects require data about Colorado, the Colorado site can act as a database server. Thus, distributed components are appropriate for open and distributed GIS environments since they can provide efficient and flexible client/server applications. Distributed GIS components and applications can freely interact and interoperate on the Internet.

The following section will provide an in-depth review of the development history and major features of the three types of distributed-component technologies, DCOM/.NET, CORBA, and Java platform. These distributed-component frameworks will provide a fundamental support for the deployment of high-level distributed GIServices architecture.

3.4.1   DCOM and .NET

The software giant Microsoft has been involved with the development of distributed-component frameworks for several years. The early development of distributed-component framework is called DCOM. After 2000, Microsoft initiated a more powerful framework for distributed components called .NET.

DCOM technology is an extension of the Component Object Model (COM), which supports interoperability and reusability of distributed components under Microsoft's operating systems, such as Windows 95/98 and Windows NT. Many programmers consider COM and DCOM as a single technology that provides a range of services for distributed-component interaction. COM is designed for a process running on a single machine and DCOM is designed for processes operating across heterogeneous networks. The COM/DCOM technology is also closely related to other Microsoft technologies, including OLE and ActiveX. In order to clarify the relationships between COM, DCOM, OLE, and ActiveX technology, which are usually confusing to the public and nonprogrammers, the following section will give a brief introduction to the development history of DCOM and its related technology (Table 3.2).

3.4.1.1   DCOM Development History   The original idea of COM/DCOM technology comes from the clipboard function created by Apple in the late 1970s (Grimes, 1997). The COPY, CUT, and PASTE tools provided users a friendly way to share documents between different programs. In 1990, the release of Microsoft Windows 3 extended the clipboard idea and the PUSH concepts developed by Apple. Then Microsoft introduced its own way to exchange data between applications, called DDE, which allowed different Windows applications to communicate with each other via a message-based protocol.

Internet GIS Showcase: The Community Health Information Profile (CHIP)

CHIP is a British system that makes up-to-date information and statistics available to anyone interested in health and health services. CHIP provides easy on-line access to health inequities indicators and information about activities that promote good health.

The information available on CHIP comes from agencies in multiple sectors and is designed to serve all of Salford, Trafford, and Manchester. It uses maps to help users search for information, display statistics, and pinpoint local activities street by street. CHIP used Autodesk's MapGuide as the Internet GIS software.

Part of the CHIP data is from the 1991 census, but the majority is current data, taken from service records of various kinds. Data maintenance agreements with data providers ensure that the data are regularly updated. The information can be displayed as maps or tables, which are then downloadable into other applications.

CHIP can help customers find the location of activities that can help to improve health. Based on extensive citizen input, most people have a very broad view of the sort of facilities that are relevant to health—not just doctors and dentists, but also recreation, education, voluntary, and community sector activities. The activities included in CHIP reflect that wide view and span all sectors.

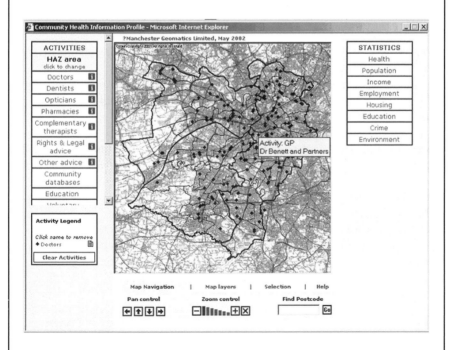

Source: http://www.healthprofile.org.uk.

TABLE 3.2 Development History of DCOM and .NET

Year	Technology Development
1990	DDE (Dynamic Data Exchange) with Windows 3.0
1991	OLE 1.0 for compound documents
1993	OLE 2.0 + COM for compound software
spring 1996	ActiveX (next generation of OLE)
summer 1996	DCOM (the distributed version of COM) with Windows NT 4.0
2000	.NET

In 1991, Microsoft released OLE 1.0, which modified the major functions of DDE and added an API on top of the DDE messages. The major improvement of OLE 1.0 is the ability to link and embed documents within applications. (Microsoft, 1996, p. 1):

OLE is a technology that enables an application to create compound documents that contain information from a number of different sources. For example, a document in an OLE-enabled word processor can accept an embedded spreadsheet object. Unlike traditional cut and paste methods where the receiving application changes the format of the pasted information, embedded documents retain all their original properties. If the user decides to edit the embedded data, Windows activates the originating application and loads the embedded document.

The linking function of OLE allowed applications with embedded documents to be linked together dynamically. If the original data were changed, the embedded contents would automatically be updated and vice versa. Figure 3.14 shows an example of compound documents that includes graphics, pictures, sound clips, and an embedded Excel document.

In 1993, the release of OLE 2.0 extended the capability of OLE beyond the compound document to compound software (Brockschmidt, 1994). The popular use of OLE 2.0 generated a shift of Microsoft software development from an application-centered paradigm to a document-centered paradigm. The document-centered paradigm allows users to move documents between many different applications without even noticing the movements among different applications. Almost every recent Microsoft package, including Office 97, Visual Basic, Visual C++, and Excel relies on OLE 2.0 technology. OLE 2.0 provides more comprehensive architecture and communication protocols to allow programmers to design applications under Microsoft's operating systems, such as Windows 98 and Windows NT.

The COM was originally designed in 1993 to specify interface interactions and communication protocols between OLE 2.0 components. COM provides the underlying support for OLE components to communicate with other OLE

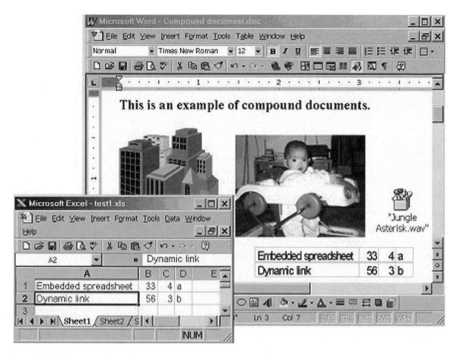

Figure 3.14 Example of compound documents in Microsoft Word 97.

components (Brockschmidt, 1994): "A straightforward way to think about COM is as a packaging technology, a group of conventions and supporting libraries that allows interaction between different pieces of software in a consistent, object-oriented way. COM objects can be written in all sorts of languages, including C++, Java, Visual Basic, and more, and they can be implemented in DLLs or in their own executable, running as distinct processes" (Chappell and Linthicum, 1997, p. 58). COM's language-independent feature means that components written in different languages can interoperate via standard binary interfaces.

ActiveX developed in 1996 is the next generation of OLE and extends the use of COM/DCOM to Web applications. ActiveX is a lean, stripped-down version of OLE, optimized for size and speed so it can execute in browser space. Actually, ActiveX loosely defines a group of Microsoft technologies, including ActiveX control, ActiveX scripting, ActiveX documents, ActiveX containers, and so on. The name ActiveX was coined in December 1995 by Microsoft (Grimes, 1997). Based on marketing considerations, Microsoft decided to repackage the related OLE technology and sell it as ActiveX technology, targeting future markets of Internet applications and becoming a major competitor to Java technology. ActiveX allows COM architecture to execute on a Web browser as buttons, list boxes, pull-down menus, and animated graphics. ActiveX has been widely used by corporate Management Information System (MIS) and independent software vendors (Knapik and Johnson, 1998).

The release of DCOM technology was packaged with Windows NT 4.0 in mid-1996. The original design of COM assumed that components and their clients were running on the same machine. DCOM extends the COM technology to communicate between different computers on a LAN, a WAN, or even the Internet (Microsoft, 1998). DCOM also includes a distributed security mechanism, providing authentication and data encryption (Chappell and Linthicum, 1997).

Figure 3.15 illustrates the relationships between OLE, ActiveX, COM, and DCOM. In general, COM and DCOM represent low-level technology (interface negotiation, licensing, and event management) that allows components to interact, whereas OLE and ActiveX represent high-level application services (linking and embedding, automation, compound documents) that are built on the top of COM/DCOM technology.

3.4.1.2 DCOM Architecture and Interfaces The architecture of DCOM is established on client machines with a remote object proxy and on server machines with a COM stub (Figure 3.16). Network communication is accomplished through Microsoft DCE RPC, which is an extension of Open System Foundation's DCE RPC specification (Grimes, 1997). DCOM uses the method of marshalling to format and bundle the data in order to share it among different components (Orfali and Harkey, 1997).

Marshalling begins as a COM client calls for its remote object proxy on the local machine. The object proxy then passes the calls over the network to a COM stub on the server machine, which marshals the parameters and passes them to the server applications. When the call is completed, the server COM stub marshals return values and passes them to the object proxy on the client machines, which returns them to the client-side applications (Orfali and Harkey, 1997). Beside the low-level communication of objects, the architec-

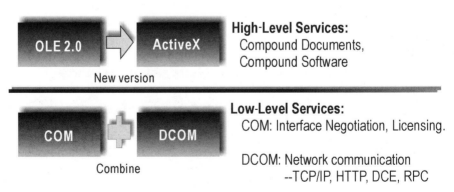

Figure 3.15 Relationships between OLE, ActiveX, COM, and DCOM.

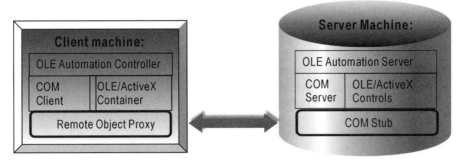

Figure 3.16 Architecture of DCOM.

ture of DCOM also incorporates a high-level object management scheme by using OLE automation controllers and servers (Figure 3.16). OLE automation controllers and servers provide the ability for distributed components to expose functions and commands for other components to access and facilitate the development of programming tools and macrolanguages, which can operate across applications (Chappell and Linthicum, 1997).

The architecture of DCOM specifies the communication mechanism and object management between clients and servers. The actual operations and executions of DCOM objects are accomplished by using the software interfaces between DCOM objects. A DCOM interface is a collection of function calls and defined as a binary-type API based on a table of pointers, called a virtual table or vtable. An interface of DCOM will be given a name starting with a capital "I," such as IUnkown, IClassFactory, and IDispatch. Each DCOM interface has a unique interface identifier, called Interface ID (IID), which is automatically generated by DCOM. Figure 3.17 illustrates a hypothetical DCOM object, MapObject, with three basic interfaces, IUnKown, IDisplay, and IZoomIn.

The *IUnkown* interface is the most important interface of DCOM, which is used for runtime interface negotiation, life cycle management, and aggre-

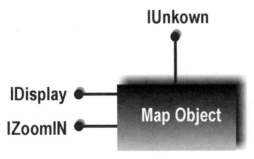

Figure 3.17 Interface example in map object under DCOM framework.

gation. The IDisplay interface and IZoomIn interface are the function calls for the MapObject. For example, a GIS application can call the IDisplay interface to display the map on a defined window as the following statement: MapObject.Idisplay (Mapextent, Window's name). If the GIS application needs to zoom in on a specific area, the IZoomIn interface will be called as MapObject.IZoomIn(X1,Y1,X2,Y2). With the use of DCOM interfaces, software programmers can easily manipulate the behaviors of MapObject for different types of GIS tasks in their applications.

The binary interfaces of DCOM are created by using Microsoft Interface Definition Language (Microsoft IDL), which describes the interfaces' methods and their arguments. Beside the use of Microsoft IDL, DCOM technology provides another type of language for DCOM automation, called DCOM Object Definition Language (ODL). DCOM automation allows client programs to dynamically invoke methods of DCOM objects in order to allow clients to dynamically discover the methods and properties (Orfali and Harkey, 1997).

3.4.1.3 Advantages and Disadvantages of DCOM

The major advantage of DCOM technology is the popularity of Microsoft's operating system (Windows 95/98, Windows NT 4.0, and Windows 2000), desktop applications (Word, Excel, PowerPoint, Access, Internet Explorer, and so on), and programming tools (Visual Basic, Visual C++, and Visual J++). All of Microsoft's products are based on and will be based on DCOM technology. Thousands of PC-related applications and software developed by other companies are also based on Microsoft's DCOM technology, such as ESRI's MapObjects and InterGraph's GeoMedia. For Microsoft Windows-based applications, DCOM is more feasible and more popular for developing distributed components than other compatible technologies, such as CORBA and Java platforms.

The second advantage is that the DCOM technology is designed from the evolution from DDE to OLE to ActiveX. The design of DCOM results from extensive implementation experience instead of being designed from pure theory, as in the case of CORBA. Its core concepts and functions have been revised, changed, and extended over almost 10 years. DCOM technology has been adopted and implemented in thousands of application programs.

The third advantage of DCOM is the language-independent interface design based on Microsoft's binary interface structure. Software programmers can develop DCOM components or ActiveX controls in any languages, including Visual Basic, C++, or even Java. Moreover, Microsoft's J++ development tools provide an integration of Java and DCOM, which allows Java programmers to write Java application with DCOM easily. The multilanguage development ability will attract more involvement in DCOM programming and development.

However, there are some disadvantages to the DCOM technology. The first drawback is that it is not based on a pure object-oriented (OO) implementation. For example, DCOM objects do not support multiple inheritance, which will limit the extensibility of DCOM object development effort. Software programmers have to manually aggregate different COM components by using a complicated software packaging approach in order to compromise the limitation.

The second problem with DCOM technology is the complexity of DCOM interfaces (Brockschmidt, 1994; Vckovski, 1998). Once created and defined, the interfaces will exist forever to ensure backward compatibility with future DCOM applications. As a result, hundreds of component interfaces and functions have been specified in different DCOM objects, which increases the difficulty of understanding and developing DCOM applications.

The third problem with DCOM technology is the inadequate support from other platforms, such as UNIX and Macintosh. Currently, Microsoft is working to make DCOM and some other parts of the ActiveX family available on other operating systems. Microsoft has provided ActiveX support for Macintosh (Chappell and Linthicum, 1997). DCOM implementation on all major UNIX platforms, such as Solaris, Linux, and HP/UX, is also available from a third-party company, Software AG (Microsoft, 1998). The main problem of DCOM with non-Window platforms remains a lack of popular applications. Most software companies develop DCOM applications on the Windows platform rather than on UNIX and Macintosh because of the marketing considerations.

The fourth problem is the compatibility with other distributed-component frameworks, such as the Java Virtual Machine (VM). Currently, only Microsoft's own Java VM (with IE 4.0 or later versions) can run DCOM components or ActiveX controls. Non-Microsoft Web browsers such as HotJava and Netscape Communicator are not able to run DCOM components and ActiveX controls. In general, the DCOM applications have become a Microsoft-dependent technology, which does not easily cooperate with other software companies and is not a fully interoperable framework for distributed components.

In general, Microsoft's DCOM technology is closely connected with current PC desktop applications. However, the proprietary DCOM technology embedded in many Microsoft products is not compatible with other platforms, such as UNIXs or Macs. The specifications of DCOM is actually against the original principle of distributed computing: bridging the heterogeneous platforms and environments. To solve these problems, Microsoft initiated a new type of distributed-component framework in 2000, called .NET (dot NET), which will be discussed in the next section.

3.4.1.4 .NET (Next Generation Distributed-Component Framework)
.NET is a next-generation distributed-component framework developed by Microsoft that can enable software developers to build "blocks" applications and to exchange data and services across heterogeneous platforms and environments. Different from DCOM technology, .NET provides a very comprehensive (and complicated) cross-platform framework where different

component applications can interoperate with one another through the Internet (Pleas, 2000).

The framework of .NET is a collection of many different component technologies, programming languages, and communication protocols. The actual framework of .NET is shown in Figure 3.18.

First of all, .NET provides very flexible development tools and languages for software programmers to create distributed components. The generic language supported by .NET is called C#, which is a pure object-oriented languages developed by Microsoft. The features of C# are very similar to Java languages developed by Sun Microsystems that include inheritance, encapsulation, abstraction, and polymorphism, although Microsoft claims that they are totally different languages. C# is the best language for developing .NET application but not the only language according to Microsoft's statement. .NET can allow programmers to have other languages selections, such as Visual Basic, Java, Jscript, and Perl. Moreover, different languages can be combined together for a single application under the .NET framework. For example, a GIS application can utilize Java Swing for the design of graphic interface, C# for buffering function, and Perl scripts for text attribute query function.

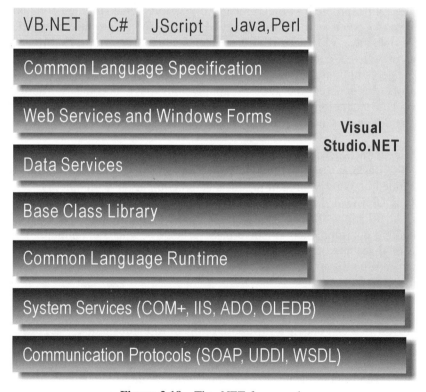

Figure 3.18 The .NET framework.

Since many different languages can be used and combined together to develop .NET applications, the goal of Common Language Specification (CLS) is to ensure the interoperability across these different languages. CLS specifies the language compliant types for .NET and the conformance specification for different language compilers. The use of CLS can ensure that distributed components compiled by different languages can work together without breaking the whole application or service. Under CLS, all data types (e.g., strings, integer) defined in different .NET languages are consistently defined and interoperable.

The next stage of the .NET framework is Web services and Windows forms. The design of Web services and Windows forms is to provide a high-level component framework and user interface tools for all .NET applications. Web services, a collection of distributed-component technologies based on the Extensible Markup Language (XML), include Simple Object Access Protocol (SOAP), Universal Discovery, Description, and Integration (UDDI), and Web Service Definition Language (WSDL). The design of Web services can allow GIS users to define, search for, and invoke distributed components remotely. WSDL is used to define the functions and services of .NET components, which is similar to the role of IDL in DCOM. UDDI is a searchable registry database that stores the metadata of .NET components (generated by WSDL) and allows other programs to search/access these components. The function of UDDI is very similar to the metadata repository in the CORBA framework. SOAP is an XML-based RPC, which can allows other programs to invoke .NET components and applications remotely. Windows forms provide various client-side user interface tools for .NET applications and self-managing capabilities for .NET applications.

The goal of data services is to provide a robust XML-based data access model for .NET applications. Microsoft created a new type of DTD (Document Type Definition) for .NET called XML Schema Definition (XSD) schemas. XSD is used to provide a standardized XML format to allow interoperable XML data processing, data manipulation, and database management.

The next level of .NET framework is Base Class Library (BCL). The BCL includes fundamental class libraries for .NET applications, including standard Input–Output (I/O) controls, security manipulation, and basic mathematic operations. The goal of BCL is to ensure that all .NET applications can be written by different languages and access the same core functionality defined in BCL. This feature will ensure a common implementation procedure for .NET applications and language interoperability.

Common Language Runtime (CLR) is a runtime environment that allows different .NET languages to interoperate with each other. The CLR has three main elements: the execution system, .NET Common Types System (CTS), and metadata systems. The function of CLR is very similar to the virtual machine defined in Java platforms, although Microsoft claims that they are totally different types of frameworks.

Since the framework of .NET is very complicated, Microsoft created the VisualStudio.NET to provide software programmers a comprehensive development environment/toolbox. This toolbox includes several programs and compilers for different types of languages and API documents.

System services are not part of .NET core technology but belong to the level of operating systems and hardware communications. Several techniques will be used to support .NET framework, such as transaction services in COM+, database connection in OLE DB, and directory services in Microsoft Active Directory.

The lowest level of .NET framework is the communication protocol. This level is closely related to the Web service level and provides the communication channels for Web services. The .NET protocols used in this level include SOAP, UDDI, and WSDL.

In general, the design of .NET provides a very comprehensive, flexible framework for distributed Web services, and Microsoft will play a significant role in the development of distributed components in the future. Many Microsoft applications, such as Microsoft Office and Windows, will use .NET to provide global Web-based services for their users. But, the .NET project is still under development. Furthermore, many technologies used in the .NET framework are proprietary to Microsoft, and this may become a serious problem for other distributed-component frameworks such as Java and CORBA.

3.4.2 Common Object Request Broker Architecture

CORBA is another distributed-component framework developed and standardized by OMG (http://www.omg.org). CORBA provides a standardized interface model and object framework for solving network computing problems in a distributed heterogeneous environment.

3.4.2.1 CORBA Development History   The development of CORBA has been in progress for over 10 years and has been dominated by OMG. OMG is a nonprofit consortium founded in May 1989 by eight companies: 3Com Corporation, American Airlines, Canon, Data General, Hewlett-Packard, Philips Telecommunications N.V., Sun Microsystems, and Unisys (Yang and Duddy, 1996). In 1998, OMG included over 800 member companies internationally. The main goal of OMG is to promote theories and practices of object technology in distributed computing, including reusability, portability, and interoperability. The direction of OMG does not focus on developing new computing technologies, but rather relies on existing technologies offered by member companies. OMG's members may propose specifications based on OMG's Requests for Proposals (RFPs) under different commercially available computing technologies. The proposed specification will be reviewed and voted by the OMG board of directors to decide whether the specification is formally accepted or not (Vinoski, 1997; Yang and Duddy, 1996). Essentially, OMG is a standards organization.

OMG released the first specification of CORBA 1.1 to the computer industry in 1991, following the standardized Object Management Architecture

(OMA). Later, OMG released CORBA 2.0 in 1994, CORBA 2.2 in 1998, and CORBA 3.0 in 2002, which adds QOS control and CORBA component architecture. The specification of CORBA defines an IDL and APIs, which enable client/server object interaction within a specific implementation of an ORB (Orfali and Harkey, 1997). The architecture of CORBA differs from DCOM in that it does not distinguish between clients and servers, as discussed below.

3.4.2.2 CORBA Architecture and Interfaces   CORBA's architecture is based on OMA, a high-level conceptual infrastructure for distributed-computing environments proposed by OMG. OMA provides the means to build interoperable software systems in heterogeneous network computing environments.

The reference model of OMA has been consistently modified since it was published in 1990. The 1996 version of the OMA reference model added a new category, domain interface, and introduced the object frameworks category (Thompson et al., 1997). Figure 3.19 shows that the OMA reference model consists of an object request broker and four software interface categories (application interface, domain interface, common facilities, and object services).

Object services are used for the management of distributed object programs and the discovery of other available services. Two examples include the naming service, which allows clients to find objects based on names, and the trading service, which allows clients to find objects based on their properties. Other object services specify software life-cycle management, security, transactions, event notification, and so on (Vinoski, 1997).

Common facilities provide standardized interfaces to common application services, such as system management, data interchange, printing, and user

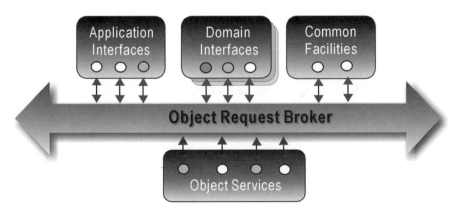

Figure 3.19   OMA reference model interface categories (Vinoski, 1997). (Figure reprinted with permission from IEEE, © 1999 IEEE.)

interface. They are oriented toward end-user applications. An example of such a facility is the Distributed-Document-Component Facility (DDCF) that permits interchange of objects based on a document model, for example, facilitating the linking of a spreadsheet object into a report document. Other types of common facilities include the printing facilities, database facilities, electronic mail facilities, and user interfaces (Vinoski, 1997; Yang and Duddy, 1996).

Domain interfaces are oriented toward specific task domains. One of the first OMG domain interface categories is Product Data Management (PDM), which focuses on the manufacturing industry domain. Other types of OMG domain interfaces will soon be issued for telecommunications, medical, and financial application domains. In Figure 3.17, multiple boxes are shown for domain interfaces to indicate the existence of many separate application domains (Vinoski, 1997).

Application interfaces are developed specifically for a given application. Because they are application specific and because the OMG does not develop applications (only specifications), the interfaces are not standardized. However, if over time it appears that certain broadly useful services emerge out of a particular application domain, they might become candidates for future OMG standardization (Vinoski, 1997). For example, a GIS vendor can develop its own application interfaces for a specific GIS product in the framework of OMA and utilize other types of interfaces for the purpose of system management or object services.

In addition to the reference model, OMA also defines an object model and framework. An OMA object is an encapsulated entity using OO modeling techniques. The object model of OMA defines common object semantics for specifying the externally visible characteristics of objects in a standard and implementation-independent way. A client-side object can request services from a target object (a server) through a software interface, which is specified in OMG IDL. The request includes an object reference of the service provider, which is a unique object identifier. The design of each object reference protects the content from the client-side intervention (Vinoski, 1997). In general, each object has its own types of interfaces in order to provide their functionality and communicate with other types of objects (OMG, 1998).

Essential to CORBA is the design of ORB. The main function of ORB is to deliver requests from clients to target objects. In general, ORB is middleware that maintains client-server relationships for the application programmers. The protocol for client/server interaction is defined through a single implementation language-independent specification, the IDL. The IDL can be defined underneath different programming languages, such as C++, Java, and SmallTalk.

The IDL provides operating-system-independent interfaces to all the services and components that reside on a CORBA bus. Programmers can use the IDL to specify, for example, a component's attributes, the parent classes it inherits from, the exceptions it raises, the typed events it emits, and the methods it supports (Orfali et al., 1996). Current CORBA specification 2.2, released in 1998, provides several language mappings for the IDL, including C, C++, Smalltalk, COBOL, and Java (OMG, 1998; Vinoski, 1997).

Figure 3.20 illustrates the CORBA architecture. A CORBA client can use the dynamic invocation interface or an IDL stub to make a request to the server-side objects. The client can also directly interact with the ORB for some functions. The object implementation (server-side) object receives a request as an up-call either through the IDL-generated skeleton or through a dynamic skeleton (Schmidt and Vinoski, 1995). The ORB locates the appropriate implementation code, transmits parameters, and transfers control to the object implementation interface through an IDL skeleton or a dynamic skeleton. In performing the request, object implementation may obtain some services from the ORB through the object adapter. When the request is complete, control and output values are returned to the client (Orfali and Harkey, 1997; Vinoski, 1997).

3.4.2.3 CORBA 3.0   OMG released CORBA 3.0, which embraces several new capabilities to CORBA 2.0 specifications, in 2002. There are three major new categories created in CORBA 3.0:

- Java and Internet integration
- QOS control, and
- CORBA component architecture.

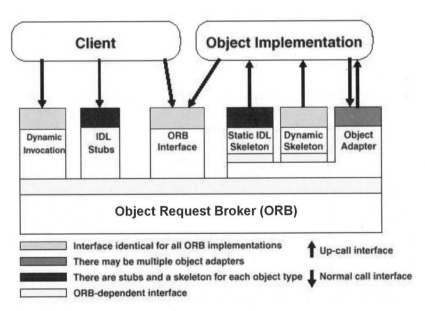

Figure 3.20   CORBA architecture (OMG, 1998). (Figure reprinted with permission from Object Management Group.)

First, due to the popular use of Java technologies, the development of CORBA specification began to focus on the integration of Java and CORBA technologies. The addition of Java functions will be a great improvement over CORBA functionality. For example, the language mapping between Java and IDL can allow Java RMI to invoke CORBA objects and to use the Internet Inter-ORB Protocol (IIOP) to access Java beans. Also, several new services, such as Interoperable Name Services and Firewall Specification, can improve the performance and the security of CORBA applications.

Second, the new CORBA 3.0 provides the control of QOS. Different types of CORBA objects and clients can be ordered by their priority and create more efficient and reliable Web services. CORBA 3.0 also specifies real-time resource control to create a predicable real-time access environment.

Finally, the new version of CORBA specification defines the setting of CORBAcomponents. There are three elements of CORBAcomponents, the CORBAcomponent container, enterprise Java beans, and multiplatform software distribution format. The design of CORBAcomponents will create flexible and Java-enhanced Web services frameworks for CORBA applications.

3.4.2.4 Advantages and Disadvantages of CORBA   The design of CORBA provides a scalable and flexible framework for distributed client/server components and for the Internet and the intranet. CORBA follows comprehensive OMA guidelines with the full range of object services, common facilities, domain interfaces, and application interfaces. CORBA developers can create a sophisticated, well-organized object set whose elements can interact dynamically via the ORB. Well-defined object categories can facilitate communication between objects. CORBA implementation procedures can help programmers to conquer the most critical challenges in distributed-network environments, such as monitoring object life cycles, global naming procedures, transaction services, licensing, and security problems.

A second advantage is that CORBA provides a pure object-oriented concept for modeling, including encapsulation, inheritance, and polymorphism, in the object implementation and language-mapping methods. At the same time, CORBA objects can be implemented using traditional procedure languages, such as C, FORTRAN, and COBOL.

The third advantage is the extensibility for future development of distributed objects/components. CORBA has been in development for almost 10 years, which is much longer than the other competitors, DCOM and Java. CORBA specifications create innovative design and concepts for distributed network environments, such as self-describing, self-managing objects. Other distributed-component technologies, such as Java platform and DCOM, have followed the same concepts from the original CORBA and OMA specifications. Indeed, the development of CORBA illustrates the future direction for distributed network environments and distributed computing. However, there are still some drawbacks of CORBA.

The first problem with CORBA development is that the desktop integration with Microsoft Windows-based environments is difficult. Although CORBA implementations support a wide range of mainframe and workstation UNIX platforms, CORBA provides only limited support for Windows NT applications and other Windows-based environments. Most Windows-based environments use DCOM technology, which as stated above is a closed architecture.

The second problem is a marketing issue. CORBA is not a free technology. Users have to purchase the development tools and implementation frameworks to develop CORBA objects, whereas DCOM and Java technology provide free download for programmers to develop their applications. Therefore, the marketing strategy limits the popularity of CORBA objects and applications.

The third problem is the slower development process compared with other distributed-component technologies. Java and DCOM evolve very quickly with many new functions released every year. The main reason for the slow evolution of CORBA is that all major changes and modifications must be approved by the OMG members in hundreds of different software companies. Although the democratic approach of CORBA can ensure the standardization of methods, it also slows the progress of CORBA relative to its competitors, which are developed by a single software company. This presents a classic trade-off in standards development, namely the dynamic tension between institutional consensus and market competition.

To summarize, CORBA has a comprehensive, extensible, well-defined architecture to support complicated applications in distributed, heterogeneous network environments. The implementation of CORBA also supports both new and legacy languages and applications, which is a very important feature for the integration of distributed data and programs. The main goal of CORBA is to allow distributed business applications to work together seamlessly across a network. Many programmers prefer CORBA technology because of its innovative design and well-defined architecture. However, the integration of desktop computers will be a critical issue for the future success of CORBA development. One possible solution for the desktop integration is to let Java technology bridge the gaps between CORBA applications with desktop PCs. The Java/CORBA integration has emerged as a new direction for CORBA development (Orfali and Harkey, 1997).

3.4.3 Java Platform

In contrast to DCOM and CORBA, the original development of the Java platform is as a programming language instead of in support of distributed-object frameworks. However, with the rapid growth of Java applications, the Java language has developed its own component framework, called JavaBean, with the architecture specifications for distributed computing. Currently, many Java-related technologies are already beyond the scope of programming language. Its original developer, Sun Microsystems, called all related Java technologies and specifications an integrated Java platform: "The Java programming language platform provides a portable, interpreted, high-performance, simple, object-oriented programming language and supporting runtime environment" (Gosling and McGilton, 1996, p. 11).

The original goal of Java is to meet the challenges of application development in the context of heterogeneous, network-based distributed environments (Gosling and McGilton, 1996). The key to Java's power is its "write once, run anywhere" software model. The Java runtime environment translates Java byte codes into a virtual machine that runs on any supported platform (Hamilton, 1996). With its powerful cross-platform capability, many software vendors and organizations have launched their projects to explore the potential of the Java language and on-line applications (Halfhill, 1997).

3.4.3.1 *Java Development History*   The Java language was developed at Sun Microsystems in 1990 by James Gosling as part of a research project to develop software for consumer electronics devices (Anuff, 1996; Harmon and Watson, 1998; Lemay and Perkins, 1996). The original name of the new language was called Oak. The purpose of Oak was to provide an object-oriented programming language and a software platform for smart consumer electronics, such as cable boxes and video game controls. In 1991, a team called Green Project from Sun Microsystems began to work on Oak. Sun renamed the Oak language as Java and introduced it to the public in 1995. Java technology has become one of the most important developments in Internet history.

In 1997, Sun released Java 1.1, which includes many important new features and functions for distributed computing, including Java beans, internationalization, new event model, jar files, object serialization, reflection, security, JDBC, and Remote Method Invocation (RMI). One of the most important new features of Java 1.1 is the Java bean for creating reusable, embeddable software components, which are similar to the Microsoft's ActiveX model. Two other significant features in Java 1.1 for distributed computing are the RMI API and JDBC, which allow a Java program to invoke methods of remote Java objects or communicate with remote Database Management Systems (DBMSs) directly (Weber, 1997).

In late 1998, the Java 2 platform was released and provided more advanced network-centered functions and APIs. The new content included the Java version of ORB for the integration of CORBA, Java 2D APIs, Java foundation classes, and Java servlets for enterprise server-side applications (Flanagan, 1999; Horstmann and Cornell, 1998).

In 2001, Sun grouped Java technologies into three different editions:

- Java 2 Micro Edition (J2ME),
- Java 2 Standard Edition (J2SE), and
- Java 2 Enterprise Edition (J2EE).

The reason for providing three different editions of the Java platform is to extend the capability of the Java framework to different types of computing environments, including mobile/pocket devices, desktops/workstations, and enterprise servers.

3.4.3.2 Java Language and Architecture The Java language is a pure object-oriented language designed to enable the development of secure, high-performance, and highly robust applications on multiple platforms in heterogeneous, distributed networks (Gosling and McGilton, 1996). From the computer programming perspective, Java looks like C and C++ while discarding the overwhelming complexities of those languages, such as typedefs, defines, preprocessor, unions, pointers, and multiple inheritance (Gosling and McGilton, 1996).

The design of the Java language draws on the best concepts and features of previous object-oriented languages, primarily from Eiffel, SmallTalk, Objective C, and C++. Java also incorporates garbage collection and dynamic links from Lisp and Smalltalk, interface concepts from Objective C and OMG's IDL, packages from Modula, concurrency from Mesa, and exceptions from Modula-3 (Harmon and Watson, 1998, p. 62).

The architecture of the Java platform is illustrated in Figure 3.21. There are two procedures for the implementation of Java applications, a compile-time environment (server side) and a runtime environment (client side). The

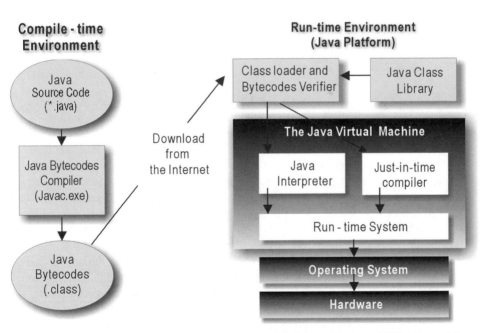

Figure 3.21 JAVA platform architecture. (Modified from Kramer, 1996, Figure 3.)

compile-time environment can be constructed by using the Java Development Kit (JDK) provided by Sun and includes a Java compiler (Javac.exe), a Java interpreter (Java.exe), a Java debugger (jdb.exe), and several standardized Java libraries. Programmers can use the Java compiler to generate a Java class from a text-based Java source code to a Java byte-code format and put the class on the server-side machine. Then, the Java class is ready for download by client machines.

The runtime environment is comprised of three components: class loader, Java class library, and Java VM. When a client requests a Java class, the client-side VM will download the Java class via the class loader and combine it with other required Java classes from the library. Then, the Java class will be interpreted or compiled into the actual machine codes in the runtime system, which can be executed under the client-side operating system and hardware environment.

Besides the mobile class download functions, the Java platform also supports RMIs on objects across different Java VMs by using the RMI. By using RMI, Java programmers can create a remote Java class with object serialization and create client stubs and server skeletons for the communication between clients and servers. The implementation of RMI is very similar to the procedure of CORBA object implementation (Orfali and Harkey, 1997).

Key Concepts

Java applications *are stand-alone programs. They provide full access to the entire local machine resources. A* **Java applet** *is a specific kind of application that can only run from within a Web browser that contains the Java VM. A* **Java servlet** *is a server-side Java program that extend the capabilities of the server.*

Three types of Java programs are Java applications, Java applets, and Java servlets. Java applications are stand-alone programs. They do not need to be embedded inside a HTML file or use any Web browser to execute the programs. Java applications can provide full access to the entire local machine resources, such as writing files and changing database contents. Also, Java applications run faster than Java applets because the applications do not need to deal with browsers and have full control of the local client environment. A Java applet is a specific kind of application that can only run from within a Web browser that contains the Java VM. In contrast to a Java application, Java applets must be included as part of a Web page in HTML format. Java applets are designed for WWW and can be dynamically downloaded via the Internet. In order to protect the Web users and prevent possible damage to the local machines, Java applets execute within a closed, secure Web browser environment and have only limited access to the memory, data, and files on the local machine. More recently, server-side Java programs, called Java servlets, become more and more important for distributed computing environment and the Internet. Java servlets can let a user upload an executable program to the network or server. These servlets can actually be linked into the server and extend the capabilities of the server. By interacting with server-side applications, a Java servlet can share the loads between servers and clients. The results will reduce server load and provide the balance of functionality on server and client machines.

Most programmers and technology consultants are very optimistic about the future development of Java technology. The main reason is that Java language is truly designed for the distributed network environment, such as the Internet and intranet. In the future, Java technology will embrace more new functions and APIs in order to cope with the rapid development of network technology.

3.4.3.3 *Three Kinds of Java: J2SE, J2EE, J2ME*   In 2001, Sun Microsoft created three different editions of Java platforms; J2ME, J2SE, and J2EE (Figure 3.22). The J2SE focuses on general, personal, client-side applications. J2SE is the original edition of Java 2 platform. Many useful APIs developed in J2SE are closely related to the applications of Internet GIS and on-line mapping, such as Java advanced imaging APIs, Java 2D APIs, and Java 3D APIs.

J2EE focuses on the server-side, business, IT-based applications. Software programmers can use J2EE Java bean architecture to create reusable distributed components for different Web services applications and Business-to-Business (B2B) applications. Several key technologies of the Enterprise edition include the integration of Java and XML, the Java servlets technologies, and J2EE Java bean.

J2ME focuses on the applications of small/wireless devices, such as cellular phones, pagers, smart cards, and PDAs. For example, the Java card technology can allow very tiny Java programs to run on smart cards and other

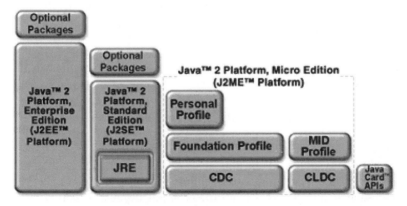

Figure 3.22   Three editions of Java platforms. (*Source:* http://java.sun.com/java2.)

hand-held devices with very limited memory size (between 128K and 512K). J2ME can also allow different Java applications to specify their own "profiles," which are the minimum sets of APIs for particular devices or applications. By utilizing the profiles, the size of Java VMs and programs could become very small and be able to be embedded inside a small device. J2ME also introduces two kinds of Java VMs with their configurations. Connected Device Configuration (CDC) utilizes a portable, full-feature Java 2 VM called C Virtual Machine (CVM). Connected Limited Devices Configuration (CLDC) utilized another type of VM called K Virtual Machine (KVM), which is a minimum-footprint Java runtime machine for tiny, resource-constrained devices. Moreover, J2ME also provides Mobile Information Device Profile (MIDP) to create and specify APIs for wireless communications and mobile devices.

In general, the development of these different Java platforms indicates that distributed-component technologies like Java can be applied in many different Web services and mobile devices. The key issue in the development of Java applications is that the new framework can provide more flexible, efficient services compared to traditional computer programs. Currently, the Java platform is facing a serious challenge from the .NET frameworks developed by Microsoft, although the programming environments of the Java 2 platform are more open, easier, and more flexible compared to the proprietary .NET architecture. It is difficult to foresee who is going to eventually win the distributed-component framework market.

3.4.3.4 Advantages and Disadvantages of Java Platform
In general, Java provides a simple and creative way to develop, manage, and deploy distributed client/server applications. It also provides an easy way to quickly distribute and update applications and programs via the Internet. From a distributed computing perspective, there are four main advantages of Java technology.

The first advantage of Java is to provide a dynamic component framework of Java applets and servlets. The use of Java applets and servlets in Web applications can facilitate a more dynamic and efficient interaction between client and server. Therefore, the Java platform can provide a truly distributed computing environment with the balance of server-side/client-side processes.

The second advantage of Java technology is the similarity between the Java language and C++ language. The similar syntax and statements encourage more and more software engineers to develop powerful Java applications without too much difficulty. Programmers with C++ experience can shift to Java programming very quickly. Therefore, Java programming becomes more and more popular due to its similarity to C++.

The third advantage of Java is its robust performance with the cross-platform capability. Traditional programming languages such as C++ cannot provide such a robust, cross-platform program because "their designs primarily support programmer-directed memory allocation and de-allocation, pointers and pointer arithmetic, multiple inheritance and procedural features such as functions, structures, union, typedefs, defines, and pre-processor directives, including macros" (Knapik and Johnson, 1998, p. 279). Since Java gets rid of many problematic designs and functions in traditional programming languages, the execution of Java programs becomes more robust and reliable across different platforms.

The fourth advantage is the dynamic binding feature for Java with the downloadable Java applets framework: "Imagine a multi-media word processor written in Java. When this program is asked to display some type of data it has never encounter before, it might dynamically download a class from the network that can parse the data, and then dynamically download another class (probably a Java bean) that can display the data within a compound document. A program like this uses distributed resources on the network to dynamically grow and adapt to the need of its user" (Flanagan, 1997, p. 5). The dynamic download for new classes will facilitate the sustainable growth of Java applications in the heterogeneous network environments.

On the other side, Java technology still has some weakness. First, the Java platform does not provide a standardized distributed-object infrastructure, such as CORBA's OMA. Many different software companies develop unique Java libraries and applications with nonstandard frameworks. Without the standardized categories, open architecture and integration between different packages and libraries will remain very difficult.

Second, the performance of Java byte-code programs is slower than genetic machine-level binary-code programs written in C++ or other languages. Sun provides some solutions for improving the Java program performance, such as the Just-in-Time (JIT) compiler and Java chips. However, the general performance of Java applets and applications is still slower than traditional programs.

To summarize, Java is a simple, object-oriented, distributed, interpreted, robust, secure, architecture-neutral, portable, high-performance, multi-threaded, and dynamic binding language (Anuff, 1996). Java technology is still evolving and changing. The great success of Java technology changes the nature of the Internet and the WWW, and Internet GIS. In the future, Java technology may extend territory to provide smart electronic devices such as interactive TVs, smart air conditioners, or smart microwaves or palm-size GPS applications (Horstmann and Cornell, 1998).

3.4.4 From Client/Server, Web Client/Server, to Distributed-Component Models

Client/server, Web client/server, and distributed-component models are similar architectures that rely on networks for the communication between the client and the server. They are interrelated and are not necessarily easy to separate. They all use the same client/server architecture model. However, there are major differences among these three approaches.

First, the Web client/server model relies on the Web browser as a front-end user interface, while client/server and distributed-component applications may use their own GUI other than the Web browser. Second, both the Web client/server and distributed-component architecture rely on TCP/IP as the transport service, but the Web client/server applications have to use the HTTP protocol as the communication glue. The client/server applications, on the other hand, may use other different protocols.

3.5   DEPLOYMENT OF THE DISTRIBUTED GISERVICES ARCHITECTURE

Based on the concept of the distributed-component model, the distributed GIServices scenario of the prior Wal-Mart site selection example can be established dynamically on several GIS nodes (Figure 3.23). Figure 3.23 illustrates the major differences between traditional GISystems solution using the client/server architecture and Internet GIServices solution using the distributed-component model. The Internet GIServices can relocate or access both geodata sets and GIS components (programs) remotely from a user's machine. The software architecture of distributed systems implemented in the distributed GIServices provides more flexible and scalable GIS capabilities for different applications. This user scenario also illustrates two types of client/server architecture:

1. the Web Client/server architecture and
2. the distributed-component framework (Web services architecture).

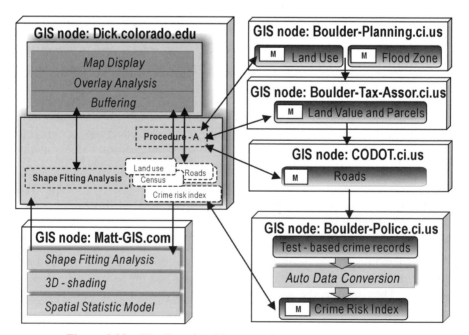

Figure 3.23   Distributed architecture of Wal-Mart site selection.

The Web client/server architecture is illustrated in the first step of this scenario, while Dick, the GIS spatial analyst, can access the land use datasets from the planning department of Boulder County. Boulder County created a Web server for multiple users to access and download their GIS datasets (Figure 3.24). The Web client/server architecture allows distributed clients to access a centralized server remotely. The client-side components are usually platform independent, requiring only an Internet browser to run. However, each client component can access only one specified server at a time. The software components on client machines and server machines might be different and not interchangeable. For example, if the Web server in the Boulder planning department is using the ESRI ArcIMS feature server, Dick needs to download a specific ESRI Java viewer in order to access the land use data sets. The ESRI Java viewer is not compatible with Autodesk MapGuide server and MapGuide plug-in viewer. Also, the relationships between clients and servers are one server to many clients. The Web client/server is not scalable and the services may be limited by the numbers of clients and applications provided on the servers.

The second type of distributed GIS architecture is the distributed-component framework. More recently, the IT industry described this type of framework as "Web services." In this scenario, Dick can download a "shape-fitting analysis" GIS program from Matt's Linux server and also send out procedure A to the Police Department servers to perform remote GIS oper-

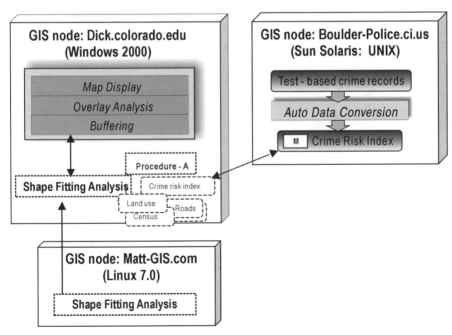

Figure 3.24 Traditional client/server architecture.

ations and data conversions on a UNIX machine (Figure 3.25). Distributed-component frameworks provide a more flexible, scalable, cross-platform solution for Internet GIS. These functions (interoperable programs in different platforms) can only be achieved in distributed-component frameworks or Web services.

In general, distributed-component frameworks allow software programs to travel, access, and interact with multiple and heterogeneous systems and platforms without the constraints of traditional client/server relationships (Montgomery, 1997). A distributed GIServices architecture permits dynamic combinations and linkages of geodata objects and GIS programs via networking. The client/server relationships in distributed-component frameworks are many clients to many servers.

3.6 SUMMARY

This chapter introduced several kinds of client/server computing models and reviewed the development of different distributed-component frameworks: DCOM/.NET, CORBA, and Java. In general, the progress of network technology provides a modern hardware/software infrastructure for Internet GIS. The concepts of distributed systems and open systems facilitate the shift of GIS architecture from a centralized system to distributed services. The in-depth description of three distributed-component frameworks illustrates the possible choices of technical frameworks for distributed GIServices. The understanding of these technologies should help the GIS community and software designers recognize the potential capabilities and the technical

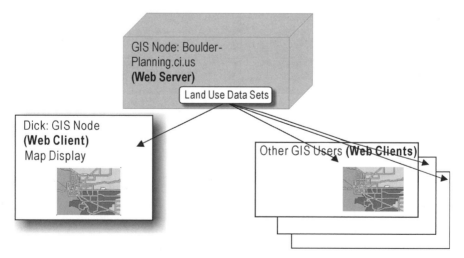

Figure 3.25   Distributed-component framework (Web services).

limitations of distributed GIServices. There are constraints on these technologies, such as vendor dependency, complex software specifications and design, and lack of integration between different component frameworks. To deploy a distributed GIServices architecture, the GIS community has to confront the limitations and the drawbacks of distributed-component technology and utilize the potential capabilities of distributed-component technologies such as dynamic binding, self-managing components, and remote method invocations. The following discussion illustrates the major considerations of adopting distributed-component technologies from the GIServices-oriented perspective.

First, selecting the right component technology for distributed GIServices is extremely difficult. The selected technology should provide a robust, secure, and efficient communication mechanism via the Internet/intranet. Security and stability will become major considerations for distributed GIServices because many geospatial data sets and services are valuable and critical, and networking brings opportunities for viruses, hacker attacks, network traffic jams, and so on. The integration of legacy systems must be another criterion for GIServices because many valuable GIS programs providing essential services reside in legacy systems. The best example of this is federally produced public domain GIS data such as census data. Current distributed-component technologies usually provide certain approaches to integrate legacy systems, such as object-wrapping and middleware solutions. However, these approaches may reduce the performance of legacy systems or simply cannot be applied in some specific cases.

The third criterion is the future development of these technologies. Many people think that a superior technology will guarantee successful adoption in the future. However, many cases in the computing industry do not support such an assumption, such as the failures of the NeXt operating system, OpenDoc, and IBM's OS2. The best technology does not automatically ensure continued development. Support from software vendors, marketing strategies, and users' feedback will also decide the future development of component technologies. Thus, to choose an appropriate distributed-component technology, one should consider not only its technical features and implementation details but also the actual users' experiences, vendor support, and marketing strategies.

Furthermore, customizing these technologies for distributed GIServices is a major task for the GIS community. Distributed-component technologies are designed not specifically for GIServices but for general information services. Many requirements and functions of GIServices are not considered in the original design of generic distributed-component technologies. For example, the complexity of geodata models and functions, the huge volume of geospatial databases and remote sensing images, and the visualization requirements of geospatial information are not taken into account. Adequate GIS functions above the low-level technical frameworks are essential for the successful implementation of distributed GIServices.

Third, integrating different distributed-component technologies is essential for providing truly distributed GIServices. Inevitably, the future development of Internet/mobile GIServices will have to tolerate heterogeneous techniques and frameworks because there is no perfect distributed-component technology for all kinds of GIServices. A sound technology should be able to integrate and migrate different frameworks. Although most distributed-component technologies have proposed solutions for integrating other technologies, only a few successful cases demonstrate that these approaches really work. Software vendors are not willing to integrate their technology with others because of marketing considerations. The GIS community should push these vendors for a true integration of distributed-component technologies because it will not happen automatically. It is dangerous for the GIS community to just wait and see what happens.

To summarize, maximizing the capability of GIServices by using distributed-component technologies is the main goal of distributed GIServices. Traditional GISystems do not provide users with flexible and dynamic services. The future development of distributed GIServices should provide innovative GIS functions and services instead of mimicking the original functions of GISystems. Putting traditional GISystems on-line is not equal to distributed GIServices. The GIS community should invent new services and functions specifically for distributed GIServices, such as digital libraries, distance learning, cyberspace navigating, network-based decision support systems, and virtual tourism. Innovative GIServices and functions will energize the development of distributed GIServices to a higher level of functionality and provide users with more comprehensive services.

This chapter has provided a comprehensive introduction of current software development and component technologies. These distributed-component frameworks, such as .NET, Java, and CORBA, will be able to provide truly interoperable geospatial services for wired and wireless Internet GIS applications. The next chapter will take a look at the technology evolution of Web mapping.

WEB RESOURCES

Descriptions	URL
Java 2 software	http://java.sun.com/java2
.NET framework	http://www.microsoft.com/net
CORBA3	http://www.omg.org/technology/corba

REFERENCES

Anuff, E. (1996). *The Java Sourcebook*. New York: Wiley.

Hollis, B. S., and Lhotka, R. (2001). *VB.NET Programming with the Public Beta*. Birmingham, United Kingdom: Wrox Press.

Brockschmidt, K. (1994). *Inside OLE 2.* Redmond, Washington: Microsoft Press.

Chappell, D., and Linthicum, D. S. (1997). ActiveX Demystified. *BYTE,* 22(9), pp. 56–64.

Dixon, R. (1996). *Client/Server and Open Systems.* New York: Wiley.

Flanagan, D. (1997). *Java in a Nutshell: A Desktop Quick Reference,* 2nd ed. Sebastopol, California: O'Reilly & Associates.

Flanagan, D. (1999). *Java in a Nutshell: A Desktop Quick Reference,* 3rd ed. Sebastopol, California: O'Reilly & Associates.

Gosling, J., and McGilton, H. (1996). *The Java Language Environment,* A White Paper. Sun Microsystems. URL: http://www.Java.sun.com/docs/white/langenv, May 10, 2000.

Graham, I. (1994). *Object-Oriented Methods,* 2nd ed. Workingham, England: Addison-Wesley.

Grimes, R. T. (1997). *Professional DCOM Programming.* Chicago, Illinois: Wrox Press.

Halfhill, T. R. (1997). Today the Web, Tomorrow the World. *BYTE,* January 1997, 22(1), pp. 68–80.

Hamilton, M. A. (1996). Java and the Shift to Net-Centric Computing, *Computer,* August pp. 31–39.

Harmon, P., and Watson, M. (1998). *Understanding UML: The Developer's Guide.* San Francisco, California: Morgan Kaufmann Publisher.

Horstmann, C. S., and Cornell, G. (1998). *Core Java 2,* Vol. 1: *Fundamentals.* Englewood Cliffs, New Jersey: Prentice-Hall.

Knapik, M., and Johnson, J. (1998). *Developing Intelligent Agents for Distributed Systems: Exploring Architecture, Technologies and Applications.* New York: McGraw-Hill.

Kramer, D. (1996). *The Java Platform, A White Paper,* Sun Microsystems, URL: http://java.sun.com/docs/white/index.html, May 1996.

Lemay, L., and Perkins, C. L. (1996). *Teach Yourself Java in 21 Days.* Indianapolis, Indiana: Samsnet.

Malek, M. (1995). Opening Keynote Address: Omniscience, Consensus, Autonomy: Three Tempting Roads to Responsiveness, In *Proceedings of the 14th Symposium on Reliable Distributed Systems.*

Microsoft. (1996). *OLE Concepts and Requirements Overview.* Redmond, Washington: Microsoft Online Library. URL: http://support.microsoft.com/support/kb/articles/Q86/0/08.ASP, May 11, 2000.

Microsoft. (1998). *DCOM Architecture,* White Paper. Redmond, Washington: Microsoft Press.

Montgomery, J. (1997). Distributing Components. *BYTE,* April 1997, 22(4), pp. 93–98.

Newton, H. (1996). *Newton's Telecom Dictionary,* 11th ed. New York: Flatiron Publishing.

Object Management Group (OMG). (1998). *The Common Object Request Broker: Architecture and Specification,* 2.2 ed. Framingham, Massachusetts: OMG.

Oracle (1992). Distributed Databases. In *ORACLE 7 Server Concepts Manual.* Redwood Shores, California: Oracle, Chapter 21, pp. 21.1–21.6.

Orfali, R., and Harkey, D. (1997). *Client/Server Programming with Java and CORBA.* New York: Wiley.

Orfali, R., Harkey, D., and Edwards, J. (1996). *The Essential Distributed Objects Survival Guide.* New York: Wiley.

Orfali, R., Harkey, D., and Edwards, J. (1999). *Client/Server Survival Guide,* 3rd ed. New York: Wiley.

Pleas, M. (2000). Microsoft .NET. *PC Magazine,* December 5, pp. IP01–IP08.

Rumbaugh, J., Blaha, M., Premerlani, W., Eddy, F., and Lorensen, W. (1991). *Object-Oriented Modeling and Design.* Englewood Cliffs, New Jersey: Prentice-Hall.

Schmidt, D. C., and Vinoski, S. (1995). Object interconnections: Comparing Alternative Programming Techniques for Multi-Threaded Servers. *Column 5. IGSC++ Report Magazine,* Feb. 1995. URL: http://www.cs.wustl.edu/~eschmidt/report-doc.html.

Schroeder, M. D. (1993). A State-of-the-Art Distributed System: Computing with BOB. In S. Mullender (Ed.) *Distributed Systems.* Wokingham, England: Addison-Wesley, Chapter 1, pp. 1–16.

Seltzer, L. (1998). NT 5.0 Preview. *PC Magazine,* 17(20), pp. 100–130.

Shan, Y.-P., and Earle, R. H. (1998). *Enterprise Computing with Objects: From Client/Server Environments to the Internet.* Reading, Massachusetts: Addison-Wesley.

Taylor, D. A. (1992). *Object-Oriented Information Systems: Planning and Implementation.* New York: Wiley.

Thompson, C., Linden, T., and Filman, B. (1997). *Thoughts on OMA-NG: The Next Generation Object Management Architecture.* URL: http://www.omg.org/docs/ormsc/97-09-01.html, May 11, 2000.

Vckovski, A. (1998). *Interoperable and Distributed Processing in GIS.* London: Taylor & Francis.

Vinoski, S. (1997). CORBA: Integrating Diverse Applications within Distributed Heterogeneous Environments. *IEEE Communication,* February 1997, 35(2), pp. 46–53.

Weber, J. (Ed.). (1997). *Special Edition: Using Java 1.1,* 3rd ed. Indianapolis, Indiana: Que Corporation.

Yang, Z., and Duddy, K. (1996). CORBA: A Platform for Distributed Object Computing. *ACM Operating Systems Review,* April, 30(2), pp. 4–31.

CHAPTER 4

TECHNOLOGY EVOLUTIONS OF WEB MAPPING

Technology does not stand still, even in this field. It is very likely that new methods will become available in the near and distant future . . . and allow you to include features not possible today.
—Brandon Plewe (1997, p. 253)

4.1 INTRODUCTION

The development of distributed GIS is following the progress of computer technologies and telecommunication networks. As we mentioned in the first chapter, it evolved from centralized mainframe GISystems to personal desktop GIS to distributed GIServices that include the applications of wired Internet GIS and wireless mobile GIS. Along with the progress of distributed GIS applications, the technologies adopted by distributed GIS are also changing constantly.

The technology evolution of distributed GIS is shown in Figure 4.1. It started with static map publishing and evolved to static Web mapping, to interactive Web GIS and to distributed GIServices. Static map publishing distributes maps on the Web page as static map images in graphic formats like Portable Document Format (PDF), GIF, or JPEG. It relied on the early stage of Web technology—a giant URL-based HTTP server—to hyperlink ready-made maps on the Web. Maps are usually part of the HTML document to enrich the contents of the document. Users cannot interact with the maps or change their display format in any way.

The second stage is static Web mapping. It involves the use of HTML forms and the CGI to link the user input on the Web browser with GIS or

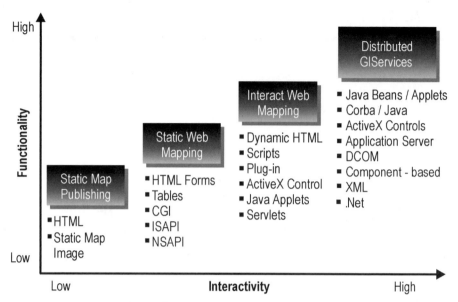

Figure 4.1 Evolution of distributed GIS.

mapping programs on the servers. Users make requests from the Web browser using customized HTML forms. The request is then sent to the CGI through an HTTP server to invoke GIS or mapping engines. The GIS or mapping engines create the map based on the user's request and generate an image map on the fly. The new image is sent via HTTP back to the user on the Web browser. However, the drawback of the static Web mapping technologies is that the performance of HTTP with CGI is slow, cumbersome, and stateless. Several variations of CGI were developed to improve the performance of CGI, such as Netscape's NSAPI, Microsoft's ISAPI and ASP, NeXT/Apple's WebObjects, Javasoft's servlets, and fast CGI. But the interaction between the user and the maps on the Web browser is still limited. The HTTP form is text based and allows limited user input. Users cannot define or draw a circle or a square on the image maps.

The third stage is the interactive Web mapping, where more interactivity and intelligence are added to the Web client side by using scripts like dynamic HTML and/or client-side applications like plug-ins, ActiveX controls, and Java applets. Some user queries can be processed on the client side without sending requests to the servers. But this approach still requires HTTP connections and the Web servers to mediate between software objects running on the client-side machines and the servers that store these objects.

The fourth stage is the distributed GIServices where GIS components on the Web client side can directly communicate with other GIS components on the server without going through an HTTP server and CGI-related middleware. Distributed GIServices rely on the communication between CORBA/ Java ORB or Microsoft's SOAP on the client side and the CORBA/IIOP and server-side Java or .NET/COM + technology in the Microsoft world (we will discuss these in detail late in the next chapter).

Key Concepts

Distributed GIServices *is a broad term for network-based geospatial information services. There are two major application of distributed GIServices, wired Internet GIS and wireless mobile GIS. This book uses the two terms* Internet GIS *and* distributed GIServices *interchangeably. Internet GIS emphasizes the aspect of physical networks and distributed GIS focuses on the distributed access mechanisms of information services.*

Distributed GIServices *refers to a specific software framework where GIS components on the Web client side can directly communicate with other GIS components on the server.*

This chapter and the next will cover the underlying technologies that support the evolution of the distributed GIServices. This chapter starts with the early development of static map publishing on the webpage. It then introduces the static Web mapping technology, including HTML forms, CGI, servlets, and ASPs. It follows with the description of interactive Web mapping that covers dynamic HTML and client-side applications such as plug-ins, ActiveX controls, and Java applets. The technologies that constitute distributed GIS will be discussed in the next chapter, which includes the CORBA/Java ORB, CORBA/IIOP, COM+, XML, and Document Object Model (DOM).

4.2   STATIC MAP PUBLISHING

A static map publishing refers to embedding maps as graphic images like GIF, JPEG, and Portable Network Graphics (PNG) inside an HTML page. The map images are usually used as a visual presentation to illustrate the points inside the HTML text. But the map image itself is not intelligent. That is, the map image is a static image displayed on the Web browser. The user cannot click on it to zoom to a certain area or get more information. A static map publishing does not support feature data at the client side and does not have map-rendering tools. It is a very thin client application that only supports ready-made map images on the Web browser. To publish a static map image, you can save a ready-made map as a graphic map image format and embed it inside an HTML page. Figure 4.2 illustrates a graphic map example showing a map of a regional park.

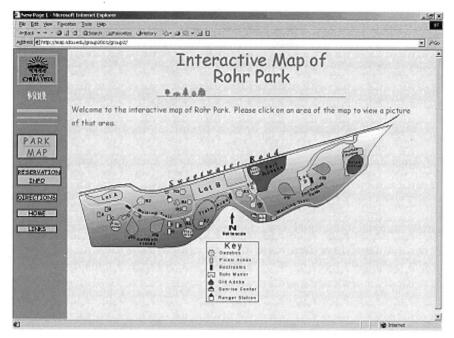

Figure 4.2 Example graphic map image. (Map generated by R. Thornberry, C. Cronk, and K. Hess at San Diego State University.)

In addition, Acrobat's PDF file is another popular method to publish maps on the Web. Figure 4.3 is an example of a PDF map that can be embedded inside an HTML document.

Static map publishing also includes a clickable map. That is, the whole map image is divided into different parts. If you click on one part, additional information on this part of the map will be displayed. For example, if you have a U.S. map and you click on a state on the U.S. map, information about that state or even the separate state map will be displayed. Additional information and maps for that state are separate HTML files or graphic image files that are stored in the Web server as separate files. Figure 4.4 illustrates the park example, which can allow users to click on different playgrounds or picnic area to show the actual photos.

Both the embedded static map images and the clickable maps are simple static map images. They both use the simple Web publishing technology, and no additional technologies are needed. So we treat them in the same category. We will now discuss how this static map publishing works on the Web.

4.2.1   Embedding Map Images in HTML Documents

To embed map images inside the HTML document or as separate static image files, you need to first make a map as one of the many graphic image formats,

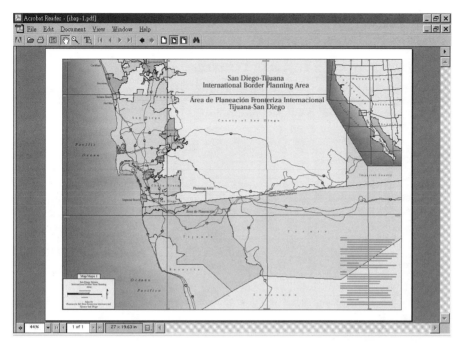

Figure 4.3   Example PDF map, San Diego–Tijuana International Border Planning Area. (Map generated by A. Perry and K. Wells at San Diego State University.)

such as GIF, JPEG, and PNG or PDF. You then embed these map image files inside the HTML document using the ⟨IMG⟩ tag or element in the HTML.

The ⟨IMG⟩ element in HTML includes an attribute "SRC" to indicate the file name and its location. For example, ⟨IMG SRC="/maps/USAmap.gif"⟩ tells the Web browser that this is an image file "USAmap.gif" and it is located at the maps directory on the server. There are other parameters associated with the ⟨IMG⟩ tag. For example, the ALLGN parameter tells the browser to place the map images at a certain place on the Web page. The ALT parameter displays the alternative text for nongraphic browsers. The ALT information is important for complying with the ADA (Americans with Disabilities Act) requirements. Here is a simple example:

```
⟨IMG SRC=''/maps/USAmap.gif'' ALLGN=''center'' ALT=''A
USA map''⟩
```

The map image could also be linked with other map images or HTML pages. For example, in the case of

```
⟨A HREF=''About_USAmap.html''⟩ ⟨IMG SRC=''/maps/
USAmap.gif''⟩
```

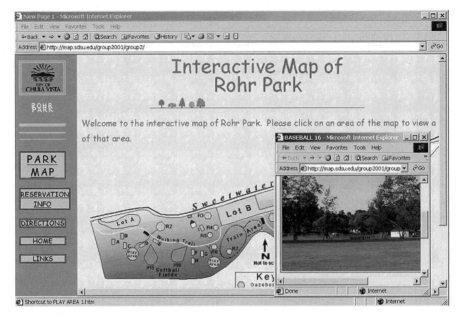

Figure 4.4 Clickable static Web maps. Users can click on the location of playgrounds or picnic areas to show the actual photos. (Map generated and designed by R. Thornberry, C. Cronk, and K. Hess at San Diego State University.)

when the user clicks on the USAmap image on the browser, the server will return an HTML page (About_USAmap.html) to the user.

4.2.2 Clickable Maps

A clickable map refers to a map that links to separate information about different parts of the map image. For example, if on the U.S. map you click on the state of Wisconsin, the map Wisconsin will be displayed as a new Web page; if on the Wisconsin map you click on Milwaukee, the city of Milwaukee map will show up. Here the terms *map image, clickable map,* and *imagemap* are often confusing. A map image is a geographic map in a graphic image format such as GIF or JPEG. A clickable map is a static map image but can be clicked and linked with other HTML or image files. An imagemap is simply a clickable image file that has hot links under different areas of the image, not necessarily a geographic map.

The clickable map can be created using imagemaps in the HTML page. Clickable images or imagemaps are similar to the static on-line GIF images. They are simply static images. The only difference is that clickable maps have hot spots or links assigned to them. Hot spots are areas of the map image that link to certain URLs.

4.2.3 Architecture of Static Web Publishing

Static map publishing uses the simple client/server architecture model as shown in Figure 4.5. It is a simple two-tier client/server model. The client is a Web browser such as Netscape or Internet Explorer, while the server is an HTTP server (or Web server), and the glue is the HTTP. The Web browser handles the presentation element for users to request information and for information to be displayed. The Web server receives users' requests and sends out the file in a user-requested URL. Therefore, this Web client/server model is simply a huge file server that serves files from URLs to all browsers.

4.2.3.1 *The Client: Web Browser* The client in this early Web stage is the simple Web browser with no client-side plug-ins or Java applets. The sole purpose of the Web browser is to interpret the contents of the HTML documents that were sent by the Web server and display them graphically. The Web browser also helps navigate from one page to another using the embedded hypertext links. The Web browser is incapable of interpreting any other documents or data formats except for HTML documents. The use of early Web browsers is similar to the use of a 3270 terminal, a simple display monitor. All the contents are prepared on the server side, and there is little intelligence on the client side. The partition point for this early Web model is at the Web browser, as shown in Figure 4.6. It is a thin-client and thick-server application. As mentioned before, this partitioning that is defined by a protocol (HTTP, in this case) is not flexible.

The client gets resources from the server by clicking a URI (Uniform Resource Identifier). The URI provides a global naming scheme to identify the names and the location of resources on the Web. It identifies the address of a resource and how to access it. A typical URI consists of four parts, as shown in Figure 4.7: the protocol scheme, the server name or domain name, the port number, and the location of target resources.

The protocol scheme specifies the type of protocol to be used to access the resources on a server. URI supports the following Internet protocols: HTTP, FTP, Gopher, Wide-Area Information Server (WAIS), News, and Mailto. A URI to identify a file for downloading would require the "ftp" protocol, such as this:

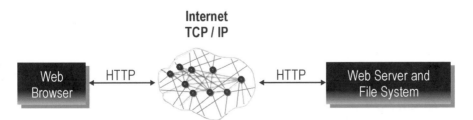

Figure 4.5 Static two-tier Web client/server model.

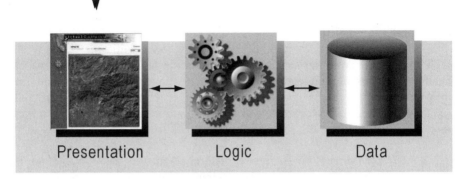

Figure 4.6 Partition point for static Web client/server model.

ftp://www.GIScompany.com/downloadfiles/street.shp

This would result in a download of the shapefile "street.shp" from the server www.GIScompany.com under the directory of "downloadfiles."

The domain name is the server name to identify the Web server site. It could be a registered domain name such as www.yahoo.com or a numeric IP address such as 129.79.82.108.

The port number is to identify the program that runs on a server. It is specified after the server name using a colon (:) as the separator. If no port number is specified, the browser will direct the request to a well-known port that is associated with a particular program. For example, 80 is usually a port number for accessing HTTP and 21 is a reserved port number for FTP.

Key Concepts

URI *contains information on the specific location of target resources. It typically consists of four parts: the protocol scheme, the domain name, the port number and the location of target resources.* **Protocol scheme** *specifies the type of protocol to be used to access the resources on a*

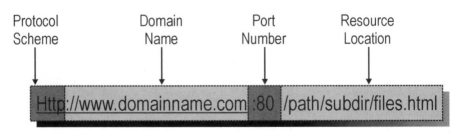

Figure 4.7 Structure of a typical URI.

server. The **domain name** *is the server name to identify the Web server site. A* **port number** *identifies the program that runs on a server. The* **location of target resources** *specifies the path and the name of resources: document, images, titles, and so on.* **URL** *is an informal term associated with popular schemes: http, ftp, mailto, etc.*

Finally, the URI contains information on the specific location of target resources. It is a hierarchical description of a file location on the computer. This usually includes a file directory, subdirectories, and files names. For example, /path/subdir/meeting.html indicates that the HTML file meeting.html is stored at the /path/ directory and /subdir/ subdirectory. The URI-supported Web resource can be HTML documents, image files, video clips, and programs such as a CGI or Java applet.

A URI for a program such as a forms-handling CGI script written in Practical Extraction and Reporting Language (PERL) might look like this:

```
http://www.getcomments.com/cgi-bin/comments.pl
```

It should be noted that URL is an informal term associated with popular URI schemes: http, ftp, mailto, telnet, and so on. It is no longer used in W3C's technical specifications (http://www.w3.org/addressing).

4.2.3.2   The Glue: HTTP   The glue, or middleware, between the Web client and Web server is the HTTP that is used by the Web to communicate between the Web client and the Web server. Like RPC, HTTP is a request–response oriented protocol. In fact, the HTTP is a *stateless* RPC on top of TCP/IP. That is, for each call from a client, the HTTP establishes a connection between the client and server; the server then fulfills a client request and hands over a reply to the client. The connection is then broken and the server forgets everything it sent out.

Key Concepts

HTTP is a **stateless** *RPC on top of TCP/IP. A* **cookie** *is a text file stored on the client machine to keep the state.*

To keep the state, a text file called a cookie is stored on the client machine. For example, e-commerce applications often use cookies to store user-selected items and to remember a user's personal profile so that the next time the user visits the site the information is automatically retrieved. Cookies contain attributes that tell the browser to what servers to send them. But this is not sufficient to support state-oriented client/server conversation, especially the map-based graphic user interface in GIS. For example, the user cannot draw a rectangle on the Web browser because it requires clicking on two points, while the stateless nature of the HTTP does not allow the server to remember the first click once the mouse moves to the second point. Therefore, some client-side applications such as plug-ins, Java applets, and ActiveX controls are developed to enhance the capability of the Web browsers (We will discuss them later in the chapter).

How exactly does the HTTP help clients to communicate with the HTTP server? The answer is through a negotiation process of describing the data type by both the Web browser and the Web server. The HTTP allows Web browsers to inform their server about the type of files they could understand; in return, the server in its response informs the client about the type of data it sends out. Web browsers and HTTP servers use the Internet's MIME (Multipurpose Internet Mail Extensions) data representations to describe and negotiate the contents of the message. MIME is an extension of the original Internet e-mail protocol that specifies a standard to exchange different kinds of data files on the Internet, including audio, video, images, application programs, and others.

An HTTP request from the client consists of a request method (GET or POST), a URI, header fields, and a body (which can be empty). The GET method asks the server to send a copy of the file to the client. The POST method allows a client to send a form's data to the specific URI at the server. When you request a URI in a Web browser, the GET method is used for the request. When you send an HTML form, either GET or POST can be used. With a GET request the parameters are encoded in the URI; with a POST request they are transmitted in the HTTP message body.

The current common protocol is HTTP 1.0, which sets up a new connection for each client request and creates a separate TCP connection to download each URI. This may cause some communication problems while providing multi-tasking services. The newest version of HTTP is 1.1, which has been adopted by some advanced Web servers, such as Apache and Netscape Fasttrack, and some mobile application servers. The new HTTP 1.1 allows persistent connection. That is, HTTP 1.1 keeps the connection with the HTTP server open for multiple request–response interactions. This means that many images embedded in the same HTML document can be downloaded consecutively without breaking and reestablishing the link with the server. The second advance of HTTP 1.1 is pipelining; that is, the client can send multiple requests to the server before waiting for a response. In HTTP 1.0, the client has to wait for the response before sending another request. This means that HTTP 1.1 will allow users to click on two points in a map before sending the request to the server. Therefore, HTTP 1.1 offers the potential capability of drawing a box on a map on a Web browser. Lastly, HTTP 1.1 provides cache validation commands to help clients maintain a consistent local cache of documents. This can reduce the overall network traffic and improve the protocol performance.

4.2.3.3 The Server: HTTP Server   As we discussed before, A Web server is often simply referred to as an HTTP server, which is a daemon that runs continuously on the server machine. A daemon is a program that runs continuously to handle periodic service requests from other services or programs. The daemon program responds to simple requests or forwards the requests to other programs or processes. Each Web server has an HTTPD (HTTP Daemon) that is continually listening to requests that come in from Web clients and then serving pages out to clients. The role of the Web server is to listen to the client request and respond by sending a precomposed hypertext document or other documents.

4.3 STATIC WEB MAPPING

Static map publishing on the Web is simply an electronic copy of a paper map. The users can only take a look at the map images on the Web page and cannot interact with the map in any other ways. This is because the HTTP server in the two-tier Web client/server system cannot handle user requests other than serving ready-made files. To increase user interactivity, Web mapping emerged.

Web mapping refers to making maps, conducting queries, and doing some limited spatial analyses in the server while presenting the output on the standard Web browsers. The output presented on the Web browser is a copy of *static* map images that are generated by the programs in the server. So we call this kind of Internet GIS static Web mapping. The emergence of static Web mapping is the first true representation of the distributed GIServices on the Web.

4.3.1 Early History of Static Web Mapping

The early research of static Web mapping and distributed GIServices (Gardels, 1996; Plewe, 1997; Tang, 1997) has been motivated by the concepts of an open and distributed architecture. Three projects are of primary importance in the development of GIServices. They are important because these projects initiated the design of preliminary distributed GIServices frameworks and the adoption of early Internet technologies. They provided the GIS community with a glimpse of the potential of the Web and motivated the improvement of geospatial technologies. All three early examples of Internet GIS belong to the category of "static Web mapping" because the process of mapmaking occurred only in the servers.

4.3.1.1 Xerox PARC Map Viewer   The Xerox PARC Map Viewer was one of the earliest prototype of static Web mapping and was created in June 1993. Map Viewer was developed at Xerox Corporation's Palo Alto Research Center as an experiment in providing interactive information retrieval via the WWW (Putz, 1994). Map Viewer is an interactive Web application that combines the ability of HTML documents to display graphical images with the ability of HTTP servers to create new documents in response to user input (Figure 4.8). Map Viewer used a customized server module (a CGI program) written in the PERL scripting language. Map images in GIF format were generated by two separate utility programs on a UNIX server. The first program, MAP-WRITER, produced raster map images from two public domain vector map databases. The second program, RASTOGIF, converted raster images to GIF format. In subsequent work, Xerox Map Viewer was integrated with U.S Gazetteer WWW services created by Plewe (1997) to provide a text-based query function, which is essential for a complete prototype of distributed GIServices. The design of Map Viewer introduced many innovative and advanced concepts in 1994, many of which have since been adopted by other Internet GIS projects. But Xerox did not expand its effort to further the Map Viewer development, and has taken it off line now. The screen shot of the viewer is still available at http://www2.park.com/istl/projects/www94/mapviewer_example1.html.

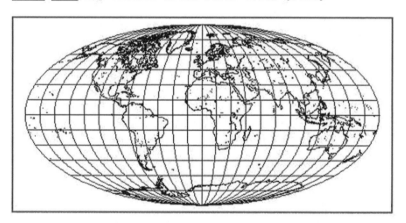

Xerox PARC Map Viewer: world 0.00N 0.00E (1.0X)

Select a point on the map to zoom in (by 2), or select an option below. Please read About the Map Viewer, FAQ and Details. To find a U.S. location by name, see the Geographic Name Server.

Options:

- Zoom In: (2), (5), (10), (25); Zoom Out: (1/2), (1/5), (1/10), (1/25)
- Features: Default, All; +borders, +rivers
- Display: color; Projection: elliptical, rectangular, sinusoidal; Narrow, Square
- Change Database to USA only (more detail)
- Hide Map Image, Retrieve Map Image Only, No Zoom on Select,
- Place mark at (0.00N 0.00E), Reset All Options

Figure 4.8 Xerox Map Viewer. (Figure reprinted with permission from Palo Alto Research Center.)

4.3.1.2 NAISMap and World Map Maker NAISMap was developed by the National Atlas Information Service (Natural Resources Canada) in September 1994. NAISMap allowed users to interact with map images (in GIF) on the Web to select map layers, order map layers, and even overlay map layers. The user could select a location from the map; the client application then passed the coordinate to the NAISMap server application, which then returned the requested map areas as a GIF image to the user at the Web client. NAISMap could provide maps in both a national view and regional view. It is an early operational and interactive web-based mapping service released on-line.

NAISMap has been operational for many years since its introduction in 1994. The current Atlas of Canada is its implementation (http://atlas.gc.ca/site/english/index.html) (Reed, 2002).

Another operational online Web mapping is the World Map Maker that was developed at the Charles Sturt University by Paul Wessel and Walter Smith in 1995 as an integrated Web interface to the GMT 3.x (Generic Mapping Tools) package and a geographic database (http://life.csu.edu.au/cgi-bin/gis/Map). With GMT, users can create maps on the Web using HTML forms and generate maps as PostScript output files. The World Map Maker is a typical implementation of HTML forms and CGI scripts. The user on the Web client specifies the mapping parameters, which are transferred to GMT commands. "The output of these commands is captured in a PERL scalar variable. The contents of this variable are then written to a temporary PostScript file, which is converted to the GIF format by the GhostScript Program" (Reed, 2002, p. 11).

4.3.1.3 GRASSLinks GRASSLinks was developed in 1995 by Huse (1995) based on her Ph.D. dissertation at the University of California at Berkeley. GRASSLinks was the first fully functional on-line GIService, connecting GRASS GIS software (from the U.S. Army Corps of Engineers) with the WWW. GRASS is a grid-based GIS package offering public domain access to environmental and geographical data. The development of GRASS-Links was supported by the Research Program in Environmental Planning and GIS (REGIS) at the University of California at Berkeley. To utilize GRASS-Links, a user only needed a Web browser to access GIS functions provided by GRASS (Figure 4.9).

The main goal of GRASSLinks was to encourage cooperation and data sharing between different environmental agencies. In traditional GIS applications, each federal/local government agency would maintain its own database as well as data obtained from other sources. GRASSLinks introduced a new model of data sharing where each agency could maintain its own data and access other agencies' data over the network as needed (Huse, 1995). GRASSLinks could perform many GIS operations, including map display, spatial query, overlay operations, reclassification, buffering, and area calculation. On-line users can save their work temporarily on the server and retrieve

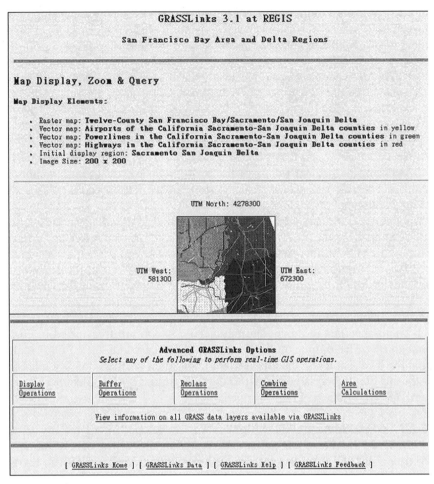

Figure 4.9 GRASSLinks and its GIS operations.

saved files later. In general, GRASSLinks demonstrates an ideal prototype for high-end distributed GIS functions and provides an example of the first true on-line GIService (Plewe, 1997).

4.3.1.4 Alexandria Digital Library Project The ADL project illustrated a digital library framework for heterogeneous spatially referenced information that can be accessed across the Internet. The ADL project was launched in 1994 concurrent with five other digital library projects (NSF, 1994). Many important collections of information, such as maps, photographs, atlases, and gazetteers, are currently stored in a nondigital form, and collections of considerable size and diversity are found only in large research libraries. ADL provides a framework for putting these collections on-line, providing search and access services to a broad class of users, and allowing both collections and users to be distributed throughout the Internet (Buttenfield, 1998; Buttenfield and Goodchild, 1996; Goodchild, 1995).

The major contribution of the ADL project was to introduce the digital library services metaphor for distributed GIServices and to extend the types of GIServices to cataloging, gazetteer searching, and metadata indexing. Another contribution of ADL was its exploration of Internet-based interface design processes. ADL utilized three different technologies for the actual implementation. The first version ran as a customized ArcView project. The second version was based on HTML and CGI programs (Figure 4.10). The final version utilized Java applets and Java applications. However, the incompatible technologies between the three prototypes caused inconsistent problems of data integration and delivery of services. The ADL user interfaces proved difficult to migrate to each new version, and the task of redeveloping their functions and interfaces was time consuming and costly.

Overall, the ADL project explored different computer technologies and frameworks, identified major tasks of digital libraries, and became the first on-line library service to provide comprehensive metadata browsing, display, and query functions for geospatial information. The recent Alexandria Digital Earth Prototype (ADEPT) is a follow-up to the ADL project. ADEPT aims to use the digital earth metaphor for organizing, using, and presenting information at all levels of spatial and temporal resolution with a specific focus on geodata and images in California.

These early examples of static Web mapping represent milestone achievements for distributed GIServices. Xerox Map Viewer provided a preliminary technical solution for distributed GIServices by using HTTP servers and CGI programs. The technical framework of Map Viewer was followed by many other static Web mapping applications. The development of Xerox Map Viewer also indicated that a single GI service—map browsing—is not sufficient and other GIS functions should be provided. GRASSLinks illustrated a comprehensive prototype that provided many traditional GIS functions, such as map browsing, buffering, and overlaying. However, both Xerox Map Viewer and GRASSLinks were only built to mimic traditional GIS functions. The ADL project introduced a new type of GIService using a digital library metaphor and provided sophisticated library functions, including collections holding, catalog searching, and metadata indexing. Notwithstanding the different functions and interfaces, these three examples adopted the same architecture in developing the applications: server-centered, three-tier Web server architecture.

4.3.2 Architecture of Static Web Mapping

Static Web mapping takes advantage of two advancements in Web technology: Web forms on the Web client side and the CGI on the server side. Web forms are created in the Web client to facilitate user input, and CGI is developed at the Web server to process the user requests. With the introduction of Web forms and CGI, the Web now essentially became a three-tier client/server

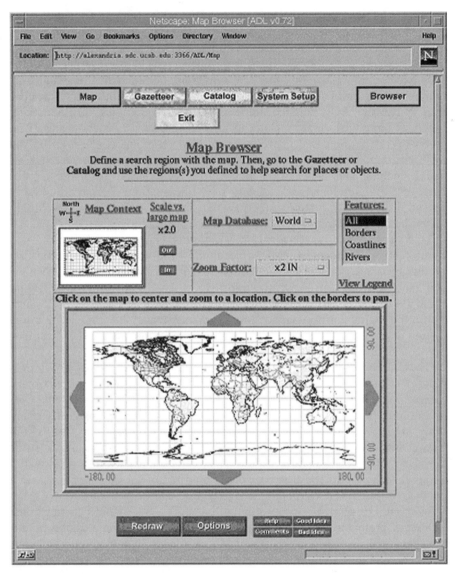

Figure 4.10 HTML/CGI version of the ADL Project.

model, as shown in Figure 4.11. The first tier is the Web client with the function of displaying HTML and forms. The second tier is the HTTP server coupled with a CGI. The third tier consists of traditional application servers such as map servers and DBMS servers.

The Web client is still a simple Web browser with the capability of handling HTML and Web forms. A Web form is an HTML page with data entry fields for user input. The user inputs are collected by the Web browser, which

Internet GIS Showcase: Digital Libraries

"**Digital libraries** basically store materials in electronic format and manipulate large collections of those materials effectively. Research into digital libraries is research into network information systems, concentrating on how to develop the necessary infrastructure to effectively mass-manipulate the information on the Net. The key technological issues are how to search and display desired selections from and across large collections" (from the Web site of the Digital Library Initiative (DLI) Phase I http://www.dli2.nsf.gov/dlione).

The DLI was supported by the (NSF) in 1994. There were six major DLI projects funded by four NSF awards during 1994–1998:

- *University of California at Berkeley* (http://elib.cs.berkeley.edu), Environmental Planning and Geographic Information Systems
- *University of California at Santa Barbara* (http://www.alexandria.ucsb.edu), Alexandria Project: Spatially Referenced Map Information
- *Carnegie Mellon University* (http://www.informedia.cs.cmu.edu), Informedia Digital Video Library
- *University of Illinois at Urbana-Champaign* (http://dli.grainger.uiuc.edu/idli/idli.htm), Federating Repositories of Scientific Literature
- *University of Michigan* (http://www.si.umich.edu/UMDL), Intelligent Agents for Information Location
- *Stanford University* (http://www-diglib.stanford.edu/diglib/index.html), Interoperation Mechanisms among Heterogeneous Services

In 1998, the NSF announced *DLI-Phase 2,* which focused on the following issues:

- Selectively build on and extend research and testbed activities in promising digital libraries areas.
- Accelerate development, management, and accessibility of digital content and collections.
- Create new capabilities and opportunities for digital libraries to serve existing and new user communities, including all levels of education.
- Encourage the study of interactions between humans and digital libraries in various social and organizational contexts.

There are currently more than 30 university and DLI projects that are funded under DLI-Phase 2.

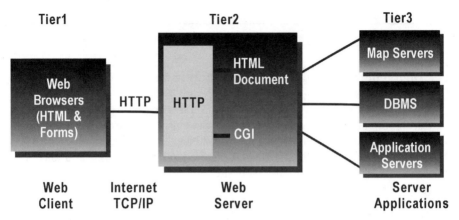

Figure 4.11 Architecture of static map publishing.

invokes a POST HTTP method and sends the user inputs to the server in an HTTP message.

The Web server receives the HTTP message but cannot respond to it because the Web server does not understand any requests other than those for an HTML or other MIME-type document. Therefore, the HTTP server passes the user requests to a back-end program specified in the URI. It uses a CGI to pass the method request and the parameters to the back-end programs.

The back-end programs are traditional server-side applications that do the actual processing. In the case of Web mapping, these back-end programs include map servers and DBMS servers. Any client/server-based GIS programs that work on the server can become a map server. The role of map servers is to fulfill the user requests and return the results to the Web server via the CGI protocol. The Web server then returns the results to the Web client. The Web server becomes middleware, connecting the Web client and back-end server applications.

This is an important and common architecture that dominates the early-stage Web mapping applications. An important characteristic of this architecture is that all user requests are processed by server-side applications. All output at the client side is merely mirrors of map images created by the server. Besides the three examples presented above, some more examples include Visa ATM locator (http://www.visa.com), MapQuest (http://www.mapquest.com), MapBlast! (http://www.mapblast.com), and many others.

4.3.3 The Client: HTML Viewers with Forms

Key Concepts

Forms *are generally used in HTML 3.2 or later versions to gather information from users for a CGI-based server application.* **METHOD**

specifies invocation method for the data to be transmitted to the CGI server application. The **GET** *method encodes the data input in the URI, while the* **POST** *method transmits the user input in the HTTP message body. The* **ACTION** *attribute specifies the URI where the data are processed.*

The client of static Web mapping is the HTML viewer with Web forms. In order for users to interact with back-end map servers, the HTML viewer needs to have two basic functions. The first one is a mechanism for users to enter text, such an address, and/or to select different options, such as selecting different display layers. The second function is to submit the user input and selections to the server. For example, the user may enter an address and then submit the request to the server to return with a map showing the location of the address. After seeing the returned map, the user may decide to zoom in (submit another request) to see a more detailed map in a larger scale.

These two functions can all be made available in Web-based *forms*. Forms are generally used in HTML 3.2 or later versions to gather information from users for a CGI-based server application. A form in HTML starts with a ⟨FORM⟩ tag and ends with the ⟨/FORM⟩ tag. The ⟨FORM⟩ tag has two mandatory attributes: METHOD and ACTION. METHOD specifies the invocation method for the data to be transmitted to the CGI server application. There are two methods, GET or POST. The GET method encodes the data input in the URI, while the POST method transmits the user input in the HTTP message body. The ACTION attribute specifies the URI where the data are processed. The URI is the name of a server and the location of the CGI program or scripts. For example,

```
⟨FORM METHOD=''POST''ACTION=HTTP://WWW.dgis.edu/cgi-
bin/geocode⟩
```

where the CGI program is the *geocode* program that resides in the cgi-bin directory at the server of WWW.dgis.edu. All CGI programs or scripts are located in the cgi-bin directory. This ACTION attribute tells the Web server that the incoming request is for a CGI program that is located at the cgi-bin directory, so the Web server will invoke the geocode program via the CGI protocol.

A form in HTML generally has three types of interface elements: the *INPUT field* for the user to enter data and submit requests, the *SELECT field* to select one option from a list of options in a dropdown list box, and the *TEXT AREA field* to enter multiple-line text input, such as comments. An HTML document can contain one or more forms, but a form cannot have another form nested within it.

A general syntax of an INPUT field is as follows:

```
⟨INPUT TYPE=''field-type'' NAME=''variable name''
VALUE=''default value''⟩
```

where INPUT indicates this is an input field; it has three properties: TYPE, NAME, and VALUE. The TYPE property indicates an input type. There are eight input types, including text, password, hidden, checkbox, radio, reset, submit, and image. NAME specifies the name of the variable (it is not the displayed name); and VALUE is the actual data value of the variable NAME. NAME and VALUE are a name–value pair to be sent to the server. The VALUE property could have a default value.

The HTML viewer with forms is a thin client; there is no restriction on the operation platform. There is also no requirement on the Web browser or the computing power of the client computer. Therefore, it is mostly applicable to the vast majority of the audience. However, there are some drawbacks of the HTML viewer. Notably, the interactivity between the user and the map image is very limited. The user cannot select a spatial feature or draw a box or a circle on the map due to the *stateless* nature of the HTTP. To improve the interactivity, DHTML or JavaScript could be used. The use of DHTML, VBscript, and JavaScript can allow the user to better interact with the map images.

4.3.4 HTTP Server with CGI

The data input from the HTML form is passed to the map server through the HTTP server and CGI. When the user fills the form and clicks on the Submit button, the data input is sent to the HTTP server, which relays the information to the CGI program. The CGI program then interacts with other applications in the server such as a map server and a DBMS server. The map server does the work and returns the results to the CGI program, which reformats the result in an HTML format and sends it to the HTTP server. The HTTP Web server then forwards the results to the Web client.

It can be seen that the CGI is an important middleware to link the Web client and server with a back-end external server application such as a map server. It is a simple language-independent standard interface that runs on top of the operating system to interface external applications with Web servers. It can be used to process user requests that involve computation or invoke other applications on the server. It works on any type of Web server and allows a server to start an external process. The CGI scripts handle the information exchange between the Web server and other server applications such as the map server (Figure 4.12). Basically, CGI is a message-handling protocol or interpreter that receives user inputs and parses them into param-

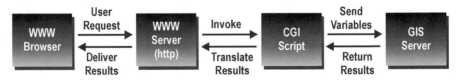

Figure 4.12 Architecture of CGI-based Internet GIS. (*Source:* Peng, 1999.)

eters of variables to be used in map servers or other GIS programs. It can invoke running map servers and/or other GIS programs, reformat output, and send it back to Web browsers.

A CGI program can be written in any language, such as C or Perl. It can access external resource managers such as files and databases. It is thus a native program. The CGI program can also connect with other applications using any of the communication middleware such as RPC and object messages. CGI programs add more functionality and interactivity to the Web page.

To illustrate how CGI works with the Web server and the map server, we will look at an example step by step [adapted from Orfali et al. (1999)]:

1. A Web user requests a map by typing a street address or a city name and clicking on the Submit request button. The Web browser collects the data within the form and assembles it into a string of name–value pairs that are separated by an ampersand (&). For example,

    ```
    Street=''2131 E. Hartford Ave.''
    &city=''Milwaukee'' &State ='' WI'' &Zip=53211
    ```

2. The Web browser makes an HTTP request that specifies a POST HTTP method, the URI of the target program in the cgi-bin directory, and the typical HTTP header. The HTTP message body or entity contains the forms data as mentioned in step 1, that is, the string of name–value pairs.

3. The HTTP server receives the HTTP request via a socket connection. The server parses the HTTP message and discovers that it is a POST for the cgi-bin program. It then starts the interaction process with the CGI.

4. The Web server sets up environmental variables to send parameters to the CGI program. Environment variables are used by the Web server to communicate with the CGI program about the environmental information such as the server name, request method, content types, content length, path and directory, and script name.

5. The HTTP server starts a CGI program by executing an instance of the CGI program specified in the URI.

6. The CGI program reads the environment variables and discovers, in this case, that it is responding to a POST method.

7. The CGI program receives the HTTP message body (i.e., those name–values and name–value strings) via the standard input pipe (stdin) and parses the string to retrieve the form data. It uses the *content_length* environment variable to determine how many characters to read in from the standard input pipe.

8. The CGI program invokes the GIS program or map server and translates the request to a format or set of variables that the map server can understand. Each request is answered in a separate process by a sep-

arate instance of the CGI program, and a new process of the GIS program or map server is launched for each request. The map server creates as many processes as the number of user requests received. More simultaneous requests require the server to create more concurrent processes.

9. The map server then processes the request by geocoding the address and making a map centered at the requested address. It then sends the output back to the CGI script.

10. The CGI program wraps the output with HTML or some other MIME-type format by writing the map server output to its standard output stream and sends the output as the HTTP response entity and the HTTP response header back to the Web server. The CGI program then returns the results to the HTTP server via its standard output (stdout).

11. The HTTP server receives the results on its standard input and concludes the CGI interaction. The HTTP server sends the results back to the Web browser. Either it can append some response header to the results it received from the CGI program or it sends it "as is" if the response header was added by the CGI program. It then breaks the connection with the Web client.

12. At the Web client, the map image is displayed at the Web browser.

The use of HTML forms and CGI to process user requests makes the Web mapping possible. The user can request its own map by specifying the layers and scales. It can also take advantage of the analysis functions of existing GIS programs. But it has four main drawbacks: low performance (a new process has to be created for every request), statelessness, platform dependence, and security concerns (Orfali et al., 1999).

First, every request has to create a new CGI process, which is time consuming and requires large amounts of server RAM. This can restrict the resources available for sharing with other server applications. Therefore, CGI applications do not scale well. When there are many simultaneous requests from Web clients, the system will perform poorly. CGI may become a bottleneck or even a failing point in the whole Web mapping system.

Second, the CGI and HTTP server is stateless. This means that every single request, even a simple zoom in or zoom out, needs to go through the whole process from the Web client to the Weber server to invoking the CGI program and map server and back to the Web client. This creates a lot of network traffic and slows down the whole process. Furthermore, the stateless nature limits the user interactivity at the Web client. In addition, the output is still a map image. Users cannot directly interact with the map.

The third drawback of this approach is that CGI is platform dependent; that is, different CGI programs have to be created for each computing platform.

Finally, CGI programs could pose a security risk to the Web server and other applications on the server because CGI is comprised of native codes that have access to other native programs. Hackers can send malicious codes through the CGI program to infect the server programs.

Two problems of the HTTP and CGI approach—slow performance and statelessness—can be mitigated. The performance of the CGI program can be improved by using the Dynamic Link Library (DLL). The slow performance of the CGI is due to two reasons: (1) the HTTP server has to create a separate process for each request received and (2) when the request is done, the process is then closed. The opening and closing of the process take time and slow down the server response. The DLL serves the function of CGI, but unlike CGI, it stays in memory, ready to service other requests until the server decides it is no longer needed. The ISAPI and the NSAPI are in the form of a DLL. ISAPI works on the Microsoft Internet Information Server (IIS), while NSAPI works on the Netscape Enterprise/FastTrack server. They are used to extend the capabilities of the Web server.

ISAPI and NSAPI DLLs reside in the same process as the HTTP server; therefore, all the resources that are made available by the HTTP server process are also available to the ISAPI or NSAPI DLLs, whereas the CGI applications run in different processes. Some benchmark programs show that loading DLLs in-process can perform considerably faster than loading them into a new process. Furthermore, in-process applications scale much better under heavy loads. Multiple ISAPI or NSAPI DLLs can coexist in the same process as the server. They are multithread-safe to handle multiple simultaneous requests.

The statelessness of the HTTP and CGI can be eliminated by using hidden fields in the HTML forms and/or cookies. A hidden field is invisible from the form but contains values that can be transmitted to the CGI program. The values in the hidden field from previous user input and kept by the Hidden field in the ⟨INPUT⟩ tag so that the user does not need to reenter them each time. For example, if the user entered a street address to request for a geocoded map, when the user received the map from the server, he or she decided to zoom in to a larger scale. How does the CGI keep the state that this user already has a map that centered at a specific address? Well, the CGI uses the hidden field to store information from previous forms to the next. So, when the user later clicks on the "Zoom in" button in the form, the previous user input, such as street address and map extent, becomes a hidden field and is sent to the CGI program through the HTTP request. The CGI program would parse these hidden fields as well as the new INPUT fields and send them to the GIS server to produce a new map. The new map would be sent to the client via the CGI and HTTP server.

Another approach to keep state information is to use *cookies*. A cookie is a small piece of text file that is stored in the client machine. It records user information such as user IDs or other basic configuration information. The role of the cookie is similar to a hidden field; the data stored on the client are similar to the value of the hidden field in the forms. Cookies communicate with the server that creates them every time the user revisits the Web site. Cookies are commonly used in e-commerce Web sites, but they are not often used to keep the state of the Web mapping.

4.3.5 Xerox Map Viewer Example

The use of HTTP plus CGI for online mapping dates back to 1989, when the first such system was designed by McDonald Dettwiler and Associates (Reed, 2002). The early client was built using Mosaic (pre-Netscape) to display map images generated from the server. But the most popular example is Xerox PARC map viewer.

The software framework of Xerox Map Viewer is another one of the earliest examples of CGI-based Web mapping application (Figure 4.13). The HTTP clients can submit their requests by sending a URI to the HTTP server. The HTTP server will parse the URI strings and then launch the CGI extension to invoke two Perl programs (MAPWRITER and RASTOGIF) located on the same machine.

Map images in GIF format were generated by the two separate utility programs on the Sun UNIX server. The first program, MAPWRITER, produced raster map images from two public domain vector map databases. The second program, RASTOGIF, converted raster images to GIF format.

After the RASTOGIF generates a new GIF image, the CGI program on the server will create a new HTML with the new GIF image link and POST the new HTML file back to the client-side Web browser.

Here are the two examples of encoded URI requests for Xerox Map Viewer. First, the URI request includes the following commands:

- border=1 (turn on the country border theme)
- lon=117.75 (longitude)

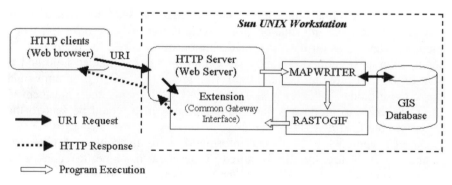

Figure 4.13 Software architecture of Xerox Map Viewer.

- lat=25.03 (latitude)
- proj=rect (projection uses rectangles)

The CGI programs located on the server-side process the request and generate a new image, such as Figure 4.14. Figure 4.15 is another example of URI requests sent by users generated by the CGI programs:

- color=1 (turn on the color display)
- db=usa (access U.S. database)
- feature=alltypes (turn on all types of features)

4.3.6 Map Servers and Other Server-Side Applications

The map server is the actual workhorse that processes the user requests. You could develop your own map server to serve a special request, such as a travel plan program. You could also use existing client/server GIS programs as map servers, such as ArcInfo, ArcView, GeoMedia, or MapInfo. The advantage of using existing GIS programs is that the GIS functions that were developed in the program can be used by the user on the Web browsers. But the existing desktop GIS programs were not designed for the Internet and thus do not scale well in the Internet and Web environment. Since each user request has to create a process, the map server has to be able to create multiple processes. This requires you to have a map server with sufficient user seats or user licenses.

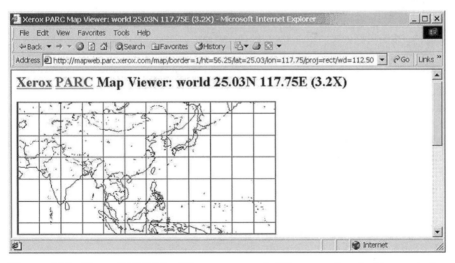

Figure 4.14 Xerox PARC Map Viewer request 1. (Figure reprinted with permission from Palo Alto Research Center.)

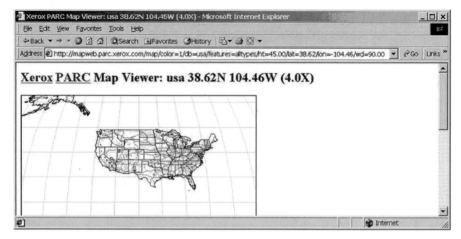

Figure 4.15 Xerox PARC Map Viewer request 2. (Figure reprinted with permission from Palo Alto Research Center.)

Besides a map server, other server applications such as DBMS could also be run to fulfill the user request. In fact, in a more complex user request, the map server has to work together with other server applications, such as a DBMS server and TP Monitor to produce the output that the user requested. TP Monitor could be used to manage and balance the load of user requests. It is essential to handle a lot of simultaneous requests.

In summary, Web mapping based on HTTP/CGI provides somewhat limited interactivity and functionality between the Web client and the map server. It can do a reasonable job in creating customized maps on-line under a low load of user requests. There is no restriction on the client-side computer platform. Anyone with a Web browser is able to make maps on demand.

However, Web mapping using HTTP/CGI as middleware does not scale well. It does not offer users the desktop GIS feel and functions. Users cannot directly interact with the maps, as they can in desktop GIS. Every interaction has to go through the forms, which is an indirect interaction. Users cannot draw a box or square or query a spatial element directly from the map. CGI has serious shortcomings. First, it is stateless and does not maintain the state between connections. This poses major obstacles in stateful GIS operations. Second, the CGI has to load the GIS program into memory for each request. This hogs a lot of server resources and creates considerable performance limitations.

4.4 INTERACTIVE WEB MAPPING

A simple HTML viewer with forms is very limited in terms of user interactivity, especially when dealing with maps and spatial objects. To create more interactive Web mapping, we need alternative viewers that can facilitate the user to interact with the spatial object and maps directly. Therefore, dynamic HTML and client-side applications such as plug-ins or help programs, Java applets, and ActiveX controls were developed to handle maps and spatial objects. These are dynamic or interactive viewers for users to directly interact with spatial objects that are interactive Web mapping applications.

One major characteristic of the interactive Web mapping applications is that they offer more interactions between the user and the client interface and more client-side processing and functionalities than the static Web mapping applications. Many current Internet GIS programs such as Arc IMS, Geo-Media Web Map, MapXtreme, and MapGuide belong in this category.

Another characteristic of interactive Web mapping is that CGI extensions are used as the middleware to mitigate the shortcomings of the CGI. These CGI extensions include Netscape's NSAPI, Microsoft's ISAPI and ASP, Apple's WebObjects, Javasoft's servlets, Allaire's ColdFusion, and many others. These CGI extensions generally perform better than the CGI scripts. The common feature of these CGI workarounds is that they all run some sort of server-side scripts (or plug-in codes) in the same address space as the Web server.

Most current interactive Web GIS programs are based on this model, that is, a dynamic viewer coupled with CGI or CGI extensions. This section will describe the characteristics of different client viewers and server-side CGI extensions.

4.4.1 Interactive Viewers

Client viewers are places for users to interact with maps and spatial objects. Different interactive viewers with various functions have been developed using different programs and technologies, from very simple HTML interfaces with forms as we discussed in the previous section to dynamic HTML and to more advanced client-side applications such as plug-ins, ActiveX controls, and Java applets. We will introduce four types of client viewers in the interactive Web mapping programs: DHTML viewer, GIS plug-ins, Java applets, and ActiveX controls.

Key Concepts

The **HTML viewer** *is static or noninteractive for Web mapping. DHTML makes plain HTML dynamic by using client-side scripting, DOM, and CSS.* **Plug-in viewers** *are software executables that run on the browser to extend the capabilities of Web browsers. A* **Java applet** *is an executable Java code that is downloadable from the server and executed on the client at runtime. A* **Java applet viewer** *displays geospatial information and handles requests.* **ActiveX viewers** *use ActiveX controls to program the viewer. An* **ActiveX control** *is a modular piece* *of software that performs tasks and communicates information to other programs and modules over the Internet via OLE.*

4.4.1.1 DHTML Viewer The HTML viewer is static or noninteractive for Web mapping. After a page and map are loaded in the browser, they become static. The only things a user can do in a static web page are:

• If there is a link, click on it.
• If there is a form or image form, fill it out and click on the submit button.

The response to either of the above is not all that quick because, in either case, the page appears after a complete round trip to the server and back, even for a simple response such as zooming into a map feature. This is where DHTML comes in.

DHTML is just plain HTML that can change even after a page has been loaded into a browser. An area of a map can change color when the mouse moves over it or a menu can drop down or a new popup window appear. Most HTML elements can be made to react to user actions after the page loads. The DHTML viewer has three major advantages over the static HTML viewer:

• The Web page and the map will respond to user actions.
• That response is immediate (without making a round trip to the server).
• No special plug-in is needed to install at the browser.

DHTML makes static HTML page dynamic by using the following:

1. *Client-Side Scripting* Client-side scripting uses JavaScript and VBScript to change HTML. VBScript works only in Microsoft Internet Explorer, while JavaScript works in other browsers as well.
2. *Document Object Model* The DOM is the hierarchy of elements that are present in the browser. This includes browser properties such as the browser's version number, window properties such as window location (the page's URI), and HTML elements such as ⟨p⟩ tags, or tables. For example, you can point to the specific check box in a specific form in a page and make it checked. Thus, by exposing the elements and their properties to scripting languages, browsers enable the user to manipulate them. The DOM also specifies the events that get triggered as a result of a user action. For example, the DOM defines an event "onMouseOver" for a link. This enables you to write a script for something to happen when a user passes the mouse over that link.
3. *Cascading Style Sheets (CSS)* CSS not only let the user specify style information in one place for an entire Web but also allow the user to set style values in such a way that they can be easily manipulated by a

scripting language. By changing the CSS properties of a page element (such as its color, position, or size), it is possible to change almost anything about the way a page looks.

4.4.1.2 Plug-In Viewer Plug-in viewers are software executables that run on the browser to extend the capabilities of Web browsers. Plug-in viewers can support both vector data and raster images. The role of plug-in viewers is to provide user interactivity with geospatial data and map images so that the user can view the maps and select features and make queries directly on the map. While plug-in viewers are small applications installed in the Web browser, GIS helper programs can be large GIS applications or existing GIS software that is located in the user's local machine. GIS software such as ArcView, MapInfo, and GeoMedia can all be GIS helper programs. When the Web browser detects a GIS data type in an HTML page, it can automatically launch the respective GIS helper program (though this function is not yet available for many existing desktop GIS programs).

Internet GIS Showcase: Brownfield Location Information System

The Brownfield Location Information System (BLIS) is Wisconsin's effort to promote the redevelopment of underused properties throughout the state. Potential redevelopers can use the site's map and query system to locate the tax-delinquent, abandoned, blighted, or hazardous site that best fits his or her needs and selection criteria. BLIS is designed to help commercial, industrial, and retail businesses locate reusable land while simultaneously assisting landowners market their sites.

To provide easy access, BLIS was developed so that users need only a Web browser that can handle HTML and JavaScript. There are no applets or extra plug-ins required. Buttons and functions on the site are intended to be intuitive, even to users who are not familiar with mapping or GIS software. Users are able to perform drag-box zooms (without a Java applet) and view the 10 best sites for their query criteria, not just sites that meet all their criteria. Also, the map size is sizable variable and will always fit the maximum available space in a user's browser.

Map clutter on BLIS was reduced by limiting which layers users are able to toggle on and off and by setting zoom thresholds where certain layers appear and disappear automatically. The incorporation of an "identify" function lessens the need for labeling all features.

BLIS uses ArcIMS technology developed by ESRI for displaying and interacting with a map.

Source: http://comgis1.commerce.state.wi.us/wiscomp/blis_start.htm.

In the case of direct feature data support, plug-in viewers can communicate with vector geospatial data. Similar to other plug-ins, GIS plug-in viewers handle GIS data from a URI that is provided as a stream as it arrives from the network. This allows a GIS plug-in to implement a progressive viewer without seeing an entire stream. Individual plug-ins can request multiple data streams simultaneously.

Figure 4.16 illustrates the working process of a GIS plug-in viewer. When the Web browser encounters a geospatial data type (most GIS data types are unknown to Web browsers) in a Web page from a server, it will look for a plug-in that is associated with that data type and then load it. If a GIS plug-in or helper program is not available in the client computer, it has to be downloaded from the Web server over the network. Once the GIS plug-in or helper program is installed, it then communicates directly with the GIS data stream from the server.

4.4.1.3 Java Applet Viewers Viewers in the form of Java applets are executable Java code that is downloadable from the server and executed on the

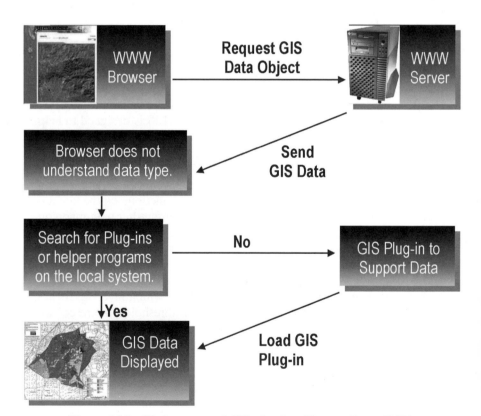

Figure 4.16 Work process of GIS plug-ins. (*Source:* Peng, 1999.)

Internet GIS Showcase: City of Nanaimo

The City of Nanaimo, Canada, has used MapGuide to create an on-line mapping tool for public use. Nanaimo's city map provides users with the ability to

find a street address on the map,

print a replacement garbage calendar,

review school catchment areas,

locate city parking lots,

locate parks and trails, and

search for common points of interest such as shopping centers, schools, and hospitals.

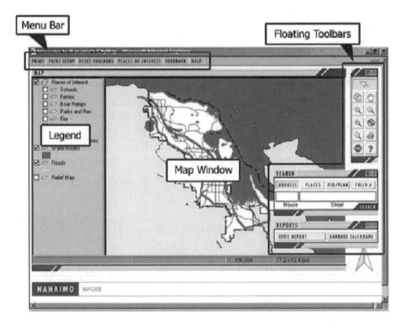

The Civic Report is a Web-based document that provides information about a selected property. The Civic Report provides

· legal description information,

· address information,

· zoning information,

· garbage calendar route information, and

· sewer service information.

Source: http://www.city.nanaimo.bc.ca/citymap.asp.

client at runtime. Java applets initially reside on the Web server. They are referenced inside an HTML document and executed by a Web browser at the client side. These applet files are downloaded and executed when a user connects to the Web site and invokes the HTML document containing a reference (the ⟨applet⟩ tag in the older version of HTML or the ⟨Object⟩ tag in HTML 4.0 or later) to the Java applet. The Java applet is then seamlessly integrated inside the Web browser. Figure 4.17 illustrates the process of loading a Java applet viewer.

Java viewers use a Java applet for displaying geospatial information and handling requests. It allows the user to interact directly with the spatial features on the map. The Java viewer usually incorporates map-rendering and data processing functions in the Java applet so that the user can render maps, make queries, and do other processing inside the viewer without going back to the map server.

Java viewers can support both feature data and map images. In the case of supporting map images, the Java applet is simply a fancy display of map images similar to the DHTML viewer. The user interacts with map images. But all logic processing such as map rendering and query processing is conducted at the server.

If the Java viewer supports streamed feature data, the data server streams the vector data to the Java viewer, and the user then interacts with the feature data. Some or all processing can be performed at the client viewer. Java viewers have the potential to support unlimited client-side processing. Feature data that are streamed to the Java viewers are temporarily cached on the client machine. The Java viewer establishes a connection channel between the Java viewer and the database server via a JDBC drive. When the user's request requires data that are not currently in the cache, the request is sent to the

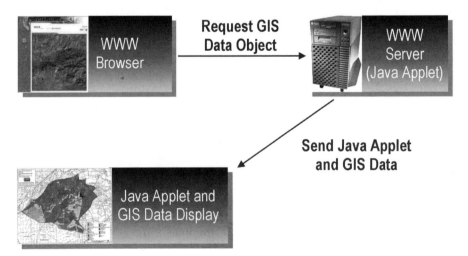

Figure 4.17 Work process of Java applets. (*Source:* Peng, 1999.)

server to either retrieve more data or process data residing on the server. The temporary cache is removed when the Java viewer is closed. So the Java viewer does not take any permanent disk space in the client's computer.

In addition to Java applets that work with a Web browser, a whole Java-based viewer, Java Beans, can also be created to work as an independent viewer or a mobile client agent.

Many GIS vendors are developing Java-based viewers, such as ESRI's ArcIMS Java viewer (Figure 4.18).

4.4.1.4 ActiveX Viewer ActiveX viewers use ActiveX controls to program the viewer. ActiveX was developed by Microsoft to "activate the Internet." It builds on the OLE standard to provide a common framework for extending the capability of Microsoft's Web browser, Internet Explorer (Chappell, 1996).

An ActiveX control is a modular piece of software that performs tasks and communicates information to other programs and modules over the Internet via OLE. It can also be used and reused by any programming language or

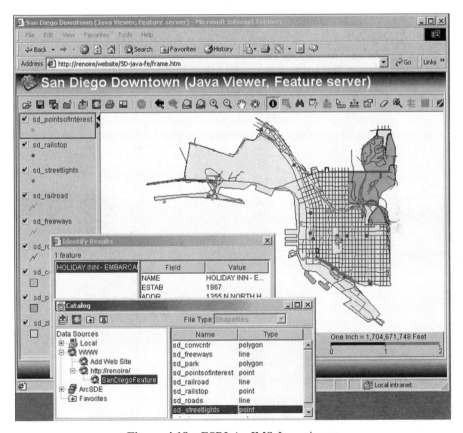

Figure 4.18 ESRI ArcIMS Java viewer.

application that supports the OLE standard (Chappell, 1996). ActiveX controls are general componentware that can plug into any application. There are many different types of ActiveX controls, each with different capabilities and functionalities. Client viewers implemented in ActiveX controls are created to handle GIS data and as a graphic interface between the user and the feature data and/or maps over the Web browser. They can be used just like plug-ins and Java applets within Web pages.

ActiveX viewers are referenced inside an HTML document and executed by a Web browser on the browser's machine. The GIS control is downloaded from a Web server when it is needed or it might already be present on the client machine if it was previously downloaded.

The ActiveX viewer can access a URI and retrieve GIS data just as a standard Web browser client does. GIS data are streamed asynchronously to the GIS control as the information arrives from the network, making it possible to implement viewers and other interfaces that can progressively display information. If the GIS control needs more data than can be supplied through a single data stream, multiple and simultaneous data streams can be requested by the ActiveX viewer. Furthermore, ActiveX viewers can be set up to allow the user to combine local data and streamed feature data from the server to process request in the same ActiveX viewer. Figure 4.19 illustrates the simple architecture of the ActiveX viewer.

While Java applets are in the form of byte codes and have to run inside a Java VM, Activex is native binary code and runs directly inside the computer's native operating system. Being a native code, ActiveX controls can take full advantage of the local computer's computing power and have direct access to all local platform functionality such as local files, local memory, and other system resources that are unavailable to a Java applet. Therefore,

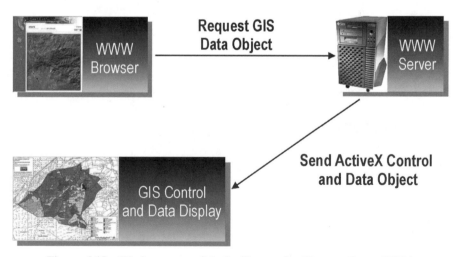

Figure 4.19 Work process of ActiveX controls. (*Source:* Peng, 1999.)

compared with a Java applet, ActiveX components have better performance (Shan and Earle, 1998).

However, this performance advantage also carries a price—portability and safety concerns. First, because ActiveX controls are compiled to the native executable format, different versions of ActiveX controls have to be made available for all platforms. That means ActiveX controls are platform dependent. This is in contrast with the Java applets, which are platform independent. One Java applet code (theoretically) can be run on all platforms. Second, because ActiveX controls are able to access to local files and other local resources, they could pose greater danger to the users' local computers. Someone could write a vicious ActiveX code that could erase all local files. (We will discuss the safety issues in more detail in Chapter 10.)

Figure 4.20 illustrates an example of the ActiveX viewer developed by Intergraph's GeoMedia Web.

Another interesting implementation of the ActiveX controls is Citrix's WinFrame Web Client ActiveX control (http://www.citrix.com/). The WinFrame Web Client ActiveX control can allow users to access GIS data and applications remotely. WinFrame Web Client uses the techniques of Application Launching and Embedding (ALE) to run applications remotely from a Web page. Users can launch an application from a Web page by clicking on

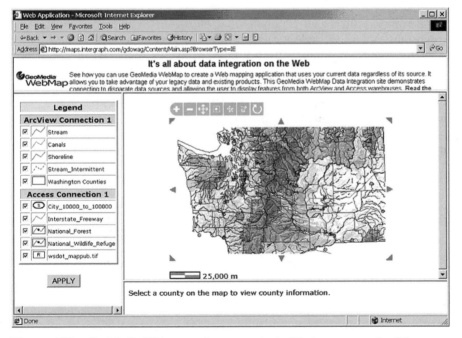

Figure 4.20 GeoMedia Web map viewer, an ActiveX viewer example. (Figure reprinted with permission from Intergraph.)

a hyperlink. The server will launch a GIS application in a window on the user's local desktop. The launched application can run on its own window or within the Web page (embedding). The users can then use this application as if it were installed and running locally on his or her own machine. They can also use the application to edit and manipulate the information, save it to a local file, or even save it back to the remote site (if the file permissions allow).

In summary, the viewers for interactive Web mapping could be as simple as a dynamic HTML page or as complex as a Java applet or ActiveX control. But regardless of the form of the viewer, a client viewer should be able to display a map, make a query, identify a data source, extract a subset of data from a data server, and have the result rendered on the screen in the form of a map image or a vector map. More advanced clients would have added functions and capabilities, such as editing, data integration, and spatial analysis.

4.4.2 Server-Side CGI Extensions

Those client viewers are linked with a map server and data server via CGI or CGI extensions. To overcome the previously discussed shortcomings of the CGI, several extensions are used to extend the capabilities of CGI in interactive Web mapping. Rather than create a separate process and close the process for each request received, as the CGI does, the server-side CGI extensions stay in the memory and are always ready to service other requests. Furthermore, unlike CGI programs that run in different processes, the CGI extensions run server-side scripts in the same address as the Web server. Therefore, all the resources that are made available by the HTTP server process are also available to CGI extensions. Loading CGI extension programs in-process can improve performance considerably compared to loading them into a new process. We have covered NSAPI and ISAPI in the last section; we will discuss servlets, ASP, and ColdFusion here.

4.4.2.1 Servlets

Key Concepts

Servlets *are modules of Java code that run in the Web server to extend the capabilities of the HTTP server.* **ASP** *is a Web server extension to receive and process user requests on the Microsoft Internet Information Server (IIS).* **ColdFusion** *is a Web application server that runs on a Web server that works with Linux, Solaris, and Windows servers.*

Similar to "applets" on the client side that are used to extend the capabilities of the Web client, servlets are modules of Java code that run in the Web server to extend the capabilities of the HTTP server (e.g., to answer client requests). Servlets are commonly used with HTTP servers, so they are often referred to as "HTTP servlets," even though they can be used with any client/server protocols. Since servlets are written in the highly portable Java language and follow a standard framework, they can be used independently of server types and operating systems.

HTTP servlets are similar to CGI scripts and usually have the following major functions (Orfali et al., 1999):

- processing and/or storing data submitted by an HTML form;
- providing dynamic contents (e.g., returning the results of a database query to the client);
- managing state information on top of the stateless HTTP (e.g., managing many concurrent requests for the same map services);
- initiating a connection to a database and maintaining its connection across requests;
- passing a client request to another servlet, a feature called servlet chaining; and
- providing an interface between Web users and a legacy (mainframe) application and its database.

Compared to CGI, servlets have several advantages (Orfali et al., 1999):

- While CGI runs in a separate process, a servlet does not run in a separate process. This removes the overhead of creating a new process for each request every time, thus improving the performance.
- Similar to ISAPI and NSAPI DLL, a servlet stays active in memory between requests, while a CGI program needs to be loaded and started for each request. This is another way to increase responsiveness and performance.
- There is no need to create multiple instances to respond to multiple requests. Only a single instance is needed to answer all requests concurrently. This saves memory and allows a servlet to easily manage persistent data. Servlets are multithread safe.
- A servlet can be run by a servlet engine or servlet container in a restrictive sandbox just like an applet runs in a Web browser's Java VM, which increases the server security.

Because of these features, servlets become a pretty good alternative to the CGI programs, especially for simple applications. Servlets are better than CGI at accepting form input, interacting with a single database, and dynamically generating an HTML response page. They provide functions to easily extract the HTTP name–value pairs and compose a dynamic HTTP response.

Servlets do a good job for a simple CGI-like request–response system. However, the servlet is just a little better than CGI as Web middleware. It is still very primitive. Since servlets use a generic API, they have a set of predefined methods. Therefore, you have to do your own marshaling and unmarshaling of parameters. Also, servlets do not support typed interfaces; you have to create your own command formats. Servlets do not fit well with the distributed-object system. They cannot take advantage of object interfaces and do not have the features that many scalable server-side component technologies provide, such as transaction (Orfali et al., 1999).

4.4.2.2 Active Server Page

4.4.2.2 Active Server Page   An ASP is another server-side feature or Web server extension to receive and process user requests on the Microsoft IIS. It is used to replace the CGI scripts on the Web server. An ASP is essentially an HTML page that includes one or more scripts (small embedded programs). These scripts are processed on a Microsoft Web server before the page is sent to the user. The user accesses the ASP Web page on the Web browser; the user requests are then sent to the ASP scripts on the server. The script in the Web page at the server then accesses data from a database and builds or customizes a page on the fly before sending it back to the user at the browser. In other words, the server-side ASP script simply creates a regular HTML page or ASP file by processing the user requests and/or extracting data from the database on the server.

ASP scripts can be written in either VBScript or Jscript. User requests can be fulfilled using ADO program statements. Scripts can reside in either the server side or the client side. Client-side scripts create more interactivity while scripts on the server side are more versatile and have no limitation on browsers.

4.4.2.3 ColdFusion   ColdFusion was developed by the Allaire Corporation, which has merged with Macromedia, to be an alternative to Perl and other CGI technologies. It is a Web application server that runs on a Web server that works with Linux, Solaris, and Windows servers. Similar to ASP, the ColdFusion Web application server works with the HTTP server to process user requests for Web pages. When a user sends a request from a Web browser for a ColdFusion page, the ColdFusion application server executes the scripts or programs the page contains.

ColdFusion can create and modify variables just like other languages. It has some built-in functions for performing some complicated tasks. ColdFusion applications can access databases using Microsoft's OLE DB, ODBC, or drivers that access Oracle and Sybase databases. Just like ASP, ColdFusion uses standard SQL to link Web pages and Web applications with the back-end data servers to retrieve, store, format, and present information dynamically.

ColdFusion uses its proprietary markup language CFML (ColdFusion Markup Language) to make web programming. CFML encompasses HTML and XML and is tag based. A JIT compiler turns the CFML into Web pages to be served to the Web client. CFML has 70+ CFML tags and over 200 custom functions. It also offers tools similar to those at the server side CGI extensions to extend the server-side functions. ColdFusion can be coordinated with distributed applications that use CORBA or Microsoft's DCOM to interact with other network applications. ColdFusion also has tags to embed COM, CORBA objects, and Java applets/servlets.

In addition to the middleware or application server between the Web server and the map server, there are other services on the server, including catalog services, load balance services, and state services. Catalog services keep track of where the data are in a distributed environment; load balance services balance the load of different server functions; while the state services keep the state of user requests.

These CGI-like middleware and other services are connected with mapping servers and database servers. It is the mapping server that fulfils the user request and makes maps. The middleware is simply a translator that receives requests from the Web browser and forwards the requests in a proper format to the map server for process.

Notice that in the interactive Web mapping applications, although the client side could use the Java applets and ActiveX controls, the middleware is still CGI or CGI extensions. This Java-to-CGI client/server approach is still the traditional Web client/server architecture. It is different from the distributed GIS, the next phase of distributed GIS, as we will discuss in the next chapter.

Although interactive mapping programs, whether static or interactive, are very popular and well recognized in the GIS community, there are common problems: The performance is slow, the functions are limited, and they are proprietary and not interoperable. Different Web mapping programs were developed in different database frameworks and using different technologies. The heterogeneous techniques and software programs prevented the integration and sharing of information among these Web mapping programs. Furthermore, it is difficult to migrate technology from one platform and one stage to another, as demonstrated in the ADL project.

The problem will get worse as more vendors start developing Web mapping programs. OGC has been making efforts to develop a set of standards to guide the development of Web mapping programs so that they can be interoperable. Therefore, OGC developed Web Map Server (WMS) implementation information specifications based on some Web mapping testbeds or pilot programs, which represents the first effort to standardize the implementation of the Web mapping programs.

4.5 OPENGIS WMS IMPLEMENTATION INTERFACE SPECIFICATIONS

Key Concepts

The OpenGIS **WMS implementation interface specifications** *provide guidelines for current Web map servers with the specifications of HTTP contents and URI communication syntax. Its specifications also lay out*

Internet GIS Showcase: Mason County, Kentucky, PVA Project

This demonstration model was created to showcase the benefits of an Internet/intranet-based mapping system. The local Property Valuation Administration (PVA) office, with cooperation from the State of Kentucky, provided the digital, graphical, tabular, and geographic data for this project

The application combines the PVA database information with digital maps and images. The process is fulfilled by incorporating a dynamic HTML (Web page) editor. The Web page design program used in this application was ColdFusion, from Allaire Corporation. ColdFusion allowed the developer to easily integrate the database information into an HTML (Web page) format. Also, ColdFusion was used to build dynamic queries from the database and relate them to the MapGuide viewer, a Web browser client.

This site combines the ColdFusion pages with the viewer to have a complete page showing the PVA tracts and database entries from simple queries. In the next stage, actual lot survey drawings, descriptions, photos, and so on, will be attached to each parcel.

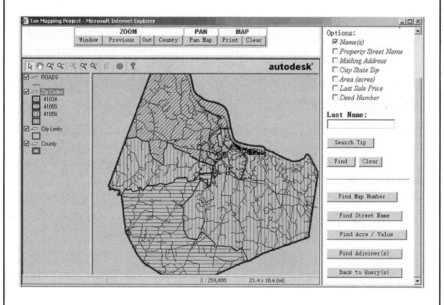

Source: http://hq.carlsonsw.com

the major tasks of Internet map servers, which can be applied in the architecture of distributed GIServices.

4.5.1 Background and Overview

One of the main goals of OGC is to come up with a set of specifications to be used as a guide to Internet GIS design for different software vendors so that their designed systems can be communicated or interoperated with each other. OGC's interoperability programs cover a range of areas, from geospatial data to geospatial processing. Web mapping is one of them. OpenGIS specifications result from common understandings and the consensus of the GIS vendor industry as well as from experiences learned from different testbeds or pilot programs.

OGC's Web mapping activities started with a WWW mapping framework by Doyle (1997). A task force of OGC was formed to come up with a consensus position on the WWW mapping Special Interest Groups (SIG) that is described in "User Interaction with Geospatial Data" by Cuthbert (1997). This document presents an abstraction for the display of geospatial data. It provides a common set of terms that can be used to describe a variety of software implementations. Based on the basic ideas from Doyle (1997) and Cuthbert (1997) as well as from "A Web Mapping Scenario" by Gardels (1998), OGC sponsored the Web Mapping Testbed (WMT) initiative. The WMT initiative invited GIS software vendors as well as governmental agencies to design pilot Web mapping systems to test implement the ideas in the WWW mapping framework and OGC consensus position papers. The WMT demonstration was made in September 1999. The OpenGIS WMS interface implementation specification was subsequently published in April 2000.

A request for a Quotation and a call for participation in the OGC Web Mapping Testbed Phase II (WMT-2) were made in April 2000 to further test and expand the Web mapping specifications. The goal of WMT-2 was to rapidly develop interface specifications that lead to Standards-based Commercial-Off-The-Shelf (SCOTS) implementations of software that support use and exploitation of geospatial data and images over the WWW. WMT-2 builds upon the framework of specifications that have already been adopted or will soon be adopted by OGC. WMT-2 efforts will help refine existing OGC specifications and may create new specifications. Ultimately, this initiative will lead to standardized geospatial tools from multiple vendors that satisfy requirements for Web mapping (OGC Project Document 00-028). The OpenGIS WMS implementation interface specifications provide guidelines for current WMSs with the specifications of HTTP contents and URI communication syntax. The WMS specifications also lay out the major tasks of Internet map servers that can be applied in the architecture of distributed GIServices.

The major content of the OpenGIS WMS specifications focuses on how to describe a Web map server and map services with standardized URI syntaxes and semantics. A URI is a short string that identifies resources in the Web.

The format of URI strings indicates the syntax and semantics of formalized information for location and access to resources via the Internet.

The OpenGIS WMS specifications standardize the syntax and semantic contents of the URIs for WMSs and focus on the three major tasks. In general, "a standard web browser can ask a Map Server to do these things just by submitting requests in the form of Uniform Resource Locators. The content of such URIs depends on which of the three tasks is requested" (OGC, 2000, p. 9). The WMS implementation interface specification indicates that a WMS should be able to (OGC, 2000, p. 9)

1. produce a map (as a picture, as a series of graphical elements, or as a packaged set of geographic feature data),
2. answer basic queries about the content of the map, and
3. tell other programs what maps it can produce and which of those can be queried further.

4.5.2 WMS Architecture

Besides the specification of three major WMS tasks, the WMS specifications also identify four main processing stages in a WMS: filter service, display element generator, render service, and display service. The concept of four processing stages is derived from Cuthbert (1997), who describes geospatial data visualization from data to a map as a flow line with four processes, as shown in Figure 4.21:

1. the *selection* of geospatial data to be displayed,
2. the *generation of display elements* from the selected geospatial data,
3. the *rendering* of display elements into a rendered map, and
4. the *display* of the rendered map to the user.

These four processes can be considered as service components. Each service component becomes a client of the other component, and each has interfaces that can be invoked by clients of that service. Depending on the location of these service components on the network, the whole process can be described as systems with "thin," "medium," and "thick" clients (Figure 4.22).

- If only the *rendered map* is carried over the Internet to the client, it would be a thin-client system with virtually no client-side capabilities. A typical example of this is displaying rendered maps as GIF files.
- If the *display elements* are carried over the Internet to the Web browser, it would be a medium-client system, which allows for a limited client-side processing, such as panning and zooming and selection.

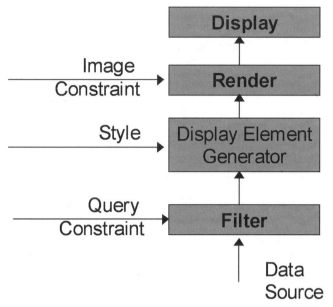

Figure 4.21   Map display process. (Figure reprinted with permission from OGC.)

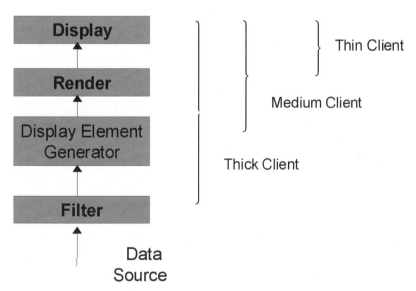

Figure 4.22   System partitions. (Figure reprinted with permission from OGC.)

• If the *geospatial data and display element generator service* is carried over the Internet to the Web client, it would be a thick-client system with unlimited client-side capabilities.

OGC later abandoned the use of "thin," "medium," and "thick" client in its WMS interface implementation specification because of some very imprecise definitions of "thin" and "thick" client used in the marketing literature. Instead, OGC uses the kind of information presented at the Web client to categorize the Web mapping services. This led to the three "cases": namely the "picture case," the "graphic element case," and the "data or feature case" (OGC, 2000).

In the picture case, what travels across the Internet to the Web browser in response to a client's request is essentially a picture of a map in such format as a GIF, JPEG, or PNG. The map image was constructed by a map server and was transported to the Web client.

In the graphic element case, what travels to the Web client is a packaged set of individual elements, typically already in a projected reference system and with already defined symbolization for geographic features. Some graphic element formats include Scalable Vector Graphics (SVG) and Web Computer Graphics Metafile (CGM). For example, a freeway might be a thick red polyline, a lake could be a blue polygon, and so on. Some of the graphic elements could themselves be pictures like a bitmap or predrawn fragment of a map, so the graphic element case may also include the picture case as a subset.

Finally, the data or feature case provides the ability to send geographic feature data from the server to the client. These feature data can be processed and manipulated directly on the Web client using display element generators and map-rendering tools. XML was tested in WMT Phase I to encode OpenGIS simple features, which resulted in an OGC specification of GML, which can be used to transport data from the server to the client.

4.5.2.1 WMS Specifications for the Picture Case The WMS implementation specification covers primarily the picture case. As mentioned before, in the picture case, only pictures of maps are presented at the Web client; all other map rendering and data selections are conducted in the server, as shown in Figure 4.23. To simplify things, we can consider the server as one unit and focus on the functions of the server, as shown in Figure 4.24.

There are three functions that a map server could or should have in order to answer the user's request. First, the map server should be able to provide users with maps at the Web browser. Second, it needs to (may not have to) provide users with information about the maps, including information about the specific areas of the map and specific layers of the map. Lastly, the map server should be able to provide information about what interfaces a map server supports and what map layers it can serve. These functions are supported by three WMS interfaces: map interface, feature information interface,

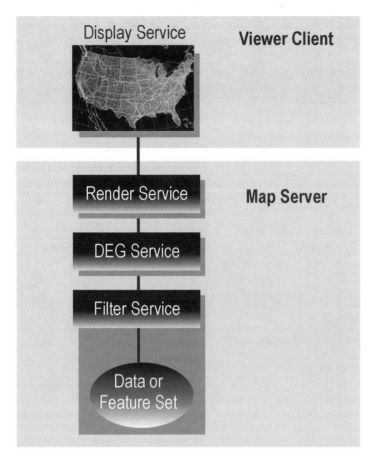

Figure 4.23 WMT Implementation of the picture case. (Figure reprinted with permission from OGC.)

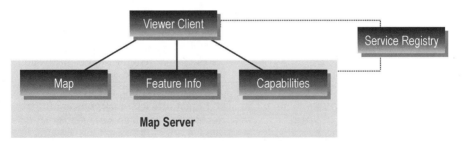

Figure 4.24 Functions of the server. (Figure reprinted with permission from OGC.)

and capabilities interface. Individually, they are sometimes informally referred to as GetMap, GetFeatureInfo, and GetCapabilities.

Map Request (GetMap) Interfaces The design of map request interfaces focuses on the display and production of Web-based map services: "To produce a map, the URI parameters indicate which portion of the Earth is to be mapped, the coordinate system to be used, the type(s) of information to be shown, the desired output format, and perhaps the output size, rendering style, or other parameters" (OGC, 2000, p. 9). The parameters of map request interfaces include the map layers, picture format, picture size, background color, and so on. Table 4.1 illustrates the parameters used in the map request interfaces.

Feature Request (GetFeature) Interfaces The feature request interfaces identify the request mechanisms for map contents and feature attributes.

TABLE 4.1 Map Request Interfaces

URL Component	Description
http://server_address/path/script?	URL prefix of server (Section 6.2.5.1.1)
WMTVER=1.0.0	Request version, required (Section 6.2.5.1.2)
REQUEST=map	Request name, required
LAYERS=layer_list	Comma-separated list of one or more map layers, required
STYLES=style_list	Comma-separated list of one rendering style per requested layer, required
SRS=srs_identifier	Spatial reference system (SRS), required
BBOX=xmin,ymin,xmax,ymax	Bounding box corners (lower left, upper right) in SRS units, required
WIDTH=output_width	Width in pixels of map picture, required
HEIGHT=output_height	Height in pixels of map picture, required
FORMAT=output_format	Output format of map, required
TRANSPARENT=true_or_false	"TRUE"|"FALSE": If TRUE, then the background color of the picture is to be made transparent if the image format supports transparency; optional; default=FALSE
BGCOLOR=color_value	A hexadecimal red-green-blue color value (0xrrggbb) for the background color; optional; default=0xFFFFFF
EXCEPTIONS=exception_format	The format in which exceptins are to be reported by the map server; optional; default=0xFFFFFF
Vendor-specific parameters	Section 6.2.5.1.5

Source: OGC, 2000, p. 23, with permission from OGC.

To query the content of the map features, the URI parameters indicate what map (layer) is being queried and which location on the map is of interest (X, Y coordinates). Table 4.2 indicates the elements of feature request interfaces.

Capabilities Request (GetCapabilities) Interfaces The capabilities request interfaces are used to provide extensive map services, such as catalog services or metadata queries, in addition to the basic map display and attribute query (Table 4.3). For example, to ask a map server about its holdings, the URI parameters can be included in the capabilities requests, such as "Database=Colorado+California." However, current OpenGIS WMS specifications do not specify the exact contents of the GetCapabilities interfaces. The WMS specifications only suggest the possible use of GetCapabilities interfaces and leave the detailed design of the interfaces and contents to software vendors with their vendor-specific parameters.

Figure 4.24 introduces another element, the service registry. It is a component that delivers information about available services of different map servers to any client. The service registry identifies services through a search of metadata across map servers by invoking the capabilities request. The service registry also provides interface for publishing service descriptions to a publisher client. In WMT Phase I, a service registry was constructed using the OpenGIS catalog services specification (OGC, 1999).

TABLE 4.2 Feature Request Interfaces

URL Component	Description
http://server_address/path/script?	URL prefix of server (Section 6.2.5.1.1)
WMTVER=1.0.0	Request version; required (Section 6.2.5.1.2)
REQUEST=feature_info ⟨**map request copy**⟩	Request name; copy of map request parameters that generated the map for which information is desired (Section 6.2.8.2)
QUERY_LAYERS=layer_list	Comma-separated list of one or more layers to be queried
INFO_FORMAT=output_format	Return format of feature information; optional; default=MIME
FEATURE_COUNT=number	How many features to return information about; optional; default=1
X=pixel_column	X coordinate in pixels of feature (measured from upper left corner=0)
Y=pixel_row	Y coordinate in pixels of feature (measured from upper left corner=0)
Vendor-specific parameters	(section 6.2.5.1.5)

Source: OGC, 2000, p. 30, with permission from OGC.

TABLE 4.3 Capabilities Request Interfaces

URL Component	Description
http://server_address/path/script?	URL prefix of server (Section 6.2.5.1.1)
WMTVER=1.0.0	Request version; required (Section 6.2.5.1.2)
REQUEST=capabilities	Request name; required
Vendor-specific parameters	Section 6.2.5.1.5

Source: OGC, 2000, p. 22, with permission from OGC.

In a distributed Internet environment there are many map servers across the Internet. Therefore, a "cascading map server" should be used to aggregate the capabilities of the individual map servers into one logical "place." The cascading map server can function as both a client and a server. It is a client to access many other map servers, while it is a map server to other Web browser clients. Furthermore, a cascading map server can also perform additional services. For example, a cascading map server can convert many different graphics formats (e.g., PNG, JPEG) into GIF format. This would allow any viewer clients to display any output from different map servers. Similarly, a cascading map server might perform coordinate transformations on behalf of other servers.

4.5.2.2 WMS Specifications for the Graphic Element Case The graphic element case is the medium-client model that the client-side machines can provide both display and render services (Figure 4.25). The servers will process the geodata from the GIS databases and generate well-defined geodata objects with associated symbols and colors. AutoDesk's MapGuide is one example of the graphic element case. The advantage of the graphic element case is that the combination of render and display services can allow more interactive user manipulation of map features, such as the vector-based highlights/selections and dynamic graphic display elements. In the graphic element case, map users can create a new graphic element on the client side and send it back to the server for updating (such as the map notes function in MapGuide). The response time and display performance is faster and better than the picture case, especially in the zoom-in, zoom-out types of display functions. However, map users have to download specialized Web plug-ins, ActiveX controls, or Java applets besides the regular Web browsers in order to see the graphic elements. The implementation of the WMS is more difficult than the picture cases because the graphic element case needs to modify the functions of the HTTP servers and add a middleware on the server, such as a Java servlet engine or CGI, for communication between Web servers, GIS databases, and client-side viewers.

Internet GIS Showcase: Internet GIS at Oregon State University

An exciting development with regard to Internet GIS is its use as a portal for GIS functionality as well as data distribution (e.g., Xue et al., 2002). Oregon State University (OrSt) researchers involved in developing Internet GIS applications along these lines include faculty and graduate students from the Departments of Geosciences, Entymology, Forest Science, Soil and Crop Science, Computer Science, and Bioengineering, the Marine Resource Management Program, and the Northwest Alliance for Computational Science and Engineering. Designs and applications range from the simple presentation of data via Web mapping to more complex signal analysis, real-time scientific collaboration and the incorporation of environmental models and decision support. For example, entymologists at the OrSt Integrated Plant Protection Center are developing a "public access GIS" using GRASSLinks (ippc2.orst.edu/glinks/) in order to integrate weather and climate data with soils, topography, insect distribution, and other environmental layers for the purposes of phenological (biology related to climate), population, and disease risk modeling. Their applications also include a real-time, multiresolution climate mapping expert system as well as the serving of national ecoregion GIS layers and maps in collaboration with EPA.

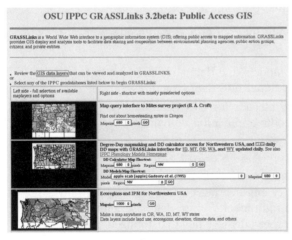

While the infrastructure for ready access to data and the resulting maps via Internet GIS (i.e., linking data to data) is desired and needed, OrSt researchers, in collaboration with geologists and computer scientists at the University of Oregon and computer scientists at the Evergreen State College (Olympia, Washington), are building the Virtual Research Vessel (oregonstate.edu/dept/vrv), an experimental linkage of Internet GIS to additional database support, tool composition, and numerical models. Major objectives include the refinement of numerical simulations, better exploration of relationships between observations of the seafloor made with various instruments and vehicles, and the quantitative evaluation of scientific hypotheses. In this regard, Internet GIS is viewed as a preliminary step toward widespread data access rather than as a final solution. Better support for analysis, modeling, and decision support within or connected to Internet GIS should move users beyond the "data-to-data" mode toward "data-to-models" and "data-to-interpretation" modes (Wright et al., 2003).

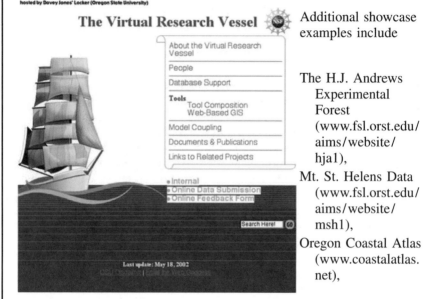

The Virtual Research Vessel

About the Virtual Research Vessel

People

Database Support

Tools
    Tool Composition
    Web-Based GIS

Model Coupling

Documents & Publications

Links to Related Projects

• Internal
• Online Data Submission
• Online Feedback Form

Search Here!   [Go]

Last update: May 18, 2002

Additional showcase examples include

The H.J. Andrews Experimental Forest (www.fsl.orst.edu/aims/website/hja1), Mt. St. Helens Data (www.fsl.orst.edu/aims/website/msh1), Oregon Coastal Atlas (www.coastalatlas.net),

Yolo County Sediment and Soil Analyses (yolo.een.orst.edu/yolo.net), and

Virtual Oregon (virtual-oregon.nacse.org).

In addition, the following Internet GIS research questions are being considered:

- How should data models and data structures for Internet GIS differ from conventional GIS data structures?
- Are there standard metrics for GIS functionality that should be developed for specific application domains?
- What are the appropriate measures of performance for Internet GIS?
- What are the primary barriers to the usability of most Internet GIS sites? Usability engineering techniques are being investigated and deployed and multilevel Web-to-database interfaces are being developed for Internet GIS to enable customized access for meeting the needs of very different user groups.
- To what end should Internet GIS be developed? It is normally best used with broadband access in order to get satisfactory results. Yet, according to McGovern (2001), who cited statistics from NetValue, only 11% of American, 5% of German, 4% of French, and 3% of British households had such access in 2001.
- Given the predominance of Windows-based systems, what is the future of UNIX/Linux open-source GIS for the Web, and how can these toolkits be exploited and proliferated?
- Should Internet GIS be an "add-on" to standard Web browsers or cross-platform languages such as Java?

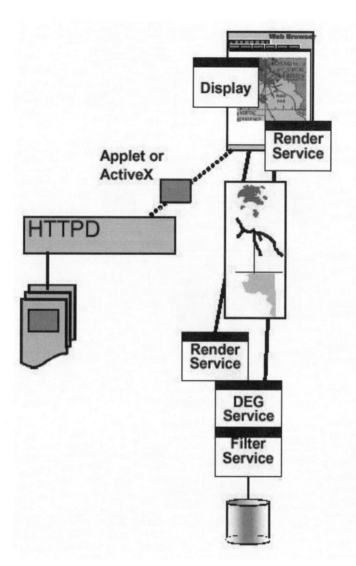

Figure 4.25 Graphic element case. (Figure reprinted with permission from OGC, 2000, p. 15.)

4.5.2.3 WMS Specifications for the Data Case The data (feature) case is the thick-client architecture where the client-side machines can perform display, render, and Display Element Generation (DEG) services (Figure 4.26). The servers will only be responsible for communicating GIS databases and the client-side map viewers. The communication between the client-side map viewers and servers may use XML or GML to specify the geodata elements and map display properties. All the map tasks, such as projections and

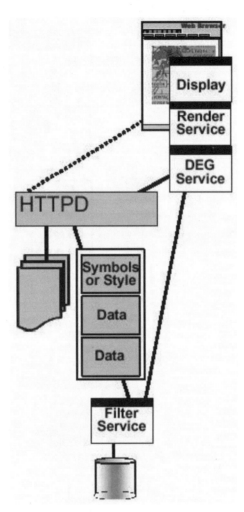

Figure 4.26 Data case. (Figure reprinted with permission from OGC, 2000, p. 15.)

symbol selections, will be performed locally in a viewer. ESRI's ArcIMS feature service is one of the examples of the data or feature case. The advantage of the data case is that it allows users to have the most freedom in manipulating geographic data items. Users can change the symbols and colors of map features locally without sending requests to the servers. Also, users can display both the Web-based map features with the data layers from local machines in local hard drives. Since the client viewer already has all the display capabilities, map users may use the client viewer to perform basic GIS operations, such as buffering and overlay operations. However, the map users may need to pay client-side software license fees in the data case arrangement because such powerful client-side map browsers can be used as regular GIS software packages.

In general, the three case examples have their own advantages and disadvantages. Currently, the picture case is the most popular framework adopted by the GIS industry. However, the picture case only provides limited map display functions and less user interactions. Along with the progress of Web mapping and information technologies, the data case and the graphic element case may become more popular than the picture case in the future. The WMS implementation interface specifications (version 1.0) only focus on the picture case (thin clients) with the standardization of URI syntax and semantic contents. The next version of WMS may focus on the graphic element cases and the data cases with the adoption of GML applications.

The three cases in the WMS specifications demonstrate that different types of GIServices may need to adopt different types of software architecture. However, the software models proposed in the OpenGIS WMS specifications do not provide an approach for dynamically changing the architecture of Web map services. For example, the software framework in the picture case will not be able to upgrade to the graphic element case or data case if client-side map users ask for a higher level of map services or want to change their map applications. The ad hoc WMS specifications do not provide a flexible mechanism for migrating a software framework from one case to another.

One possible solution for providing an upgradable software framework for WMSs is to adopt the dynamic GIServices architecture proposed in this book. By adopting the dynamic framework proposed in Chapter 5, the WMS software framework can be easily upgraded from the picture case to the graphic element case or the data case by relocating the map service elements. Figure 4.27 illustrates such a dynamic architecture for Web map services, where each service element can be freely moved or relocated among client or server-side machines. This dynamic architecture will be able to provide a flexible software architecture for Web map services.

Figure 4.27 illustrates that different map users can access the same server that provides map services in either the picture case (scenario A) or the graphic element case (scenario B). For example, scenario A could be that a map user wants to display road maps in Boulder, Colorado, and the client machine only requires display services (the picture case is the best choice). Scenario B could be that a map user wants to find out the top 10 cities in the United States with the highest population growth rate. This scenario may require advanced map query capabilities and more flexible map display functions. Thus, the client machine could dynamically download a render service element from a server to the client machine (the graphic element case). By introducing the GIS component container and the dynamic GIServices architecture, map users can download different types of map service components based on their needs from servers to clients or vise versa. The dynamic change of the architecture will provide more flexible, upgradable, and user-oriented Web map services for users.

The WMS interface specifications deal primarily with the interfaces on the Web server. It is part of a more general Web mapping architecture that involves many distributed map servers on the Internet.

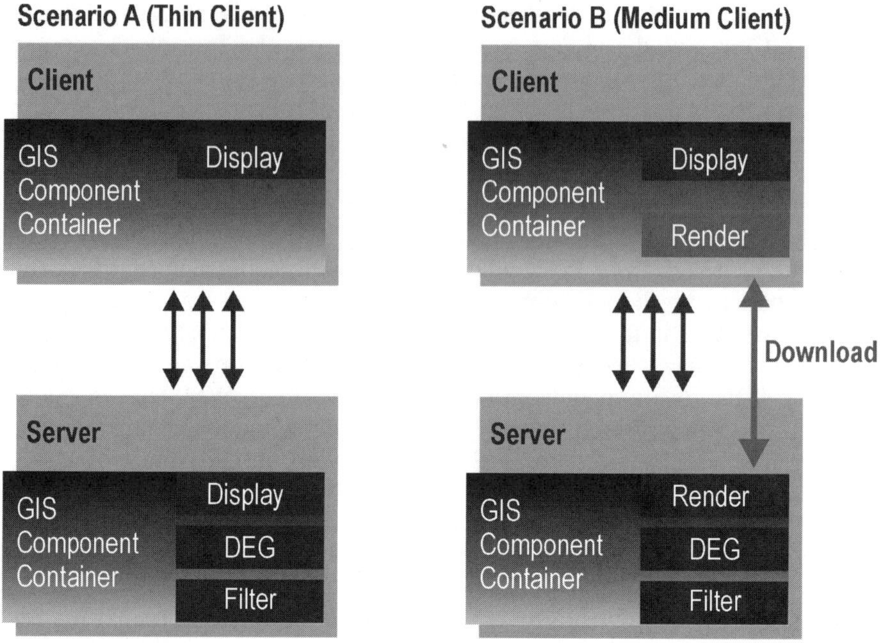

Figure 4.27 Dynamic architecture for Web map services.

Web mapping is still a preliminary GIS program that has very limited functions. To turn the Web mapping program into a truly distributed GIS, we need to rely on the more advanced distributed-component technology and its standards. In the next two chapters, we will introduce these more advanced distributed-component technologies and their emerging standards.

WEB RESOURCES

Descriptions	URL
Xerox Map Viewer	http://mapweb.parc.xerox.com/map (it is offline now)
GRASSLinks	http://www.regis.berkeley.edu/grasslinks
ADL	http://www.alexandria.ucsb.edu
DLI Phase I	http://www.dli2.nsf.gov/dlione
DLI Phase II	http://www.dli2.nsf.gov
OGC	http://www.opengis.org
ISO/TC 211	http://www.isotc211.org
ESRI ArcIMS Java viewer	http://www.esri.com/software/internetmaps/index.html
GeoMedia ActiveX viewer	http://www.intergraph.com/gis/gmwm

Visa ATM locator	http://www.visa.com
Mapquest	http://www.mapquest.com
Mapblast!	http://www.mapblast.com

REFERENCES

Buehler, K., and McKee, L. (Eds.) (1996). *The Opengis® Guide: Introduction to Interoperable Geoprocessing*. Wayland, Massachusetts: Open GIS Consortium.

Buehler, K., and McKee, L. (Eds.). (1998). *The Opengis® Guide: Introduction to Interoperable Geoprocessing and the Opengis Specification*, 3rd ed. Wayland, Massachusetts: Open GIS Consortium. URL: http://www.opengis.org/techno/guide.htm, May 11, 2000.

Buttenfield, B. P. (1998). Looking Forward: Geographic Information Services and Libraries in the Future. *Cartography and Geographic Information Systems*, 25(3), pp. 161–171.

Buttenfield, B. P., and Goodchild, M. F. (1996). The Alexandria Digital Library Project: Distributed Library Services for Spatially Referenced Data. In *Proceedings of GIS/LIS'96*, Denver, Colorado. Bethesda, Maryland: American Society for Photogrammetry and Remote Sensing, pp. 76–84.

Chappell, D. (1996). *Understanding ActiveX and OLE*. Redmond, Washington: Microsoft Press.

Cook, S., and Daniels, J. (1994*). Designing Object Systems: Object-Oriented Modeling with Syntropy*. Englewood Cliffs, New Jersey: Prentice-Hall.

Cuthbert, A. (1997). *User Interaction with Geospatial Data*. OpenGIS Project Document 98-060. Wayland, Massachusetts: Open GIS Consortium.

Doyle, A. (1977). *WWW Mapping Framework*. OpenGIS Project Document 97-007. Wayland, Massachusetts: Open GIS Consortium.

Gardels, K. (1996). The Open GIS Approach to Distributed Geodata and Geoprocessing. In *Proceedings of the Third International Conference on Integrating GIS and Environmental Modeling*, Santa Fe, New Mexico, National Center for Geographic Information and Analysis (NCGIA), CD-ROM.

Gardels, K. (1998). *A Web Mapping Scenario*. OpenGIS Project Document 98-068. Wayland, Massachusetts: Open GIS Consortium.

Goodchild, M. F. (1995). *Alexandria Digital Library: Report on a Workshop on Metadata*, Santa Barbara, California. URL: http://alexandria.sdc.ucsb.edu/public-documents/metadata/metadata_ws.html, May 11, 2000.

Huse, S. M. (1995). GRASSLinks: A New Model for Spatial Information Access in Environmental Planning. Unpublished Ph.D. dissertation, University of California at Berkeley, Department of Landscape Architecture, Berkeley, California.

ISO/TC 211 Chairman. (1998). *Draft Agreement between Open GIS Consortium, Inc. and ISO/TC 211*. ISO/TC 211-N563.

McGovern, G., 2001. The technology productivity paradox, *New Thinking*, 6(42), http://www.gerrymcgovern.com/nt/2001/nt_2001_10_29_productivity.htm. Accessed 22 July 2002.

National Science Foundation. (1994). *NSF Announces Awards for Digital Libraries Research.* NSF PR 94-52. NSF: Washington, DC.

Open GIS Consortium (OGC) (1998). *The OpenGIS Abstract Specification,* Version 3. Wayland, Massachusetts: Open GIS Consortium. URL: http://www.opengis.org/ techno/specs.htm, May 11, 2000.

Open GIS Consortium (OGC) (1999). *The OpenGIS Abstract Specification,* Version 4.0. Wayland, Massachusetts: Open GIS Consortium, URL: http://www.opengis.org/ public/abstract/99-113.pdf.

Open GIS Consortium (OGC) (2000). *OpenGIS Web Map Server Interface Implementation Specification,* Revision 1.0.0. Wayland, Massachusetts: Open GIS Consortium.

Open GIS Consortium (OGC) (2001). *OpenGIS Web Map Server Interface Implementation Specification,* Revision 1.1.0. Wayland, Massachusetts: Open GIS Consortium. URL: http://www.opengis.org/techno/specs.htm, September 11, 2001.

Orfali, R., Harkey, D., and Edwards, J. (1999). *Client/Server Survival Guide,* 3rd ed. New York: Wiley.

Peng, Z.-R. (1999). An Assessment Framework of the Development Strategies of Internet GIS. *Environment and Planning B: Planning and Design,* Vol. 26(1), pp. 117–132.

Plewe, B. (1997). *GIS Online: Information Retrieval, Mapping, and the Internet.* Santa Fe, New Mexico: OnWord Press.

Putz, S. (1994). Interactive Information Services Using World Wide Web Hypertext. In *Proceedings of the First International Conference on the World-Wide Web,* Geneva, Switzerland: CERN (European Organization for Nuclear Research) URL: http://www94.web.cern.ch/WWW94/PrelimProcs.html, May 11, 2000.

Reed, C. (2002). *Prior Art and Invention Related to Web Mapping,* Version 1. Unpublished manuscript.

Rowley, J. (1998). *Draft Business Case for the Harmonisation between ISO/TC 211 and Open GIS Consortium, Inc.* Resolution 47. ISO/TC 211-N472.

Shan, Y.-P., and Earle, R. H. (1998). *Enterprise Computing with Objects: From Client/ Server Environments to the Internet.* Reading, Massachusetts: Addison-Wesley.

Tang, Q. (1997). Component Software and Internet GIS. In *Proceedings of GIS/ LIS'97,* Cincinnati, Ohio, pp. 131–135.

Wright, D. J., O'Dea, E., Cushing, J. B., Cuny, J. E., and Toomey, D. R. (2003). Why Web GIS May Not Be Enough: A Case Study with the Virtual Research Vessel, *Marine Geodesy,* 26(1–2).

Xue, Y., Cracknell, A. P., Guo, H. D., 2002. Telegeoprocessing: The Integration of Remote Sensing, Geographic Information System (GIS), Global Positioning System (GPS) and Telecommunication, *Int. J. Remote Sensing,* 23(9): 1851–1893.

CHAPTER 5

FRAMEWORK OF DISTRIBUTED GEOGRAPHIC INFORMATION SERVICES

> *Geographic information systems are evolving to support a new, network-based architecture. This architecture is multiparticipant, collaborative, and will allow organizations to openly share and directly use GIS information from many distributed sources at the same time.*
> —Jack Dangermond (2001)

5.1 INTRODUCTION

The previous chapter introduced the technology of interactive Web mapping, which provide only half of truly distributed GIS. Client-side applications such as Java applets and ActiveX controls and dynamic HTML are designed mainly for graphic display of maps rather than truly providing GIS operations and analysis. There is very limited functionality in existing Web mapping programs, which do not offer much interactivity and flexibility for complicated GIS modeling and processing.

In addition, the architecture models for static Web mapping and interactive Web mapping require a CGI or CGI-like middleware between the Web client and the GIS and application server. The middleware approach adds the overhead of interactions between the Web client and a GIS server. What if we have a direct communication between the GIS server and the client? This is the idea of a distributed GIS.

So what is distributed GIS? Distributed GIS refers to a distributed platform of accessing and processing geospatial data using distributed GIService components on the Internet. It relies on mobile client components or downloadable clients that communicate directly with objects and data on the server across the network. The mobile client can access both the server resources from which it originated and other server resources on the Internet. The mobile client component is an interoperable component that is constructed using standard distributed technology. Distributed GIS will have the look, feel, and functionality of desktop GIS but work seamlessly with remote object and data over the Internet.

5.1.1 Basic Requirements of Distributed GIS

The architecture of distributed GIS represents a dramatic departure from the Web mapping applications. In interactive Web mapping, although client-side applications such as Java applets and ActiveX controls offer much interactivity between the user and the graphic user interface, the variations of CGI or different CGI extensions at the server side are simply Band-Aid solutions for distributed GIS. From a technical perspective, the middleware with CGI and its extensions between the Web client and the map server cannot provide a truly distributed GIS. To qualify as true distributed GIS, a system has to have the following characteristics:

1. *It is composed of distributed components;* each component has its own functions. For example, a buffer is a component, a "point in polygon" overlay is another component, and so on.
2. *The component is distributed.* That is, the components could reside in different computers or GIS nodes but interact directly with each other.
3. *The component is mobile.* Although components reside in different computers, they can be retrieved and downloaded into other computers on demand.
4. *The components are open and interoperable.* Once the components migrate to other computers, they can be assembled and interoperated with other components that may be downloaded from yet another computer. To be interoperable, the components have to be constructed according to standards.
5. *The components are searchable* and mechanisms are available to purchase and use the components from service providers. A service catalog is needed to advertise the availability and functions of all components.
6. *Data are distributed.* Distributed GIS can access any data located anywhere on the Internet. Standard metadata and/or data repository is provided to connect distributed GIS data on the Internet.
7. *Data are interchangeable.* This means that data from different sources can be integrated. Mechanisms are needed to integrate data with different spatial reference systems, different semantics, and different formats.

Are there currently any distributed GIS products? No, not at this time. But distributed GIS is the vision shared by many users, government agencies, standards organizations, and even GIS vendors. It is the vision that is most likely to become reality in the next few years. In fact, the two standards organizations, the ISO and OGC, are working hard to lay the ground rules and set standards to make this happen.

5.1.2 Why We Need Distributed GIS

Why do we need a distributed GIS? There are several reasons:

1. Distributed GIS will allow GIS components to merge into the mainstream of information technology to handle geospatial data. So GIS components can be easily merged with other programs such as word processors or spreadsheets.
2. Distributed components are open and independent from different operating systems, hardware, network environments, vendors, and applications.
3. Users do not need to make a big commitment to buy expensive GIS software. They simply pay (or rent) as they use.
4. Users do not have to lock into one GIS software vendor as they do now.
5. GIS software vendors can focus their efforts on developing high-quality GIS components and other applications.
6. The development of the distributed component shifts the software paradigm from a monolithic, feature-heavy approach to a flexible, modularized, and plug-and-play approach. This modularized, reusable software framework can improve the cycles of program development and efficiency of software engineering and result in much more powerful GIS technologies.
7. The standard software component approach would encourage innovation from small GIS software vendors, which are more likely to develop highly specialized niche components to meet specific needs.

To develop this future generation of distributed GIS, we need distributed components, distributed architecture, different ways of implementation, and standards. Therefore, this chapter will follow this logic and cover topics ranging from distributed *components* to distributed *architecture* to system *implementations*. The next chapter will discuss emerging *standards*.

5.1.3 Components, Interfaces, and Services

Before we talk about the general architecture in distributed GIS, it is necessary to differentiate the concepts of component, interface, and service. Based on the ISO TC 211 document *Geographic Information—Services,* component, interface, and service are defined as follows (ISO/TC 211/WG 4, 2000, OGC WMT 2 Request for Quotation (RFQ)):

- *Component*—"a physical, replaceable part of a system that packages implementation and conforms to and provides the realization of a set of interfaces" (p. 11).
- *Interface*—"a named set of operations that characterize the behavior of an element" (p. 10).
- *Service*—"a capability which a service provider entity makes available to a service user entity through a set of interfaces that define a behavior, such as a use case," (p. 10). Service is sometimes synonymous with *capability* and *function* in this chapter.

For example, in the case of distributed GIS, a client element is a component that is used to interact with the user. This client component has a service (capability or function) and an interface. The service defines the roles or capabilities of the component. For instance, the client component offers services for a user to zoom, pan, or query. The interface implements the capabilities or functions of a service through a set of operations. For example, the zoom service offered in the client component could be implemented as an HTML click button or a pull-down menu in the Java applet.

There could be many components in a distributed GIS application. Components communicate with each other through interfaces. Each component can be located in the client side or the server side. Client-side components interact with the user and provide input to server-side components, while server-side components provide service to client components. The communication between the client component and the server component is through the interface. An IDL or ODL is often used to specify a component's interface.

A component can be implemented or instantiated using different technologies. For example, clients can be constructed in HTML, Java applet, or ActiveX controls, and servers for distributed GIS can be implemented using a particular Distributed-Computing Platform (DCP) such as CORBA/IIOP or COMP+. A unique feature for distributed GIS is a mobile client component that can be downloaded and assembled on any Web browser yet can communicate directly with the server objects through an IDL- or ODL-defined interface.

Distributed components are building blocks of distributed GIS. In order to establish a distributed GIService, the major task is to create a dynamic architecture by adopting LEGO-like distributed GIS components and data objects. The term *dynamic* indicates that the GIService is constructed dynamically in the user's machine by temporarily connecting to or downloading GIS components and data objects across the network. Each time the user starts a task or service request, the user can construct the GIServices program on the fly by downloading the proper GIS components and data objects from the service providers.

5.2 BASIC COMPONENTS AND SERVICE REQUIREMENTS OF DISTRIBUTED GIS

5.2.1 What Are Distributed Components?

Under a general definition, a distributed component is "a ready-to-run package of code that gets dynamically loaded into your system to extend its functionality" (Pountain, 1997). For example, a Java applet, an ActiveX control, or even a plug-in function for the Web browser can each be called a distributed component. In principle, distributed components should be plug-and-play, interoperable, portable and reusable, self-describing and self-managing, and able to be freely combined in use (Orfali et al., 1996; Pountain, 1997).

Key Concepts

Distributed components *are building blocks of distributed GIS. In order to establish a distributed GIService, the major task is to create a dynamic architecture by adopting* **LEGO-like distributed GIS components** *and data objects. Each time the user starts a task or service request, the user can construct the GIServices program on the fly by downloading the proper GIS components and data objects from the service providers.*

In practice, distributed components are LEGO-like pieces of binary code. The LEGO metaphor refers to the well-known children's toy blocks that can be interlocked and stacked. Similar to LEGO blocks, the idea is to create software modules (GIS processes) that stack and interlock to form a dynamic GIS package. The LEGO architecture may persist only briefly for completion of a single GIS task. Then the LEGO modules are broken down, rearranged, and restacked to form a new software module for a new GIS task.

One important advantage of distributed components is the independence from different hardware, network environments, vendors, and applications. The same component can be copied, moved, and executed in different machines with different configurations. Distributed components will interact with each other or be combined together to provide integrated services to users.

Figure 5.1 shows an example of a LEGO-like module including a map display component that can be used, for example, in a word processing application or a GIS package. The word processing application is combined with several distributed components, including a graphic user interface component, a spell checker, and so on. The map display component is independent to the extent that it can be easily plugged into other packages when users

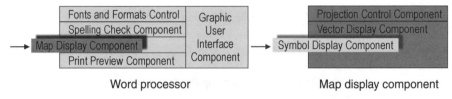

Word processor   Map display component

Figure 5.1 LEGO-like distributed GIS components.

need a map display function. Moreover, the component strategy is hierarchical. Here, the map component is made up of subcomponents, including (in this case) a projection control and a vector display control. Alternative subcomponents can be added into a map display component to extend its display functions, such as adding a symbol display control.

Another advantage of distributed-component technology is the independent operations from different software environments, database servers, and computer platforms (Figure 5.2). Distributed-component technology allows a program (component) to operate on different computer platforms and to access heterogeneous database servers via a standardized protocol. For example, a Java applet (component) can be downloaded into different computer platforms (Mac, Windows, or UNIX) and can access different types of databases, such as Microsoft SQL servers or Oracle database servers via the channel of JDBC.

5.2.2 Basic Components of Distributed GIS

The above section discussed the general framework of distributed data and service components. But it did not address the issue of how the mobile agents

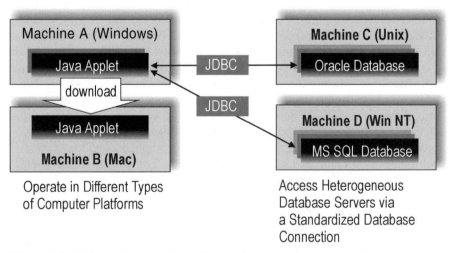

Operate in Different Types
of Computer Platforms

Access Heterogeneous
Database Servers via
a Standardized Database
Connection

Figure 5.2 Independent operations from software environments and computer platforms.

find the proper services and data across the network. If a user works on his or her client machine, how and where can he or she request service components and data objects? In order to make this dynamic construction of GIServices work, exactly which components are needed for the distributed GIS? In its development of WMS implementation specifications OGC (2000) specified four major high-level groups of components and their services:

- *Viewers and Editors*   These represent presentations of the system to allow users to view and interact with maps and the underlying data and operations. A viewer allows examination of maps and the underlying data but does not allow the user to edit the data, whereas an editor allows changes to be made to the database.
- *Catalogs*   A catalog is a collection of metadata or a repository of metadata. Information about other objects and operators can be collected into one or more catalogs. Catalogs provide a search operation that can return metadata or the names of the objects.
- *Repositories*   Repositories are collections of data. Each data or item in the repository is generally associated with a name. Therefore, a repository can find the item based on the name given. Repositories usually maintain indexes to help speed up the process of finding items by name or by other attributes of the item. Data repositories are important components in a distributed environment for efficient data discovery and management.
- *Operators*   Operators are components that can conduct some operations on the data and produce outputs based on user request. They can transform, combine, filter, or create data. The conventional GIS analysis tools such as buffer, spatial overlay, network analysis, and image processing belong to the operator category.

As OGC admits, this taxonomy is not without overlap. For example, viewers and editors can themselves be operators, as can catalogs and repositories. Catalogs are repositories, and repositories can be catalogs too.

In addition to the four groups of components, there are other components that operate on or with "data" components. All of the components and their operands have internal and external semantics. These include the following:

- Data are information about other things, or just plain information. Data can be created, stored, operated on, deleted, viewed, and so on.
- Metadata are generally considered to be information about other data, which can be stored in catalogs. Metadata in catalogs make it easier to find a specific data item.
- Names are identifiers. Names themselves are only meaningful if you know the context in which the name is valid (this is called the namespace). When a data item is stored in a repository, it can be given a name that is valid within the repository. If the repository itself has a name, the

two names together would help find the original data item. Names can refer to data or to operators.

- Relationships are links between items. These can be simple links such as WWW hyperlinks or complex, *n*-way relationships among many items. Relationships tend to link named items. Sometimes they can link to named items at a finer level of granularity. A relational database is a classic example of relationships.

These are high-level categories of distributed components that are foundations of distributed GIS. But to develop a distributed GIS, we also need architectural models to put all these component pieces together in a structured manner. Although we have discussed the thin- and thick-client models above, a more systematic discussion is needed to describe the relationship between client and server and the distributed GIS components. We now turn to this discussion.

5.3 ARCHITECTURE MODELS OF DISTRIBUTED GIS

Architecture model here refers to the architecture of a software system or a distributed GIS application. The term describes the framework of the system construct, including system components, functions of each component, the relationship between the components, and the information flow among the components. An architecture model provides the framework of what kind of system is to be built and the functions the system could have. In this section, we will discuss three architectural models: the generic distributed Web mapping architecture proposed by OGC, the restricted client/server GIS, and the open distributed GIS framework for multiple clients and servers.

OGC's *generic Web mapping architecture* provides a generic framework to construct a distributed Web mapping that allows the user on the Web browser to search, retrieve, interact, and manipulate geospatial data stored in a distributed environment across the Internet. OGC envisions that all Web mapping programs that comply with its WMS implementation specifications can interoperate with each other. In other words, all resources (data and service objects) in each Web mapping sites could be accessible and usable by any users on the Internet.

A *restricted distributed GIS framework* is a simple system in which the client can only communicate with a single designated server. The client cannot communicate with other servers across the Internet. In other words, the client can only connect to or download GIS components and data objects directly from its own designated server. There would be only one Web server, one application server, one or more map servers, and data servers that are usually (but not always) physically separated from each other yet running at the same time. These map and data servers are managed through the appli-

Internet GIS Showcase: PanCanadian Petroleum Limited

PanCanadian Petroleum Limited of Calgary, Alberta, is one of Canada's largest producers and marketers of crude oil, natural gas, and natural gas liquids. Its extensive operations, which stretch across Canada and include a variety of international interests, require extensive mapping systems. The distributed nature of the company led to adoption of a number of different mapping systems. Basemap data were inconsistent and difficult to share within the company. To fix this unworkable system, PanCanadian initiated a project called MapWiz that will allow staff members to quickly, easily, and consistently retrieve the most current basemap and environmental spatial data for any area, in any projection, and in any format needed.

The PanCanadian system was developed by Safe Software using Feature Manipulation Engine (FME) technology. Safe determined that MapWiz should be built around four main criteria:

- A *data import system* would be needed to move all the data to a central spatial database.
- A *spatial database subsystem* would then be needed to store the data while allowing import and export access.
- Once the data were in the spatial database, a *data export system* would be needed to export the data to the various projections and formats.
- Finally, all of this would have to happen *over the Web*.

Through the use of FME (this time running in the server mode), the data could be extracted from the spatial database in one of many formats and at the same time could be reprojected into a suitable datum and projection. Keeping PanCanadian's requirements in mind, the Safe team decided to create an HTML front end that users could access through any Web browser. Once the HTML front end was implemented, the browser acted as an index map viewer, allowing basic query and pan/zoom abilities. Users selected a map sheet, a destination format, and a destination projection, and a JavaScript application sent the request to FME, which translated the data, compressed it, and stored it on the local FTP site.

Source: http://www.safe.com/solutions/casestudies/
casestudy_pancanadian.htm.

cation server and can be accessed with a single interface by all clients. This model is easier to implement. In fact, many vendor-specific Web mapping programs follow this model. But this model is not interoperable and thus is less desirable from the user's point of view.

An open distributed GIS is an interoperable system, where users from any client can search, access, and retrieve GIS components and data objects from any map server and data server across the entire Internet. Geospatial data and GIS component services from any servers can be retrieved (or purchased) and displayed in any clients. This model builds on the single-site model but is made more interoperable across all sites. It is similar to the OGC's generic Web mapping architecture.

5.3.1 Distributed Web Mapping Architecture from OGC

OGC, in its request for quotation and call for participation in WMT Phase II, proposed the general Web mapping architecture shown in Figures 5.3 and 5.4.

The general idea behind this generic Web mapping architecture is that the user can search, retrieve, interact, and manipulate geospatial data stored in a distributed environment using an interoperable Web mapping system. The user interacts with client viewers and/or editors as shown in Figure 5.3. The user can do one of four things—search for data on the Internet, invoke operators

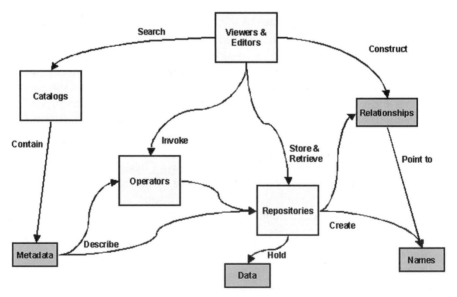

Figure 5.3 Generic spatial information management objects. (Figure reprinted with permission from OGC, 2001a.)

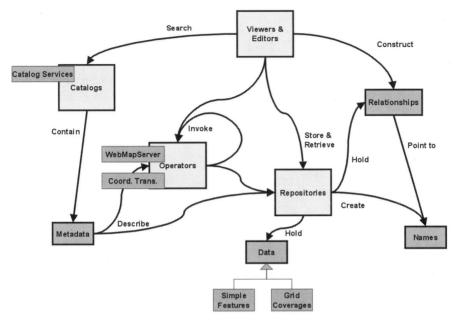

Figure 5.4   OpenGIS Specifications. (Figure reprinted with permission from OGC, 2001a.)

to conduct data analysis, retrieve and edit data, or construct object relationships:

Data Search   When a user requests geospatial data, he or she can search through a catalog that contains metadata. The metadata will direct the search to a data repository, which links to specific data sources. If the user request is a complex one, a data operation (through operators), such as a spatial query and/or spatial overlay, should be conducted.

Operations   The user can invoke operations directly from the client viewer or editor, including map rendering, spatial analysis, image processing, coordination transformation, or other operations. The underlying data are connected through the linkage with the operators.

Data Edit and Management   Through the editor client, the user can retrieve the data and metadata from the data repositories, edit the data, and store the updated data back to the data source for future usage. It can also create metadata and edit metadata. Furthermore, the user can create a name (an identifier) in a namespace (repository) to identify the specific data or operators, or create a relationship that points to a specific name of data and operators.

Construct Relationship   The user can directly construct a relationship that points to names of data and operators in the repositories. For example, the user could construct a hyperlink to link two named items in the HTML page. The user could also construct a complex GML XLink to link many named items together.

The OpenGIS specifications map specific services onto this component diagram, as shown in Figure 5.4. There are catalog services, coordinate transformation services, WMS, grid coverages, and simple features:

The *catalog service* was used in WMT 1 to collect information about map servers and their capabilities and to place the information into a searchable repository. The functions of the catalog service were extended in WMT 2 to allow clients to locate both data and services of interest in the Internet and to provide access to GML data servers via catalog searches.

The *WMS server* interface specification discussed in the previous chapter is now part of the operator in this generic Web mapping architecture. The *coordinate transformation services* are developed to transform coordinate systems from one system to another.

The data component now includes both the *simple feature* data and the *grid coverage* (raster image) data type.

This generic distributed Web mapping architecture proposed by OGC provides a grand vision of what services can be provided by the distributed GIS and how different components can link together.

5.3.2   Restricted Distributed GIS Framework

A restricted distributed GIS for a single site is a system on its own. That is, the client only communicates with its own Web server, map server(s), and data server(s). All user requests from the client are handled by a single application server (or middleware plumbing service) that manages one or more map servers and data servers. An architectural model for a distributed GIS for a single site is shown in Figures 5.5 and 5.6. The system is composed of four main components: a client, a Web server, an application server, a map server, and a data server. (We will discuss the functions of each component in detail in Section 5.5.)

Figure 5.5 shows a single-server system with only one map and data server, where all service components are stored in the map server and data objects

Figure 5.5   Simple distributed GIS.

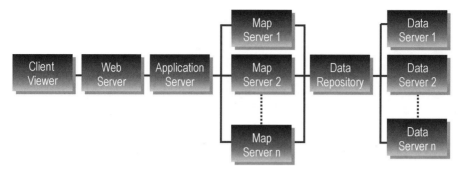

Figure 5.6 Multiserver distributed GIS.

are stored in the data server. This single-server system does not scale well to deal with a large amount of user requests (hits) from the user. It is also weak in fault tolerance. When a map server or a data server fails to function, the entire system fails. To increase scalability and fault tolerance, a multiple-server system needs to be deployed, as shown in Figure 5.6.

Figure 5.6 shows a multiserver system with multiple map servers and data servers. The multiple map servers and data servers are usually, but not necessarily always, located physically in different computers. The multiserver system has good scalability and fault tolerance. When the amount of user requests gets larger or the amount of data increases, additional map servers and/or data servers can be added.

Since many map servers and data servers are involved, in addition to the components that are available in the single-server system, the multiserver system should also include catalog services, data repository, and load balance services. A catalog service is used to keep track of what functions each map server can provide, and a load balance service is used to assign tasks to specific map servers based on the workload condition of each map server at a specific time. The purpose of the catalog service and load balance service is to make sure that the user request can be passed on to the right map server at the right time. The application server is used to provide catalog services and load balance services. Data repository is a registration service that keeps track of the types of data and the locations of all data sets. Through the data repository, the system can locate and filter the right data set from the appropriate data server.

There are two approaches to deploy multiple servers: distributed server systems and mirrored systems. In the distributed-server model, different map servers perform different functions, and different data servers contain different data sets so that the map servers and geospatial data are distributed across the servers (Figure 5.7). A catalog service is needed as a service broker dispatcher to direct user requests to appropriate servers. In the mirrored model, the multiple servers have exactly the same setup and data so that the

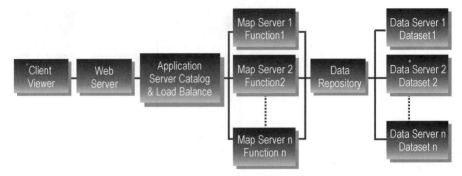

Figure 5.7 Distributed map server and data server.

additional servers act as backups if the first server is busy or unavailable (Figure 5.8).

These two approaches have their advantages and disadvantages. The distributed-server model requires less maintenance than the mirrored systems model, but it is not fault tolerant. That is, if one server is down, the whole system is then down and the viewer cannot access and display the data on that server. The mirrored systems model has more fault tolerance; that is, if one server is down, the other servers act as backup servers so that the whole system will still function. However, the mirrored systems require more maintenance than the distributed systems, especially when you need to update the servers. Therefore, a combination of the two approaches could achieve the maximal benefits.

5.3.3 Open Distributed GIS Architecture

While the client in the aforementioned distributed GIS for a single site communicates with its own Web server and map server, the client at an open or interoperable system can communicate with any Web servers and map servers

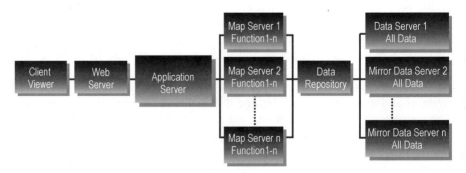

Figure 5.8 Mirrored map server and data server.

on the Internet. An open system has a repository of all distributed GIS programs that support the same protocol so that any user can take advantage of geospatial data and analysis services at any service provider across the Internet.

An open or interoperable system is technically not one system; rather it is a virtual system or a repository of distributed GIS providers that comply with the same standards. Each distributed GIS provider will have its own systems with its own Web server, map server, and data servers. But all systems will support the same protocol so that they can communicate with each other. OGC is working to develop such Web mapping protocols.

An architectural model for an interoperable distributed GIS is shown in Figure 5.9. As the figure shows, the system is a combination of distributed GIS programs in different geospatial service providers' sites. A catalog service is an important element for discovering data and analysis services from service providers over the Internet. A user from any client can discover, access, retrieve, and utilize data and GIS component services from any Web site. For example, the user can retrieve flood information from site A, land use information from site B, satellite images from site C, and analysis tools from site D. The retrieved data and analysis tool components can be integrated and displayed at one client, regardless of the format of the information and the projection and coordinate systems used for each data.

To increase data interoperability, a common format of geospatial data could be provided. GML could be such a common format. Another way to ensure interoperability is to rely on different data servers to serve the data in different formats, an approach that has been implemented in Intergraph's GeoMedia data server (see Chapter 8). The third way is to provide a data conversion tool to convert one data format to the other on demand.

When data with different projections and coordinate systems are downloaded to the client, they have to be integrated. One way to integrate them is by translating the data from one projection and coordinate system to another through a projection translator, which can reproject the geospatial data into the client's required projection and coordinate system. This translator could be placed either in the map server or at the client side. Therefore, data can be stored in a data server in a particular format in one projection or it can be transformed into other formats and projection on the fly.

The open distributed system is very scalable because it has no limit on the number of participating system services and data providers.

The client in the open distributed system could have several functions: information display, discovery, publishing and management, and so on. Each function can have its own client components and can be downloaded from servers on demand. However, a uniform user interface can also be developed to perform all these functions. Client components can communicate with any other server across the Internet, and the server components can also communicate with other server components on other sites as well.

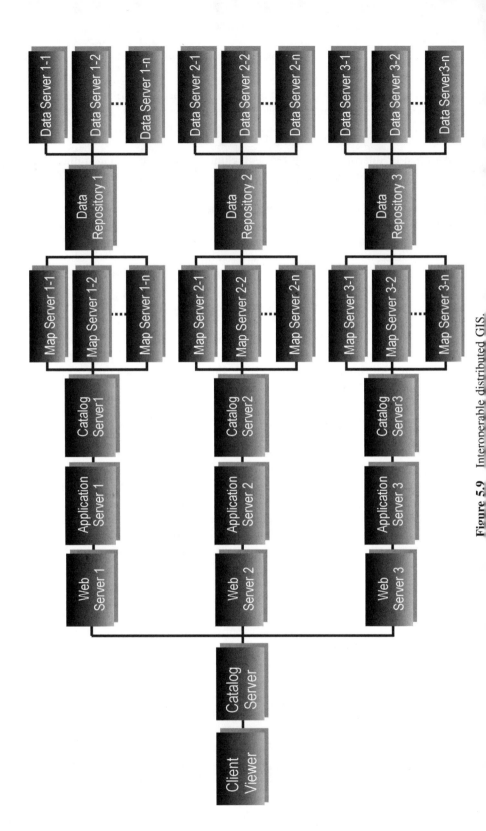

Figure 5.9 Interoperable distributed GIS.

5.4 COMPONENT FUNCTIONS IN A DISTRIBUTED GIS FRAMEWORK

Like any other applications, distributed GIS has three basic elements: presentation, business logic, and data. Presentation refers to the display of information, including maps, text, and images. Business logic includes, for example, generation of maps, map rendering, spatial analysis, and extraction of data based on certain criteria. The data component refers to data, including the spatial feature data, the properties or attributes about the spatial features, and symbologies to display spatial and nonspatial features. Each element has its own components and function requirements.

It should be noted that the client components and the presentation elements are not necessarily the same thing. A mobile client might include presentation elements, but it can also include logic elements as well. The partition point of the client/server system can be anywhere in the continuum of the whole distributed GIS application.

5.4.1 Presentation Components

Presentation components or mobile clients are important elements in distributed GIS, because the user has to rely on the presentation components to view, query, and interact with maps and other information. Presentation components in a distributed GIS should provide such services as geospatial information display, map rendering, query, and spatial analysis (OGC, 2000, WMT 2 RFQ):

- *Information Display* This is the basic function that a presentation component should provide, that is, to display geospatial information from one or many sources in the forms of maps, text (metadata), tables, and other forms. The geospatial information includes raster images, matrix (elevation), and vector (feature) data.
- *Map Rendering* The presentation component should allow the user to render the maps displayed, such as zoom and pan, and to change titles and edit text. Furthermore, the presentation component should help the user to manage the symbolization of data by providing the user flexibility of choosing a set of symbols for display.
- *Information Query and Feature Selections* The presentation component needs to be able to allow users to make queries on the map, including spatial feature selection and feature attribute query. It should allow the user to select, deselect, reorder, and prioritize the layer information being displayed.
- *Editing* An editing client is a component that allows users or data providers to manage, update, or add geospatial information at the source data server. It allows authorized users to edit and manage data in a dis-

tributed environment. The authorized user can update the data anywhere at any time. This is especially important for field data collection.

These presentation functions can be embedded into one or more mobile client components. If these are constructed as separate client components rather than as a comprehensive client component, they should be able to be assembled on demand at the Web client.

In addition, in the open and interoperable distributed GIS environment, geospatial data and analysis services may be located in different sites by data or service providers across the Internet. Therefore, in addition to providing view, query, and interactive functions with maps and other information, additional functions of the presentation components are also needed, including assisting users to search for geospatial data and analysis services over the Internet and assisting data producers to publish and manage their data sets. These more advanced functions include information discovery and data management and publishing (OGC, 2000, WMT):

- *Information and Service Discovery* This function is required for searching for geospatial data and data service providers on the Internet. A discovery client should be able to search, access, query, and retrieve (or purchase) geospatial information simultaneously from many different servers.
- *Data and Service Registration and Publishing* A mobile client component should be able, for geospatial information data providers and service providers, to register and publish available data and analysis capabilities on the Internet so that they are accessible and searchable by other users. This function can be enabled by using a data or service repository to register the data and analysis services. The registered data and services can be free or for sale.

5.4.2 Logic Components

The basic functions of business logic components of a distributed GIS are to receive user queries from the client, fulfill the queries, and return results to the clients. The function of receiving queries from the client viewers is usually provided by a Web server. Some simple queries can be fulfilled by the Web server along with third-party application servers. For example, a Web server can return a document or a ready-made map image stored at the Web server. But most spatial queries or requests for geospatial data and analysis functions cannot be fulfilled by the Web server. They have to be passed to a map server to process the request.

Therefore, there are three main components in the business logic of the distributed GIS for a single site: the Web server; the application server, or the middleware services; and a map server (Figure 5.10). The Web server is

Figure 5.10   Components of the business logic tier of distributed GIS.

responsible for receiving user requests from the clients. The middleware service or the application server is responsible in some cases (e.g., interactive Web mapping) for parsing the requests, passing them to the map server, and transporting the results from the map server to the Web server. For distributed GIS, the application server is used to manage the tasks in the map server; its functions include load balancing, managing transactions, and security. The mapping services are responsible for fulfilling user queries and/or generating maps.

Before we talk about the Web server and application server, we need to understand the server itself as hardware. Servers have been evolving from the single PC server to asymmetric multiprocessing superservers to Symmetric Multiprocessing (SMP) superservers to multiserver clusters (Orfali et al., 1999). A single PC server has a single processor; its processing power can be very powerful but can be limited in a very demanding operation with many simultaneous user requests.

If the PC server is not powerful enough, the second option is to have a superserver with multiprocessors. Now you may have two choices, an asymmetrical multiprocessing server or a SMP server. An asymmetric multiprocessing server has a hierarchical structure. That is, it divides processors into a master processor and one or more slave processors. Only one designated processor, the master, can run the operating system at any one time. The slave processors are usually dedicated by the master processor to specific functions such as disk I/O or network I/O.

In contrast, processors in the SMP server are treated equally. Every processor has the same function as any other processors. To take advantage of the processing power from multiple processors, the server application has to be multithreaded. In other words, applications have to be divided into threads that can run concurrently on any available processors. Applications that are not multithreaded cannot take advantage of the multiprocessing power offers by the SMP. SQL DBMSs and TP monitors have multithread functions. Most current servers, including UNIX, Windows NT and 2000, Netware, and OS/2, support SMP. The drawback of SMP is that processors need to share caches or memory among processors. It is thus not fault tolerant, meaning when the server fails to function, there is nothing on which to fall back or to take over. Notice that *multiprocessing* and *multiprocessor* have different meanings. Processors are hardware components while a process is an instance of a program running in a computer. Multiprocessor refers to many processors in the same server, but there can be many processes in the same processor.

The third server option is to have a multiserver cluster, that is, multiple servers working together as a single server. Multiple servers are linked together in a LAN using an Ethernet connection or a specialized high-speed LAN. Unlike SMP, multiserver clusters are fault tolerant, meaning that when one server fails, messages or requests can be automatically rerouted to an alternative server node. This is more reliable and more scalable server architecture. Now what does this have to do with the logic elements in the Web server and application server? It will help us to better understand the functions and the setup of the Web server, map server, and application server.

5.4.2.1 Web Server

The major function of a Web server, as we discussed in Chapter 3, is communicating with the Web client. It communicates with the Web client via HTTP. It constantly runs on the server, waiting for client requests. Once it receives a request from the clients, it responds to the requests in one of the following ways:

1. If the request is for an existing document or ready-made map images that are stored in the Web server, the Web server will send the document or map images to the client viewer.
2. If the request is for a mobile component such as a Java applet or ActiveX control, the Web server will send the mobile components to the Web client.
3. If the request is for an attribute data from the database, the Web server will connect directly with a third-party application server such as ASP or ColdFusion, which can fulfill the request by extracting data directly from the database.
4. If the request is for a query of a spatial data or a map, the Web server will pass the request to a map server. Once the map server processes the query or produces the map, the Web server will receive the output from the map server and pass it back to the client.

The Web server is usually but not always located physically in a separate server machine by itself.

5.4.2.2 Map Server

A *map server* is a component that generates maps, fulfills spatial queries, and delivers symbolized maps (in the form of map images or graphic element map) to a client based on the user's request. The map server can be split into smaller chunks to provide specific services. That is, each server could be specialized into one particular function or service. For example, a portrayal service is used to generate maps (as opposed to providing feature data), while a geocode service is used to locate addresses. A *portrayal service* is a service that generates symbolized maps from a feature data set (OGC, 2000, WMT 2 RFQ). The symbolized map can be either a picture (e.g., JPEG, PNG) or a graphic element (e.g., SVG) based map. These services can be located in one server with SMP or in a multiserver cluster.

A map server generally has seven basic functions (OpenGIS WMT specifications):

1. to produce a symbolized map based on user's requests;
2. to answer basic queries about the content of the map and the attribute of spatial features;
3. to extract data from a database based on the criteria on the user's request;
4. to perform spatial analysis such as buffering, feature overlay, spatial search, and network analysis;
5. to communicate with other programs about the availability of services and data, such as what maps it can produce and which of those can be queried further; and
6. to provide an interface for associating default symbology with a data set for use by clients with data publishing capability; and
7. to invoke other map servers to perform other analyses, which is sometimes referred to as a cascading map server in OGC's WMT specifications.

The output of the map server can be in one of three forms: a simple map image (i.e., GIF or JPEG); a graphic element map or a map composition comprising one or more overlays with predefined colors, styles, legends, and so on; a filtered raster and/or vector data in a form that can be manipulated (query, pan, zoom) on the client. In addition to providing links with clients, map servers can also provide mobile components for clients to download, as discussed in Section 5.2.

To better manage and coordinate these services in different map servers, application, or middleware, services are needed.

5.4.2.3 Application (Middleware) Services It can be seen from the above discussion that the Web server and the map server are important components in the business logic. The Web server receives user requests and the map server processes the requests. However, the Web server and the map server cannot directly communicate because they talk in different languages. The Web server uses HTTP and HTML, while each map server may require different types of query structure and format. Therefore, the Web server and the map server have to rely on other programs to be the translator between them. These in-between programs are called glue programs, middleware services, or application servers.

The major functions of these glue programs, or application servers, are as follows:

Internet GIS Showcase: TELUS Geomatics

TELUS Geomatics, based in Edmonton, Alberta, is helping the Canadian government develop emergency response systems. The systems are designed to help the government respond to fires, gas leaks, floods, power failures, and other such emergencies by providing accurate information quickly.

One such TELUS program is the Public Emergency Notification System (PENS), an application available in the TELUS GeoExplorer suite. The GeoExplorer suite, developed by TELUS Geomatics, uses the Autodesk MapGuide technology.

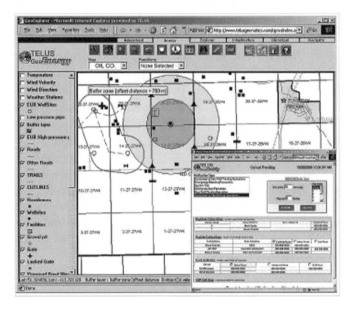

This Web-based program provides authorized subscribers with access to detailed maps containing street networks, hospital and school locations, pipeline routes, emergency contact names and numbers, wind direction, and more. Access is controlled via predefined user profile identities. The system is always available and is a valuable tool for "what if" scenario planning and actual emergency response.

The MapGuide technology is also integrated with TELUS' interactive voice response system, which can automatically make up to 50 phone calls a minute to residents and managers of an emergency situation.

Source: http://www.autodeskgovernment.com/News/
CaseStudiesfromAutodesk/Telus%20Geomatics%20Final1.pdf.

1. to establish a connection between the Web server and the map server;
2. to translate the user requests from the Web server and pass them to the map server for processing;
3. to translate the output from the map server to HTML in order for the Web server to forward to the Web browser;
4. to manage concurrent requests and balance loads among servers; and
5. to manage state, transactions and security, and so on.

A *load balance service* is a component in the application server that keeps track of the workload of each map and data server. It receives requests and assigns tasks to specific map servers and data servers based on the existing workload. When a client sends a service request, the load balance service hands it to an available prestarted application (e.g., a map server, a data server) process. A process is an instance of an application running in a computer. The map server or database server dynamically links to the services (e.g., query, spatial overlay, geocoding) called by the client, invokes the services, oversees their execution, and returns the results to the client.

There are two approaches for the load balance services: process per client and process sharing (Orfali et al., 1999). The process-per-client approach requires one process to open for each client request, while the process sharing allows application processes to be shared by many client requests. The process-per-client model allows each client request to run on a separate process, thereby eliminating interference among client requests. But this approach requires more application processes to run at the same time, which takes more CPU cache and server memory.

The process-sharing approach funnels incoming client requests to shared server processes. When one request is processed, another client can reuse the server process. If the number of incoming requests exceeds the available server processes, new processes can start automatically. The load balance service acts like a job dispatcher by keeping track of the load of the processes being used and allocating client requests to the proper server process. More sophisticated load balance services such as *TP monitors* can distribute the process load across multiple servers or multiple processors in the same server. TP monitors are software programs or server-side middleware that manages server processes, application transactions, and client/server communications.

In addition to the basic functions of a business logic tier provided for a single system such as generating maps and extracting data, the logic components in an interoperable system are also responsible for providing catalogs and registration of geospatial information and analysis services as well as other services. The function requirements discussed below are unique to interoperable distributed GIS.

5.4.2.4 Spatial Reference Services

If the user draws data from two difference sources with different spatial reference systems, the client may not display the maps properly. Therefore, a service must be provided to make the spatial reference systems compatible or to convert different spatial reference systems into a common or a standard reference system. This reference conversion system can be located in either a map server or a client (OGC, 2000, WMT 2). When it is located in the map server, the client has to send a request for a specific spatial reference system (coordinate system, projection, and datum) to the map server. The map server would then convert the geospatial data into the requested reference system before sending to the client.

If the spatial reference service is located at the client side, the client could retrieve data with different reference systems and then convert them into a common reference system. Another approach would be to send the data with a nonprojected reference system to the client. The client then projects the data to the common reference system to be displayed at the client (OGC, 2000, WMT 2).

5.4.2.5 Catalog Services In order for users to find available geospatial data and spatial analysis services in other servers on the Internet, a *catalog service* needs to provide metadata about geospatial data, features, and map servers as well as their capabilities. A catalog server is an information repository that is used to respond to queries about the availability of certain data and services. The catalog services help the middleware or application server to forward the user request to the right map server and data server. It should also be searchable on the Internet so that all Web clients on the Internet can search for the right data and map services at the right place; that is, it supports distributed searching of multiple catalogs.

The catalog may include a data set catalog to describe geospatial data, a services catalog to describe available components with certain analysis functions, and a feature catalog to describe spatial features and feature attributes. A metadata is usually developed to support data search and query. To support distributed searching, a standardized catalog service should be implemented that defines a common interface available to any client. The XML is a good medium to develop metadata for the catalog services, so that the user request can be transferred into XML, and the search results can then be translated into XML again.

A catalog server usually works together with a *type registry* (OGC 2000, WMT 2), which is a component maintained by a registration authority that describes data element types. A registry contains definitions of element types, whereas a catalog contains descriptions of instances defined using the registered elements types.

5.4.3 Data Component

A data component is usually represented by a data server. A *data server* is a component that contains and delivers data across a common interface to any client. A client application can retrieve and even modify a set of database records on the data server. The database server responds to the client's request and provides secure access to shared data. A data server contains the whole data set but delivers part of the data to a map server or a client, typically as the result of a query, usually a SQL query. Therefore, a data server is usually called an SQL engine or SQL server.

5.4.3.1 SQL Database Servers Although SQL is an ISO standard, its implementation by commercial database vendors is varied. The SQL standard has evolved from the simple and "watered-down" version of SQL-89 (established in 1989) to SQL-92 (also called SQL2, ratified in 1992) to SQL99 (published in 1999). New features are added to each evolution. However, these features have not been fully implemented by commercial DBMS products. Some features in SQL-92 are not even fully implemented by some DBMS vendors (Orfali et al., 1999). It will take a while for all features in SQL99 to be implemented in commercial DBMS products.

Different vendors adopt different SQL database server architectures. GIS users need to understand the different architectures to be able to make a rational decision when choosing their own DBMS software to work with their GIS software. There are generally three kinds of database server architectures adopted by DBMS vendors: process-per-client, multithreaded, and hybrid architectures (Orfali et al., 1999).

The process-per-client architecture provides a separate process for each client request, as shown in Figure 5.11. This means, if there are 10 client requests, the database server has to open 10 processes to communicate with the clients. The advantage of this architecture is that every client has its own process so there is no interference among clients. A large request will not hold up other requests. The disadvantage is that it consumes more memory and CPU resources of the local operating system than the alternative architectures. The performance may be affected due to process content switches and interprocess communication overhead, but this problem can be overcome by using a TP monitor to manage a pool of reusable processes. Informix, DB2, and Oracle 6 are examples of database servers that implement this architecture.

The second architecture is the multithreaded architecture. Multithreading refers to the ability of a program or an operating system process to provide services to more than one user and/or more than one request at a time. It should be noted that *multithreading* is different from *multitasking*. Multitasking refers to the capability of performing more than one computer task at a time. For example when you open your Web browser and then open Word and a spreadsheet program at the same time, the operating system is doing multitasking. Multithreaded database server programs can respond to multiple client requests from the same user or multiple users without having to have multiple copies of the database server programs running in the computer. All client requests run in the same process, as shown in Figure 5.12. The program keeps track of each client request as a thread with a separate identity. This

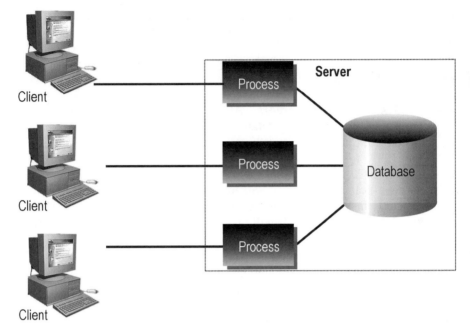

Figure 5.11 Process-per-client SQL database server. (*Source:* Orfali et al., 1999, Figure 10-2, reprinted with permission.)

architecture scheme has its performance advantage because by running all requests in the same process, there are no frequent content switches; therefore, it does not consume as much memory and CPU as does the process-per-client architecture. The disadvantage is that one client request may affect others, especially a very large request. Sybase and Microsoft SQL servers adopt this architecture scheme.

The third architecture is the hybrid architecture, which consists of a multithreaded element and a process-per-client element, as shown in Figure 5.13. Requests from multiple clients are handled by a multithreaded listener program that "listens" or receives client requests and assigns them to a dispatcher. The dispatcher places the client requests in a request/response queue. The database server then takes the requests from the queue, executes them in a shared server process pool, and returns the results back to the queue. The dispatcher dequeues the response from the queue and sends it back to the client. It is a hybrid because the listener program is multithreaded but the dispatcher and the server process are one process per client request—the processes are shared and reusable. Unlike the process-per-client architecture, this hybrid approach does not assign a permanent process to each client. But it does provide protection for each client request because each request is handled in a separate process. Therefore, it has the protection advantage of the process-per-client model and the performance advantage of the multi-

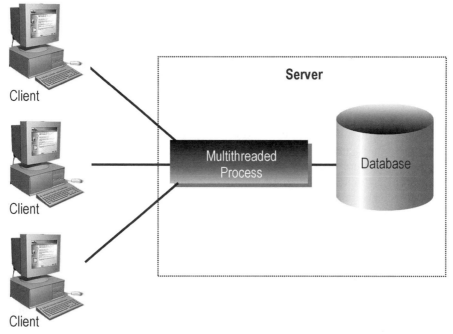

Figure 5.12   Multithreaded SQL database server. (*Source:* Orfali et al., 1999, Figure 10-3, reprinted with permission).

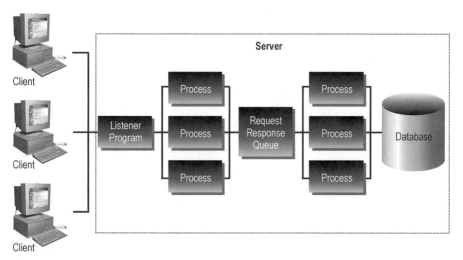

Figure 5.13   Hybrid database server architecture. (*Source:* Orfali et al., 1999, Figure 10-4, reprinted with permission.)

threaded approach by avoiding the overhead of contents switch. The downside of this approach could be the queue latency because it takes an extra step to queue and dequeue messages from the queue. Oracle 7 suffered from the queue latency issue, but that issue has been solved in Oracle 8i and Oracle 9i.

So which architecture is best for distributed GIS? It depends. If you have a lot of users accessing your site, the process-per-client architecture may not work well without a TP monitor due to performance limitations. If the user requests are many but short, the multithreaded architecture would work well. But this multithreaded architecture does not perform well with large and complicated queries and it does not provide bulletproof protection either. The hybrid architecture may provide the best balance. However, the differences in these architectures may be small when you use them to serve distributed GIS, especially when you do not anticipate a huge traffic to the server. They do matter if you want to use the database server for other purposes, such as creating a bulletproof On-Line Transaction Processing (OLTP) system.

5.4.3.2 Database Middleware A commercially available SQL server is a combination of standard SQL and vendor-specific extensions. Therefore, not all SQL servers or database servers are interoperable with each other. For example, if the data in your organization are stored in Oracle, Sybase, and Microsoft SQL servers and Informix, you need to have four different versions of SQL. To be able to access them from a single client, you need to have database middleware to mask their different SQL APIs and multiple database drivers as if they are the same database or, more accurately, federated databases.

The purpose of the database middleware is to create a standard or common SQL interface that is used by all applications so that the SQL server differences are handled by the different database drivers. There are many types of database middleware. We will discuss the three most common ones: ODBC, JDBC, and ADO.

5.4.3.2.1 Open Database Connectivity ODBC is a standard API for accessing different databases. It is used to access data in multiple database sources like Access, dBase, DB2, Excel, and Text. Applications usually gain access to a particular DBMS through a DBMS-specific SQL module or driver. Since there are many RDBMSs, there need to be many DBMS-specific SQL drivers. ODBC is similar to a SQL driver manager or dispatcher. It makes sure to plug in the correct driver to the proper database system. In more technical terms, ODBC is a SQL Call-Level Interface (SQL/CLI). CLI allows programs to submit SQL requests to different databases on the server without having to know the proprietary interfaces to the databases (Orfali et al., 1999). In other words, CLI translates calls from other programs such as a GIS program into the native database server's access language, the native SQL driver. The native SQL driver then accesses the data on the database server directly. The CLI then retrieves the results and sends them to the program that made the call.

ODBC is not a new query language. It is simply an intermediate level of procedural interface to SQL, or an SQL wrapper. ODBC adds another layer of overhead between applications and the database. Therefore, it is not nearly as fast as a native SQL driver. The ODBC specification is also controlled by Microsoft and is constantly evolving, which causes some interoperability and compatibility problems for implementation. Furthermore, ODBC is a procedural interface, not an object-oriented interface. Microsoft is now more committed to the object-oriented OLE DB; therefore, the ODBC could be replaced by OLE DB, another Microsoft-backed database access middleware.

5.4.3.2.2 Java Database Connectivity While ODBC is procedural, JDBC and OLE database ADO are object-oriented CLI. JDBC is an object-oriented API specification for connecting programs written in Java to the data in different databases. Similar to ODBC, JDBC has two API specifications, one for application interface to allow applications to access SQL services in a DBMS-independent manner. The other is a DBMS driver interface that is specific to different DBMS, the native SQL driver (Figure 5.14). Similar to ODBC or other CLIs, when an application makes a call to a database, JDBC uses a driver manager to load the proper JDBC driver to pass through the request statement to a proper native SQL driver for a specific database. It returns the results to the applications through a similar interface.

The applications access the database through JDBC in one of two ways: *direct access* or an *ODBC-bridged access*. A *direct JDBC driver* connects directly with DBMS' native SQL interface, while a *bridged JDBC driver* (JDBC to ODBC) is built on the top of an ODBC driver. Fortunately, since the structure of JDBC is similar to ODBC, the overhead of translating ODBC to JDBC is negligible, and performance is not affected much. One advantage of JDBC is the improved portability across multiple-vendor DBMSs. That is,

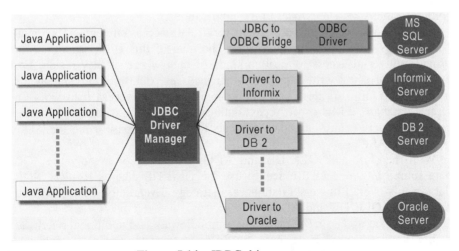

Figure 5.14 JDBC driver manager.

the JDBC driver can talk with many more databases without changing the code.

The other advantage of JDBC is the remote data access using the URI-based naming scheme. A SQL statement in the JDBC driver identifies the database as

```
jdbc:⟨subprotocol⟩⟨domain name⟩
```

For example, the following SQL statement is used to identify a database in a remote location via a JDBC-to-ODBC bridge:

```
jdbc:odbc://www.companyname.com/databasefile
```

where the subprotocol is odbc, the domain name is www.companyname.com, and the database name is databasefile. The subprotocol could also be a naming service that identifies a global name rather than a single domain name. This can take advantage of the flexibility of a networked name server and its relationship with the location of files and directories in the network.

5.4.3.2.3 OLE DB and ADO Similar to JDBC, OLE DB and ADO provide a way for applications to access data in a variety of formats in database servers. OLE DB is the low-level interface to databases. It defines a set of COM objects and interfaces for accessing data in different formats, relational or not, including SQL data, file data, spreadsheet data, directory data, and multidimensional data. ADO is an object-oriented programming interface that provides a data access API on top of OLE DB for applications. ADO is a CLI for linking calls from applications with the data from database servers. ADO allows the applications or data consumers to *connect* to a data source, performs some queries or *commands* on the data source, and returns the results of a query as a *recordset* to the applications.

OLE DB and ADO are part of an overall data access strategy—Universal Data Access (UDA)—from Microsoft. The goal of this UDA strategy is to provide universal access to various kinds of data wherever they reside. For example, you could write a program that includes ADO program statements in an HTML file to allow users of your Web site to access data from an Oracle database, a Microsoft Access database, or a spatial database. The user can make queries and make data requests from the database using the ADO code in an ASP.

With the shift with OLE DB and ADO, ODBC becomes only the SQL data source provider. ODBC has to use the OLE DB adapter to place SQL data in this overall universal data access strategy. Eventually, ODBC will be replaced by OLE DB and ADO.

It should be noted that these database middleware add additional overhead to link applications with databases in different formats. The performance is far slower than the native SQL within the database. Therefore, if you have only one database, using the native SQL will be much faster than using the ODBC, JDBC, or ADO.

5.5 DYNAMIC CONSTRUCTIONS OF DISTRIBUTED GISERVICES

The distributed GIS components and their locations in the client and server framework can be constructed dynamically. This section gives some examples of implementation.

5.5.1 Design of Dynamic GIServices Architecture

A simplified dynamic architecture for distributed GIServices is illustrated in Figure 5.15 using the Unified Modeling Language (UML). The UML nota-

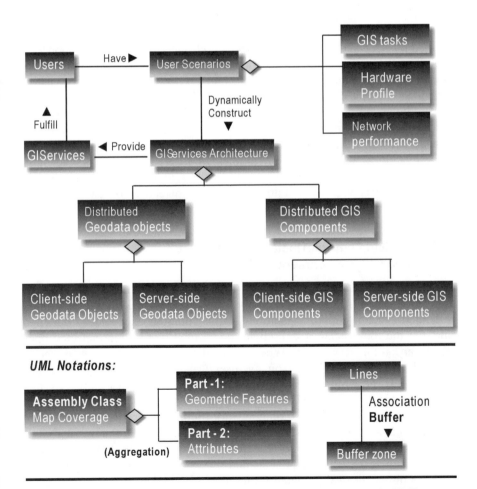

Figure 5.15   Dynamic architecture of distributed GIServices in UML.

tions use diamond shapes to represent the concept of aggregation. For example, the map coverage class is aggregated by geometric features and attributes. The line between two objects indicates the association relationship between them. The texts next to the line are association names and the arrows represent their directions. For example, lines can be buffered to create a buffer zone, which is the association between lines and buffer zone. And the arrow indicates that their association direction is from a line to a buffer zone.

Based on the UML notation Figure 5.15 identifies four major object classes in the deployment of distributed GIServices: [Users], [User Scenarios], [GIServices Architecture], and [GIServices].

There are four types of associations. [Users] have [User Scenarios]. [User Scenarios] will be used to dynamically construct [GIServices Architecture]. [GIServices Architecture] can provide [GIServices]. [GIServices] will fulfill [Users] needs.

Four types of aggregation relationships are illustrated in Figure 5.15. [User Scenarios] are composed of (or aggregated to) [GIS Tasks], [Client Machine Profiles], and [Network Performance]. [GIServices Architecture] is constructed by [Distributed Geodata Objects] and [Distributed GIS Components]. [Distributed Geodata Objects] consist of [Client-Side Geodata Objects] and [Server-Side Geodata Objects]. [Distributed GIS Components] consist of [Client-Side GIS Component] and [Server-Side GIS Component].

The dynamic construction is established by using the association between [User Scenarios] and [GIServices Architecture]. For example, different [Users] will have different [User Scenarios], which aggregate different GIS tasks, machines, and network environments. After users define the scenario, [GIService Architecture] will be constructed by combining GIS components and data objects located in either server-side machines or client-side machines.

5.5.2   Flexible Thin- or Thick-Client Model in Distributed GIS Environment

The distributed GIS model offers the flexibility of constructing a thin- or a thick-client model. In the terminology of networking, the thick-client model is defined as having major operations and calculations executed on the client side. On the other hand, a thin-client model may require that selected operations run on the server side. Whether the client-side GIS component should be thick or thin will depend on the user task and associated performance requirements. For example, it may be appropriate to use thick clients for map display services to let the GIS user take over many intuitive decisions on the graphic design, layout, and so on. Network routing or location modeling may be better off to run on the server side as complicated calculations and algorithms may be more efficiently handled by the server. The balance of functionality between client components and server components is a critical issue in the deployment of distributed GIServices.

Many research papers have argued about whether the thin-or thick-client model is the appropriate approach for distributed GIServices. This book proposes a flexible approach to solve the dilemma of the thin- or thick-client framework by dynamically rebuilding the GIS components according to different GIS tasks and network connections. The GIServices architecture is dynamically constructed by LEGO-like GIS components, based on different user scenarios, which include user tasks, client-machine profiles, networking performance, and so on. Such a dynamic approach for constructing distributed GIServices is illustrated in Figure 5.16. The arrangement of GIS components and geodata objects will be based on the criteria of the following user scenario. Different GIS tasks will require different types of GIServices architecture. For example, the GIS architecture for a map query task will focus on functions to access and query the GIS database. A Triangular Irregular Network (TIN) modeling task will require access to GIS analysis procedures.

GIS node (user machine or server machine) hardware profiles are the computer hardware specifications of the GIS nodes, including CPU speed, RAM, available hard disk spaces, and so on. Different types of GIS hardware require different design strategies in terms of thin- and thick-client models for distributed GIServices architecture.

Networking performance is the third essential factor for the deployment of GIServices architecture. Different bandwidth and connection types, such as Ethernet, ATM, DSL, and cable modem services require different types of design and thin- and thick-client deployment. A distributed GIService with a 56K bandwidth connection may require different thin- thick-client models than a 100-MB Ethernet connection.

The establishment of GIServices is collaborated by several GIS nodes, a group of network-based GIS workstations. Two kinds of elements are stored

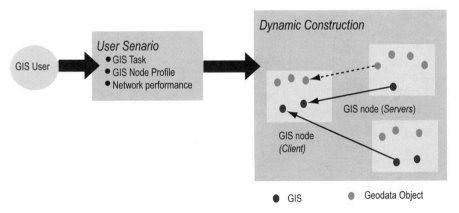

Figure 5.16 Dynamic construction of distributed GIServices by migrating and connecting geodata objects and GIS components.

in these GIS nodes in order to provide comprehensive GIServices: GIS components (programs) and geodata objects. The geodata objects and GIS components can be rearranged and linked dynamically among GIS nodes to establish dynamic, flexible, and distributed GIServices.

Figure 5.17 illustrates that the dynamic architecture of distributed GIServices is constructed instantly by migrating or connecting GIS components and geodata objects among several GIS nodes based on specific GIS tasks and scenarios. In this scenario, Mike needs to display the Colorado road map on his GIS node A machine. He will submit his GIS task to the local machine and the local GIS node A will create a dynamic service architecture by connecting or migrating GIS components and geodata sets, which are located on nodes B and C originally. After Mike's GIS task is completed, the data objects

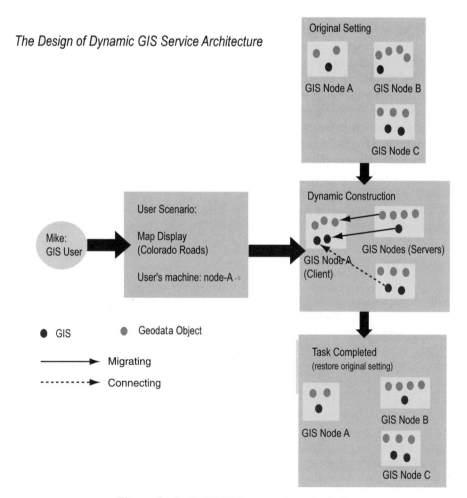

Figure 5.17 Build GIServices "on the fly."

and components are restored to their original places and wait for the next request.

5.5.3 Network Strategies for Constructing Dynamic GIServices

To construct a dynamic GIService, the user needs to either connect to or download GIS components (programs) and geodata objects from the other machine (other GIS nodes). Therefore, there are two basic strategies to accomplish dynamic construction of distributed GIServices–object migration and remote connection. There are also two basic elements to be connected or downloaded: GIS components (programs) and geodata objects.

Figure 5.18 shows two ways to construct data connection between the user's GIS node A and the remote GIS node B. The first method is the remote connection, that is, connecting the remote data objects without downloading the whole data set to the user's own machine. This is done by using SQL coupled with distributed database functions such as APIs of database connectivity and database connection middleware such as JDBC, ODBC, or OLE DB to establish the database connection between two systems.

SQL is an ISO standard for the manipulation, definition, and control of relational data. It is being used as a universal language to access, query, and

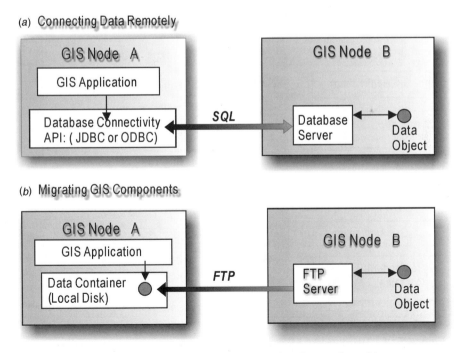

Figure 5.18 Two types of data connection for geodata objects.

manipulate all types of data in networked database servers. SQL consists of a short list of powerful and flexible commands to perform complex data operations. It can be used to manipulate, retrieve, and control sets of records from the data.

The second approach (data object migration) utilizes an FTP server to download a required data object to the targeted system and save it on the local disk or data container. Usually, the data migration approach may require both automated and manual procedures for the download and conversion of different data types into a single local GIS database. In the first case, a link is established between distributed databases; in the second case, a data object is actually transported via FTP.

Similarly, there are two approaches for accessing distributed GIS components dynamically (Figure 5.19). The first approach invokes GIS operators remotely by using distributed-component technologies and RPCs. Several technology frameworks are available for this approach, including DCOM, CORBA, and Java RMI. A GIS application will send a request to client-side component services to invoke a request for remote GIS components. The client-side service will use its client stub to build a connection with server-side server component skeleton. The server-side component services then invoke the required GIS components. The communication between the two systems uses RPC or other possible protocols, such as IIOP.

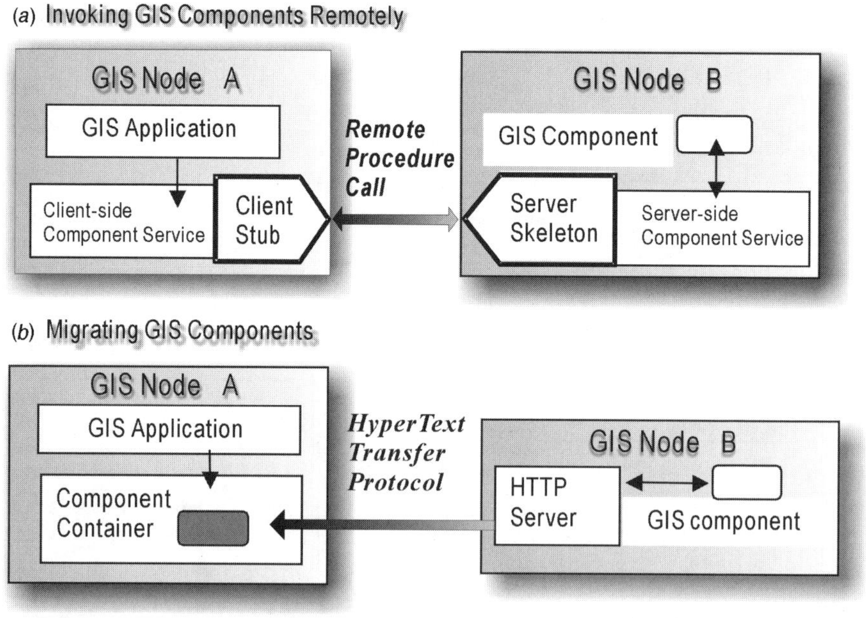

Figure 5.19 Two types of GIS component invocation for distributed GIServices.

Key Concepts

There are two basic strategies to accomplish dynamic construction of distributed GIServices: object migration and remote connection. **Remote connection** *is to connect components or objects remotely without downloading the whole software entities or data set to the targeted machine.* **Object migration** *utilizes an FTP or HTTP server to download requested components or data objects to the targeted system and save it on the local disks or data containers.*

The second approach is to actually move GIS components from one site to another. The migration process uses an HTTP server to download the required GIS components dynamically into the targeted GIS application. The downloaded GIS component is stored inside a component container, which binds with the local GIS application. This kind of approach is available from several technology frameworks, such as Java applets with a Java VM or ActiveX with an ActiveX container.

Most current Internet GIS services, such as ArcIMS by ESRI and Map-Guide by AutoDesk, have not yet utilized either strategy, due to the lack of a high-level integration framework. The following section will use two scenarios to explain how these technologies might operate in practice and to demonstrate the advantages of such a dynamic architecture for distributed GIServices.

5.5.3.1 Two Scenarios for Distributed GIS Component Access

Figure 5.20 illustrates two representative scenarios to demonstrate the advantages of dynamic GIS component access. In the first scenario, a GIS user needs to perform a road-buffering operation on a reasonably powerful PC (2.6-GHz Pentium 4 PC) with a reasonably fast network connection (10-MB Ethernet). Since the client machine has enough computing power to handle a buffering operation, the best solution is to download the GIS buffering component to the client side and to integrate it with other client-side GIS components. This is a thick-client model design. After the buffering component is downloaded, the client-side component services will initiate dynamic binding with the local GIS application and perform the buffering operation.

> *Scenario 1*   GIS task: buffering; client: 2.6-GHz PC; network: 10-MB Ethernet.
> *Solution*   Dynamic construction → a thick client.

> *Scenario 2*   GIS task: TIN modeling; client: 500-MHz PC; network: 56K modem.
> *Solution*   Dynamic construction → a thin client.

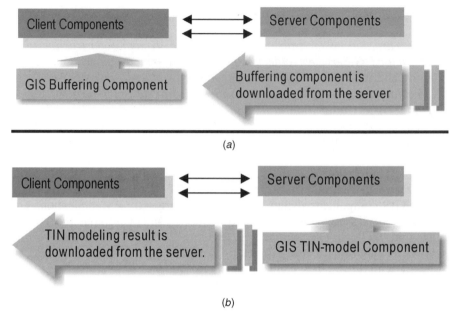

Figure 5.20   Two scenarios for GIS component access: (*a*) thin client and (*b*) thick client.

In the second scenario, a GIS user requests a TIN modeling function on a Digital Elevation Model (DEM) dataset using a less powerful PC (500-MHz PC) with a slow network connection (56K modem). Since the client machine is not powerful and the network performance is slow in downloading the TIN model component, the best solution in this case is to send a request for a TIN modeling procedure to the server machine and then perform the GIS task there. The result of the TIN modeling will be sent back to the client machine. This is a thin-client model because the major TIN modeling operations are executed on the server machine.

Under a dynamic distributed GIServices architecture, the thick- and thin-client models can be dynamically built or switched based on different scenarios. The dilemma of thin-or-thick architecture design for Web mapping no longer exists (Tsou and Buttenfield, 1998). Thus, the design of Internet GIS will be able to focus on the contents of GIS applications instead of the consideration of technical architecture.

Key Concepts

Under a dynamic distributed GIServices architecture, the thick- and thin-client models can be dynamically built or switched based on different scenarios.

In general, the requirements for dynamically migrating or connecting GIS components will include a decision-making process for choosing an appropriate architecture, a self-describing GIS component framework, and a comprehensive distributed-component service. Some distributed-component technologies, such as DCOM, CORBA, and Java can provide some low-level distributed-component services, such as object migration, global naming, life-cycle management, and object implementation. However, a dynamic GIServices architecture will also need high-level distributed-component services, including an object-oriented metadata scheme for a self-describing GIS component framework and an agent-based mechanism for decision-making processes. Besides the dynamic migration approach for GIS components, a dynamic GIServices architecture should also provide the migration capability for geodata objects. The next section will use two representative scenarios to describe how geodata object migration works in practice.

5.5.3.2 Two Scenarios for Distributed Geodata Object Access Figure 5.21 illustrates two representative scenarios for geodata object access. The two scenarios are similar to the previous examples, except that the data objects now reside on the server machine. In the first scenario, the GIS user needs to perform a road-buffering operation on a 2.6-GHz Pentium 4 PC with a 10-MB Ethernet connection. Since the [Colorado Road] data object is on the server, the best solution is to download the [Colorado Road] to the client-side

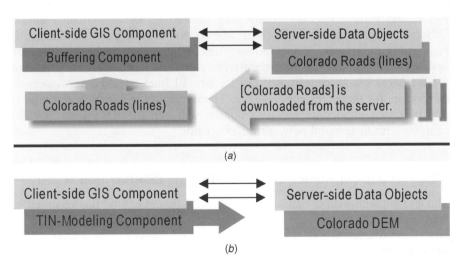

Figure 5.21 Two scenarios for geodata access: (*a*) data migration and (*b*) remote data access.

machine, then perform the buffering operation locally. This is called a data migration model.

Scenario 1   GIS task: Buffering [Colorado Roads]; Client: 2.6-GHz PC; Network: 10MB Ethernet

Solution   Data migration: Download the [Colorado Roads] to client machine.

Scenario 2   GIS task: TIN modeling; Client: 2.0-GHz PC; Network: 56K modem

Solution   Remote data access: Leaves the [Colorado DEM] data object on the server and remotely connects TIN component with the data server.

The second scenario is that a GIS user requests a TIN modeling function for Colorado DEM from a 2.0-GHz PC with a 56K modem connection. Since the [Colorado DEM] data object is huge and the available networking bandwidth (56K) is narrow, the possible solution is to leave the [Colorado DEM] data object on the server, then build the database connectivity remotely between the client-side [TIN-modeling component] and the [Colorado DEM] data object by using JDBC or OLE DB. This scenario is a remote data access model.

To summarize, the four representative scenarios mentioned above illustrate the advantages of dynamic GIServices architecture with LEGO-like GIS components and data objects. The flexibility of the distributed GIS component can provide customizable services for different users, heterogeneous platforms, and various network connections. Moreover, the design of dynamic GIServices architecture shifts from ad hoc, fixed systems to modularized, changeable component combinations. The architecture can be modified or updated later if users are facing different scenarios (e.g., users may change their network connection from ISDN lines to T1 lines or add a new category of services for their GIS applications). The dynamic combination and migration of GIS components and data objects will benefit GIS processing and analysis with distributed network environments.

5.5.4   An Implementation Example Using the CORBA/Java Model

The above-mentioned dynamic distributed GIS model can be implemented in many distributed-component model framework, particularly, the CORBA/Java model and Microsoft's COM+/.Net architecture. These two architectures are running parallel with each other; each is defining it own version of distributed computing. The distributed GIS will naturally follow these two architectural models. We will discuss only the CORBA/JAVA model as an example.

Internet GIS Showcase: The Minnesota DNR "Recreation Compass"

Minnesota is an outdoor playland of forests, parks, trails, and lakes. For travelers, locating the right spot to fish, camp, or hike can be a difficult process. To ease tourists' fact-finding efforts, the Minnesota Department of Natural Resources (DNR) developed Recreation Compass. Based on DNR GIS holdings and hosted on the University of Minnesota Mapserver, Recreation Compass provides more than 10,000 Web pages, reports, and maps. Maps and information are interlinked, allowing users to get information about items they see on the map or see a map of items they read about.

Recreation Compass allows users to search for places in three ways: by name (of a town, lake, or park), by Public Land Survey (using PLS codes such as t44r16s5), or by coordinate [Universal Transverse Mercator (UTM) or longitude/latitude]. MapServer is not a full-featured GIS system, but it does provide enough core functionality to support a wide variety of Web applications. Beyond browsing GIS data, MapServer allows users to create "geographic image maps" that can direct users to content.

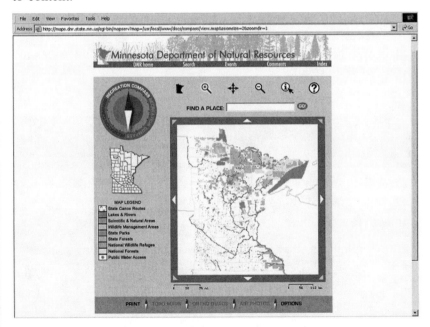

Image reprinted with permission from Minnesota DNR.

Source: http://www.dnr.state.mn.us/maps/compass.html.

Java applets running on the Web client allow the user to interact with spatial objects. But if the communication between the Java applet and the GIS server is still through CGI or CGI equivalents, Java applets are merely used to display fancy maps. However, if the Java applet communicates with objects in the GIS server through CORBA/ORBs, this would become a true distributed GIS.

Therefore, a distributed GIS in the CORBA/Java world is one in which the Java applets communicate with objects on the GIS server via object-to-object interaction methods such as ORB in a distributed-object infrastructure. ORB is an intercomponent communication middleware that establishes the client/server relationships between objects. Using an ORB, a client object on the Web browser can transparently and directly invoke a method on the server object on the same machine or different machines across the network. Therefore, ORBs can be used for intercomponent communication client to client, client to server, and server to server. ORBs run on top of IIOP, a common backbone protocol that is similar to TCP/IP with some CORBA-defined message exchanges. It is used to transport ORB requests from one ORB to the other.

Figure 5.22 shows a typical architectural model of a distributed GIS based on the CORBA/Java structure. This is the basic process of CORBA/ORB-based request and response:

1. Users on the Web browser request an HTML page that includes references to embedded Java applets.
2. The Web browser retrieves Java applets from the HTTP server in the form of byte codes.

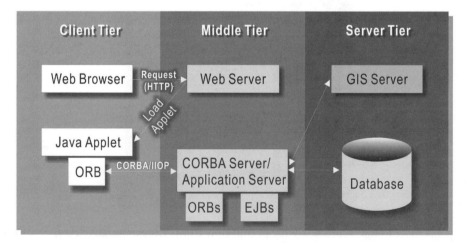

Figure 5.22   Three-tier architecture of distributed GIS using the CORBA/Java model.

3. The Web browser loads the applet and the Java VM interprets the Java byte codes. Java applets are then loaded into the memory of the client computer.

4. The Java applet communicates with the CORBA server by directly invoking the GIS objects on the server. This is done by using IDL in both the Java applet and server objects. IDL describes interfaces and structures of objects, including functions and parameters for client objects and server objects, so that the Java applets with its IDL-generated client stubs can directly invoke server objects. Once the connection or session between the Java applet and server objects [e.g., Enterprise Java Beans (EJBs)] is established, requests and responses can be exchanged until either the client or the server terminates the session.

It can be seen that this distributed GIS based on CORBA/Java does not rely on CGI or its extensions. Rather, it uses ORB-based object-to-object interactions between the client objects and server objects and allows the clients (applets or Java beans) to directly invoke object methods on a server. This is possible because CORBA allows the client objects to invoke any IDL-defined method on the server, and it allows the passage of any typed parameter from the client objects to the server, instead of just strings.

What does this mean for distributed GIS?

1. The direct object-to-object interactions mean very little overhead between a client and server, especially compared to the HTTP/CGI model. Every time an applet invokes a method in the server through a CGI, it has to restart a new instance of a program—the CORBA/Java model does not need to do this.

2. In the traditional CGI or its extension models, every client request has to invoke the CGI script. But the CORBA/Java model treats all server-side applications as regular CORBA objects that can be directly invoked from the client. The server-side applications are managed by an EJB container such as an Object Transaction Monitor (OTM) or Web application server or a map server.

3. CGI and its extensions do not maintain the state between client invocations, but CORBA does. This has important implications for distributed GIS, because now the client objects can directly invoke a wide variety of IDL-defined operations on the server. HTTP clients are restricted to a limited set of operations.

4. CORBA brings in a scalable server-to-server infrastructure to the distributed GIS. CORBA ORB provides a communication middleware to glue all server objects together. These server objects can be distributed in different servers and the ORB is used to provide load-balancing services for incoming requests. The ORB would allocate client requests to

the first available object and add more objects as the demand increases. This scalable distributed-server structure has a great advantage over the CGI applications because the CGI has to respond to all incoming requests itself and has no way to distribute the load across multiple processes or processors.

5. The CORBA/Java model also means smaller but distributed components in the Java applet on the client side. For example, instead of putting all functions on the Java applet, CORBA makes it easier to put all analysis functions like a buffer or spatial overlay as individual components and assemble them as needed inside the Java applet or beans. This makes the initial applet smaller and easier to download. This makes it possible for the client components to either access the components in the server or download the components from the server, as discussed in Section 5.2.

6. The internet-based CORBA/ORB provides a greater performance advantage over other communication methods. Based on Orfali and Harkey's (1998) client/server ping test, which measures the average response time of a Java bean client calling a server-side Java object, CORBA/IIOP has a much better performance than CGI and servlets (Table 5.1).

Table 5.1 shows that Sockets connection provides the best performance. But you cannot program directly to Sockets. So it can be used as a benchmark but cannot be used for performance comparison. The CORBA/IIOP and its counterparts DCOM and RMI/RMP perform much better than the HTTP/CGI and its extension servlets, which is about 15 times slower than the three ORB alternatives–CORBA/IIOP, RMI, and DCOM—but over 10 times faster than HTTP/CGI. HTTP/CGI is the slowest among all, about 200 times slower than the three ORB alternatives and 10 times slower than its recent extension servlets.

In summary, Java and CORBA/ORB provides revolutionary LEGO-like building blocks of distributed components for the development of distributed GIS. Java applets and Java beans provide a portable and interactive front-end component that can be directly connected with distributed objects on the server side. CORBA provides a distributed-object infrastructure over the Internet. ORBs provide a communication middleware between the Java client object and the server objects. But HTTP still plays its role in serving docu-

TABLE 5.1   Comparison of Client/Server Pings

Feature	Sockets	CORBA/IIOP	DCOM	RMI/RMP	Servlets	HTTP/CGI
Performance	★★★★	★★★	★★★	★★★	★	☆
Remote Pings	2.1 msec	3.5 msec	3.8 msec	3.3 msec	55.6 msec	827.9 msec

Source: Orfali et al. (1999), Table 28-2, p. 611, reprinted with permission.

ments and downloading HTML pages with Java applets and Java beans embedded. But once the Java applets or beans are downloaded into the client, ORBs will take over for the client/server object-to-object communications rather than relying on HTTP/CGI or its extensions.

The beauty of this CORBA/Java approach is that Java beans or CORBA beans created by multiple vendors can be assembled in both the server and the client, because they all comply with the CORBA ORB standard. CORBA and EJBs provide the architecture glue that ties these objects together so that objects from different vendors can be truly interoperable. With this distributed GIS architecture, many specialized GIS objects will potentially be created by different vendors, large or small. This can lead to a true plug-and-play GIS software program that is open and interoperable. Distributed GIS written in Java and based on CORBA offers almost all the benefits of locally executed GIS programs: responsiveness, capability to take advantage of local computing resources, graphic search, and so on. In addition, they can take advantage of computing resources from the entire global Internet. This distributed CORBA/Java model, as well as Microsoft's .Net model, holds great promise for distributed GIS, but it is up to the GIS vendors to implement it in their software design. Ultimately, it is user demand that will push the vendors to make this happen.

5.6 SUMMARY OF DISTRIBUTED GIS ARCHITECTURE

In summary, the architecture of distributed GIS has evolved from static map publishing to static Web mapping to interactive Web mapping to distributed-component GIS. But they all follow the same three-tier client/server architecture that consists of client, middleware services or application servers, and servers that include Web servers and map servers.

The client evolves from a simple HTML viewer to a dynamic HTML viewer to client-side applications such as plug-ins, Java applets, and ActiveX controls to downloadable mobile client agents such as Java beans. The interactivity and portability have been improved in this evolution process.

The middleware services or application servers progress from the CGI to CGI extensions such as ISAPI, NSAPI, servlets, ASP, and ColdFusion to CORBA/IIOP or COM+/.NET frameworks. Each progression represents an improvement in performance, scalability, and interoperability. The role of middleware or application servers also changes from linking the clients with a map server to providing services to map servers such as load balancing, transaction monitoring, and security services.

The map server has been changed from a legacy GIS program to distributed objects and components. Each function becomes a server component that can be directly invoked by the client component through an IDL or downloaded to the client.

The development of distributed GIS can also be represented by the partitioning of the client/server systems. Depending on the location of different components (presentation, logic, and data), distributed GIS can be partitioned into different categories. There are basically two types of architecture models of distributed GIS. The first architecture model categorizes distributed GIS as "thin," "medium," and "thick" clients according to the partitions of presentation, business logic, and data. If there are no data processing or logic elements such as map-rendering services on the client side and only the map images are displayed on the client, it is a thin client. If the data are streamed to the client and all map generation and rendering functions are located at the client side, the system is a thick client. If there are some map-rendering functions but no data on the client side, it is a medium client.

In this book we categorize distributed GIS based on the partitions of the distributed GIS applications, that is, on the location of different components on different computers (e.g., clients and servers). Following the tradition of system partitioning as discussed in Chapter 3, we consider presentation, logic, and data as a three-part application continuum. The partition point of a client/server system can occur in any part of the application.

We first considered the partition point in the presentation. In this case, the client is responsible for only the presentation portion of distributed GIS. All logic functions and services and data are maintained in the server. Only the rendered map is carried over the Internet to the Web client as a GIF, JPEG, or other graphic format. The map images are displayed in the HTML pages on a Web browser. The client is implemented in HTML/DHTML, and there is no client-side application such as ActiveX controls or Java applets. Users can have very limited interaction with the map images through the HTML-based client viewer. The interaction is based on a request–response scenario. The user makes a request, for example, to zoom in to the map by clicking on the Zoom link. That request goes to the web server. A communication middleware service sends the request to the map server to process the request. The updated map is returned to the client's browser embedded in the HTML page. There are virtually no client-side capabilities. Every single user request has to be transferred and processed at the map server through the Internet. This is commonly referred to as a thin-client case.

Next we consider the partition point in the logic part of the application. In this case, the partition point could be anywhere inside the logic element based on the amount of logic elements on the client side. On the one extreme, only the map-rendering service or the portrayal services are located in the client; on the other extreme, most logic elements are located at the client. In between, there would be many combinations of sharing the logic elements between the server and the client viewer. We consider several scenarios below.

In the first scenario, maps are served in the map server and presented as a map image to the Web browser. But unlike the thin case discussed above, the client has limited rendering capability, such as panning, zooming, spatial feature selection, and identification. The rendering process works with the map images and does not require access to the underlying geospatial data source. The client can be constructed in DHTML, plug-in, Java applet, or ActiveX controls.

In the second scenario, map display (graphic) elements instead of map images are transmitted to the client, so that maps can be constructed from the display elements by the portrayal service in the Web browser. Those map display elements include geographic features (points, lines, and polygons) and cartographic features or annotations (fonts, color, line and polygon styles, etc.). The map is typically already in a projected reference system. For example, a freeway might be a thick red polyline, a lake could be a blue polygon, and so on. Some graphic element formats include SVG and WebCGM, ActiveCGM from Intergraph/InterCAP, and the Drawing Web Format (DWF) from Autodesk. Since these display elements are transferred to the Web client, the user with the portrayal services at the Web client has more client-side rendering capabilities than in the first scenario. For example, the user can change colors, symbols, or styles of a geographic feature, a font of an annotation text, or a title of the map. But there is no vector data transferred to the Web client side.

In the third scenario, the geospatial data (including vector and raster data) are transferred to the Web client over the Internet, along with map-rendering or portrayal service and other analysis services. Users at the Web client interact directly with vector geospatial data as well as map-rendering function and other spatial analysis functions through a client. A data-streaming technology can stream vector data directly from the data server to the Web client. That is, the map server does not generate maps at the server; it is simply to make the initial connection between the Web client and the data server by distributing distributed components like Java applets or ActiveX controls to the client. The Java applets or ActiveX controls connect directly with the feature data server through database middleware such as ODBC or JDBC or through GML. A client would have the same or more functions as provided by the desktop GIS programs. The partitioning point in this scenario is extended to the data element, where parts of the data are transferred to the client side.

For interoperable distributed GIS, more services can be added to the client side. For example, some of the map server service that changes projection of geospatial data can be shifted from the map server to the client. Therefore, the client can request geospatial data with different projections and reproject them once they are received at the client, so that all data can be displayed in the single client.

It can be seen that the terms *thin, medium,* and *thick* client are relative. Identifying each scenario as a *thin, medium,* and *thick* client could be confusing. Furthermore, definitions of *thin* and *thick* client could be very imprecise, which makes it even more difficult to define and differentiate a very thin client, a thin client, a medium client, a thick client, or a very thick client. Any system could be a thin or thick client depending on the basis for comparison.

Obviously, there is flexibility to partition and construct the distributed GIS at any point in the client/server continuum, from a very thin client to a very thick client. A thin or thick client has its advantages and disadvantages, depending on the purpose of the system and the audience it serves.

Thin-client architecture that put all logic elements on the server side has no requirement and limitation on the client's computer. Any computer with a Web browser will be able to access and utilize distributed GIS services. A thin-client system is also easy to maintain, service, and update, but a thin-client system offers little flexibility, interactivity and responsiveness between the user and the maps and geospatial data.

Thick-client architecture that puts many logic elements on the client side, on the other hand, offers much more flexibility and interactivity in map rendering, manipulation, and analysis. The client has the potential to integrate the local data with remote data. Current GIS users will feel no difference in interacting with distributed GIS from using desktop GIS. But thick-client systems have limitations on the kind of operation systems in the user's computer. For better performance it also requires a powerful client machine and a high-speed network connection since the initial download of data and mobile clients such as Java applets or ActiveX controls takes considerable time for a slow network connection. The required downloading of client-side applications may even discourage occasional users. But as the network speeds increase over time, this concern should dissipate.

There are different technologies available to build distributed GIS, such as the CORB/Java model and the COM+/.Net model. There may be more emerging in the future. How do they interact with each other in the Internet environment? Interoperability becomes a major problem when proprietary distributed GIS programs are developed. In fact, one of the major problems of the current Internet-based mapping systems is that they do not interoperate with each other and they are not all made to be embeddable into other applications. Furthermore, as more vendors are getting into the market of Internet GIS and more distributed technologies are evolving, there is an even greater possibility of noninteroperability. Therefore, it is necessary to standardize Internet GIS programs and distributed GIServices.

WEB RESOURCES

Descriptions	URI
OGC implementation specification	http://www.opengis.org/techno/implementation.htm
ISO/TC 211	http://www.isotc211.org/
CORBA	http://www.corba.org/
Java	http://java.sun.com/
COM+	http://www.microsoft.com/com/tech/COMPlus.asp

REFERENCES

Buehler, K., and Mckee, L. (Eds.). (1998). *The OpenGIS® Guide: Introduction to Interoperable Geoprocessing and the OpenGIS Specification,* 3rd ed. Wayland, Massachusetts: Open GIS Consortium. URI: http://www.opengis.org/techno/guide.htm, May 11, 2000.

Cook, S., and Daniels, J. (1994). *Designing Object Systems: Object-Oriented Modelling with Syntropy.* Englewood Cliffs, New Jersey: Prentice-Hall.

Dangermond, J. (2001). g.net—A New GIS Architecture for Geographic Information Services, *ArcNews.* URI: http://www.esri.com/news/arcnews/spring01articles/gnet.html.

ISO/TC 211 Chairman. (1998). *Draft Agreement between Open GIS Consortium, Inc. and ISO/TC 211.* ISO/TC 211-N563.

ISO/TC 211 Secretariat. (2000). *Program of Work,* Version 8. ISO/TC 211 N 854. URL: http://www.statkart.no/isotc211/dokreg09.htm, May 11, 2000.

ISO/TC 211/WG 1. (1998a). *Geographic Information–Part 1: Reference Model.* ISO/TC 211-N623, ISO/CD 15046-1.2.

ISO/TC 211/WG 1. (1998b). *Geographic Information–Part 2: Overview.* ISO/TC 211-N541, ISO/CD 15046-2.

ISO/TC 211/WG 4. (2000). *CD 19119 Geographic Information–Services (Draft).* ISO/TC 211-N906,

Kuhn, W. (1997). *Toward Implemented Geoprocessing Standards: Converging Standardization Tracks for ISO/TC 211 and OGC,* White Paper. ISO/TC 211-N418.

Open GIS Consortium (OGC). (1997a). *OpenGIS Simple Features Specification for OLE/COM,* Revision 0. Wayland, Massachusetts: Open GIS Consortium.

Open GIS Consortium (OGC). (1997b). *OpenGIS Simple Features Specification for CORBA,* Revision 0. Wayland, Massachusetts: Open GIS Consortium.

Open GIS Consortium (OGC). (1997c). *OpenGIS Simple Features Specification for SQL,* Revision 0. Wayland, Massachusetts: Open GIS Consortium.

Open GIS Consortium (OGC). (1998). *The OpenGIS Abstract Specification,* Version 3. Wayland, Massachusetts: Open GIS Consortium. URI: http://www.opengis.org/techno/specs.htm, May 11, 2000.

Open GIS Consortium (OGC). (2000). *OpenGIS Web Map Server Interfaces Implementation Specification,* Revision 1.0.0. Wayland, Massachusetts: Open GIS Consortium. URI: http://www.opengis.org/techno/specs.htm, January 11, 2001.

Open GIS Consortium (OGC). (2001a). *OpenGIS Web Map Server Interfaces Implementation Specification,* Revision 1.1.0. Wayland, Massachusetts: Open GIS Consortium. URI: http://www.opengis.org/techno/specs.htm, September 11, 2001.

Open GIS Consortium (OGC). (2001b). *OpenGIS Geography Markup Language (GML) Implementation Specification,* Version 2.0. Wayland, Massachusetts: Open GIS Consortium. URI: http://www.opengis.org/techno/specs.htm, January 11, 2001.

Orfali, R., and Harkey, D. (1998). *Client/Server Programming with Java and CORBA,* 2nd ed., New York: Wiley.

Orfali, R., Harkey, D., and Edwards, J. (1996). *The Essential Distributed Objects Survival Guide.* New York: Wiley.

Orfali, R., Harkey, D., and Edwards, J. (1999). *Client/Server Survival Guide*. New York: Wiley.

Pountain, D. (1997). The Component Enterprise. *BYTE,* May 1997, 22(5), pp. 93–98.

Rowley, J. (1998). *Draft Business Case for the Harmonisation between ISO/TC 211 and Open GIS Consortium, Inc.,* Resolution 47. ISO/TC 211-N472.

Tsou, M. H., and Buttenfield, B. P. (1998). Client/Server Components and Metadata Objects for Distributed Geographic Information Services. In *Proceedings of the GIS/LIS '98,* Fort Worth, Texas. Bethesda, Maryland: American Society for Photogrammistry and Remote Sensing, pp. 590–599.

CHAPTER 6

STANDARDS FOR DISTRIBUTED GISERVICES

Data sharing make sense for the simple reason that there is only one Earth,
and we share it.
—K. Buehler and L. McKee (1998, p. 2)

Two major organizations that set industry standards for distributed GIServices are OGC and the Technical Committee tasked by the ISO (ISO/TC 211), both founded in 1994 (Buehler and Mckee, 1998; Rowley, 1998). The main goals of OGC are the full integration of geospatial data and geoprocessing resources into mainstream computing and the widespread use of interoperable geoprocessing software and geodata products throughout the information infrastructure (OGC, 1998). ISO/TC 211 emphasizes a service-oriented view of geoprocessing technology and a balanced concern for information, application, and systems (Kuhn, 1997). The following sections will briefly introduce the two organizations and their Internet GIS and distributed GIServices standards as well as metadata standards.

6.1 OPENGIS SPECIFICATION

In 1993, the Open GRASS Foundation (OGF) (the progenitor of OGC) proposed a comprehensive software architecture called the Open Geodata Interoperability Specification (OGIS), which supported distributed geoprocessing and geodata interoperability in distributed-network environments. The OGIS project successfully obtained support from federal agencies and commercial organizations. Later on, OGF reorganized and became the Open GIS Consortium (OGC) by adding more members from both industry and academia in 1994. OGC also renamed the specification from OGIS to the OpenGIS specification. The new OpenGIS specification defines a comprehensive software framework for distributed access to geodata and geoprocessing resources. The OpenGIS specification includes an abstract specification and a series of implementation specifications for various DCPs, such as CORBA, OLE/COM, SQL, and Java. Software developers use OpenGIS conformant interfaces to build distributed GIServices, which include middleware, componentware, and applications. These distributed GIServices will be able to handle a full range of geodata types and geoprocessing functions: "The OpenGIS Specification provides a framework for software developers to create software that enables their users to access and process geographic data from a variety of sources across a generic computing interface within an open information technology foundation" (Buehler and McKee, 1998, p. 7).

The contents of the OpenGIS specification (both the abstract and implementation specifications) are based on three conceptual models. The *OGM* provides a common data model using object-based and/or conventional programming methods. *OpenGIS services* define the set of services needed to access and process geodata defined in the OGM and provide the capabilities to share geodata with the GIS community. An *information communities model* employs the OGM and OpenGIS services in a scheme for automated translation between different geographic feature lexicons. Together, this establishes communication mechanisms among different communities of geodata producers and users (Buehler and McKee, 1998). In general terms, OGM provides a common means for digitally representing Earth and Earth phenomena, mathematically and conceptually. OpenGIS services implement geodata access, management, manipulation, representation, and sharing between information communities. The information communities model can facilitate the collaborations among different GIS research domains and applications (Buehler and McKee, 1998).

Key Concepts

The **OpenGIS specification** *includes an abstract specification and a series of implementation specifications for various DCPs. The* **OpenGIS Abstract Specification** *presents a high-level abstraction defining characteristics for designing geospatial data models and services.* **OpenGIS Implementation specifications** *focus on the actual software specifications which adopt the conceptual models proposed in the OpenGIS abstract specification.*

These three conceptual models of the OpenGIS specification are formalized in the contents of the abstract specification and the implementation specification. The two types of specifications focus on the establishment of high-level software design and technology-centered implementation. The following section will introduce the two specifications and demonstrate the OpenGIS standard in practice.

6.1.1 OpenGIS Abstract Specification

OpenGIS Abstract Specification (OGC, 1998) presents a high-level abstraction defining characteristics for designing geospatial data models and services. *OpenGIS Abstract Specification* also demonstrates the two central technology themes of OGC, which are sharing geospatial information and providing geospatial services (OGC, 1998). The contents of *OpenGIS Abstract Specification* are evolving along with the progress of information technologies. *OpenGIS Abstract Specification* embraces 16 different GIS topics (Table 6.1).

TABLE 6.1   **Contents of** *OpenGIS Abstract Specification* **(Version 4, 1999)**

Number	Topic	Technology Goals
Topic 0	Abstract specification overview	
Topic 1	Feature geometry	
Topic 2	Spatial reference systems	Prerequisites for sharing geospatial
Topic 3	Locational geometry	information
Topic 4	Stored functions and interpolation	
Topic 5	OpenGIS feature and feature collections	
Topic 6	Coverage type	Sharing geospatial information
Topic 7	Earth imagery	
Topic 8	Relations between features	Prerequisites for sharing geospatial information
Topic 9	Quality	Providing information-theoretic
Topic 10	Transfer Technology	content for information communities
Topic 11	Metadata	Prerequisites for sharing geospatial information
Topic 12	OpenGIS service architecture	Providing geospatial services
Topic 13	Catalog services	
Topic 14	Semantic and information community	Providing information-theoretic content for information communities
Topic 15	Image exploitation services (new topic added in 1999)	Supporting image exploitation, including precision measurement of ground positions and object dimensions
Topic 16	Image coordinate transformation services (new topic added in 1999)	Describing services for transforming image position coordinates, to and from ground position coordinates.

OpenGIS Abstract Specification is modified and edited at each OGC technical committee meeting (roughly twice per year). The development of the abstract specification is based on an *essential model* for geographic information representation and processing (Buehler and McKee, 1998). The essential model is an object-oriented approach developed by Buehler and McKee. OGC created the essential model as the conceptual guideline of *OpenGIS Abstract Specification*. The essential model proposed by OGC contains nine levels of abstraction (Buehler and McKee, 1998, pp. 38, 41):

1. *Real World* This is the world as it is. The real world means the collection of all facts, whether they are known by mankind or not.
2. *Conceptual World* This is the world of things people have noticed and named. The method by which the conceptual world interfaces to the real world is the extraction of the essence of a fact.
3. *Geospatial World* This is the cartoonlike world of maps and GIS, in which specific things in the conceptual world are selected to represent the real world in an abstract and symbolic way using maps and geodata.
4. *Dimensional World* This is the geospatial world after it has been measured for geometric and positional accuracy.
5. *Project World* This is a selected piece of the dimensional world (e.g., certain thematic layers in a GIS) which are structured semantically for a particular purpose, profession, discipline, or industry domain.
6. *OpenGIS Points* How points are defined, either generically or for a particular project world, in a way that all software systems can relate.
7. *OpenGIS Geometry* How geometry is constructed based on OpenGIS points in a way to which all software systems can relate.
8. *OpenGIS Features* How features are constructed from geometry, attributes, and a spatial referencing system in a way that lends itself to use in open interfaces for geoprocessing. OpenGIS features are digitally coded abstractions of real-world entities that have a geometric representation, and spatial, temporal, and other properties.
9. *OpenGIS Feature Collections* A feature collection is the unit of trade in a geoinformation-sharing transaction and the primary object of manipulation within a geospatial software processing environment. OpenGIS feature collections can be of any size and content depending on the context of the transaction.

The first five levels of the OGC essential model are based on Cook and Daniel's theory (1994), which deals with the abstraction of real-world facts, and are not modeled in software. The final four levels of the OGC essential model deal with mathematical and symbolic models of the world and thus are subject to being modeled in software (Buehler and McKee, 1998).

One disadvantage of the essential model proposed by the OGC is that the model only focuses on the design of generic data models without considering different GIS operations and distributed processes via the networks. In fact, network-based GIS processes may require different types of data models. If geospatial information is distributed across the Internet, the design of the data model will need to emphasize not only the integration of heterogeneous data formats but also the adoption of distributed-computing technologies in order to facilitate distributed GIS operations.

6.1.2 OpenGIS Implementation Specifications

OpenGIS implementation specifications focus on the actual software specifications which adopt the conceptual models proposed in *OpenGIS Abstract Specification.* The implementation specifications give explicit instructions for interoperability with other OpenGIS specification-conformant software written by other developers around the world. Application developers or software programmers are the primary users of the OpenGIS implementation specifications, which define explicit APIs or languages for accessing geodata and geoprocessing functions (Buehler and McKee, 1998).

In contrast to *OpenGIS Abstract Specification,* which is created by the OGC technical committee, implementation specifications are created by GIS software vendors, based on specific DCPs or languages such as XML. As parts of the abstract specification are completed, the OGC invites proposed implementation specifications from software vendors (Buehler and McKee, 1998). Vendors submit DCP-specific implementation specifications (such as OLE/COM, CORBA, Java, or SQL), which are reviewed by the technical committee. Once accepted, the OpenGIS implementation specifications specify in DCP-specific terms the functionality of particular OpenGIS interfaces and services.

There are several OGC implementation specifications available. For example, three implementation specifications (for CORBA, OLE/COM, and SQL) were released for simple features in March 1997 (OGC, 1997a–c). Simple features specifications implement topic 8 in *OpenGIS Abstract Specification.* These simple feature specifications provide interfaces to allow GIS software engineers to develop applications that comprise the definition of OpenGIS features and geometry (from the essential model) using several DCP-specific technologies. These standardized specifications will facilitate horizontal software integration (with graphic user interfaces, database connectivity, or task management) and vertical software integration (with different GIS software vendors and packages under the same DCP platform) (Buehler and McKee, 1998).

Another two important implementation specifications are the OpenGIS WMS interfaces implementation specification and the OpenGIS GML implementation specification.

The OpenGIS WMS interfaces implementation specification was first described as "OGC WWW Mapping Framework" in 1997. Then the WMT was proposed by the OGC WWW Mapping Special Interest Group. This group created the first version of the WMS interfaces implementation specification

(1.0.0) in 2000. That first version supported basic interoperability of simple map servers and clients but did not fully address access to simple features, coverages, data with temporal or other dimensions, and other types of geoprocessing services (OGC, 2001a). In 2001, the OGC Special Interest Group submitted the second version of the WMS (1.1.0). The new specification described many of the follow-up WMT phase 2 and the Geospatial Fusion Services Testbed. Several software vendors, such as ESRI, INTEGRAPH, and AutoDesk, began to develop OGC WMS-compliant IMS based on the WMS specifications.

The GML is an XML encoding scheme proposed by the OGC for the transport and storage of geographic information (OGC, 2001b). The GML version 1.0 was developed by OGC in 2000 and specified the XML format for both the spatial and nonspatial properties of geographic features. The original GML document is an OGC recommendation paper. But the second version of GML (2.0) became an implementation specification, which defines the XML schema syntax, mechanisms, and conventions that implementers may decide to store geographic applicatior schemas and information in GML. The GML models are based on the OGC abstract specification and the simple features mentioned above. The details of GML and its applications will be discussed in the next chapter.

6.1.3 OpenGIS Standard in Practice

As we mentioned before, the OpenGIS specification can be included in three conceptual models, the geodata model, service, and the information community model. This section will elaborate these models with actual OGC specifications.

6.1.3.1 Geodata Model The concepts of the geodata model can be found in the first topic of *OpenGIS Abstract Specification:* feature geometry. Two fundamental geographic types are recognized in this OpenGIS specification, called *features* and *coverages.* Both features and coverages can be used to map real-world elements or phenomena to an OpenGIS specification representation, which ultimately provides a common way to describe these elements in software design (Buehler and McKee, 1998).

A feature is a representation of a real-world element or an abstraction of the real world. It includes a spatial domain, a temporal domain, or a spatial/temporal domain as an attribute. Features are usually managed in groups as feature collections. A GIS thematic map layer that shows roads, for example, is a collection of features. Features, as the basic elements of geospatial information, include geometry, semantic properties, and metadata (Buehler and McKee, 1998; Gardels, 1996).

A coverage is an association of points within a spatial/temporal domain: "A coverage in the OpenGIS Specification is simply a function which can return its value at a geometric point. Scalar fields (such as temperature distribution), terrain models, population distributions, satellite images and digital aerial photographs, bathymetry data, gravitometric surveys, and soil maps can all be regarded as coverages" (Buehler and McKee, 1998, p. 42). In a coverage, a data value is associated with every location. Since coverages have all of the characteristics of features, they become a subtype of features (Buehler and McKee, 1998).

6.1.3.2 OpenGIS Services Architecture The second key concept is the OpenGIS services architecture, which creates feature collections, shares features and project schema, shares metadata, and discovers data through catalogs, traders, and standard imaging functions (Buehler and McKee, 1998). The architecture for these services basically follows the ISO Reference Model for Open Distributed Processing (RM-ODP). The specification for OpenGIS services architecture is called the OpenGIS technical reference model. The reference model includes definitions of interfaces and its behaviors that may be supported. The same services may be supported by multiple, Well-Known geospatial data Types (WKTs) (Buehler and McKee, 1998). Following the design of ISO RM-ODP, OpenGIS services include applications, shared-domain services, common facilities, distributed computing and object services, platform services, and external entities (OGC, 1998, topic 12). These services interoperate with each other according to the architecture shown in Figure 6.1.

Figure 6.1 illustrates five essential layers for the OpenGIS service architecture. The application layer contains custom-built computer programs that

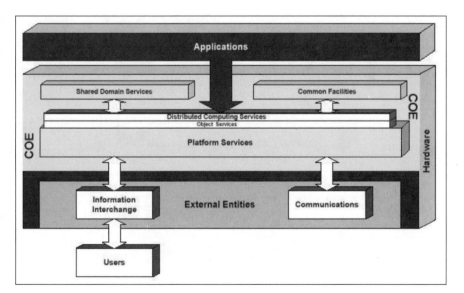

Figure 6.1 OpenGIS technical reference model. (Figure reprinted with permission from OGC, 1998, topic 12.)

allow users to perform specific tasks (e.g., a buffering program or a GPS extension). The shared-domain service layer contains computer programs that are specific to a single information domain (e.g., transportation, health care, geospatial domain). Common facilities are computer programs that provide general support across multiple domains (e.g., spreadsheets and word processors). Platform services and external entities are the software that communicate between actual hardware and operating systems and the other devices in the platforms, such as databases and modems (OGC, 1998, topic 12).

Figure 6.2 is a detail view of the shared-domain services, which contain geospatial domain services. OGC's distributed GIServices architecture is built upon the geospatial domain services: "Geospatial Domain . . . [Services] will be defined by the Open GIS Consortium to ensure that the Open GIS Services Architecture can be realized with standards-based, Commercial-Off-The-Shelf (COTS) products available from multiple vendors" (OGC, 1998, topic 12, p. 9). The intention is that, once complete, the OpenGIS standards will allow different GIS components to work together under an integrated framework. For example, a feature generalization routine written by one vendor can interoperate with a coordinate transformation routine written by another vendor, and both are transparent to the user.

Several problems need to be addressed here based on the current specification of the OpenGIS service architecture. First, the classification of geospatial domain services is quite arbitrary and ambiguous without a high-level integration. Some contents overlap with others. For example, some contents of geospatial domain access services are duplicated with feature manipulation

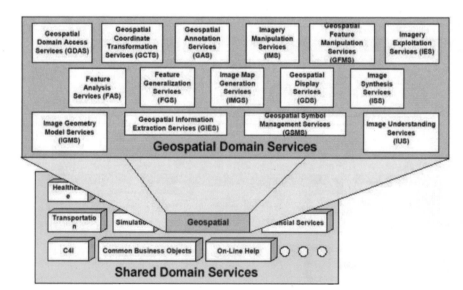

Figure 6.2 Geospatial domain services. (Figure reprinted with permission from OGC, 1998, topic 12.)

services and geospatial display services. Many similar types of services sharing the same GIS functions should be included under a single service domain, such as the integration of geospatial annotation services and geospatial symbol management services. Second, the separation between geospatial data sets and remote sensing images will cause a serious problem for the integration of true geospatial information services because many GIServices will require both types of data sets at the same time. Third, the OpenGIS service architecture does not specify any approach for dynamic binding of these geospatial domain services or business objects. The specification should address how to combine these GIServices for different applications across the Internet.

The latest version of the OpenGIS service architecture adopts ISO/DIS 19119, geographic information services. This new version is still based on ISO RM-ODP but also adopts the ISO Extended Open Systems Environment (OSE) for geographic services. OSE defines the following six classes of information technology services:

- human interaction services,
- model/information management services,
- workflow/task services,
- processing services,
- communication services, and
- system management services.

These six types of ISO information services will be discussed in detail later on.

6.1.3.3 Information Communities Model A third component of the OpenGIS standard in practice is the OGC information communities model. The model provides for automated translation between different geographic feature lexicons and establishes communications among different communities of geodata producers and users. The purpose of the information communities model is to share information between geospatial databases with inconsistent definitions and help communication between different communities that describe geographic features in different ways (Buehler and McKee, 1998).

An information community is a collection of people (a government agency or group of agencies, a profession, a group of researchers in the same discipline, corporate partners cooperating on a project, etc.) who, at least part of the time, share a common digital geographic information language and share common spatial feature definitions. This implies a common world view as well as common abstractions, feature representations, and metadata. The feature collections and geoprocessing functions that conform to the information community's standard language, definitions, behaviors, and representations belong to that information community (Buehler and McKee, 1998, p. 53).

Although the OpenGIS information communities model has not been fully developed, there are some interesting perspectives on its preliminary design. One is that intercommunity sharing of geodata will be achieved through software "semantic translators." A semantic translator can be used by an information community to filter its view (in the database sense) of the data in another information community: "The Semantic Translator will contain all of the information it needs to find and translate feature collections from the source to the target semantics" (Buehler and McKee, 1998, p. 55). Another interesting concept, called a "trader," will help an information community determine what information will be exposed and with whom it will be shared. However, current OGC documents do not specify any actual implementation approaches and functions of semantic translators and traders. Implementation details are yet to be developed.

To summarize, the primary emphasis of OGC specifications is on the interoperability of the geospatial data model instead of distributed processing. However, fully distributed GIServices cannot happen without open standards and communication mechanisms for distributed processes. The OGC's standard-building effort illustrates that the success of establishing comprehensive distributed GIServices requires collaboration among GIS community members, including the other major player, the ISO/TC 211, which sets distributed GIServices standards for the international community. The next section will introduce the ISO/TC 211 and related ISO standards.

6.2 THE ISO/TC 211 AND ITS ISO STANDARDS

ISO/TC 211 is the Technical Committee of Geographic Information/Geomatics tasked by the ISO to prepare a family of geographic information standards in cooperation with other ISO technical committees preparing related information technology standards. ISO is a worldwide federation of national standards bodies (ISO member bodies) including many different stakeholders from governments, authorities, and industry and professional organizations. ISO/TC 211 is working on a series of international standards (from ISO 19101 to 19135) that focus on geographic information. The ISO geographic information standard (formally known as the ISO 15046 standard) specifies methods, tools, and services for data management, processing, analyzing, accessing, presenting, and transferring geospatial data in digital form between users, systems, and locations. In 1999, the ISO/TC 211 had 25 active member nations from all over the world plus 16 observing member nations. In addition, there are 19 liaison organizations, of which OGC was one of the first. Two other liaison organizations that will base their future revisions on ISO/TC 211 standards include the North Atlantic Treaty Organization (NATO) through its Digital Geographic Information Working Group (DGIWG) and the maritime society through the International Hydrographic Organization (IHO) (Buehler and McKee, 1998; ISO/TC211/WG 1, 1998b; Kuhn, 1997).

Internet GIS Showcase: City of Hamilton GIS Services

This site provides a variety of Web-based services for the City of Hamilton, including a property assessment system, a street network atlas, an aerial photo viewer, and a sewer and water inventory. The site also processes customer data requests and allows for data capture and data maintenance functions.

OPAS—Online Property Assessment System. Searches can be made by intersection, by address, or by roll number.

OSNA—Online Street Network Atlas. A street segment selected from the summary page is highlighted in yellow. The map also shows "points of interest" within the city of Hamilton:

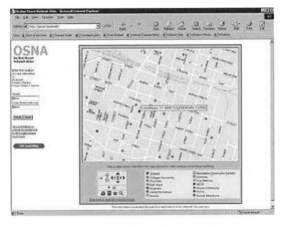

Image reprinted with permission from Intergraph.

OSWI—Online Sewer and Water Inventory. Users can retrieve a map of a selected asset (asset in red). Clicking on any of the infrastructure assets displayed on the map will return associated data:

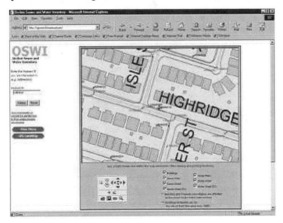

Image reprinted with permission from Intergraph.

Source: http://www.intergraph.com/gis/customers/internal/hamilton.asp.

ISO standards proposes a standard framework for the description and management of geographic information and geographic information services. The main goals of the ISO standard are to "increase the understanding and usage of geographic information; increase the availability, access, integration, and sharing of geographic information; promote the efficient, effective, and economic use of digital geographic information and associated hardware and software systems; and contribute to a unified approach to addressing global ecological and humanitarian problems" (ISO/TC 211/WG 1, 1998a, p. V).

The ISO standard framework is based on five major areas that incorporate information technology concepts to standardize geographic information (Figure 6.3).

1. The *framework and reference model* identifies how components fit together. The reference model provides a common basis for data sharing and communication.

2. *Geographic information services* define the encoding of information in transfer formats and the methodology for cartographic presentation of geographic information. Services also include satellite positioning and navigation systems.

3. *Data administration* focuses on the description of quality principles and quality evaluation procedures for geographic information data sets. Data

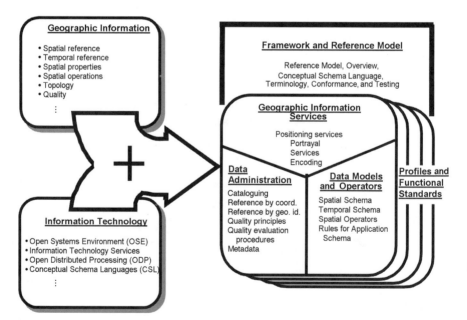

Figure 6.3   Integration of geographic information and information technology in the ISO standard [ISO 19101: 2002 (E) Fig. 2]. (Figure reprinted with permission from ISO.)

administration also includes the description of metadata, together with feature catalogues.

4. *Data models and operators* are concerned with the underlying geometry of the globe and how geographic or spatial objects may be modeled (as points, lines, surfaces, and volumes).

5. *Profiles and functional standards* consider the technique of putting together packages/subsets of the total set of standards to fit individual application areas or users. For example, different countries may have different profiles for their own geospatial datasets. This supports rapid implementation and penetration in user environments. Equally important is the task of absorbing existing de facto standards from the commercial sector and harmonizing them with profiles of the emerging ISO standards (ISO/TC 211/WG 1, 1998a).

Key Concepts

The ISO standard framework is based on five major areas: the **framework and reference model, geographic information services, data administration, data models and operators,** *and* **profiles and functional standards.** *The ISO framework incorporates information technology concepts to standardize geographic information services. ISO/ TC 211 created five* **working groups (WGs)** *and identified 34 parts of the ISO geographic information standard in order to cover the full range of standardization issues.*

Based on this framework, ISO/TC 211 created five WGs in 1995 and identified 20 parts of the ISO geographic information standard initially in order to cover the full range of standardization issues that need to be addressed. Working group 1 is tasked with the framework and reference model, providing an overview, and defining terminology for the standards, a conceptual scheme language, and methods for testing conformance. Working group 1 also involves the integration of imagery and grid data standards.

Working group 2 focuses on the geospatial data models and operations, which include the spatial schema and temporal schema for geographic information, the rules for application schema, classification of geography objects and their relationships, and the spatial operators for access, query, management, and processing geographic information. Working group 3 focuses on geospatial data administration, including the feature cataloguing methodology, the guidelines for spatial referencing by both coordinates (geodetic reference) and geographic identifiers (indirect spatial reference), geographic data quality principles and evaluation procedures, and the definition of metadata schema. Working group 4 is tasked with the geospatial services, including positioning services, portrayal definition, and encoding rules for geographic information, and the service interface and the relationship to the open system environment model. Working group 5 focuses on the profiles and functional standards and will define the guidelines for profiles and functional standards for other international standardization as well as try to harmonize between these standards and the ISO/TC 211 standards (ISO/TC 211 Chairman, 1998; ISO/TC 211 Secretariat, 2000).

ISO/TC 211 has initiated 34 different geographic information standard projects (the 19102 project was deleted) as follows:

- 19101 (*15046-1): Geographic information—reference model
- 19102 (15046-2): Geographic information—overview (project deleted, see resolution 192—Adelaide)
- 19103 (15046-3): Geographic information—conceptual schema language
- 19104 (15046-4): Geographic information—terminology
- 19105 (15046-5): Geographic information—conformance and testing
- 19106 (15046-6): Geographic information—profiles
- 19107 (15046-7): Geographic information—spatial schema
- 19108 (15046-8): Geographic information—temporal schema
- 19109 (15046-9): Geographic information—rules for application schema
- 19110 (15046-10): Geographic information—feature cataloguing methodology
- 19111 (15046-11): Geographic information—spatial referencing by coordinates
- 19112 (15046-12): Geographic information—spatial referencing by geographic identifiers
- 19113 (15046-13): Geographic information—quality principles
- 19114 (15046-14): Geographic information—quality evaluation procedures
- 19115 (15046-15): Geographic information—metadata
- 19116 (15046-16): Geographic information—positioning services
- 19117 (15046-17): Geographic information—portrayal
- 19118 (15046-18): Geographic information—encoding
- 19119 (15046-19): Geographic information—services
- 19120 (15854): Geographic information—functional standards
- 19120/Amendment 1: Geographic information—functional standards
- 19121 (16569): Geographic information—imagery and gridded data
- 19122 (16822): Geographic information/geomatics—qualifications and certification of personnel
- 19123 (17753): Geographic information—schema for coverage geometry and functions
- 19124 (17754): Geographic information—imagery and gridded data components

- 19125-1: Geographic information—simple feature access, Part 1: common architecture
- 19125-2: Geographic information—simple feature access, Part 2: SQL option
- 19125-3: Geographic information—simple feature access, Part 3: COM/OLE option
- 19126: Geographic information—profile, FACC data dictionary
- 19127: Geographic information—geodetic codes and parameters
- 19128: Geographic information—WMS interface
- 19129: Geographic information—imagery, gridded and coverage data framework
- 19130: Geographic information—sensor and data models for imagery and gridded data
- 19131: Geographic information—data product specifications
- 19132: Geographic information—location-based services possible standards
- 19133: Geographic information—location-based services tracking and navigation
- 19134: Geographic information—multimodal location-based services for routing and navigation
- 19135: Geographic information—procedures for registration of geographic information items

(*Note:* the ISO geographic information standards were known as the ISO 15046 standards. The ISO/TC 211 renamed the whole series of standards in 2001 and added more topics in the standards.)

6.2.1 ISO 19101: Reference Model

The core concept of ISO geographic information standards is illustrated in the ISO 19101 Geographic Information reference model, which is a guide to ensure an integrated and consistent approach to structuring other ISO geographic information standards. The reference model plays the same role as in *OpenGIS Abstract Specification* for the OpenGIS standard. This reference model uses concepts derived from the ISO/International Electrotechnical Commission (IEC) OSE approach for determining standardization requirements and the IEC ODP reference model and other relevant ISO standards and technical reports (ISO/TC 211/WG 1, 1998a).

Similar to the three conceptual models (OGM, OpenGIS services, and information communities) in *OpenGIS Abstract Specification,* the reference model also has four conceptual components: conceptual modeling, the domain reference model, the architectural reference model, and profiles. Their relationships and major tasks are summarized as follows.

- *Conceptual modeling* is used to describe and define services for transformation and exchange of geographic information. Conceptual modeling is the process of creating an abstract description of some portion of the real world or a set of related concepts. ISO's conceptual modeling module specifies the languages (EXPRESS, ISO IDL, Object Modeling Technique), approaches (conceptual schema, conceptual schema languages, and conceptual formalism), and principles (the 100% principle, the conceptualization principle, the Helsinki principle) for the standardization of conceptual modeling and the integration of ISO standards (ISO/TC 211 /WG 1, 1998a).

- The *domain reference model* provides a high-level representation and description of the structure and content of geographic information. The domain reference model includes a general feature model, which defines what kinds of descriptive information shall be recorded about features and the relationships that exist between features and this information. The domain reference model encompasses both the information and computational viewpoints, focusing most closely on the structure of geographic information in data models and operations, and the administration of geographic information (ISO/TC 211/WG 1, 1998a).

- The *architectural reference model* describes the general types of services that will be provided by computer systems to manipulate geographic information and enumerates the service interfaces across which those services must interoperate with each other. This model also provides a method of identifying specific requirements for standardization of geographic information that is processed by these services. Standardization at these interfaces enables services to interoperate with their environments and to exchange geographic information (ISO/TC 211/WG 1, 1998a).

- *Profiles* combine different parts of the ISO geographic information standard and specialize the information in these parts in order to meet specific needs. Profiles and functional standards facilitate the development of GIS and application systems that will be used for specific purposes (ISO/TC 211/WG 1, 1998a).

To summarize, the ISO 19101 reference model provides a comprehensive development framework for the ISO geographic information standards. The contents of the reference model are very similar to the model in *OpenGIS Abstract Specification.* For example, conceptual modeling is related to the first four levels of OGC's essential model. The scope of the domain reference models is similar to the last four levels of OGC's essential model and some parts of the OGM. The contents of the architecture reference model and profiles are overlapped with the OpenGIS services model and the OGC's information communities model.

Key Concepts

The contents of the ISO/TC 211 reference model are very similar to the contents in OpenGIS Abstract Specification. Different from the OpenGIS implementation specifications, ISO/TC 211 did not specify the actual implementation specifications for different platforms and the private software vendors. Instead, ISO/TC 211 defines a high-level data model for the public sector, such as governments, federal agencies, and professional organizations.

The reference model provides a general understanding of the underlying principles and requirements of the ISO standard, the detailed presentation of system implementation approaches, and data standard conformance. The following sections will provide a more practical overview of the reference model and related works of ISO/TC 211.

6.2.2 ISO Geospatial Data Model

Different from the OpenGIS implementation specifications, ISO/TC 211 did not specify the actual implementation specifications for different platforms and the private software vendors. Instead, ISO/TC 211 defines a high-level data model for the public sector, such as governments, federal agencies, and professional organizations. The geospatial data model is described in the view of the domain reference model (Figure 6.4).

The domain reference model defines a high-level view of the geospatial data model, that includes four major components: data set, application schema, metadata data set, and geographic information services.

Data set consists of features, spatial objects, positions, and coverages. Features define feature attributes, feature relationships, and feature functions. Spatial objects describe the spatial aspects of features. Position describes the spatial object's location by using units of measure provided by reference systems. Coverages combine the associate values of attributes to individual positions within a defined space or geographic area. For example, a coverage contains the values of one or more attributes of a geographic location over a region of interest.

The *application schema* provides a description of the semantic structure of the data set and identifies the spatial object types and reference systems. Data quality elements and data quality overview elements are also included in the application schema. The metadata data set allows users to search for, evaluate, compare, and order geographic data. It describes the administration, organization, and contents of geographic information in data sets. The structure of *metadata data set* is standardized by ISO 19115—Metadata. *Geographic information services* define how to implement software programs operating on geographic information. These services reference information in the metadata

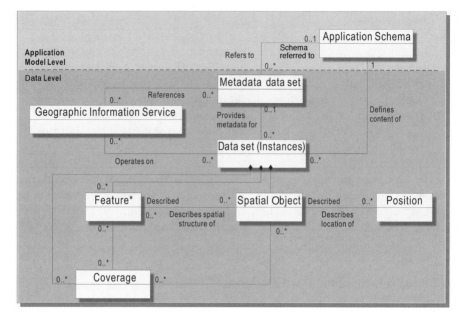

Figure 6.4 High-level view of domain reference model [ISO 19101: 2002 (E), Fig. 5]. (Figure reprinted with permission from ISO.)

set in order to perform retrieval operations correctly as well as manipulation operations such as transformation and interpolation (ISO/TC 211/WG 1, 1998a).

6.2.3 ISO Standard in Practice

There are many ISO standards and models proposed by ISO/TC 211. One of the most important components for the actual implementation in practice is the definition of software architecture, the *architecture reference model*. Under the 19101—Reference Model, the architectural reference model defines a structure for geographic information services and a method for identifying standardization requirements for those services. This model provides an understanding of what types of services are defined in the different parts of ISO geographic information standards and distinguishes these services from other information technology services (ISO/TC 211/WG 1, 1998a).

GIS developers and GIS users can use the architectural reference model to establish standardized geographic information services. The architectural reference model is shown in Figure 6.5. The model shows application systems and services residing at different computing sites linked by a network. Services are capabilities provided for manipulating, transforming, managing, or presenting information. Service interfaces are boundaries between applica-

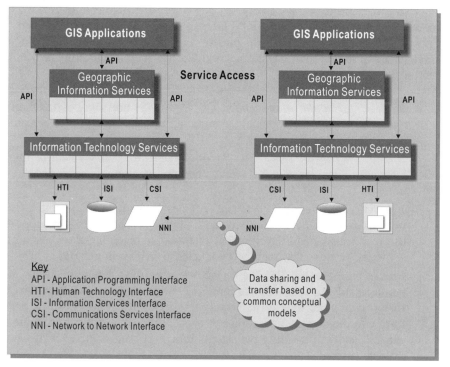

Figure 6.5 ISO architectural reference model. [ISO 19101: 2000 (E), Fig. 12.] (Figure reprinted with permission from ISO.)

tions, external storage devices, communications networks, and human beings across which services are involved.

The architectural reference model identifies four general types of interfaces in order to enable the interoperability of GIS in distributed-computing environments: API, Human Technology Interface (HTI), Information Services Interface (ISI), and Communication Services Interface (CSI). The API is the interface between services and application systems, which is used to invoke geographic information services. The CSI is the interface for accessing data transport services and communicating across a network via Network-to-Network Interface (NNI). The HTI allows the end user to access the computing system, which includes graphic user interfaces and keyboard specifications. The ISI is a bridge across heterogeneous database services and allows persistent storage of data (ISO/TC 211/WG 1, 1998a).

The architectural reference model defines these service interfaces in order to enable a variety of applications with different levels of functionality to access and use geographic information. GIS and software developers will use these interfaces to define and implement geographic information services (ISO/TC 211/WG 1, 1998a).

Beside the specification of service interfaces, the architectural reference model also identifies six classes of generic services for geographic information (ISO/TC 211/WG 1, 1998a, p. 28). The six categories of technology services are based on the OSE framework that was mentioned in the previous section:

1. *Model/Information Management Services* (MS) are services for management of the development, manipulation, and storage of metadata, conceptual schemas, and datasets.

2. *Human Interaction Services* (HS) are services for management of user interfaces, graphics, and multimedia and presentation of compound documents.

3. *Workflow/Task Services* (WS) are services for support of specific tasks or work-related activities conducted by humans. These services support use of resources and development of products involving a sequence of activities or steps that may be conducted by different persons.

4. *Processing Services* (PS) are services that perform large-scale computations involving substantial amounts of data. Examples include services for providing the time of day, spelling checkers, and services that perform coordinate transformations. A processing service does not include capabilities for providing persistent storage of data or transfer of data over networks.

5. *Communication Services* (CS) are services for encoding and transfer of data across communications networks.

6. *System Management Services* (SS) are services for the management of system components, applications, and networks

However, there are several potential problems in the ISO/TC 211 architecture reference model. First, the four types of service interfaces (API, CSI, HTI, and ISI) are too generic and emphasize the computational view instead of service view. Moreover, the definitions of ISO interfaces are not compatible with the current computer industry and may cause software development problems. For example, the CSI should be closed mapped into the network communication protocols, such as TCP/IP or SOAP. The HTI should be defined under the user interface design category, and the ISI should be closed mapped to the GIS database connectivity.

Second, with the interface standardization proposed by the architecture reference model, it is difficult to achieve software interoperability. Different computer languages have their own APIs, which are not compatible with others. Also, different component technologies (DCOM, CORBA, and Java platform) also have their own interface frameworks, which is also quite different from the architecture reference model. The new ISO 19119 standards did try to match the architecture reference model to several industrial platforms, such as CORBA, Microsoft DNA/COM+, and the J2EE/EJB framework in the Annex D section. However, it is difficult to actually implement the architecture reference model by using current available technologies. Moreover, the standardization of user interface (HTI) is problematic because different GIS applications have different GIS tasks, which will require unique and application-oriented user interface design. The standardization of interfaces is only reasonable for the database connectivity (ISI) and network communication protocol (CSI).

Third, the six generic classes of geographic information services are too ambiguous to guide the implementation of GIServices components. From a GIS processing perspective, it is really difficult to distinguish the differences between WS and PS because each GIS task always involves complex computing and substantial amounts of data. Also, it is difficult to separate the HS from the WS and PS because the design of HS is highly dependent on the fundamental features of WS and PS. In fact, geographic information services are unique and should have their own domains and classifications. Thus, the strategy of ISO/TC 211—mapping the geographic information service domains into general information service architecture—is not appropriate because the uniqueness of geographic information services requires specialized design and considerations.

6.3 COMPARISON BETWEEN OGC AND ISO/TC 211

The above sections reviewed the major concepts and models developed from both OGC and ISO/TC 211. In general, ISO is an international organization and its members are mainly from the public sector, including national standards bodies and organizations. For example, the U.S. national standard body is ANSI. On the other hand, OGC's members come mainly from the private sector, including software vendors and GIS companies, such as ESRI, ERDAS, INTERGRAPH, and AutoDesk. Since the backgrounds and resources of OGC and ISO/TC 211 members are quite different, the strategy and emphasis of open and interoperable GIServices frameworks are not really compatible. Basically, OGC focuses on both abstract definitions of OpenGIS frameworks and the technical-oriented implementations of data models and (to a lesser extent) services. ISO/TC 211 mainly focuses on the high-level definitions of the GIS standards from an institutional perspective. Although both OGC and ISO/TC 211 were formed in 1994, the early development and activities of the two organizations were independent processes. Until early in 1997, there was a strong case to reassure the market that the two activities were compatible and would produce a consistent family of standards (Rowley, 1998). Both OGC and ISO/TC 211 are currently dedicated to harmonizing their collective works, models, and standards.

In general, the contents of the OpenGIS specification and the ISO standards have significant overlaps but adopt different frameworks in their data models and architectures. Table 6.2 lists the major areas of overlaps in the programs of OGC and ISO/TC 211.

In order to build a close working relationship between ISO/TC 211 and OGC, both organizations adopted some actions and modified their work programs. First, OGC is currently a Class A external liaison with ISO/TC 211, which means OGC's experts can provide their knowledge and contribute to the setting of the ISO 14056 standard. However, OGC does not have a voting seat in ISO/TC 211 to participate in the decision-making processes. On the other hand, Olaf Østensen, Chairman of ISO/TC 211, holds a voting seat on OGC's management committee. Second, the two groups have formally com-

TABLE 6.2 Areas of Overlap between ISO/TC 211 and OGC

ISO/TC 211 Standards	OGC Equivalent Standards
19101: Reference model	Essential model
19102: Overview	Open GIS guide
19103: Conceptual schema language	Up to RFP authors
19104: Terminology	Draft document including ISO/TC 211 terms
19105: Conformance and testing methodology	Project document 97-200
19106 Profiles	RFP is profiling; submissions define profiles
19107: Spatial subschema	Abstract specification provides general feature model
19108: Temporal subschema	None
19109: Rules for application schema	Domain working groups
19110: Feature cataloguing methodology	None
19111: Spatial referencing by coordinates	Project document 97-017
19112: Spatial referencing by Geographic Identifier (GID)	None
19113: Quality principles	Quality topic in abstract specifications
19114: Quality evaluation procedures	None
19115: Metadata	Request for Information (RFI) results, scenarios
19116: Positioning services	Abstract specification, topic 2
19117: Portrayal of geographic information	WWW mapping SIG
19118: Encoding	Well-known structures, transfer technology RFP
19119: Services	Abstract specifications, topic 12
19120: Functional standards (formerly called spatial operators)	Abstract specifications, topic 4

Source: Modified from Kuhn, 1997, p. 8.

mitted to work closely together to converge and match their respective work plans to avoid duplicate or divergent work in their efforts (Kuhn, 1997).

In 1998, a formal cooperation agreement was generated in the ISO/TC 211-N472 document, which indicates the cooperation tasks between OGC and ISO/TC 211 through planning, coordination, and quality control activities. These tasks emphasize that the cooperation should lead to conformance with a single industry reference model, provide confirmation that TC 211 standards remain relevant and conformant with market-driven requirements, provide confirmation that OGC technology is conformant with TC 211 standards, and provide the opportunity for stable OGC interface specifications to be transposed into ISO endorsed documents (Rowley, 1998).

One example of the cooperation between ISO/TC 211 and OGC is the metadata standard. ISO/TC 211 proposed a joint project with OGC to demonstrate mutual cooperation and the feasibility of implementing the emerging ISO metadata standard (ISO WD 15046-5: geographic information, Part 15: metadata). In the future, OGC will adopt the metadata standard developed by ISO/TC 211 and create several scenarios based on the ISO metadata model.

In general, ISO has broader goals and is working at a level of abstraction above OGC, so the two efforts complement each other, and both are necessary. ISO's work is not likely to result in immediate implementation-level specifications, so it is in both organizations' mutual interest to see that OGC's implementation specifications fit into the ISO framework as implementation profiles (Kuhn, 1997). According to the document of ISO/TC 211-N563, the general principles of cooperative agreement between ISO/TC 211 and OGC are as follows (ISO/TC 211 Chairman, 1998, p. 3):

1. The OGC produces publicly available industry specifications through an open, consensus-based process with international participation by hundreds of individuals and organizations.
2. ISO/TC 211 wishes to adopt suitable industry specifications as ISO deliverables.
3. Both organizations desire to harmonize their procedures. Initial technical development work relevant to this agreement is done primarily in OGC with provisions for ISO participation. Once the technical work is stable and the editorial state is satisfactory, final editorial and independent assessment technical work, eventually resulting in an international standard, is done primarily in the ISO/TC 211 with provisions for OGC participation.
4. Both organizations desire that OGC implementation specifications be adopted as ISO standards or deliverables as quickly as is feasible and with only minimal changes based on an agreed-upon set of criteria.

Collaboration between ISO and OGC is essential for the future development of distributed GIServices. However, due to the differences in their backgrounds, members, and development strategies, many potential problems reside in their development agendas and working items. OGC is a highly commercial-oriented organization and is supported by GIS vendors. On the other hand, ISO is a nonprofit organization and relies on the international body and academic support.

In the future, both OGC and ISO will have significant impacts for the development of a National Information Infrastructure (NII) in the United States and many other countries. ISO works closely with other federal organizations, such as FGDC and ANSI, in order to deploy a standardized geographic information infrastructure for nationwide adoption. The focus of OGC is more marketing oriented for the future development of GIS software and applications. Therefore, resolution of the diversified perspectives between the two organizations and pursuit of an integrated framework for distributed GIServices architecture will be the most important issues for both OGC and ISO. In the next few years, GIS is most likely to be integrated with other information technologies and become one of the family members in the IT industry. However, without the comprehensive deployment infrastructure, the integration of GIS and IT may be delayed or even prevented. Thus, the GIS community should encourage the cooperation between ISO and OGC in order to provide a comprehensive infrastructure to harmonize different types of ITs. Collaborative work on a GIS metadata standard is a good start for both OGC and ISO. The GIS metadata standard will be used to support many distributed GIS applications.

The OpenGIS specification and the ISO standard illustrate efforts made by the GIS community to try to solve the problem mentioned above. The approaches and specifications provided by both organizations are very feasible and provide the GIS community with a promising future. The early development of specifications and standards in both organizations were isolated and had no connections between each other. Fortunately, both organizations recently realized the importance of integration and cooperation with each other and have since worked together for the harmonization between ISO/ TC 211 and OGC. However, based on the documents and specifications from both organizations, there are several potential problems in their model designs, which were mentioned in the previous sections. Examples include the lack of high-level classification of GIServices, the needs of dynamic combination and integration for distributed GIS components, and the ambiguous definition of semantic translators and traders in OGC's information communities model. The main problem in the OGC and ISO/TC 211 specifications and proposals is the lack of identifying distributed GIS processing and designing an integration framework for heterogeneous GIS components.

The technology evolution of Internet GIS from static map publish to static Web mapping to interactive Web mapping and distributed GIServices demonstrated that different programs utilize different kinds of computer and network technologies that may not be compatible with each other. Therefore, the goals of ISO standards and OGC specifications are to formalize these technologies and geodata models and provide a truly interoperable, exchangeable geographic information services via the Internet.

Internet GIS Showcase: Civic Center

Civic center is a generic Internet GIS program marketed to municipal governments. Its goal is to make key property information available to all local government staff, not just GIS staff. Civic Center is designed to be accessible from all municipal workstations and even a few home computers, such as for the mayor or city councilors.

Civic Center uses Autodesk's Mapguide and offers the following functions:

• Display property ownership and other relevant parcel data.

• Locate a specific property.

• Link documents and blueprints to a property.

• Create buffers and mailing lists.

• Generate form letters.

Civic Center interface.

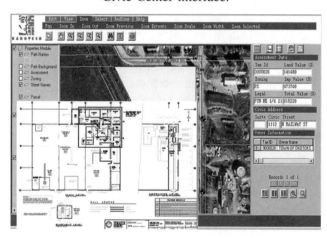

Blueprint of selected property. Image reprinted with permission of Kanotech Information Systems Ltd.

Source: http://mapguide.kanotech.com/eproperties/framespage.cfm

6.4 DEVELOPMENT OF GEOSPATIAL METADATA STANDARDS

6.4.1 Introduction to Geospatial Metadata and FGDC Metadata Standards

The development of geospatial metadata standards is essential to distributed GIServices. Metadata can provide users and systems with the descriptions in accessing, archiving, and operating geodata objects and components and make them self-describing and self-managing in distributed-network environments. The major uses of metadata include organizing and maintaining an organization's investment in data, providing information to data catalogs and clearinghouses, and providing information to aid data transfers (FGDC, 1995).

A conceptual model for metadata includes description, history, and findings (Gardels, 1992). Description focuses on the generic feature of the data. History refers to the derivation, update, and processing chronology of the data sets. Findings consist of aspects such as precision, consistency, and accuracy. The conceptual model indicates a wide range of potential metadata functionality. In the development of distributed-network environments, the use of metadata plays a key role for the interoperability of heterogeneous systems and data models.

The development of metadata in GIS applications began at the federal level with the work of the Spatial Data Transfer Standard (SDTS) committee in the 1980s (Moellering, 1992). The goal of SDTS is to provide a common ground for data exchange by defining logical specifications across various data models and structures (Fegeas et al., 1992). The original concepts of metadata were described in the data quality component of SDTS but without a detailed specification (Wu, 1993). Fifteen years later, on June 8, 1994, a Content Standard for Digital Geospatial Metadata (CSDGM) was approved by the FGDC. The standard includes eight major components: identification, data quality, spatial data organization, spatial reference, entity and attribute, distributed information, and metadata reference information. Hundreds of fields need to be filled to complete a comprehensive, standardized metadata record. There are three types of metadata elements inside the FGDC content standards: *Mandatory* elements must be provided in all kinds of geospatial data; *mandatory-if-applicable* elements must be provided if the data set exhibits the defined characteristics; and *optional* elements could be provided at the discretion of the metadata provider (FGDC-STD-001-1998). Figure 6.6 illustrates the framework of the FGDC content standards published in 1994 (FGDC, 1995).

One unique feature of CSDGM is the design of compound metadata elements. Figure 6.7 illustrates section 2 of the CSDGM, Data Quality Information, which includes three compound elements and three basic data elements. A data element is a logically primitive item of data. Compound elements are composed of other compound elements or data elements. For example, the "attribute accuracy" is a compound element which includes a

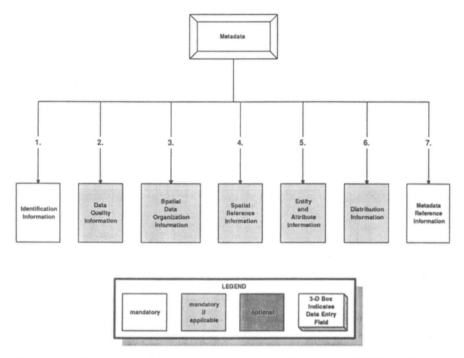

Figure 6.6   FGDC Metadata Framework, June 8, 1994 version (FGDC, 1995). (Prepared by S. Stitt, Technology Transfer Center, National Biological Survey, in conjunction with FGDC Standards Working Group. Figure reprinted with permission from FGDC.)

data element (attribute accuracy report) and a compound element (quantitative attribute accuracy assessment).

In 1995, the ADL project designed its own metadata scheme by extending the FGDC metadata standard and combining it with the USMARC standard, a U.S. national metadata scheme for libraries (Smith, 1996). Neither FGDC nor USMARC standards can fully represent digital spatial data or map materials. Therefore, the ADL project created a spatial data extension to add additional fields for specialized items that are not found in either standard. Figure 6.8 illustrates an example of ADL metadata records.

After the FGDC published the first version of the content standards in 1994, many federal and local governments and agencies adopted the FGDC metadata standards. In 1998, FGDC released the second version of the content standards, which modifies some production rules for easy implementation of metadata. The new version also added two new functions for the CSDGM: the definition of profile and user-defined metadata extensions.

The first significant enhancement of the new FGDC metadata standards is the ability of the metadata producer to "profile" the base standard by defining

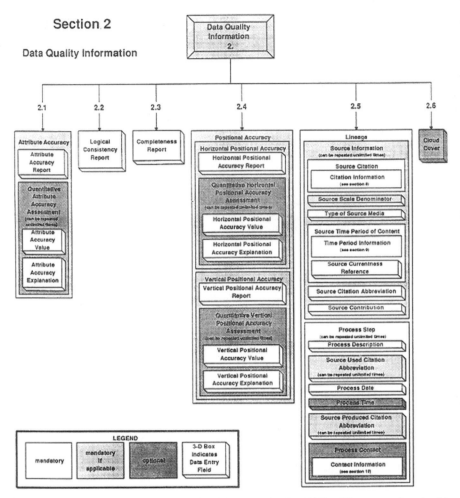

Figure 6.7 Data quality section of CSDGM (FGDC, 1995). (Figure reprinted with permission from FGDC.)

a subset of the metadata entities and/or elements that are used by a specific discipline. A profile is a subset of the standard metadata elements that describes the application of the FGDC metadata standard to a specific user community. For example, the biological research community can define its own profiles for the biological/ecological data sets, such as vegetation, land uses, and habitats (FGDC, 1999). Profiles are formalized through the FGDC standards process or may be used informally by a user community (FGDC, 1998). Actually, the use of Profile in the FGDC metadata standards is adopted from the metadata specification by the ISO/TC 211, which will be mentioned in the next section.

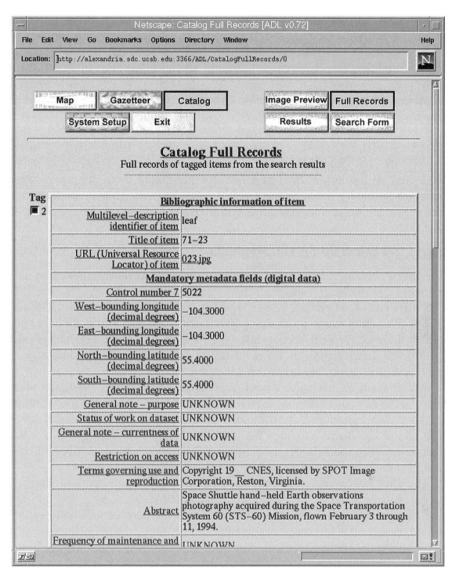

Figure 6.8   Example of metadata records in ADL project.

Another significant improvement of the FGDC content standard is the ability of the metadata producer to extend the base standard, through user-defined extensions. A specific research discipline can define a set of metadata entities for its specific applications. Extended elements to the standard may be defined by a data set producer or a user community. For example, a remote sensing community can define the metadata extensions for remote sensing research (FGDC, 2000).

Although the concepts of profiles and extensions are very similar, the main difference between profiles and extensions is that the extensions emphasize the new metadata elements outside the original standards but the profiles focus on the modification of original standards and change their attributes from mandatory to optional or vise versa. Profiles could also include extended elements, which must be formally documented following the hierarchical structure of the standard.

Currently, one of the major metadata international standards is the ISO 19115 metadata standard (previous published as ISO 15046-15) created by the ISO TC 211. The ISO metadata standards proposed a conceptual framework and an implementation approach for geospatial metadata which were developed partially based on the 1994 FGDC standards. The detailed description of the ISO metadata standards will be mentioned in the next section.

6.4.2 ISO Standard for GIS Metadata

The ISO 19115 metadata standard is one of the most comprehensive (but also the most complicated) metadata scheme for distributed GIServices. The ISO GIS metadata standards are currently adopted and used by both the OGC and FGDC. The framework of geospatial metadata specified by ISO/TC 211 includes three conceptual levels: data level, application level, and metamodel level. Each level highlights different aspects of the metadata model and its relationship to geographic data sets (Figure 6.9).

The element in the highest level (metamodel level) of ISO metadata standards is the metadata schema language, which is used to describe a conceptual metadata schema and an application schema at the application model level. The metadata schema provides the metadata element definitions for a metadata set. A metadata set describes the administration, organization, and content of a data set at the data level (ISO 19101: 2002).

The ISO 19115 metadata schema includes three major scopes. The first scope is the mandatory and conditional metadata sections, metadata entities, and metadata elements. These sections include the core or minimum set required to serve the full range of metadata applications, that is, data discovery, determining data fitness for use, data access, data transfer, and use of digital data. The second scope is the optional metadata element, which allows for a more extensive standard description of geographic data. The third scope is a method for extending metadata to fit specialized needs (ISO/TC 211/WG 3, 1998).

Similar to the FGDC metadata attributes specifications, the design of mandatory, conditional, and optional items in the ISO 19115 metadata standard allows the implementation of metadata standards that are more flexible, dynamic, and easily adopted in a distributed-network environment.

Based on the scope of ISO metadata standards, ISO/TC 211 identifies two levels of conformance for metadata elements. Conformance level 1 is the minimum metadata required to identify a data set uniquely. This level of

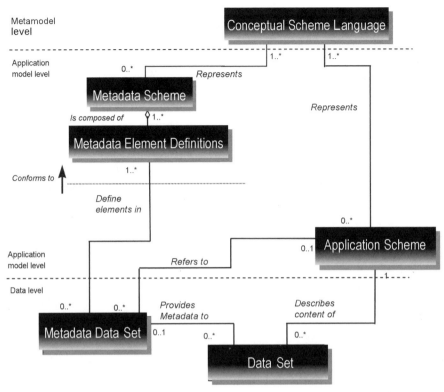

Figure 6.9 Details of ISO/TC 211 metadata relationships (ISO 19101: 2002). (Figure reprinted with permission from ISO.)

conformance shall be used only for the purpose of cataloging data sets and data set series and to support data clearinghouse activities facilitating data discovery. Thus, only one type of metadata schema is specified on this level, which is called cataloguing information. The cataloguing information schema includes 60 metadata elements, including title, initiative identification information, responsible party information, data set extent, category, abstract, and metadata date (ISO/TC 211/WG 3, 1998).

Conformance level 2 provides the metadata required to document a data set completely. This level of conformance fully defines the complete range of metadata required to identify, evaluate, extract, employ, and manage geographic information. This level also specifies a method for extending metadata to accommodate user-defined requirements. Eight information sections and five supporting entities are specified on this level. Information sections include identification information, data quality information, lineage information, spatial data representation information, reference system information, feature catalogue information, distribution information, and metadata reference information. Supporting entities include citation information entity, responsible party information entity, address information entity, extent information entity, and on-line resource information entity (ISO/TC 211/WG 3, 1998). Over 300 metadata elements are defined in conformance levels 1 and 2. Each metadata element has a descriptor indicating whether a metadata element shall always be present or not. The descriptor has three types of values: mandatory (M) means the element shall be present, conditional (C) means the element shall be present if the data set exhibits the characteristics defined by that element, and optional (O) means the element may be present or not. The values of descriptors can be modified by individual communities, nations, or organizations that will develop a community profile of the ISO standard. A community profile is used for customizing the value of descriptors to meet the actual needs of specific communities. The community profile provides a flexible and customizable framework of the ISO metadata standard. Another flexible approach proposed by the ISO metadata standard is the metadata extension capability specified in the extent information entity of conformance level 2. The extent information entity schema provides the rules for defining and applying additional metadata to meet special user needs. Users can add a new metadata element or a new metadata entity type or even a new metadata section based on their needs. The design of community profiles and user extensions was adopted by the second version of the FGDC metadata standards in 1998.

Another significant feature of the ISO metadata standard is to provide a language-based implementation framework for metadata structure and encoding. ISO/TC 211 suggests that metadata software will support the input and output of metadata entries using Standard Generalized Markup Language (SGML), as defined by ISO. Each metadata entry will be encoded as a SGML document entity including a SGML declaration, a base DTD, and the start and end of a base document element. The same format is also used in the XML DTD. Although the metadata implementation method using SGML and XML is not mandatory for the ISO metadata standard, the metadata encoding using XML DTD format will become a major advantage for the future development and implementation of metadata data sets, especially for Web-based applications. Most research indicates that XML will replace HTML and become the main language used by Web applications in the future.

In general, the use of metadata can facilitate the identification, interoperability, and autotransfer functions in distributed GIServices. A comprehensive metadata structure is essential for the future development of open and distributed GIServices. However, the complicated metadata standards may undermine the real use of metadata and their implementation procedures. Since different types of geospatial data sets have unique data structures and formats, the contents of metadata should represent their unique data features. In fact, the construction of metadata should be flexible and have alternative methods for different data types because metadata are both data oriented and application oriented. Current metadata standards by FGDC are sophisticated but also restrictive. An alternative approach is to establish the metadata exchange mechanisms instead of enforcing the standardization of metadata structure (Gardner, 1997). The metadata standard proposed by ISO/TC 211 illustrates a flexible framework for the construction of geospatial metadata that has great potential to become the de facto metadata standard in the future. Besides the argument between establishing a standardized format and a standardized exchange mechanism, current metadata research also focuses on its potential functionality, such as machine-readable features, error propagation, and data lineage (FGDC, 1995; Lanter and Surbey, 1994; Wu, 1993). The use of metadata will become more and more important for bridging the heterogeneous frameworks in distributed-network environments.

The development of geospatial metadata is closely related to the standardization of the GIS data model. Establishing a standardized metadata schema has become one of the major tasks in setting up GIS data standards. The early development history of metadata standards, such as SDTS and FGDC's metadata standard, indicates that setting up a rigid, ad hoc metadata standard may not be well accepted by the GIS community and GIS users. The ISO metadata standard solves the inflexible problem of metadata structure by proposing an extension capability and customizable profiles to meet the various needs from different GIS users. The ISO metadata standard has received support from OGC, FGDC, and other GIS community members. However, there are still potential problems for the design of ISO metadata standard. First, the contents of the metadata standard are only designed for geospatial data sets without considering the distributed GIS components. A comprehensive metadata scheme for distributed GIServices should consider both geospatial data sets and GIS components and the interactions between GIS operators and data sets. The metadata for distributed GIS components should be designed and specified in order to provide self-managing, self-describing GIS components, which can be freely combined and used with various data sets and provide comprehensive GIServices. Second, the ISO metadata standard does not specify how to protect the contents of metadata from distributed-network environments and how to ensure the connections between metadata and geodata. The contents of metadata need to reside in a safe place, where the metadata will not be modified without authorization or lose connections with data. One possible solution is to encapsulate metadata elements into data objects and protect the contents of metadata automatically by the object-oriented modeling mechanism. The deployment of both geospatial metadata and GIS component metadata and metadata encapsulation will be discussed in detail in the next section.

6.4.3 Object-Oriented Metadata Framework

Metadata standards developed by ISO/TC 211 (Kuhn, 1997; OGC, 1998) and FGDC demonstrated the need for extensions, such as adding to an existing data element or adding a new metadata element. The major problem of current metadata schemes is the detachment of metadata from data in the database

management and data model. The detachment of metadata and data jeopard-
izes the availability of metadata when geospatial data sets are frequently
moved, downloaded, or modified in the dynamic network environment. It is
quite possible to lose metadata during data processing and copying. An
object-oriented metadata scheme for both GIS data sets and components could
solve the problem of the formalization of geospatial data sets and GIS
operators in distributed-network environments (Tsou and Buttenfield, 1998).
The optimal scheme of metadata for distributed information should embed
metadata within the data object itself. Figure 6.10 demonstrates these two
different metadata schemes.

   Traditional metadata schemes emphasize the establishment of a standardized format and adopt traditional relational database concepts, where each
metadata item is represented as an individual record. However, ad hoc approaches to the metadata issues do not scale and thus cause interoperability
problems (Baldonado et al., 1997). The standardization of metadata formats
may undermine their application to services, because it is impossible to design
a single standard for heterogeneous geospatial data processing methods. For
example, a single standard would be inadequate to simultaneously describe
both a TIN data model and a raster data model without lots of extraneous
fields. Likewise, a single service-based metadata model designed to describe
both interpolation and buffering would be both cumbersome and inefficient.
Thus a single standard for metadata likely will not be feasible. Encapsulated
metadata schemes adopt object-oriented modeling techniques and embed
metadata as encapsulated items within the data itself. The contents of metadata can be designed from an operational metadata scheme that facilitates
dynamic interaction between geodata objects and GIS components (programs).

   There are several advantages to establishing an object-oriented metadata
model. First, metadata objects provide a flexible approach to construct metadata by object-oriented modeling. The adoption of object-oriented modeling
methods will permit a flexible storage where metadata are tailored to the type
of object they describe. Second, when a user moves or copies geodata objects,

Figure 6.10   Two metadata schemes (relational and object oriented).

metadata will automatically be exchanged. Users will never worry about where to find the metadata for their data objects. Third, encapsulation of metadata information will protect the metadata from exterior environments. Only authorized programs (with electronic access-key codes) can access the metadata information. Moreover, when a new geodata object is generated, the new metadata can inherit parent metadata information and then add new metadata information for itself. For example, if a subset area is clipped from a satellite image, the new metadata object could inherit the information about image resources, sensor types, and resolution from the original image and add new spatial boundary coordinates for the new metadata.

In a comprehensive metadata object scheme, each data object should be able to automatically generate its own metadata and encapsulate it into the data object in the process. Each geodata object must have metadata in distributed geographic information environments and metadata objects can be retrieved from data objects, saved in a repository, and be accessed by other application programs. The metadata scheme can apply in both geodata and GIS components, which will be discussed next.

Key Concepts

A **metadata scheme** *is the abstract framework for the development of metadata models. For example, an operational metadata scheme will focus on the operational functions of metadata. A remote-sensing metadata scheme will focus on the remote-sensing applications.*

A **metadata schema** *(schemata) provides the actual metadata element definitions for a metadata dataset, which describes the administration, organization, and content of data sets. Usually, metadata schema is associated with XML schema and related specifications.*

6.4.3.1 Design of Operational Metadata for Geodata Objects
Operational metadata for geodata should facilitate the access of geodata objects, remote GIS database connectivity, and the migration of geodata objects from one location to another. Two categories of metadata elements for geodata described below (Figure 6.11) are essential for distributed geodata objects.

GIS-operation metadata bridge the connections between geodata objects and GIS components. For example, in order to interact with a specific GIS component, [Map Display], GIS operation metadata should include, for example, scale, projections, map units, map extension, and cartographic symbol mapping. Other examples of GIS operation metadata include topology for buffering, and map accuracy for overlaying.

Data connectivity metadata connect the geodata objects with remote systems to facilitate the migration of geodata objects. Data connectivity metadata should identify the acceptable connectivity mechanism, such as JDBC, ODBC, and the preferred types of database engines, such as Oracle, Informix,

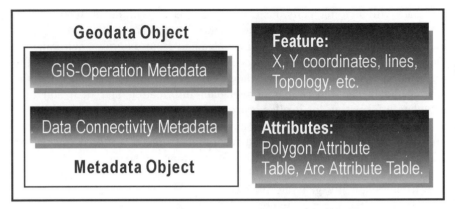

Figure 6.11 Content of encapsulated metadata for geodata objects.

or Microsoft Access databases. Data connectivity metadata should also include the types of migration, such as COPY (duplicate the data object on the client and leave the original object on the server) or MOVE (duplicate the data object on the client and delete the original object on the server), LIFE-CYCLE (does the geodata object temporarily or permanently reside in the client machines?), and data transform methods for heterogeneous database environments.

Figure 6.11 also illustrates the concepts of encapsulated metadata for geodata objects. Under a comprehensive object-oriented scheme, the two elements of metadata should be wrapped inside data objects and the encapsulation will protect the critical contents of metadata from outside interventions. With the help of the operational metadata, distributed geodata objects will become more accessible, self-describing, and self-managing. Therefore, distributed GIS components and client/server machines will be able to handle distributed geodata objects more automatically and more efficiently.

6.4.3.2 Design of GIS Component Metadata An operational metadata scheme can also be applied to GIS components. For on-line GIServices, different GIS components will be developed and designed for specific user tasks and functions. The contents of GIS component metadata need to facilitate dynamic interactions with geodata objects and plug-in functions for client/server machines. Metadata embedded in a GIS component should include two major parts (Figure 6.12): system integration metadata and GIS-operation metadata.

System integration metadata describe available functions, methods, and behaviors of GIS components for system controls, plug-in functionality, and collaborations with other GIS components. System integration metadata will also include the types of GIS components (DCOM, CORBA, or Java bean) and the component migration methods. Currently, many distributed-

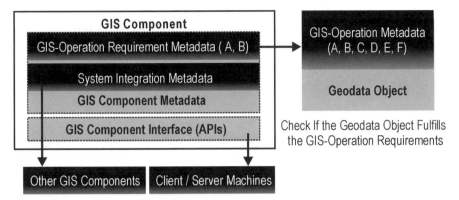

Figure 6.12   Contents and functions of GIS component metadata.

component frameworks such as CORBA and DCOM have already included some of the system integration metadata in their design.

GIS operation requirement metadata specify the data requirements for specified GIS operations. For example, a [Map Display] GIS component will embed a GIS operation requirement metadata item that specifies the data requirements for the map display function, such as map units, projection method, symbols, and scales. The [Map Display] component will check if the available geodata object meets the requirements of a map display operation by accessing the GIS operation metadata inside the geodata objects. If the requested metadata are not available, the [Map Display] component cannot operate on the geodata object.

With the collaboration between two types of metadata, distributed GIS components will become reusable, modularized, self-describing, and self-managing. To summarize, the use of operational GIS component metadata is the key to interoperability and the plug-and-play function for distributed GIS components.

The operational metadata design emphasizes three important concepts for the future use of metadata. First, the design changes the traditional functions of metadata from a descriptive type of information into processing-oriented, operational, machine-readable metadata contents. The processing-oriented metadata scheme will facilitate distributed GIS processing, accurate map display and overlay analysis, and automatic data conversion and exchanges via the networks. Second, encapsulation of metadata into data objects will protect the metadata from being lost in the network environment and prevent accidental intervention for critical metadata content. Third, the operational metadata scheme can be applied to both geodata objects and GIS components.

In contrast to the traditional metadata design, which can only be applied to data objects, the metadata sharing between GIS components and geodata objects will improve the efficiency of GIS operations and analysis. The GIS components and geodata objects with encapsulated metadata will be able to communicate with each other in distributed-network environments.

Currently, XML is the most popular tool for the actual implementation of metadata, which involves metadata modeling, data-type definitions, and language specifications.

6.5 DISCUSSION

This chapter has reviewed the two major standards for distributed GIServices and metadata development. The following will discuss the major problems and tasks of developing distributed GIServices standards based on previous reviews.

The OpenGIS specification and the ISO 15046 standard illustrate efforts made by the GIS community to try to solve the problem mentioned above. The approaches and specifications provided by both organizations are very feasible and provide the GIS community with a promising future. The early development of specifications and standards in both organizations were isolated and had no connections between each other. Fortunately, both organizations recently realized the importance of integration and cooperation between each other and have since worked together for the harmonization between ISO/TC 211 and OGC. However, based on the documents and specifications from both organizations, there are several potential problems in their model designs, which were mentioned in the previous sections. Examples include the lack of high-level classification of GIServices, the needs of dynamic combination and integration for distributed GIS components, and the ambiguous definition of semantic translators and traders in OGC's information communities model. The main problem in OGC and ISO/TC 211 specifications and proposals is the lack of identifying distributed GIS processing and designing an integration framework for heterogeneous GIS components.

The development of geospatial metadata is closely related to the standardization of the GIS data model. Establishing a standardized metadata schema has become one of the major tasks in setting up GIS data standards, such as SDTS and ISO 15046. The early development history of metadata standards, such as SDTS and the FGDC metadata standard, indicates that setting up a rigid, ad hoc metadata standard may not be well accepted by the GIS community and GIS users. The ISO metadata standard solves the inflexible problem of metadata structure by proposing an extension capability and customizable profiles to meet the various needs from different GIS users. The ISO metadata standard has received support from OGC, FGDC, and other GIS community members. However, there are still potential problems for the design of an ISO metadata standard. First, the contents of the metadata standard are only designed for geospatial data sets without considering the distributed GIS components. A comprehensive metadata scheme for distributed

GIServices should consider both geospatial data sets and GIS components and the interactions between GIS operators and data sets. The metadata for distributed GIS components should be designed and specified in order to provide self-managing, self-describing GIS components that can be freely combined and used with various data sets and provide comprehensive GIServices. Second, the ISO metadata standard does not specify how to protect the contents of metadata from distributed-network environments and how to ensure the connections between metadata and geodata. The contents of metadata need to reside in a safe place, where the metadata will not be modified without authorization or lose connections with data. One possible solution is to encapsulate metadata elements into data objects and protect the contents of metadata automatically by the object-oriented modeling mechanism.

To summarize, the development and standardization of distributed GIServices domains provides a high-level service framework for the future directions of GIServices architecture. To establish a truly interoperable, distributed GIServices architecture, both the main IT industry and the GIS community should play essential roles in the setting of these standards, including cross-platform environments, communication protocols, and collaboration with different GIS components and data sets.

WEB RESOURCES

Descriptions	URI
OGC Implementation Specification	http://www.opengis.org/techno/implementation.htm
ISO/TC 211	http://www.isotc211.org/

REFERENCES

Baldonado, M., Chang, C. K., Gravano, L., and Paepcke, A. (1997). The Stanford Digital Library Metadata Architecture. *International Journal on Digital Libraries,* 1, pp. 108–121.

Buehler, K., and McKee, L. (Eds.). (1998). *The OpenGIS® Guide: Introduction to Interoperable Geoprocessing and the OpenGIS Specification,* 3rd ed. Wayland, Massachusetts: Open GIS Consortium. URI: http://www.opengis.org/techno/ guide.htm, May 11, 2000.

Cook, S., and Daniels, J. (1994). *Designing Object Systems: Object-Oriented Modelling with Syntropy.* Englewood Cliffs, New Jersey: Prentice-Hall.

Federal Geographic Data Committee (FGDC). (1995). *Content Standards for Digital Geospatial Metadata Workbook,* Version 1.0. Reston, Virginia: FGDC/USGS.

Federal Geographic Data Committee (FGDC). (1998). *Content Standards for Digital Geospatial Metadata.* FGDC-STD-001-1998. Reston, Virginia: FGDC/USGS.

Federal Geographic Data Committee (FGDC). (1999). *Content Standards for Digital Geospatial Metadata Part 1: Biological Data Profile.* FGDC-STD-001.1-1999. Reston, Virginia: FGDC/USGS.

Federal Geographic Data Committee (FGDC). (2000). *Content Standards for Digital Geospatial Metadata: Extensions for Remote Sensing Metadata (Public Review Draft).* Reston, Virginia: FGDC/USGS.

Fegeas, R. G., Cascio, J. L., and Lazar, R. A. (1992). An Overview of FIPS 173, the Spatial Data Transfer Standard. *Cartography and Geographic Information Systems,* 19(5), pp. 278–293.

Gardels, K. (1992). A (Meta-) Schema for Spatial Meta-data. In *Proceedings of Information Exchange Forum on Spatial Metadata.* Reston, Virginia: FGDC (Federal Geospatial Data Committee), pp. 83–98.

Gardels, K. (1996). The Open GIS Approach to Distributed Geodata and Geoprocessing. In *Proceedings of the Third International Conference on Integrating GIS and Environmental Modeling.* Santa Fe, New Mexico: National Center for Geographic Information and Analysis.

Gardner, S. R. (1997). The Quest to Standardize Metadata. *BYTE,* November 1997, 22(11), pp. 47–48.

ISO/TC 211 Chairman. (1998). *Draft Agreement between Open GIS Consortium, Inc. and ISO/TC 211.* ISO/TC 211-N563.

ISO/TC 211 Secretariat. (2000). *Program of Work,* Version 8. ISO/TC 211 N 854. URI: http://www.statkart.no/isotc211/dokreg09.htm, May 11, 2000.

ISO/TC 211/WG 1. (1998a). *Geographic Information—Part 1: Reference Model.* ISO/ TC 211-N623, ISO/CD 15046-1.2.

ISO/TC 211/WG 1. (1998b). *Geographic Information—Part 2: Overview.* ISO/TC 211-N541, ISO/CD 15046-2.

ISO/TC 211/WG 3. (1998). *Geographic Information—Part 15: Metadata.* ISO/TC 211-N538, ISO/TC 211-N538, ISO/CD 15046-15.

ISO/TC 211/WG 4. (2000). *CD 19119 Geographic Information—Services (Draft).* ISO/TC 211-N906.

ISO 19101: 2002, Geographic Information—Reference Model.

Kuhn, W. (1997). *Toward Implemented Geoprocessing Standards: Converging Standardization Tracks for ISO/TC 211 and OGC,* White Paper. ISO/TC 211-N418.

Lanter, D. P., and Surbey, C. (1994). Metadata Analysis of GIS Data Processing: A Case Study. In *Proceedings of the 6th International Symposium on Spatial Data Handling,* Edinburgh, UK, T. Waugh and R. Healey (Eds.). London, U.K.: Taylor Francis, pp. 314–324.

Moellering, H. (1992). Opportunities for Use of the Spatial Data Transfer Standard at the State and Local Levels. *Cartography and Geographic Information Systems,* Special Issue, 19(5), pp. 332–334.

Open GIS Consortium (OGC). (1997a). *OpenGIS Simple Features Specification for OLE/COM,* Revision 0. Wayland, Massachusetts: Open GIS Consortium.

Open GIS Consortium (OGC). (1997b). *OpenGIS Simple Features Specification for CORBA,* Revision 0. Wayland, Massachusetts: Open GIS Consortium.

Open GIS Consortium (OGC). (1997c). *OpenGIS Simple Features Specification for SQL,* Revision 0. Wayland, Massachusetts: Open GIS Consortium.

Open GIS Consortium (OGC). (1998). *OpenGIS Abstract Specification,* Version 3. Wayland, Massachusetts: Open GIS Consortium. URI: http://www.opengis.org/techno/specs.htm, May 11, 2000.

Open GIS Consortium (OGC). (2001a). *OpenGIS Web Map Server Interfaces Implementation Specification,* Revision 1.1.0. Wayland, Massachusetts: Open GIS Consortium. URI: http://www.opengis.org/techno/specs.htm, September 11, 2001.

Open GIS Consortium (OGC). (2001b). *OpenGIS Geography Markup Language (GML) Implementation Specification,* Version 2.0. Wayland, Massachusetts: Open GIS Consortium. URI: http://www.opengis.org/techno/specs.htm, January 11, 2001.

Orfali, R., and Harkey, D. (1998). *Client/Server Programming with Java and CORBA,* 2nd ed., New York: Wiley.

Orfali, R., Harkey, D., and Edwards, J. (1996). *The Essential Distributed Objects Survival Guide.* New York: Wiley.

Orfali, R., Harkey, D., and Edwards, J. (1999). *Client/Server Survival Guide.* New York: Wiley.

Rowley, J. (1998). *Draft Business Case for the Harmonisation between ISO/TC 211 and Open GIS Consortium, Inc.,* Resolution 47. ISO/TC 211-N472.

Smith, T. R. (1996). *The Meta-Information Environment of Digital Libraries. D-Lib Magazine,* July/August URI: http://www.dlib.org/dlib/july96/new/07smith.html.

Tsou, M. H., and Buttenfield, B. P. (1998). Client/Server Components and Metadata Objects for Distributed Geographic Information Services. In *Proceedings of the GIS/LIS '98,* Fort Worth, Texas. Bethesda, Maryland: American Society for Photogrammetry and Remote Sensing, pp. 590–599.

Wu, C. V. (1993). Object-Based Queries of Spatial Metadata. Unpublished Ph.D. dissertation, State University of New York at Buffalo, Geography Department, Buffalo, New York.

CHAPTER 7

GEOGRAPHY MARKUP LANGUAGE

GML is new and exciting technology that will drive the future of spatial information on the Internet.
—Ron Lake (2000)

Chapter 2 illustrated the fundamental Internet infrastructure and the hardware components of networking. Chapter 3 provided the foundation of the evolution of software development. Chapters 4–6 discussed the technology evolution of distributed GIS from the system architecture perspective. This chapter will focus on the data, specifically on the spatial information discovery, display, and exchange over the Web. It will introduce the development of XML and GML. This chapter will start with a brief discussion of HTML and its shortcomings and follows with a discussion of the basic elements of XML and GML.

7.1 INTRODUCTION: PROBLEMS OF SPATIAL INFORMATION DISCOVERY, DISPLAY, AND EXCHANGE OVER THE WEB

The current dominant application of the Internet is the WWW. Internet GIS needs to rely on the Web, at least initially, to access remote data and process components. The Web provides an easy and friendly user interface to display information. Anyone who has an Internet connection can access on-line data and information at any time. However, there are some problems with the traditional Web-based technology that uses HTML. HTML is a Web-based language which can embed texts, images, and multimedia information in a single, integrated format. HTML documents can be interpreted and displayed in a Web browser. The original design of HTML focuses on the display of information rather than information search, discovery, and exchanges. The lack of information query and search functions is the major limitation of HTML documents, especially for geospatial data and maps. We will first look at some examples of the problem.

7.1.1 Problems

7.1.1.1 Spatial Information Discovery There are plentiful goespatial data on the Internet; it is just a matter of finding them. One way to look for them is to use search engines. For example, if you want to find all data about Mississippi River, search for the key word "Mississippi River." Well, to your surprise, you will find hundreds of results about "Mississippi River," but most of them are not what you looked for. They could include books about the Mississippi River, cruise lines on the Mississippi River, and the history of the Mississippi River. The real database about Mississippi River may not be shown in your search results at all.

7.1.1.2 Spatial Data Display on the Web HTML was originally created to code texts and images only. It cannot handle vector-based geospatial data with geometry features like points, lines, and polygons. Even when you have the data coded as text inside the HTML, the Web browser can only display it as text and numbers. The Web browser cannot convert HTML-based geospatial information into maps.

7.1.1.3 Data Integration and Exchange Assume you are lucky enough and have found some data sets about the Mississippi River. But these data were created for different applications, based on different spatial data models, in different formats, and in different spatial reference systems. For example, the Mississippi River in one data set may be represented by a linear feature; it may be a polygon in another data set using different spatial reference systems. And, quite often, these data are usually available in different formats. As an end user, how could you integrate these data in a coherent manner?

Furthermore, truly Internet GIS or distributed GIServices require real-time data exchanges among different data servers over the Internet. Except in the case of an Intranet, real-time data exchange among Internet GIS programs is not currently feasible.

7.1.1.4 Causes of the Problem The major causes of those aforementioned problems are twofold: HTML is a limited medium and geospatial data are nonstandardized. Specifically, the problems include the following:

1. HTML has a weak and flat internal structure that makes it difficult to find the correct information on the Web.

2. HTML is only a coding mechanism for presentation of text and images on the Web browser. It cannot be used to describe the content of the text and images.

3. The HTML tags are fixed and nonexpandable. That is, it is impossible to use HTML to code and correctly display geospatial vector data on the Web.

4. Different geospatial data models are used to model the physical world.

5. Different semantics are used to describe different elements of the real-world spatial objects.

6. Proprietary data structures are implemented in different GIS vendors.

For example, "Mississippi River" becomes a linear feature in some databases, a polygon in other applications, and a routable network in some other applications. Furthermore, "Mississippi River" could be a spatial feature, an attribute of a feature, or both. This makes data searches on the Internet very difficult. People compensate for the Web's current shortcomings in representing information by spending more time in their searches. But more time cannot solve the problem for a spatial data search that seeks a particular data type, a particular geographic area, and a particular layer of spatial data that may be buried inside the Web servers somewhere on the Internet. Furthermore, this process would become prohibitive for distributing computing that requires spontaneous information exchanges.

To fully understand these problems, their causes and the possible solutions, we now look at the structure and appearance of HTML.

Key Concepts

Information display *focuses on the efficiency and mechanism of showing information via computer output devices, such as CRT screens or printers. To display information appropriately, software programmers need to consider the size of display devices, color capability, typography, sharpness, projection methods, and so on.*

Information discovery *focuses on searching and finding requested information among large set of data collections, such as distributed databases or Web documents. Information discovery is different from data mining, which is the mechanism of finding out previously unknown relationships among the data.*

Information retrieval *is the topic of accessing relevant documents as opposed to exactly matching items stored in separated machines. The mechanisms of information retrieval focus on how to communicate and share information among computers, including text documents, bibliographic data, images, and multimedia.*

Information exchange *is a general term indicating the sharing of information among specific groups or users via information exchange systems or networks.*

7.1.2 Limitations of HTML

HTML is a markup language. A markup language is a set of markup codes and rules to be added to a document to indicate its structure and appearance. An application program such as a Web browser processing the document can interpret these markup codes and rules. A markup language is totally different from programming languages, because you cannot use markup languages to create a series of instructions for a computer to execute.

Any markup needs rules on how to interpret the markup codes. Without rules, anyone can create codes that nobody else can understand. For example, a ⟨h1⟩ could mean "level 1 heading" in one markup language (e.g., HTML); it could mean "horizontal alignment" in another markup language.

The rules could be in the form of a fixed set of tags that have fixed meanings, as HTML tags do. Or the rules could be setup such that as long as the application program knows the rules, it is able to interpret the codes correctly, as is the case with SGML, which was issued as an international standard in 1986 to standardize the markup language for a document. With SGML the developers can develop their own markup tags. SGML does not specify a set of markup tags; it simply establishes the rules on how tags are created. SGML separates the logical content of a document (semantics) from its presentation. This makes documents more versatile to allow document-processing programs to interpret the documents more efficiently.

Key Concepts

Markup languages *are the languages which specify different markup codes and rules for the display of documents. For example, Adobe's PostScript files and Microsoft's Rich Text Format (RTF) are the early examples of markup languages. HTML, XML, and SGML are other examples of markup languages.*

Programming languages *are the languages or executable instructions for computer programming and software applications. There are various types of programming languages, such as Visual Basic, Java, and C++.*

However, SGML itself does not specify semantics or a tag set. It leaves this job to the applications (e.g., browsers) that process SGML documents. In this sense SGML is really a metalanguage for describing markup languages (DuCharme, 1999). In other words, SGML only provides a facility and rule to define tags and the structural relationships between them. But this provides great flexibility and extensibility for any SGML document authors to mark up any documents, whether the documents are text, images, or vector GIS data. In short, SGML provides several features or capabilities to mark up a document (Castro, 2001; DuCharme, 1999):

- *Extensibility*   Authors of SGML document can define their own tags and attributes to mark up a document by specifying their syntax and semantics.
- *Structure*   Structure in SGML is explicit. That is, documents can contain other documents, with arbitrary nesting structure.
- *Validation*   Any SGML document can be verified if its grammar and structure conforms to its required grammar and structure.

SGML has been the standard, vendor-independent way to maintain repositories of structured documentation for more than a decade (DuCharme, 1999). But the flexibility and extensibility it offers also carry a price if it is to be used on the Web: The document processing program (a Web browser) must be very sophisticated in order to understand and correctly interpret all different kinds of tags in different documents. Therefore, SGML is too complicated and not well suited to serving documents over the Web. A simplified version, HTML, was developed in 1991 to present documents over the Web.

HTML simplifies SGML rules by creating a fixed set of tags to mark up the document. Instead of allowing individual authors to develop their own tags, HTML standardized a fixed set of HTML tags. By dealing with these finite numbers of standard tags, the Web browser can easily and correctly interpret those tags on the Web.

An HTML document is thus a structured document with different components or "elements" marked by HTML tags (DuCharme, 1999). The HTML tags mark up a document (text and images) as headers, paragraphs, lists, hypertext links, and other structural elements. The Web browsers interpret the tags to display the document according to its internal rules about how each tag should look.

Each HTML tag starts with a set of angle brackets (⟨ ⟩) and ends with a slash after the opening angle bracket (⟨/ ⟩). For example, the following markup codes indicate a first-level header element (⟨h1⟩) and the second-level header element (⟨h2⟩):

```
<h1> Internet GIS </h1>
<h2> Chapter 1. Introduction </h2>
```

Some elements have attributes. For example, an ⟨a⟩ element in the HTML has an "href" attribute to inform browsers the Web URL of the link's destination, as shown in the following example.

```
<a href=''http://www.w3.org''> World Wide Web
Consortium</a>
```

The text between the start and end tags of an element describes the element's *content* (e.g., the "Internet GIS" between ⟨h1⟩ and ⟨/h1⟩). Some elements have no content text and are called "empty" elements, such as the "img" element shown below:

```
<img src=''worldmap.gif'' align=''right''>
```

In this example, "img" is an empty element because it does not have any content text or end tag. Its attribute "src" identifies a file that has a picture for the browser to display. The "align" is an optional attribute to specify the alignment of the image in the browser window.

Unlike SGML, HTML mixed document structure and appearance together in its tag system. For example, the tag ⟨h1⟩ means "first-level header" from the structure point view, but it may also mean a certain font face and font size, such as "24-point bold Times Roman" from the appearance viewpoint and will be interpreted by the browser as such. Therefore, the HTML tag ⟨h1⟩ mixes both the structure and appearance.

The Web browser does not require strict hierarchy structure. For example, a Web browser can interpret and display the following HTML document without any problem, even though it has no head element and no paragraphs of text and it has section headings in a meaningless order:

```
<html><body>
<h3> Milwaukee</h3>
<h2> Wisconsin </h2>
<h3> Madison </h3>
<h1> United States </h1>
<h3> Los Angeles </h3>
<h3> San Francisco</h3>
<h2> California </h2>
```

This is because the HTML standard does not have any requirements on the order of the heading, and you can place the heading and the contents inside the ⟨body⟩ in any order you please. This means that you do not need a h1 with the h2s inside it and h3s inside the h2s. A well-structured HTML document should ideally be written this way, but the HTML standard does not require it.

This flat and loose structure allows Web page developers flexibility to create Web pages with minimal effort and with the simplest of tools. However, the lack of structure makes it more difficult to search and reuse the information inside the Web page. For example, logically, if a user wants to find all information about California (under header ⟨h1⟩), information about San Francisco (under header ⟨h2⟩) should also be included, because San Francisco (⟨h2⟩) is part of California (⟨h1⟩). However, the flat structure of HTML does not allow the user to extract the information this way. The user has to search for information on every subelement within the *body* element. In addition, the loose and flat internal structure also means that a valid HTML document may not make sense at all when you consider the semantics of the elements.

It can be seen from the above discussion that HTML is a presentation technology only. It helps the Web browser to display the document based on the markup tags used in different parts of the document. But the markup tags do not indicate anything about the information contents inside the tags. In other words, HTML was never intended as a general means of defining metadata (DuCharme, 1999). For example, HTML can identify ⟨h2⟩Wisconsin⟨/h2⟩ as a level 2 header and the Web browser interprets it as a specific font size and font style. But we do not know from the tag ⟨h2⟩ whether Wisconsin is a state, a river, or a street. The lack of a metadata standard and rigid structure makes it difficult to search for the exact information on the Web.

HTML has fixed tags set and each of them has a definite meaning. The Web developers cannot develop their own tags. Therefore, an ⟨h1⟩ is always a first-level heading, but you cannot create a new tag ⟨state⟩ to indicate Wisconsin is a state. For GIS users, we cannot create a new tag ⟨line⟩ or ⟨polygon⟩ to represent linear and polygon features. This lack of creativity or extensibility makes it difficult to present and display spatial features and GIS data using HTML.

Furthermore, although the HTML standard tags are approved and maintained by the World Wide Web Consortium (W3C), not all tags in use are standardized. Some tags are created by browser vendors to be browser specific and are not part of the W3C's HTML standard. For example, Internet Explorer has some of its own tags, which can only be interpreted by Internet Explorer and cannot be interpreted by Netscape. Netscape has it own tags too.

In short, HTML uses tags like ⟨h1⟩, ⟨h2⟩, and ⟨p⟩ to present both structure and appearance of a document. However, these tags do not reveal anything about its contents; they simply tell the browser how to display the contents inside the tags. This is not very useful for Internet GIS, because we cannot use HTML tags to specify spatial features like a line, a point, or a polygon or whether "Mississippi River" is a name of a place, a cruise line, or a restaurant name.

Because of these limitations, HTML is not enough for Web developers to develop versatile Web documents, especially for discovering, describing, and exchanging spatial information on the Web. Other markup languages should be used. This markup language should be able to allow users to define the contents of the information inside the tags, should have rigid hierarchical structure, and should be able to separate structure and appearance. SGML meets all these requirements, but it is too complicated to be implemented on the Web. XML was created to fill the void.

7.1.3 Current and Potential Solutions

To overcome the limitations of HTML, the current solution for spatial information discovery and data exchange is to utilize standardized metadata, and the current solution for spatial information display on the Web is to display map images or use Web browser plug-ins, applets, or ActiveX controls. The standardized metadata allow the data provider to unambiguously indicate what each piece of information and data element mean. The Web browser add-ins allow the Web browsers to interpret spatial information and data on the Web browser. But the more promising solution would be to use XML for information discovery, exchange, and display over the Web browser.

7.1.3.1 Standardized Metadata   Metadata are data about data. They represent a formal way to describe what a piece of information means. Anyone can compose metadata to describe the information about the geospatial data, such as the date of data creation, the accuracy, the georeference system used, the content of the data, and so on. But in order for the metadata to be searchable and to facilitate data exchange, a metadata standard should be created. The FGDC established a metadata standard to standardize the way to describe geospatial data (see Chapter 6).

In addition to being used to describe the content and format of data, metadata are also commonly used to describe distributed objects, including the behaviors of an object and the arguments necessary to activate each behavior.

Metadata allow for more efficient data search over the Internet, but they do not describe the structure of the data. For example, if we search for data about the Mississippi River, the metadata may not describe the specific data model used to construct the data. Therefore, the metadata approach is limited in assisting data exchange.

To facilitate efficient data exchange, a formal rule of shared content must be established. That is, both the data providers and the data consumers must agree that the data conform to the content standard, including spatial data models, spatial feature representations, spatial reference systems, data formats, and so on.

For example, two or more parties agree to use a shared-context standard, a specific data model, a certain spatial reference system, and a predefined data format. The parties would know exactly how to interpret the data. The data standard can be published over the Internet, and more data providers and users could follow the same standard or extend the standard. The shared-context standard can greatly improve data exchange, but it does not provide complete spontaneous data exchange (DuCharme, 1999).

7.1.3.2 XML Approach   The XML approach toward interoperability or spontaneous information exchange over the Internet is through the combination of metadata and shared-context standard. Unlike current metadata, which typically describe the data set as a whole, XML adds metadata down to the spatial feature level through a series of tags. The syntax for adding these tags is very similar to that of HTML. For example, to indicate "Mississippi River" as a linear feature, you simply write ⟨linear⟩ Mississippi River ⟨/linear⟩. As you can see, XML is significantly different from HTML, because XML tags now can be used to describe the contents of the information inside the tags.

The shared-context rule is added through DTDs or *application schemata*. A DTD is a set of rules or declarations that govern the structure of an XML document. It specifies the allowable order, structure, and attributes of tags for a particular type of document. An XML document has to conform and reference to the DTD. The DTD is a public document and allows anyone to access it. Application schema or XML schema is another way to specify the structures of a document type.

GML is an XML application to encode for geographic features. GML is designed to support interoperability by providing metadata through some common basic geometry tags that are supported by all systems and by providing shared context through application schemata. In addition, GML also provides a common spatial data model (features/properties) to support data exchange and interoperability.

The following section will introduce the concepts of XML and its application in geospatial information: GML.

7.2 WHAT IS GML?

GML is "an XML encoding for the transport and storage of geographic information, including both spatial and nonspatial properties of geographic features" (OGC, 2001b). It is a recommended standard to encode or mark up spatial and nonspatial information in XML format by OGC. GML 2.0 is currently (February 2001) an OGC-endorsed "adopted specification." [The materials presented here are based on OGC (2001a).]

GML is designed to support *interoperability* among different data models and feature representations by providing a common data model, a set of basic geometry tags to describe the spatial features, and a mechanism for creating and sharing application schemata.

7.2.1 GML Is a Means to Encode Geospatial Data

GML is a markup language that is based on the XML standard to construct structured spatial and nonspatial information to enable data sharing and interchange over the Web. It provides a standard way to encode spatial features, feature properties, feature geometries, and the location of the feature geometries based on a standard data model: the "simple feature" data model.

GML offers great flexibility to design your own markup tags or elements to describe spatial features and geometries. But to promote goespatial sharing and exchange, GML provides standards to describe features, geometry, and feature associations by developing three schemata: the geometry schema (geometry.xsd), the feature schema (feature.xsd), and the XLink (XML Linking Language) schema (xlinks.xsd). As long as your GML documents follow the rules defined in these base schemata, your GML-encoded document and data can be understood and exchanged among different users and heterogeneous systems. This is a validation mechanism to validate your GML document with the standard GML schemata.

GML also provides a mechanism to explicitly encode the spatial feature's spatial reference systems, including the main projection and geocentric reference frames in use today (Lake, 1999), as are found at the European Petroleum Survey Group (EPSG) Web site. Furthermore, GML provides flexibility for users to define units and reference system parameters as well as special local coordinate systems.

This standardized encoding of spatial features, geometries, feature associations, and spatial reference systems provides a foundation of data sharing and interchange over the Web and ensures information can be exchanged more easily and efficiently.

7.2.2 GML Is a Means to Transport Geospatial Data over the Web

Because different organizations (or even different units of the same organization) may not use a single set of data and GIS programs, data transport and exchange are inevitable. The traditional way of exchanging data among a diverse group of users is to send the data set to the other party, sometimes along with the metadata to describe the data set. Often, however, there is no metadata attached. The party who receives the data has to transform the data into a format that his or her GIS program can handle. Some information gets lost and even distorted in the process of data transport and transformation. This is especially problematic for real-time access by a third-party GIS processing program in the Internet environment.

When the GML-coded geospatial data are transported, all the markup elements that describe all spatial and nonspatial features, geometry, and spatial reference systems of the data are also transported to the recipient. Based on the standard markup elements, the receiving party knows exactly what each data component means and how to extract it so that nothing gets lost or distorted in the transport and translation process. This is particularly important for real-time data access and transport in the Internet environment.

However, one problem of GML in transporting GML data over the Internet is its size. The detailed description of GML feature elements causes the GML-coded data to be very bulky. For example, a 5.5-Mb shapefile becomes 13 Mb when it is transformed into GML. The large size of the GML will become a problem for data transporting over the Internet. Ways to compress the GML data should be studied and, we believe, will become available in the near future.

7.2.3 GML Is a Means to Store Geospatial Data

GML provides a mechanism [using XLink and XML Pointer Language (XPointer)] to develop a multidirectional association among different features and feature properties—we will discuss how this is done later in this chapter). This is ideal for distributed data that are built and maintained locally but are accessible and interoperable globally. For example, a city can build and maintain a GML database according to the requirements of GML. This database can be accessible and seamlessly integrated with data from other cities and from counties and states, because they are using the same GML coding schemata. Within a city or a large organization, data from different departments can then be integrated seamlessly; for example, a bus route segment is associated with specific streets or a type of land use is associated with property information and environmental information that is built, stored, and maintained in different departments of the city governments. If all data collected from different department for different purposes are stored in GML and interconnected with XLink and Xpointer, these data would be readily accessible and integrated with other data.

However, you do not have to store your data in GML, and you do not need to convert all your existing data into GML either. If your data are stored in other formats, you can convert the data into GML for data transport purpose, but you need a GML data conversion program to convert the vector GIS format into a GML database. There are already such conversion programs available; more are expected in the near future.

7.2.4 GML Is NOT a Presentation Language for Data Display

GML is a description language for geospatial data representation, but it is not a presentation language for data display. This means that GML has a strict rule about how to describe the geospatial data and has a strict hierarchy internal structure, but it says nothing about how to present or display the data. GML separates the data content and presentation.

In order to display the GML data, you need to have other programs to interpret the GML contents using certain graphic symbols, line styles, and area patterns so that a point feature will display as a point and a linear feature will display as a line on the map. This process is usually called *map styling* (Lake, 2000). The styling program is called a map styler or style engine. The XML Transformation Language (XSLT) is an example of a styler. It can be used to style XML data into XML graphical display format such as W3C SVG, Vector Markup Language (VML), or the Web 3D Consortium's X3D. To view an SVG, VML, or X3D graphic files, you need to have a graphic data viewer or renderer like an SVG viewer in Internet Explorer 5.0 or above, the Adobe SVG viewer, the Java applet SVG viewer, or SVG and X3D libraries.

OGC is developing a Web feature service to get data directly from GML for other applications, such as using SVG or VML technology to display spatial data on the Web as vector maps and using WML to display spatial data as maps on wireless devices.

7.2.5 GML Is NOT a Programming Language for Data Processing

GML is based on XML. Like HTML, XML is only a markup language. It is not a programming language that can execute operation and express behavior. Therefore, you cannot rely on GML for buffer analysis, spatial overlay, image processing, and network analysis. You have to rely on other languages such as C++, Visual Basic, or Java, existing desktop GIS programs for data processing.

However, you could use other XML-based technologies like Extensible Stylesheet Language (XSL), XSLT, Xpath, and XML Query Language (XQL) for GML data query and manipulation. For example, XSLT can call functions in other programming languages, such as VBScript or Java, to perform needed computation and manipulation on the GML data. The GML-coded spatial and nonspatial data enable vector mapping and even spatial analysis on standard Web browsers and establish the foundation for Internet GIS.

To illustrate what is a GML document and how it works, we look at a simple yet complete example:

Example 1. A Simple GML Document

```
<?XML Version=''1.0''?>
<mke:street, fid=''3490''>
 <mke:streetName>Broadway</mke:streetName>
 <mke:speedLimit>45</mke:speedLimit>
 <mke:numberLanes>4</mke:numberLanes>
  <gml:centerLineOf>
   <gml:LineString srsName=''EPSG:4326''>
    <gml:coordinates>5.5,80.0 60.5,130.5</gml:
     coordinates>
   </gml:LineString>
  </gml:centerLineOf>
 <mke:majorRoad/>
</mke:street>
```

The first line of this document is a processing instruction: ⟨?XML . . .?⟩. It is simply the XML declaration that identifies that the document is an XML document and adheres to the version 1.0 XML standard.

The second line is an element tag called "street" that begins with the tag ⟨mke:street⟩ and ends with ⟨/mke:street⟩ at line 12. The "mke:" is a namespace prefix, indicating the element "street" belongs to the "mke" namespace. This street element has a unique feature identifier (fid) "3490." Other elements from line 3 to line 6 are child elements or subelements of the element ⟨mke:street⟩. These elements have a strict hierarchy structure. "Broadway" is a content of the element ⟨mke:streetName⟩.

Notice ⟨mke:majorRoad/⟩ is an empty element, which does not have any content. The empty element uses the modified syntax that ends with a trailing /⟩. This modified syntax indicates to a program processing the GML document that the element is empty and no matching end tag should be expected.

The geometry of the feature street is a linestring as indicated by the ⟨gml: centerLineOf⟩ and ⟨gml:LineString⟩ elements. Based on the spatial reference system named "EPSG:4326," the location of this street feature is identified by the ⟨gml:coordinates⟩ element.

It can be seen from this simple example that GML documents are very similar to HTML documents. The contents of a document are marked up by GML elements or tags. But there are several important differences between a GML document and an HTML document.

7.3 CHARACTERISTICS OF GML

Compared with HTML, GML has its own unique characteristics that make it dramatically different. Some of the characteristics are also those of XML, including extensible elements, hierarchy structures, data validation, separation of structure and presentation, and separation of contents and relationships (Castro, 2001; DuCharme, 1999; Lake, 1999, 2000, 2001).

There are many other ways to encode geospatial data, including the Canadian Council on Geomatics Interchange Format (CCOGIF), Map And Chart Data Interchange Format (MACDIF), State Access Inspection Fitting (SAIF), DLG, and SDTS (Lake, 1999). Why do we need another standard? Well, GML is based on the XML technology and has some unique characteristics and advantages. It is text based, open, and nonproprietary; it is extensible in terms of elements and schemata and thus offers a lot of flexibility; it has strict internal structure and construct rules; it offers schemata to validate GML documents to ensure interoperability; and it separates structure and presentations as well as contents and relationships (Lake, 1999).

7.3.1 Text-Based, Open, and Nonproprietary

GML documents are text based and intuitive and thus can be easily read and understood by human beings (Lake, 2001). Anyone can open GML documents by a simple Notepad or any other word processor. Therefore, GML documents can be easily edited, maintained, and updated. GML standard is also open. There are no proprietary data structures and data formats for GML documents. This makes it easy for any other programs to communicate, extract, and exchange, the data.

In addition, since GML is text based, it can readily integrate geospatial data with a wide variety of nonspatial data types, including text, business transactions, graphics, audio, voice, and more. This capability would greatly enhance the value and accessibility of geospatial information. For example, you can easily insert a map in a financial report or vice versa.

7.3.2 Extensible Elements

GML offers great flexibility to create your own markup tags (called *element* in GML) to construct a GML document. For example, you can create a ⟨state_name⟩ *element* to indicate the content inside the ⟨state_name⟩ ⟨/state_name⟩ as a name of a state. This extensibility of GML elements provides a foundation to clearly define the semantics of the content inside the element context. So we could use such useful names as ⟨point⟩, ⟨linestring⟩, and ⟨polygon⟩ to define our spatial features. This is in great contrast with HTML where only a finite predefined set of markup tags can be used.

7.3.3 Strict, Expandable, and Enforceable Hierarchy Structure

A GML document can have a strict hierarchy structure to clearly identify the relationships among the document's content elements. The child element or subelement is always the subset of the parent element. This is in contrast with HTML's flat and loose structure. The structure of GML can be displayed in a variety of XML editors, which makes viewing and navigating GML data very easy (Lake, 1999). Furthermore, by using XML editors, it is easy to edit GML documents without corrupting the data structure.

GML also uses GML schemata or DTD to specify and validate the rules that the hierarchy structure in a particular document must follow. GML encodes data according to the rules in DTD or GML schemata. A validation parser can verify that a given GML document (officially called a GML document instance) complies with the DTD or GML schemata. GML document instances would be invalid if they do not comply with the DTD or GML schemata.

GML 2.0 adopts three XML schemata (feature schema, geometry schema, and XLink schema) to establish standards to encode spatial feature properties, geometries, and spatial feature associations. These schemata are standardized, expandable, and public. Anyone can expand them and build shareable application schemata for other industries that use spatial data such as transportation, telecommunications, utilities, forestry, tourism, and location-based services.

7.3.4 Separation of Structure from Presentation

In HTML, markup tags are used to represent both structure and appearance. An ⟨h1⟩ tag means the first-level header in terms of structure but it also means a font type and a display size in terms of presentation. It tries to do two things at the same time using the same tags. It ends up doing neither task very well (Lake, 1999). In GML, however, the document structure and document presentation are independent. The markup *element* in a GML document describes the document's structure and semantics. It indicates the relationship between elements, that is, which elements are associated with which other elements.

However, the markup element does not indicate how the document should be displayed. For example, it does not indicate how to display a point, whether as a green dot or a red circle or a purple star. In other words, GML is a structural and semantic markup language, not a presentation markup language. In our simple Example 1, there is nothing there to indicate how to display or format the contents of each element. Rather, the elements simply describe what each content is. Specifying display methods is the task for a styling engine such as XSL, XSLT, or CSS, and displaying GML data as maps requires the use of graphic renderers such as an SVG viewer, a VML processor, or X3D libraries.

With the separation of structure from presentation, GML is used to focus on the structure of the document without considering the presentation styles. With these independent presentation methods, readers could choose different presentations to render the document or spatial data as they wish for the same GML-coded document. For example, some readers may like to see roads in the red color in some circumstances; others may prefer a brown color, while still others may want to see the road as yellow (see Figure 7.1). This is in great contrast with an HTML document, in which readers are limited to the

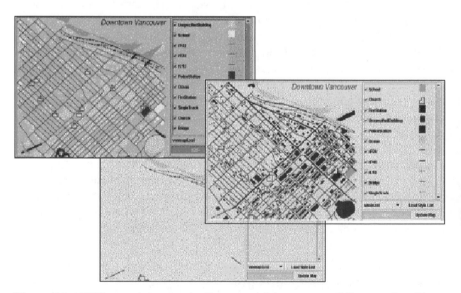

Figure 7.1 GML mapping example: Same spatial data—many different styles. (Courtesy of Ron Lake, Galdos Systems Inc. Figure reprinted with permission from Galdos Systems Inc.)

presentation imposed by the document author and have no flexibility of changing the appearance of the document once it is published on the Web.

7.3.5 Separation of Content from Relationships

The capability of linking one Web page to another is the foundation of the HTML-based Web. To represent relationships among elements in different documents, HTML uses hot links (a link reference in the source page and an anchor or bookmark in the target page) inside the HTML document to link one document with another document. In other words, HTML embeds or hardwires relationships within the document content structure, which makes the links difficult and expensive to maintain. If the linked document is relocated or deleted and the information in the source document has not changed, there will be broken links. You may have experienced numerous such broken-link messages in your own Web browsing. Furthermore, the HTML-based link can only associate two resources (the source and target pages) and in one direction (from source to target). The link point is also limited to only one point in the source page and one point in the target page. These limitations pose particular barriers to link different spatial features in geospatial objects, because one spatial object is usually associated with many other objects; for example, one point may be associated with many lines and polygons.

In GML, the relationship between documents is not maintained inside the XML document content. Rather, relationships with other documents can exist and be maintained outside the content of documents. In other words, the document content and document relationships can be separate. The advantage of this is that the GML document content does not need to change when its linked documents change. Furthermore, the linkage can be established using only read access to the target database. This maintains the relationship of different databases while as the same time the autonomy of each individual database remains. XLink and Xpointer make this sophisticated document linkage possible.

With XLink's and Xpointer's sophisticated linking capability, GML can enable distributed spatial datasets to link together to allow local maintenance and development but with global access and sharing. The data are maintained locally at the source to keep them current. But the changes in local data will not affect the linkage with related data. There will be no broken linkages, as is the case with HTML. Furthermore, with XLink, GML data can easily be mixed with nonspatial data including text, video, and imagery.

GML's links are also much more complex and flexible. It provides a mechanism for linking one source with multiple targets and in both directions (from source to the targets or from targets to the source). GML's linkage is much more fine grained than HTML's. GML linking can associate single GML elements or even element fragments within one GML document or between

GML documents. This means that one feature can associate with other features in the same database or other features across the Internet (Lake, 2001).

GML's linking capability has a great advantage over the current GIS data's binary data structure, which makes them very difficult to integrate with one another (Lake, 1999). With a binary data structure, you have to understand the file structure or database schema in order to link the databases together, even for a simple task of adding a text document or attribute data with a separately developed and maintained spatial database. This is particularly problematic for real-time data exchange and integration on the Internet environment for Internet GIS applications. But with GML, integrating spatial and nonspatial data is much easier.

Furthermore, GML is encoded based on clearly identified coordinate systems and is easily transformed to other coordinate systems by using coordinate conversion systems.

7.3.6 Interoperable GML-Based Services

GML itself is only a document and data description language. It does not have behavior and computation capabilities. However, GML is based on XML, so other technologies that work with XML will also work with GML. This will include XST, XSLT, Java, Javascript, C++, Visual Basic, SVG, VML, or X3D. For example, XSLT, along with Xpath/XQL, can be used to conduct spatial and nonspatial queries from GML documents. XSLT can also be used to transform an XML document that complies with one DTD or schema to a newer document type that has richer functionality or change from XML format to WML format to display in wireless devices. Furthermore, XSLT can work together with Java or Visual Basic to build XSLT extension functions to perform tasks such as string manipulation and mathematical computation (Lake, 1999). Therefore, many GIServices can be created with specific functions based on GML, such as routing, tracking, spatial analysis, and spatial information discovery. GML provides a STANDARD means to define I/O arguments.

Furthermore, since GML is nonproprietary and open, any client can talk to any server. This will enable developers, especially small developers not associated with any larger GIS vendors, to develop and provide their own geographic information services on the Web. This will enable the development of nonproprietary Web feature servers, image/map annotation, map styling, and spatial analysis on the Web and on the desktop.

This has been the high-level discussion of GML. With that in mind, we now discuss in more detail how to use GML to represent the spatial objects in the real world and how to encode them and link them.

7.4 MARKUP BASICS OF GML

GML, like XML, has basic rules or specifications to govern the markup of a GML document. These rules, such as the placement of tags, the validity of element names, and the attributes of elements, are strictly enforced by the rules in XML. GML documents must satisfy these rules, or grammars, in order for an XML parser program to interpret and display the documents. These GML documents that satisfy the XML specifications are said to be *well formed*. We will discuss well-formedness rules below. (Most of the following materials are derived from W3C's XML specifications. See http://www.w3 .org/XML.)

7.4.1 Basic Building Blocks: Elements, Attributes, and Values

Similar to HTML, the basic building blocks of GML include elements, attributes, and values. The *element* is the fundamental unit of the GML content. An element consists of an *element name* and *element content*. An element name is any name specified by the author to describe the semantics of element contents. But it must start with a letter, an underscore, or colon and cannot start with numbers. The three letters, upper- or lowercase, x, m, and l are reserved by the W3C. The nonempty element always starts with a start tag (⟨ ⟩) and ends with an end tag (⟨/⟩). For example, in the element ⟨River⟩ Mississippi ⟨/River⟩, ⟨River⟩ is a start tag with the element name "River" and ⟨/River⟩ is the end tag. A GML document is case sensitive, so ⟨River⟩ is different from ⟨river⟩.

The text between the start tag and end tag (the "Mississippi" in this case) is called the element's *content*. The content could be data, text, mixed data and text, as well as other elements; it could also be empty. Here are some examples of different element contents:

Data content (number and text): ⟨speedlimit⟩ 55 ⟨/speedlimit⟩.

Mixed content: ⟨address⟩ 2131 E. Hartford Ave. ⟨/address⟩.

Empty content: ⟨OneWay/⟩.

The last example, ⟨OneWay/⟩, is an empty element. It does not contain any text or data. The empty element is represented by an empty-element tag, that is, angle brackets with a trailing slash as shown in the example.

There can be only one element at the top level. This element is called the *root element* or *document element*. Every GML document must contain one root element. The root element contains all of the other elements in the document. Elements have a parent–child relationship. Elements within an element are child elements. For Example 1,

```
<?XML Version=''1.0''?>
<mke:street, fid=''3490''>
<mke:streetName>Broadway</mke:streetName>
<mke:speedLimit>45</mke:speedLimit>
<mke:numberLanes>4</mke:numberLanes>
  <gml:centerLineOf>
```

```
<gml:LineString srsName=''EPSG:4326''>
<gml:coordinates>5.5,80.0 60.5,130.5</gml:
coordinates>
</gml:LineString>
</gml:centerLineOf>
<mke:majorRoad/>
</mke:street>
```

The ⟨mke:street⟩ is a root element or document element. It is also called the *parent element* or *out element,* and other elements within the ⟨mke:street⟩ element are ⟨mke:street⟩ element's *child elements, inner elements,* or *subelements.* (The prefixes "mke:" and "gml:" are namespaces; we will discuss namespaces in Section 7.7.4.)

A well-constructed element name should provide some information about the element contents. But it may not be enough to accurately describe the metadata associated with an element. Therefore, *attributes* are usually used to further describe the purpose, metadata, and other information about the element. Attributes allow consumers of the document, such as database access programs, to use the element content more effectively.

An attribute is contained in the open tag of an element. It consists of a pair of *attribute names* and an *attribute value.* The attribute value is always bounded by quotation marks. The author can specify as many attributes for an element as necessary to adequately describe it. Attribute names follow the same rule of naming element names. That is, they must start with letters, underscore, and colon. Similarly, like element names and element contents, attribute names and values are also case sensitive. For example, in the element ⟨street fid="3490"⟩, fid="3490" is the attribute of the element ⟨street⟩. Similarly, srsName="EPSG:4326" is the attribute of the element ⟨gml:LineString⟩.

With this brief introduction of GML elements and attributes, we now turn to the criteria of a *well-formed* document. In general, the following criteria must be met for a document to be well formed (Castro, 2001):

- There must be one root element.
- All nonempty elements have start tags and end tags; that is, a closing tag such as ⟨/mke:street⟩ is required for every element.
- All empty elements have the correct empty-tag syntax (i.e., with a trailing slash): ⟨mke:oneway⟩ ⟨/mke:oneway⟩ is equivalent to ⟨mke:oneway/⟩.
- Attribute values must be enclosed in quotation marks.
- Parent and child elements must be properly nested. Child elements must be inside the parent element. Furthermore, if you start element A, then start element B, you must first close element B before closing element A.

Example 1 above is a well-formed XML document.

7.4.2 XML Declaration and Comments

An XML document generally starts with an *XML declaration,* which contains a processing instruction and the XML version. The XML declaration always appears before the root element. The XML declaration gives special information to software that may process the document that this is an XML version 1.0 document and should be interpreted as such. It starts with ⟨? and ends with ?⟩. The XML declaration could have one or more attribute–value pairs, as in ⟨?xml version= "1.0" ?⟩.

Comments in GML documents start with ⟨! and end with —⟩. You can put comments anywhere in the document after the XML declaration.

GML documents have to be read by an XML parser. The parser is responsible for dividing the document into individual elements, attributes, and other pieces. It parses the contents of the GML document to the application piece by piece. The application is typically a Web browser that displays the document to the reader. But the XML parser can also parse the contents of the GML document to other applications, such as a word processor, a database server, or other GIS programs.

With this brief introduction of XML encoding as being used in GML, we now turn to how GML models and encodes the real-world spatial features or objects.

7.5 SIMPLE FEATURE DATA MODEL USED IN GML

GML 2.0 is based on the simple feature model that is a simplification of the more general model given in *OpenGIS Abstract Specification* (OGC, 1998), which describes the real-world phenomenon as a set of features, including spatial and nonspatial features. The spatial feature is associated with a geographic location. The nonspatial feature is not necessarily associated with a location.

A nonspatial feature, such as a person, a household, or a car, is described by a name and a set of feature properties. The *property* used here is the same as *attribute* in the conventional GIS terminology. The property is used in GML rather than the conventional attribute to avoid confusion with the attributes of XML elements. Each feature property is defined by a feature name, a feature type and value, or as a {name, type, value} triple in GML. Each feature has its own type definition, which determines the number of properties a feature may have as well as their names and types. For example, a person may have a special feature type called "personType," which may contain a name property whose value must be "string," a weight property whose value must be "float," and an income property whose value must be "integer." In this example, the names of the properties (name, weight, and income) and their types ("string," "float," and "integer") are defined in the feature type "personType." Properties with simple types (e.g. integer, string, float, boolean) are collectively referred to as *simple properties* (OGC, 2001b). Furthermore, GML 2.0 (OGC, 2001b) allows features to have complex or aggregate nongeometric properties such as dates, times, and addresses. Such *complex properties* may themselves be composed of other complex and simple properties.

In the simple features geometry model, spatial features are simplified as simple geographic features "whose geometric properties are restricted to 'simple' geometries for which coordinates are defined in two dimensions and the delineation of a curve is subject to linear interpolation" (OGC, 2001b). For spatial features, there are two general types: simple features and feature collections. *Simple features* include the traditional geometric features of points, line strings, and polygons. A *feature collection* is a collection of other simple features such as multipoint, multiline string, and multipolygon collections or a collection of points, lines, and polygons. A feature collection is a simple feature itself; therefore, it has its own properties in addition to other features it contains. But simple features are not sufficient to explicitly model more sophisticated topology such as a true circle. Future versions of GML will hope to address this limitation and provide more elaborate geometry models.

Geometric types are used to describe spatial features such as the location and shape of a geographic feature, just as they are also used to describe nonspatial features, such as property names and simple property types (e.g., integer, string, float, Boolean). For example, a lake feature type can be described by a simple property "name" whose value can be expressed as "string." It could have another simple property "depth" whose value can be expressed as "float."

Like simple property types, geometric types must have a name and a definition of a type. For example, a road may be represented by line strings. To describe this geometric type, we can name this geometric property Line-StringType or CenterLineOf and use a string of (x, y) coordinates to represent its value. A feature could have multiple geometric properties as well. For instance, the road feature may have a geometric property of line strings, and/or a polygon property. For each geometry type, a spatial reference system must be specified, because for different spatial reference systems, the coordinates of the geometry would be different.

GML 2.0 defined seven GML geometry classes, including point, line string, polygon, multipoint (a collection of point geometries), multiline string (a collection of line strings), multipolygon (a collection of polygons), and multigeometry (a collection of other geometry classes).

The simple feature model and its associated properties provide a conceptual framework of a common data model to describe the real-world phenomena. Based on this framework, it is possible to define a spatial feature in GML. Since XML offers great flexibility to define your own elements, you could define the spatial features anyway you want by defining your own rules. However, this is not better than the current proprietary data structures dominant today. In order for GML to be interoperable, we need rules to define features and their properties and to specify which elements and attributes are allowed or required. This is done in GML by developing DTDs or XML schemata. The roles of DTDs or GML schemata are similar but the implementation is different.

In GML 1.0, a set of declarations or a collection of rules were developed in DTDs to specify the allowable structure of an GML application. There were three DTDs: GML feature DTD (gmlfeature.dtd), GML geometry DTD (gmlgeometry.dtd), and GML spatial reference system DTD (ebcsdictionary .dtd). The DTDs list all legal markup tags and specify where and how the markup tags may be included in a document. They include feature element structures such as the allowed order and nesting of tags, attribute values, types and defaults, and so on.

In GML 2.0 GML schemata are used to replace the DTDs in GML 1.0 due to the shortcomings of DTDs (see Section 7.6.3). An XML schema is very similar to a DTD. It defines the rules that an XML document type should follow, such as what elements could be included in a document and in what order and what attributes each element could contain. But there are major differences between DTD and XML schemata, as we will discuss later in this chapter. The most important advantage of XML schemata over DTDs is that XML schemata provide a rich set of primitive data types (e.g., string, Boolean, float, date, time) to describe element types and attribute types, and they further confer flexibility of creating derived and user-defined data types and substitution groups (OGC, 2001b).

GML DTDs and GML schemata are important tools for keeping individual GML documents consistent across GML applications. You can compare or validate a particular GML document instance with GML DTDs or GML schemata. If a particular GML document matches all the rules specified in the DTDs or schemata, as well as the criteria for well-formed documents, it is called *valid*. Documents that do not match are invalid. Validity depends on the schema; a document could be valid or invalid depending on which schema you compare it to. In GML, any GML documents must be compared with GML schemata. It should be noted that a GML document has to be *well formed*, but it does not have to be *valid* in order to work with a XML parser.

7.6 DOCUMENT-TYPE DEFINITIONS IN GML 1.0

A DTD defines a special tag: the document type. It defines what tags can go in the document, what tags can contain other tags, the number and sequence of the tags, the attributes the tags can have, and, optionally, the values those attributes can have. But it does not specify the meaning of the tags. It is a formal set of grammar that defines the rules of certain XML implementations. A typical DTD needs to declare or define four types of rules or four kinds of declarations: elements, attributes, entities, and annotations. We will only discuss the element and attribute declarations in DTD as an example.

7.6.1 Element Declarations

Every element in a GML document must be defined in a DTD in order for the GML document to be valid. If you want to add new elements in your GML document, you also have to add their definitions to the DTD or to create a new DTD. Each element declaration begins with ⟨!ELEMENT and ends with ⟩. It contains the element name and a content model surrounded by parentheses. The content model defines what subelement or child element an element may contain. A typical element definition looks like this:

```
<!ELEMENT Feature (description?, name?, boundedBy?,
property*, geometricProperty*)>
```

This DTD defines the element named "Feature." The content model indicates that the feature element may contain zero or one of the subelements "description," "name," and "boundedBy" and may contain zero or more of the subelements "property," and "geometricProperty." The commas between subelement names indicate that they must occur in sequence. The special characters after each subelement name indicates the number of subelements that an element may contain. The default value is exactly 1. Other characters have the following meanings:

?   Permits zero or one of the subelements
*   Permits zero or more of the subelements
+   Permits one or more of the subelements

7.6.2 Attribute Declarations

Besides element declaration, all the elements' attributes must be declared in DTD. The attribute declaration identifies what elements may have attributes and what kinds of attributes the elements may have. It starts with the ATTLIST declaration and typically has four consecutive parts:

- *Element Name*   This refers to the element to which the attribute list applies. It is immediately after the ATTLIST keyword.
- *Attribute Name*   This is the attribute name that identifies the attribute to the element.
- *Attribute Type*   This indicates the type of attribute allowed. There are 10 possible attribute types.
- *Default Value*   This indicates the default value for each attribute. There are four possible default values: #IMPLIED (optional attribute),

#REQUIRED (required attribute), #Fixed (fixed attribute value), and the literal "value" (the actual default attribute value).

A typical attribute list declaration looks like this:

```
<!ELEMENT Feature (description?, name?, boundedBy?,
property*, geometricProperty*)>
<!ATTLIST Feature
typeName CDATA #REQUIRED
identifier CDATA #IMPLIED >
```

In this example, the "Feature" element has two attributes: "typeName" and "identifier." Both of them must have an attribute type of text (character data) as indicated by CDATA. But "typeName" is required while "identifier" is optional.

7.6.3 Shortcomings of DTD

DTDs have some shortcomings in specifying rules and structures of XML document types (Castro, 2001). First, DTDs are not written in XML document syntax. They cannot be parsed with an XML parser. Therefore, you need applications and tools that are able to process both an XML document and the XML DTD syntax.

Second, DTDs provide no mechanism for specifying what kinds of information or data type a given element or attribute can contain. Therefore, you are not able to validate automatically whether a value is an integer, float, or string.

Lastly, all the declarations in a DTD are global, which means all the declarations apply to the whole document. You cannot specify a "local" element with the same name as the "global" element, even when they appear in separate context. Therefore, a GML schema is used to specify the structures of document types in lieu of DTDs in GML 2.0.

7.7 GML SCHEMATA IN GML 2.0

To overcome the shortcomings of DTD, the GML schemata are developed in GML 2.0 and later version, because GML schemata have a few advantages over the DTDs (Castro, 2001; Lake, 2001):

- The GML schema is written in XML and can be parsed using the XML parser.
- The GML schema can define a system of *data types* to specify the types of data an element can contain. The data type could be explicitly specified

Internet GIS Showcase: Hong Kong City Map

This Web site, developed by Telecom Directories Limited, is a source of useful information about a wide array of activities and locations in Hong Kong. It can map the location of buildings, streets, or services; it can identify dining and tourist destinations on a map; it can track weather and traffic conditions; and it can provide demographic data in the city. The site, which uses ArcIMS and ArcSDE GIS, is an excellent resource for residents and visitors alike.

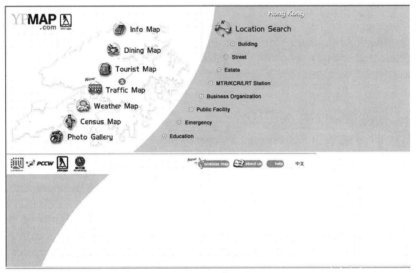

Fig. 1 The site map interface.

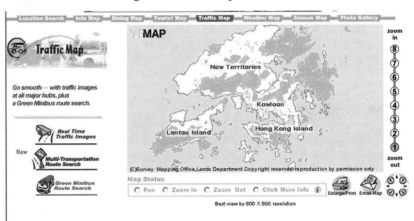

Fig. 2 The traffic map interface.

Source: http://www.hkcitymap.com/eng.

as an integer, a time, a string, and so on. This is in contrast with text type in DTD, where the element that contains only text content is specified as #PCDATA, where #PCDATA could be a string, a number, a date, a name, or anything else. You could not define #PCDATA exactly as an integer, date, or string. In an XML schema, you could specify exactly what data type you want the element to contain, with no ambiguity.

- GML schemata can define local elements and global elements. Global elements apply throughout the GML document while local elements apply only to a specific context within a GML document.

- Furthermore, GML allows you to define recurring blocks of elements or attributes once and then reuse this definition repeatedly later.

There are three base schemata in GML 2.0: feature schema (feature.xsd), geometry schema (geometry.xsd), and XLink schema (xlink.xsd). The schemata were created to offer a validation mechanism to ensure individual GML applications to be compliant to the simple feature data model.

The feature schema (feature.xsd) provides a set of rules, or the general feature–property model, to specify simple geographic features (including feature collections) and common feature properties such as fid (a feature identifier), name, and description. The geometry schema (geometry.xsd) provides rules and components to construct simple feature geometries by specifying the XLink attributes. The XLink schema (xlinks.xsd) provides a mechanism for linking different features. These three schemata are interrelated. The definitions and declarations contained in the geometry schema are included in the feature schema by using the ⟨include⟩ element. The definitions and declarations contained in the XLink schema are "imported" in the geometry schema by using the ⟨import⟩ element.

These three schemata provide *base schemata* that can be used in any individual GML documents. Any GML document instance must create its own application schemata by deriving, extending, or restricting the base schemata. An *application schema* declares the actual feature types and property types for a particular application or domain using components from these standard GML schemata (OGC, 2001b). For example, you could develop an application schema to describe features in your environmental applications; you may create another application schema for the transportation application. The definitions of features and properties defined in specific applications are distinguished from one another using namespaces.

A GML schema is simply a GML document with the .xsd extension. It is usually defined by an XML schema namespace as in the following example:

```
<?XML version=''1.0'' ?>
<schema xmlns=http://www.w3.org/2000/10/XMLSchema>
<mke xmlns:ex=http://www.w3.org/2000/10/XMLSchema-
  instance
```

```
ex:schemaLocation=''http://www.dgis.org/xsd/mke.xsd''>
<annotation>
  <documentation> This is a simple XML schema example.
  You can add anything here for documentation purpose.
  </documentation>
</annotation>
</schema>
```

The first line of this example is a normal XML declaration. The second line is the default namespace for the "schema." The third line identifies the namespace prefix ex and the fourth line identifies the location of the actual schema file mke.xsd. The ⟨annotation⟩ and ⟨documentation⟩ elements add more information about the schema for documentation purpose.

7.7.1 Element and Attribute Declarations in an XML Schema

In an XML schema, all elements in an XML document are divided into two categories: a *simple-type* element and a *complex-type* element. A simple-type element describes the content type of an element and contains only text. A complex-type element describes the structure of the document and contains attributes and/or other elements.

The text in simple-type elements can have different data types such as integer, date, and string. You could use these predefined simple data type or build your own data types in a GML schema. Simple-type elements cannot contain attributes. Complex-type elements, on the other hand, could contain both attributes and other elements.

7.7.2 Simple Element Types

The declaration of simple element types in an XML schema is very simple. We now look at a simple example. For the two elements in an XML document

```
<streetName>Broadway</streetName>
<speedLimit>45</speedLimit>
```

their corresponding XML schema is

```
<element name=''streetName'' type=''string''>
<element name=''speedLimit'' type=''integer''>
```

This XML schema forces the element "streetName" to be a string data type and the element "speedLimit" to be an integer data type. If you define these two elements in GML codes as

```
<streetName> 10 </streetName>
<speedLimit> 45 MPH </speedLimit>
```

the streetName element is valid because 10 could be a string, but the speedLimit element is not valid because its content is not an integer data type.

There are many predefined or built-in simple types, including string, decimal, Boolean, date, time, URI reference, and so on. The entire list can be found at http://www.w3.org/TR/xmlschema-2/#built-in-datatypes.

In addition, you can also build custom simple types yourself. To define a custom simple type, you begin the definition by using the keyword *simpletype* followed by an optional name attribute that identifies the new custom simple type. If there is a name attribute, it is called *named type*. If there is no name attribute specified, it is called *anonymous custom type*. Anonymous custom types are used when you just use this type once to set a particular element. The difference between a named type and an anonymous type is that a named type can be used more than once, whereas the anonymous type can only be used for the element in which it is contained (Castro, 2001). The next element following the simpletype element is a restriction element and a base attribute. The value of the base attribute is the simple type upon which your custom type is based. For example,

```
<simpleType name= ''zipcodeType''>
<restriction base= ''string''>
<pattern value= ''\d{5}(-\d{4})?''/>
</restriction>
</simpleType>

<element name= ''zipcode'' type= ''zipcodeType''>
```

This example created a new data type named "zipcodeType." It is based on the string simple data type and limits the contents of the element as five digits followed by a hyphen and four additional digits. The "?" indicates that the hyphen and four additional digits is optional. The new data type can then be used to define element "zipcode."

Therefore, based on the new custom data type, the following two XML codes are valid:

```
<zipcode> 53201 </zipcode>
<zipcode> 53201-0413 </zipcode>
```

The custom data types can also be used to limit the content of an element or attribute to a set of acceptable values. For example,

```
<element name=''DataAllowed''>
<simpleType>
<restriction base=''string''>
    <enumeration value=''string''/>
    <enumeration value=''integer''/>
    <enumeration value=''time''/>
    <enumeration value=''date''/>
    </restriction>
</simpleType>
</element>
```

In this example, the custom data type "dataAllowed" could be any of the list of the simple data types: string, integer, time, and date. But if the element were defined as decimal or Boolean, it would be invalid.

Here is another GML schema example:

```
<attribute name=''show''>
<simpleType>
<restriction base=''string''>
    <enumeration value=''new''/>
    <enumeration value=''replace''/>
    <enumeration value=''embed''/>
    <enumeration value=''other''/>
    <enumeration value=''none''/>
    </restriction>
</simpleType>
</attribute>
```

In this example, there are five possible values for the attribute named "show," including new, replace, embed, other, and none. Any other value is undefined and thus invalid.

In addition, custom simple types allow you to specify as many restrictions (or facets) as necessary to define your new custom types (Castro, 2001):

- You could specify a pattern for the content of an element to match exactly in order to be valid.
- You could specify a range of acceptable values (highest, lowest, or both) for the content of an element or attribute.
- You could specify the limits of the length of the content value of an element, such as exactly 20 characters or at least 5 characters or at most 50 characters.
- You could limit the number of digits of a numerical value.
- You could predefine the default value or a list of available values of an element's content.

7.7.3 Complex Element Types

In contrast to simple element types, complex types refer to elements that contain other elements and/or attributes. In GML schema, you always use the keyword *complexType* to start the complex element type declaration followed by the name attribute to label the complex type, as in the following example:

```
<complexType name= ''ComplexTypeLabel''>
```

where the "complexType" element indicates this is a complex-type element called "ComplexTypeLabel."

There are four complex element types: (1) elements that contain other elements only—*element-only* elements; (2) elements that contains text only—*text-only* element; (3) elements that contain both text and other elements—*mixed-content* elements; and (4) *empty elements* that contain no text or other elements but may contain attributes.

7.7.3.1 Element-Only Type Element-only type refers to elements that contain elements and/or attributes only but no text. The purpose of declaring element-only type is to require the child elements within the elements to follow a certain structure. These structures require that the elements follow a certain order, that certain elements are grouped together, and so on.

For example, to define an XML schema for the XML document

```
<streetAddress>
<houseNumber>1234</houseNumber>
<dirPrefix>S.</dirPrefix>
<streetName>Broadway</streetName>
<dirSuffix/>
<cityName>Anytown</cityName>
<state>AnyState</state>
</streetAddress>
```

one could define the XML schema as

```
<element name= ''streetAddress'' type= ''Ex:
 AddressType'' />
<complexType name= ''AddressType''>
<sequence>
<element name= ''houseNumber'' type= ''integer''/>
<element name= ''dirPrefix'' type= ''string''
minOccurs= ''0'' maxOccurs=''1''/>
<element name=''streetName'' type=''string'' />
```

```
<element name = ''dirSuffix'' type= ''string''
minOccurs=''0'' maxOccurs=''1''/>
<element name=''cityName'' type=''string'' />
<element name=''state'' type=''StateType'' />
</sequence>
<element ref= ''streetAddress'' MinOccurs= ''0'' />
</complexType>
```

In this example, the first line defines the "streetAddress" element with an "AddressType" type in the Ex namespace. The second line indicates that this is an element with a complex type that is labeled "Address." The third line ⟨sequence⟩ indicates that the following elements inside the "streetAddress" should be listed in the order they appear. Lines 4–9 indicate the sequence of the elements and their individual data types. Notice that "StateType" itself is a named complex type. It is quite common for a complex type to include another complex type. The minOccurs and maxOccurs attributes indicate the minimal and maximal amount of time the element may occur for the document to be considered valid.

The element "streetAddress" is a globally declared element because it is declared at the top level of the schema. A globally declared element must be explicitly referenced in order to appear in the XML document. In XML schema, the globally declared elements, once defined, can later be referenced using the ref=keyword, as in ⟨element ref="streetAddress" MinOccurs= "0" /⟩.

Besides the sequence, the element-only type includes other means of defining a structure of elements, for example, creating a set of choices (using ⟨choice⟩), unordered groups (the elements could appear in any order, using ⟨all⟩), or defined groups (grouping the elements together for easy reference, using ⟨group⟩):

```
<complexContent>
 <extension base=''gml:AbstractGeometryType''>
  <choice>
   <element ref=''gml:coord''/>
   <element ref=''gml:coordinates''/>
  </choice>
</complexContent>
```

In this example, you could choose to use either the "gml:coord" element or "gml:coordinates" element, two elements that are defined in GML to describe the coordinates of point, line strings, and polygons.

7.7.3.2 Text-Only Elements Text-only elements refer to certain restricted text elements. It is represented by the keyword *element* ⟨*simpleContent*⟩:

```
<ComplexType name=temporalType />
 <simpleContent>
  <extension base=''string''>
   <attribute name=''year'' type=''year'' />
  </extension>
 </simpleContent>
</ComplexType>


<element name= ''landUse'' type = ''temporalType''
```

In this example, the first line indicates that this is a complex-type element. The ⟨simpleContent⟩ indicates this is a text-only element. The ⟨extension base="string"⟩ indicates that the new complex type is based on the simple-type definition (string). The element "extension" means expanding on the simple type, while "restriction" (not shown) means to limit the base simple type with additional facets. The ⟨attribute⟩ elements indicate the attribute names and types. The last line is to declare the element "landUse" that uses the complex-type definition. Based on this text-only complex type, the following XML document would be valid:

```
<landUse year= ''1960''> residential </landUse>
```

7.7.3.3 Empty Elements   Empty elements refer to elements that have no content between the opening and closing tags, but empty elements could have attributes:

```
<ComplexType name=temporalParcelType />
 <complexContent>
  <extension base=''string''>
   <attribute name=''year'' type=''year'' />
   <attribute name=''parcelID'' type=''integer'' />
  </extension>
 </complexContent>
</ComplexType>


<element name= ''landUse'' type =
 ''temporalParcelType''>
```

Based on this defined complex type, the following XML document would be valid. Notice there is no content between the opening tag ⟨landUse⟩ and the closing tag ⟨/landUse⟩:

```
<landUse year= ''1960'' parcelID= ''123890''> </
 landUse>
```

7.7.3.4 *Mixed Elements*   Mixed elements contain both other elements and text. They are useful mainly in text-based documents. Mixed elements are declared by a Boolean attribute mixed="true" as in the following example:

```
<ComplexType name=parcelDesc mixed= ''true''/>
 <sequence>
  <element name =''parcelName'' type= ''string''>
  <element name =''parcelLocation'' type= ''string''>
 </Sequence>
  <attribute name=''year'' type=''year'' />
  <attribute name=''parcelID'' type=''integer'' />
</ComplexType>

<element name= ''parcelDescription'' type =
''parcelDesc''
```

Based on this defined mixed type, the following XML document would be valid. Notice that the "parcelDescription" element contains text, elements, and attributes:

```
<parcelDescription year= ''1960'' parcelID= ''123890''>
 The <parcelName> Lakewoods </parcelName> is located at
 <parcelLocation> 123 N. Lake drive, anytown, anystate <
 /parcelLocation>
</parcelDescription>
```

Besides these four basic complex-type elements, you could also derive more complex types based on predefined (or old) complex types by expanding or restricting the old complex type. You could also declare attributes. For example, you could define the type of attributes, make the attributes required, or set up a default attribute value:

```
<attribute name=''decimal'' type=''string''
 use=''default'' value=''.'' />
<attribute name=''character'' type=''string''
 use=''required'' />
```

In this example, the first line indicates the attribute "decimal" is automatically set to the default value ("."). The second line indicates that the attribute "character" is required for the document to be valid.

7.7.4 Namespaces

Assume you are searching for geospatial information from different XML documents, but two documents may use the same element name to refer to different geospatial objects. For example, "Route" in a transit scheduling application means a bus service serving a particular area. But "Route" in a street network application may mean a traverse of a particular path from a trip origin to a destination. If you want to combine information from these two applications, there would be a problem of *naming collision*. Such a naming collision could cause potentially serious interoperability problems in information exchanges.

One solution to this naming collision problem is to create a label or an XML namespace to distinguish an element name in one GML application from the other. Therefore, "mke:street" will be different from "lax:street." In other words, the "street" element is associated with separate namespaces in different GML applications. Namespaces have two purposes in GML:

- They avoid naming collision by differentiating element names from different GML applications.
- They help XML parsers and processors to sort out all related elements and attributes from a single GML application for different treatments. Elements in the same namespace may be treated the same, while elements in other namespaces may be treated differently. For example, the transformation language XSLT relies on namespaces to distinguish between XML objects that are data and those that are instructions for processing the data. Anything with an XSL: namespace prefix is treated as the instruction elements, while anything without a namespace prefix is treated as data in the transformation process (Ray, 2001).

Namespaces are implemented by attaching a prefix or namespace name to each element and attribute name. For example: "mke:street" is a qualified element name with a namespace. "mke" is a namespace prefix or namespace name and "street" is a local element name. Each namespace prefix is usually mapped to a URI. This is because a namespace must be unique and permanent. The URI is unique because no two URIs can be the same, and it is permanent because the URI seldom changes. For example, the following is an example of a namespace that defines "street" in the "mke" namespace:

```
<mke:street xmlns:mke=http://www.dgis.org/xmlns/mke/
1.0>
```

where "mke:street" is a qualified element name and "mke" is a namespace prefix. xmlns is an attribute inside the element; it is also the namespace declaration keyword to indicate to the XML parser that this attribute is a namespace declaration. The URI and its path are unique for the namespace. The URI usually belongs to the organization that maintains the namespace. It may point to a DTD or XML schema, but this is not required. In fact, a URI could refer to no specific file in a URI. The URI is used here purely for the purpose of formality to provide additional information about the namespace, such as who owns it and what version it is.

A namespace can apply to a single element or subelement; it could also apply to a root element. When a namespace applies to the root element, it applies to all its contained child elements. This namespace is called a default namespace. A default namespace is declared by omitting the colon (:) and the namespace prefix from the "xlmns" attribute, such as

```
<mke xmlns = http://www.dgis.org/xmlns/street/1.0>
```

where "mke" is a root element and the URI is a default namespace name. The default namespace for the root element will apply for all elements in the document unless other namespaces are declared for some elements.

However, even after you declare a default namespace, you could also label specific individual elements in a document with a different namespace. You do this by declaring a special prefix for the namespace and then using that prefix to label the individual elements:

```
<mke xmlns=''/examples''
 xmlns:gml=''http://www.opengis.net/gml''
 xmlns:xlink=''http://www.w3.org/1999/xlink''
 xmlns:xsi=''http://www.w3.org/2000/10/XMLSchema-
 instance''
 xsi:schemaLocation=''/examples city.xsd''>
  <street fid=''3490''>
   <streetName>Broadway</streetName>
   <speedLimit>45</speedLimit>
   <numberLanes>4</numberLanes>
    <gml:centerLineOf>
     <gml:LineString srsName=''EPSG:4326''>
     <gml:coordinates>5.5,80.0 60.5,130.5</gml:
     coordinates>
     </gml:LineString>
    </gml:centerLineOf>
   <majorRoad/>
  </street>
```

where the element "centerLineOf" is only identified with the namespace "gml." The element "street" is an unprefixed element that belongs to the default namespace "mke." Note that an XML processor considers the prefix an integral part of the element's name. Therefore, you have to close the element with the prefix attached, such as ⟨/gml:CenterLineOf⟩. The combination of prefix and default namespace for specifying namespaces provides much flexibility to avoid naming collisions.

When you add a namespace prefix before an element, the XML consider the whole prefix:element as one element name. Therefore, if you use DTD, you have to change your DTD to reflect that by declaring each prefixed element. In addition, you have to declare attribute "xmlns" or "xmlns:prefix" in the DTD. But you do not have to do that if you compare your GML document with the GML schema. In other words, the GML schema directly supports namespace while DTD does not. This is a major reason that the XML schema is overtaking DTDs to become the dominant XML validation choice (Castro, 2001).

7.7.5 Referencing to Different XML Schemata Using Namespace

As we mentioned before, there are three base schemata in GML: feature schema, geometry schema, and Xlink schema. There is also a main application schema that is used to validate the GML document instance. How do we reference elements in the three base schemata in the application schemata?

We do this by associating each element in a GML application document instance with a namespace, or by populating a namespace with elements that the schema contains. Once every element is associated with a namespace, it becomes a namespace-prefixed element and then can be referenced by other schemata and XML documents.

In an XML schema, we could associate a globally declared element (at the top level) with a namespace by declaring a targetNamespace= "URI", where URI is the namespace with which you want to associate the components defined in this schema. Once the association is established, namespace-prefixed elements can be used and referenced in other documents, and those documents can be validated by looking at the schema(s) that is (are) identified with the target namespace (Castro, 2001).

We could also add locally declared elements or attributes to the target namespace by using the attribute elementFormDefault= "qualified" or attributeFormDefault= "qualified". We could also add locally declared elements individually to associate with the target namespace by using the attribute Form= "qualified" or use Form= "unqualified" to prevent a particular locally declared element from being associated with the target namespace.

We could import top-level schema elements and attributes from other schemata with a different target namespace:

```
<schema targetNamespace=''http://www.opengis.net/
examples''
xmlns=''http://www.w3.org/2000/10/XMLSchema''
xmlns:ex=''http://www.opengis.net/examples''
xmlns:xlink=''http://www.w3.org/1999/xlink''
xmlns:gml=''http://www.opengis.net/gml''
elementFormDefault=''qualified'' version=''2.01''>
```

```
<!— import constructs from the GML Feature and Geometry
schemata —>
<import namespace=''http://www.opengis.net/gml''
schemaLocation=''feature.xsd'' />
</schema>
```

In this example, the first line declares that the globally declared element schema is associated with the target namespace. It means that all globally declared elements (top level, including xmlns, xmlns:ex, xmlns:xlink, xmlns: gml) in this schema are now associated with the target namespace. The second line declares the default namespace for the schema element. It means that all default element data types, simple or complex, come from the default namesapce http://www.w3.org/2000/10/XMLSchema. The third line is the declaration of the namespace of "ex"; it indicates that any element associated with the namespace ex is defined in or belongs to the namespace http://www.opengis.net/examples. The same is true for namespaces xlink and gml, as shown in lines 4 and 5. Line 6 adds all the locally declared elements to the target namespace (http://www.opengis.net/examples) and the version of the schema.

To exclude the locally declared elements from adding to the target namespace, you could use elementFormDefault="unqualified". Similarly, to add all the locally declared attributes to the target namespace, use attributeFormDefault="qualified", and to exclude the locally declared elements from adding to the target namespace, use attributeFormDefault="unqualified".

The ⟨import⟩ line imports schema components from other schemata, in this case, the schema file feature.xsd located at the namespace http://www .opengis.net/gml as indicated by "schemaLocation." ⟨Import⟩ is used when you want to combine two schemata with different target namespaces. But if the target namespaces are the same, you can use "include" to add another schema into the schema of interest. If the included schema has no target namespace specified, it is assumed that that schema would have the same target namespace as the schema of interest. For example, the GML feature schema uses the ⟨include⟩ element to bring in the definitions and declarations contained in the geometry schema:

```
<?xml version=''1.0''>
<!— File: feature.xsd —>
<schema targetNamespace=''http://www.opengis.net/gml''
 xmlns:gml=''http://www.opengis.net/gml''
 xmlns=''http://www.w3.org/2000/10/XMLSchema''
 elementFormDefault=''qualified'' version=''2.06''>
<!— include constructs from the GML Geometry schema —>
<include schemaLocation=''geometry.xsd''/>
```

Where the ⟨include⟩ element includes the geometry schema geometry.xsd in the feature schema feature.xsd so that components (e.g., elements, attributes, data types) from both feature schema and geometry schema could be used to describe GML documents.

7.7.6 Encoding Features without Geometry in GML 2.0

Encoding aspatial features in XML is simple. Here is an example to describe a street:

```
<?XML Version=''1.0''?>
<street>
 <streetName>Broadway</streetName>
 <speedLimit>45</speedLimit>
 <numberLanes>4</numberLanes>
 <majorRoad/>
</street>
```

The corresponding XML schema (or application schema) to support this XML coding is

```
<element name=''Street'' type=''ex:StreetType'' />
<complexType name=''StreetType''>
 <sequence>
  <element name=''StreetName'' type=''string''/>
  <element name=''streetLimit'' type=''integer''/>
  <element name=''numberLanes'' type=''integer''/>
  <element name=''majorRoad'' type=''Boolean''
  MinOccur=''0''/ MaxOccur=''1''>
 </sequence>
</complexType>
```

Now, we introduce GML and consider the ⟨street⟩ element as a feature type in GML. Since now ⟨street⟩ is a feature in GML, it can extend or inherit the "AbstractFeatureType" type in the feature schema (feature.xsd). The application schema becomes

```
1 <element name=''Street'' type=''ex:StreetType''
   substitutionGroup=''gml:_Feature'' />
2
3 <complexType name=''StreetType''>
4  <complexContent>
5   <extension base=''gml:AbstractFeatureType''>
6    <sequence>
```

```
7      <element name=''streetName'' type=''string''/>
8      <element name=''streetLimit'' type=''integer''/>
9      <element name=''numberLanes'' type=''integer''/>
10     <element name=''majorRoad'' type=''Boolean''
       MinOccur=''0''/ MaxOccur=''1''>
11     </sequence>
12     </extension>
13   </complexContent>
14   </complexType>
```

In this example, "Street" is a feature type and "*streetName*" is a property. The first line is a global declaration of the feature "Street." It indicates that the feature "Street" is of the "StreetType" type in the "ex" namespace. You could define the "ex" namespace in a URI. The "substitutionGroup" attribute indicates that if "ex:StreetType" is a defined abstract feature type (abstract feature type always starts with an underscore in GML), then a ⟨Street⟩ element can appear wherever the (abstract) "gml:_Feature" element is expected. Substitution groups promote a specific type (e.g., "ex:StreetType" in this case) as a more general supertype (e.g., "gml:_Feature" in this case).

Notice that the "StreetType" data type is now a complex type and is extended from the "gml:AbstractFeatureType." Therefore, all predefined attributes in "gml:AbstractFeatureType" can be used by the "StreetType." Now what is "gml:AbstractFeatureType" and what attributes does it have? We will look at the declaration in the feature schema (from GML 2.0 feature.xsd):

```
1 <complexType name=''AbstractFeatureType''
   abstract=''true''>
2   <annotation>
3    <documentation>
4      An abstract feature provides a set of common
properties. A concrete feature type must derive from
this type and specify additional properties in an
application schema. A feature may optionally possess an
identifying attribute ('fid').
5    </documentation>
6   </annotation>
7   <sequence>
8    <element ref=''gml:description'' minOccurs=''0''/>
9    <element ref=''gml:name'' minOccurs=''0''/>
10   <element ref=''gml:boundedBy'' minOccurs=''0''/>
11      <!— additional properties must be specified in an
application schema —>
```

```
12   </sequence>
13   <attribute name=''fid'' type=''ID''
     use=''optional''/>
14   </complexType>
```

It can be seen that the "gml:AbstractFeatureType" is an abstract feature that provides a set of common properties including "gml:description," "gml: name," and "gml:boundedBy." It has an optional attribute feature identifier (fid). Since "Street" extends "gml:AbstractFeatureType," the predefined "gml:description" property and the "fid" attribute can be used to describe "Street," as shown in the following:

```
<?XML Version=''1.0''?>
 <street fid='A9490''>
 <gml:description>Highway 44</gml:description>
 <streetName>Broadway</streetName>
 <speedLimit>45</speedLimit>
 <numberLanes>4</numberLanes>
 <majorRoad/>
 </street>
```

How do we encode the location information or geometry of this street? We now turn to the GML geometry schema.

7.8 GEOMETRY SCHEMA IN GML 2.0

GML provides mechanisms to encode eight geometric features as geometry elements in XML according to the OGC simple features model:

- Point,
- LineString,
- LinearRing,
- Polygon,
- MultiPoint,
- MultiLineString,
- MultiPolygon, and
- MultiGeometry.

These geometry types have their formal names in GML as shown in Table 7.1 and as defined in GML 2.0. To ensure all the markups are consistent and interoperable, the geometry schema (geometry.xsd) was created to set up the

TABLE 7.1 Basic Geometric Properties

Formal Name	Descriptive Name	Geometry Type
boundedBy	—	Box
pointProperty	Location, Position, CenterOf	Point
lineStringProperty	CenterLineOf, EdgeOf	LineString
polygonProperty	ExtentOf, Coverage	Polygon
geometryProperty	—	Any
multiPointProperty	MultiLocation, MultiPosition, MultiCenterOf	MultiPoint
multiLineStringProperty	MultiCenterLineOf, MultiEdgeOf	MultiLineString
multiPolygonProperty	MultiExtentOf, MultiCoverage	MultiPolygon
multiGeometryProperty	—	MultiGeometry

Source: OGC (2001a).

encoding rules. In addition, GML also defined the ⟨coordinates⟩ and ⟨coord⟩ elements to encode coordinates and a ⟨Box⟩ element for defining map extents.

7.8.1 Encoding Coordinates

GML has created two ways to encode the coordinates of any geometry class instance using either the ⟨coord⟩ or ⟨coordinates⟩ element. The ⟨coord⟩ element uses a sequence of ⟨coord⟩ element (e.g., X, Y, Z) to identify the values of the coordinates. The ⟨coordinates⟩ element uses a single string of coordinates separated by commas. Both approaches can convey coordinates in one, two, or three dimensions. Between the two elements, the ⟨coord⟩ element may have a slight edge since it allows the XML parser to validate basic coordinate types and enforce constraints on the number of (X, Y, Z) tuples that appear in a particular geometry instance.

The geometry schema (geometry.xsd) defines the ⟨coord⟩ and ⟨coordinates⟩ elements as follows:

```
<element name=''coord'' type=''gml:CoordType'' />

<complexType name=''CoordType''>
 <sequence>
  <element name=''X'' type=''decimal''/>
  <element name=''Y'' type=''decimal'' minOccurs=''0''/>
  <element name=''Z'' type=''decimal'' minOccurs=''0''/>
 </sequence>
</complexType>
```

```
<element name=''coordinates'' type=''gml:
CoordinatesType''/>

<complexType name=''CoordinatesType''>
 <simpleContent>
  <extension base=''string''>
   <attribute name=''decimal'' type=''string''
   use=''default'' value=''.''/>
   <attribute name=''cs'' type=''string''
   use=''default'' value='',''/>
   <attribute name=''ts'' type=''string''
   use=''default'' value=''&#x20;''/>
  </extension>
 </simpleContent>
</complexType>
```

The gml:CoordinatesType indicates the coordinates can be conveyed by a single string. They are separated by commas, and successive coordinate tuples are separated by a space character (#x20). The delimiters are specified by several attributes: "decimal," "cs," and "ts."

Once ⟨coord⟩ and ⟨coordinates⟩ are defined in the geometry schema (geometry.xsd), we can use them to describe the location of spatial features or geometry elements in GML.

For the same point with the coordinates (10.0, 2.0), you can encode it in GML as follows using either the ⟨coord⟩ or ⟨coordinates⟩ element:

```
<Point>
 <coord>
  <X>10.0</X>
  <Y>2.0</Y>
 </coord>
</Point>
```

or

```
<Point>
 <coordinates>10.0,2.0</coordinates>
</Point>
```

7.8.2 Geometry Elements

The values of coordinates for a feature geometry are always associated with SRS. Different SRS have different coordinate values for the same location. Therefore, all coordinate values must specify to which SRS they refer. Ideally, the value of coordinates would change if the SRS changes. But GML 2.0

does not address the details of defining and transforming SRS. This is left to a proposed XML-based specification for handling coordinate reference systems and coordinate transformations (OGC 00-040).

However, GML 2.0 uses the attribute *srsName* to ensure that the coordinates of different geometry types are from the same SRS. The value of the attribute *srsName* is a URI reference that may point to the definition of the SRS. As long as the value of *srsName* is the same, it is assumed the coordinates are derived from the same SRS.

The other optional attribute of the geometry types is the *gid* attribute that represents a unique identifier for geometry elements. The *gid* is an ID-type attribute, which means that it must be a legal XML name. Since XML names cannot start with a digit, the *gid* usually starts with a common letter or an underscore followed by digits. Furthermore, the *gid* must be unique within the GML document. No other ID-type attribute can have the same value, and each element can have no more than one ID type attribute.

7.8.3 Primitive Geometry Elements

There are five primitive geometry elements in GML 2.0: Point, Box, LineString, LinearRing, and Polygon.

The *Point* element is used to encode instances of the point geometry class. It is defined in the geometry schema (geometry.xsd) as the "PointType" as follows:

```
<complexType name=''PointType''>
 <annotation>
  <documentation>
  A Point is defined by a single coordinate tuple.
  </documentation>
 </annotation>
 <complexContent>
  <extension base=''gml:AbstractGeometryType''>
   <sequence>
    <choice>
     <element ref=''gml:coord''/>
     <element ref=''gml:coordinates''/>
    </choice>
   </sequence>
  </extension>
 </complexContent>
</complexType>
```

The "pointType" is an extension of the "gml:AbstractGeometryType." It could have a single ⟨coord⟩ element or a ⟨coordinates⟩ element containing exactly one coordinate tuple. The *srsName* attribute is optional since a point element may be contained in other elements that specify a reference system.

Similar considerations apply to the other geometry elements. The *gid* attribute is also optional. The following is an example of a GML encoding of a point element:

```
<Point gid=''P1'' srsName=''http://www.opengis.net/gml
/srs/epsg.xml#4326''>
 <coord>
  <X>56.1</X>
  <Y>0.45</Y>
 </coord>
</Point>
```

The *Box* element is used to encode extents. It is defined in the geometry schema (geometry.xsd) as the "BoxType" as follows:

```
<complexType name=''BoxType''>
 <annotation>
  <documentation>
   The Box structure defines an extent using a pair of
   coordinate tuples.
  </documentation>
 </annotation>
 <complexContent>
  <extension base=''gml:AbstractGeometryType''>
   <sequence>
    <choice>
     <element ref=''gml:coord'' minOccurs=''2''
     maxOccurs=''2''/>
     <element ref=''gml:coordinates''/>
    </choice>
   </sequence>
  </extension>
 </complexContent>
</complexType>
```

The "BoxType" is also an extension of the "gml:AbstractGeometryType." It must have either a sequence of *two* ⟨coord⟩ elements or a ⟨coordinates⟩ element containing exactly two coordinate tuples. The ⟨coord⟩ element is constructed from the minimum values measured along all axes, and the ⟨coordinates⟩ element is constructed from the maximum values measured along all axes. A value for the *srsName* attribute should be provided, since a Box cannot be contained by other geometry classes. Here is an example of a GML-encoded Box instance:

```
<Box srsName=''http://www.opengis.net/gml/srs/
 epsg.xml#4326''>
```

```
<coord><X>0.0</X><Y>0.0</Y></coord>
<coord><X>100.0</X><Y>100.0</Y></coord>
</Box>
```

A *LineString* is "a piece-wise linear path defined by a list of coordinates that are assumed to be connected by straight line segments" (OGC, 2001b). A LineString is not necessarily a closed path. In fact, it usually is not. At least two coordinates are required as indicated by minOccurs="2" in the following definition of "LineStringType" (geometry.xsd):

```
<complexType name=''LineStringType''>
 <annotation>
  <documentation>
   A LineString is defined by two or more coordinate
   tuples, with linear interpolation between them.
  </documentation>
 </annotation>
 <complexContent>
  <extension base=''gml:AbstractGeometryType''>
   <sequence>
    <choice>
     <element ref=''gml:coord'' minOccurs=''2''
     maxOccurs=''unbounded''/>
     <element ref=''gml:coordinates''/>
    </choice>
   </sequence>
  </extension>
 </complexContent>
</complexType>
```

Here is an example of a GML-encoded LineString instance:

```
<LineString srsName=''http://www.opengis.net/gml/srs/
epsg.xml#4326''>
 <coord><X>0.0</X><Y>0.0</Y></coord>
 <coord><X>20.0</X><Y>35.0</Y></coord>
 <coord><X>100.0</X><Y>100.0</Y></coord>
</LineString>
```

A *LinearRing* is a closed LineString. But it requires that the last coordinate be coincident with the first coordinate and that there be at least four coordinates (the three to define a ring plus the fourth duplicated one):

```
<complexType name=''LinearRingType''>
 <annotation>
```

```
<documentation>
A LinearRing is defined by four or more coordinate
tuples, with linear interpolation between them; the
first and last coordinates must be coincident.
</documentation>
</annotation>
<complexContent>
<extension base=''gml:AbstractGeometryType''>
 <sequence>
  <choice>
   <element ref=''gml:coord'' minOccurs=''4''
   maxOccurs=''unbounded''/>
   <element ref=''gml:coordinates''/>
  </choice>
 </sequence>
</extension>
</complexContent>
</complexType>
```

The *srsName* attribute is not needed because a LinearRing is used in the construction of Polygons (which specify their own SRS). Here's an example of a GML-encoded LinearRing instance:

```
<LinearRing srsName=''http://www.opengis.net/gml/srs/
epsg.xml#4326''>
 <coord><X>0.0</X><Y>0.0</Y></coord>
 <coord><X>20.0</X><Y>35.0</Y></coord>
 <coord><X>100.0</X><Y>100.0</Y></coord>
 <coord><X>50.0</X><Y>65.0</Y></coord>
 <coord><X>0.0</X><Y>0.0</Y></coord>
</LinearRing>
```

A *Polygon* is a connected surface that has an outer boundary and zero or more inner boundaries. The outer and inner boundaries of the polygon are a set of LinearRings. The "PolygonType" is defined in the geometry.xsd as follows:

```
<complexType name=''PolygonType''>
 <annotation>
  <documentation>
  A Polygon is defined by an outer boundary and zero or
  more inner boundaries which are in turn defined by
  LinearRings.
  </documentation>
 </annotation>
```

```
<complexContent>
 <extension base=''gml:AbstractGeometryType''>
  <sequence>
   <element name=''outerBoundaryIs''>
   <complexType>
    <sequence>
     <element ref=''gml:LinearRing''/>
    </sequence>
   </complexType>
   </element>
   <element name=''innerBoundaryIs'' minOccurs=''0''
   maxOccurs=''unbounded''>
   <complexType>
    <sequence>
     <element ref=''gml:LinearRing''/>
    </sequence>
   </complexType>
   </element>
  </sequence>
 </extension>
</complexContent>
</complexType>
```

Although not specificly defined in the definition of the "PolygonType," it is implicit that "the LinearRings of the interior boundary cannot cross one another and cannot be contained within one another" (OGC, 2001a). But the ordering of LinearRings and whether they form clockwise or counterclockwise paths are not important.

A following example of a Polygon instance has two inner boundaries and uses coordinate strings:

```
<Polygon gid=''_98217'' srsName=''http://
www.opengis.net/gml/srs/epsg.xml#4326''>
 <outerBoundaryIs>
  <LinearRing>
   <coordinates>0.0,0.0 100.0,0.0 100.0,100.0 0.0,100.0
   0.0,0.0</coordinates>
  </LinearRing>
 </outerBoundaryIs>
 <innerBoundaryIs>
  <LinearRing>
   <coordinates>10.0,10.0 10.0,40.0 40.0,40.0 40.0,10.0
   10.0,10.0 </coordinates>
  </LinearRing>
 </innerBoundaryIs>
```

```
<innerBoundaryIs>
 <LinearRing>
  <coordinates>60.0,60.0 60.0,90.0 90.0,90.0 90.0,60.0
  60.0,60.0 </coordinates>
 </LinearRing>
 </innerBoundaryIs>
</Polygon>
```

7.8.4 Geometry Collections

A geometry collection is a collection of geometry elements to form a new geometry element. There are a number of homogeneous geometry collections that are collections of the same primitive geometric elements and are predefined in the geometry schema (geometry.xsd). A *MultiPoint* is a collection of Points; a *MultiLineString* is a collection of LineStrings; and a *MultiPolygon* is a collection of Polygons. Here is an example of the multiLineString geometry element defined in the geometry schema:

```
<complexType name=''MultiLineStringType''>
 <annotation>
 <documentation>
   A MultiLineString is defined by one or more
 LineStrings, referenced through lineStringMember
 elements.
  </documentation>
 </annotation>
 <complexContent>
  <restriction base=''gml:GeometryCollectionType''>
   <sequence>
    <element name=''lineStringMember''
    maxOccurs=''unbounded''>
     <complexType>
      <sequence>
       <element ref=''gml:LineString''/>
      </sequence>
     </complexType>
    </element>
   </sequence>
  </restriction>
 </complexContent>
</complexType>
```

Where "MultiLineStringType" is derived from "gml:GeometryCollectionType," which refers to a geometry collection that must include one or more geometries, referenced through geometry member elements

("LineStringMember" in this case). There must be at least one "Line-StringMember" element but it could have unlimited members. The srsName attribute can only occur on the outermost geometry collection and must not appear as an attribute of any of the enclosed geometry elements. Here is an example of a "MultiLineString" instance with three members from the GML 2.0 Recommendation Paper (OGC, 2001b):

```
<MultiLineString srsName=''http://www.opengis.net/gml/
srs/epsg.xml#4326''>
 <lineStringMember>
  <LineString>
   <coord><X>56.1</X><Y>0.45</Y></coord>
   <coord><X>67.23</X><Y>0.98</Y></coord>
  </LineString>
 </lineStringMember>
 <lineStringMember>
  <LineString>
   <coord><X>46.71</X><Y>9.25</Y></coord>
   <coord><X>56.88</X><Y>10.44</Y></coord>
  </LineString>
 </lineStringMember>
 <lineStringMember>
  <LineString>
   <coord><X>324.1</X><Y>219.7</Y></coord>
   <coord><X>0.45</X><Y>4.56</Y></coord>
  </LineString>
 </lineStringMember>
</MultiLineString>
```

In addition, the *MultiGeometry* element is a collection of any geometric elements, including any of the primitive geometry elements such as Points, LineStrings, Polygons, MultiPoints, MultiLineStrings, MultiPolygons, and even other geometry collections. The MultiGeometry element uses a generic "geometryMember" property to include the next geometry element in the collection. An example of a heterogeneous MultiGeometry instance appears below:

```
<MultiGeometry gid=''c731''
srsName=''http://www.opengis.net/gml/srs/
epsg.xml#4326''>
 <geometryMember>
  <Point gid=''P6776''>
   <coord><X>50.0</X><Y>50.0</Y></coord>
  </Point>
 </geometryMember>
```

```
<geometryMember>
 <LineString gid=''L21216''>
  <coord><X>0.0</X><Y>0.0</Y></coord>
  <coord><X>0.0</X><Y>50.0</Y></coord>
  <coord><X>100.0</X><Y>50.0</Y></coord>
 </LineString>
</geometryMember>


<geometryMember>
 <Polygon gid='' 877789''>
  <outerBoundaryIs>
   <LinearRing>
    <coordinates>0.0,0.0 100.0,0.0 50.0,100.0 0.0,0.0
    </coordinates>
   </LinearRing>
  </outerBoundaryIs>
 </Polygon>
</geometryMember>
</MultiGeometry>
```

7.8.5 Encoding Features with Geometry

The purpose of this predefined set of geometry properties in GML 2.0 is to consistently describe geographic features. GML 2.0 also offers flexibility for users to define their own geometry properties based on those predefined geometry properties. For example, in our "street" example, we can add the geometry properties to describe its location by adding a linear property called "centerLineOf," which is one of the predefined descriptive names that can substitute for the formal name "LineStringPropertyType":

```
<?XML Version=''1.0''?>
<street>
 <streetName>Broadway</streetName>
 <speedLimit>45</speedLimit>
 <numberLanes>4</numberLanes>
 <majorRoad/>
 <gml:centerLineOf>
  <gml:LineString>
   <gml:coordinates>1.0 1.0, 3.0 4.0, 5.5 6.0</gml:
   coordinates>
  </gml: LineString >
 </gml:centerLineOf>
</street>
```

The corresponding XML schema (or application schema) to support this XML coding is

```
<element name=''Street'' type=''ex:StreetType'' />
<complexType name=''StreetType''>
 <sequence>
  <element name=''StreetName'' type=''string''/>
  <element name=''streetLimit'' type=''integer''/>
  <element name=''numberLanes'' type=''integer''/>
  <element name=''majorRoad'' type=''Boolean''
  MinOccur=''0''/
  MaxOccur=''1''>
 </sequence>
</complexType>
```

which is based on the following application schema fragment:

```
<element name=''Street'' type=''ex:StreetType''
substitutionGroup=''gml:_Feature'' />

<complexType name=''StreetType''>
 <complexContent>
  <extension base=''gml:AbstractFeatureType''>
   <sequence>
    <element name=''streetName'' type=''string''/>
    <element name=''streetLimit'' type=''integer''/>
    <element name=''numberLanes'' type=''integer''/>
    <element name=''majorRoad'' type=''Boolean''
    MinOccur=''0''/ MaxOccur=''1''>
    <element ref=''gml:centerLineOf''/>
   </sequence>
  </extension>
 </complexContent>
</complexType>
```

Alternatively, you could define a user-defined geometry property such as "streetLocation" and declare it globally as shown below:

```
<element name=''Street'' type=''ex:StreetType''
substitutionGroup=''gml:_Feature'' />
<element name=''streetLocation'' type=''gml:
LineStringPropertyType'' substitutionGroup=''gml:
lineStringProperty''/>

<complexType name=''StreetType''>
 <complexContent>
  <extension base=''gml:AbstractFeatureType''>
   <sequence>
    <element name=''streetName'' type=''string''/>
```

```
<element name=''streetLimit'' type=''integer''/>
<element name=''numberLanes'' type=''integer''/>
<element name=''majorRoad'' type=''Boolean''
MinOccur=''0''/ MaxOccur=''1''>
<element ref=''ex:streetLocation''/>
</sequence>
</extension>
</complexContent>
</complexType>
```

In this case, the "streetLocation" becomes a global property to identify the location of the ⟨street⟩ instance. It has the "*gml:LineStringPropertyType*" and it can substitute for the "gml:lineStringProperty" element. You can then reference it in a type definition as in ⟨element ref="ex:streetLocation"/⟩.

7.9 XLINK SCHEMA IN GML 2.0

One of the main reasons for the Web's popularity is its hyperlink capability. A hyperlink establishes a one-way link with another element in the same document, an element in a different document, or an entirely separate document. HTML implemented this by using the ⟨a⟩ element and the *href* attribute:

```
<a href=''dgis.html''>What's Distributed GIS?</a>
<a href =''#XML''>What's XML?</a>
```

In this example, ⟨a href="dgis.html⟩ links to an entire document named dgis.html, while ⟨a href="#XML⟩ links to a section marked ⟨a name="XML"⟩ in the same document.

But hyperlink in HTML is very limited. For example, it does not offer multidirectional links. Except for a display of the linked element, we cannot control how the link will perform when clicked, and so on. XLink goes beyond simple HTML links. Coupled with XPointer, XLink offers ways to define links that are much more flexible than HTML's ⟨a href=⟩ links and do much more than just offer simple hypertext links.

XLink provides a mechanism for defining relationships between elements and offers multiple ways to represent those relationships. Its companion, Xpointer, provides ways to point to specific elements, character strings, vector graphic elements, or even individual characters of an XML document. XPointer describes how to address a resource, while XLink describes how to associate two or more resources. A *resource* is any addressable (or URI-referenced) unit of information or service, including files, images, documents, programs, and query results, or a portion of a resource.

XLink describes syntax for defining relationships between objects or the "participating resources" of a link (i.e., resources associated with a link). According to the XLink specification, XLink "uses XML syntax to create structures that can describe links similar to the simple unidirectional hyper-links of today's HTML, as well as more sophisticated links" (http://www.w3.org/TR/xlink/). More sophisticated links enable us to specify links with multiple target documents. GML uses XLink to link two or more geographic features.

7.9.1 Basic Concepts in XLink

An important concept in XLink is the link. An *XLink link* is defined in the W3C XLink specification as "an explicit relationship between resources or portions of resources" (http://www.w3.org/TR/xlink/). It is made explicit by an XLink *linking element*, an XML element that describes an XLink link. (The discussion here about XLink is derived from the W3C-recommended *XLink Version 1.0*, June 2001, available at http://www.w3.org/TR/xlink.)

When we use or follow an XLink link, we say we follow a *traversal*. Traversal always involves a pair of resources or portions of resources, starting from the *starting resource* and ending at the destination or *ending resource*. Information about how to traverse a pair of resources (e.g., the direction of traversal and possibly application behavior information) is called an *arc*. An arc has directions. If a link has two arcs that specify the same pair of resources but with different starting and ending resources, that link is *multidirectional*.

Xlink resources are divided into local resources and remote resources. According to the XLink specification, "a *local resource* is an XML element that participates in a link by virtue of having as its parent, or being itself, a linking element." A *remote resource* is any resource or resource portion that participates in a link by virtue of being addressed with a URI reference, even if it is in the same XML document as the link, or even inside the same linking element. You can see that the same resource could be local or remote depending on how it is referenced. A local resource is specified "by value" and a remote resource is specified "by reference (URI)."

With the definition of local and remote resources, we can define the directions of an arc. Any arc could generally have three directions: inbound, outbound, or third-party. An arc that starts from a local resource and ends with a remote resource goes *outbound,* that is, away from the linking element. The HTML ⟨a⟩ element is one example of an outbound arc. Similarly, the reverse direction of an arc is *inbound*. That is, its ending resource is local but its starting resource is remote. If an arc starts from a remote resource and ends with another remote resource, it is a *third-party* arc. Inbound and third-party links are more difficult to discover than the outbound links and are usually contained in a *link database* or *linkbase* for easier link discovery.

7.9.2 XLink Element Types and Attributes

There are six XLink elements or XLink element types, including simple, extended, locator, arc, resource, or title. Two of them (simple and extended) are considered linking elements. The other four provide various pieces of information that describe the characteristics of a link.

The two linking elements—simple and extended—are used to describe the *existence* of a link, either a simple link or an extended link. A *simple link* is an outbound link with exactly two participating resources from a local re- source to a remote resource, which is very similar to a standard HTML ⟨a href⟩ link. An *extended link* describes multipaths in a collection of arbitrary numbers of participating resources. It includes inbound and third-party arcs as well as outbound arcs. The structure of an extended link is much more complex than the simple links, and thus extended links have more function- ality as well.

The other four XLink elements are used to further describe the *character- istics* of a link. The locator-type elements address the remote resources par- ticipating in the link. The arc-type elements provide traversal rules among the link's participating resources. The title-type elements provide human-readable labels for the link. The resource-type elements supply local resources that participate in the link. For example, you can use locator, arc, resource, and title as child elements to further describe the extended links.

The XLink global attributes include *type, href, role, arcrole, title, show, actuate, label, from,* and *to.* The use of these global attributes makes the elements in the XLink namespace or other namespaces recognizable as XLink elements.

The *type* attribute indicates the XLink element type ("simple," "extended," "locator," "arc," "resource," "title," or "none"); each type has its own mean- ing and behavior as described above. When the value of the type attribute is "none," the element has no XLink-specified meaning and no XLink-specified relationship with any XLink-related content or attributes.

The *href* attribute is a locator attribute that provides the location of a remote resource. Its value must be a URI reference. It may be used on simple-type elements and must be used on locator-type elements.

The *role, arcrole,* and *title* are semantic attributes to describe the meaning of resources within the context of a link. The value of the *role* or *arcrole* attribute must be a URI reference. The value of the *title* attribute is optional, but it must be a string to describe the meaning of a link or resource.

The *show* and *actuate* attributes are behavior attributes. They signal what should happen to the traversal and ending resources, whether to display in the same browser and open a new window, and so on. Specifically, the *show* attribute indicates the desired presentation of the ending resource on traversal from the starting resource. The values of the show attribute must be one the following: "new," "replace," "embed," "other," and "none":

- "new" indicates that the ending resource should load it in a new window, frame, pane, or other relevant presentation context.

- "replace" indicates that the ending resource should load the resource in the same window, frame, pane, or other relevant presentation context as the starting resource.

- "embed" indicates that the ending resource should load its presentation in place of the presentation of the starting resource.
- "other" indicates that the behavior of an application traversing to the ending resource is unconstrained. Other markup presented in the link should be used to determine the appropriate behavior.
- "none" indicates that the behavior of an application traversing to the ending resource is unconstrained, and no other markup is present to help the application determine the appropriate behavior.

The *actuate* attribute is used to "communicate the desired timing of traversal from the starting resource to the ending resource (W3C, 2001)." Its values must be one of the following: "onLoad," "onRequest," "other," and "none":

- "onLoad" indicates that an application should traverse to the ending resource immediately upon loading the starting resource.
- "onRequest" indicates that an application should traverse from the starting resource to the ending resource only on request, for example, when a user clicks on the presentation (e.g., a hyperlink) of the starting resource.
- "other" indicates that the behavior of an application traversing to the ending resource is unconstrained. Other markup presented in the link should be used to determine the appropriate behavior.
- "none" indicates that the behavior of an application traversing to the ending resource is unconstrained, and no other markup is present to help the application determine the appropriate behavior.

The last three attributes (*label*, *from*, and *to*) are traversal attributes. The value of a *label*, *from*, or *to* attribute must be an NCName, that is, a valid XML element name type, except that colons are not permitted. If a value is supplied for a *from* or *to* attribute, it must correspond to the same value for some label attribute on a locator- or resource-type element that appears as a direct child inside the same extended-type element.

7.9.3 Simple Links

A simple link can only link two resources, one local and one remote, and the direction of that traversal arc is one way: from the local resource to the remote resource. It is thus always an outbound link. The local resource is always an XML element, or sometimes called a link element. The remote resource must be identified by a URI and could be many things, such as another element in another XML document, a group of elements, a whole XML document, or other non-XML document.

Any link that can be implemented in the HTML ⟨a href⟩ hyperlink is an example of a simple link in XLink. For example, if you want to show the

Internet GIS Showcase: California Department of Transportation Intelligent Mapping Server

This system, developed by the Intergraph Corporation, creates custom maps for the California Department of Transportation by dynamically combining numerous data. Users can query and locate, change map projections, combine numerous data, and even submit their own dynamic segmentation data to create custom maps.

The application is targeted to anybody who is interested in information about streets, highways, bridges, and public transportation systems in California. GeoMedia WebMap has been used to combine a variety of information sources into a single map view. GeoMedia WebMap serves live ARC/INFO, ArcView, Microstation, MGE, and Oracle Spatial Cartridge GIS data sets, stored in at least 11 different surveyor's coordinate systems without translation or data conversion.

California State Highway System
Click on an area of the state for a more detailed map.

Caltrans Districts

1. Northwest California
2. Northern California
3. North Central California
4. San Francisco Bay Area
5. Central Coast
6. South Central California
7. Los Angeles & Ventrua Co.'s
8. San Bernardino & Riverside Co.'s
9. Eastern Sierra Nevada
10. Central California
11. San Diego & Imperial Co.'s
12. Orange Co.

Roadside Rest Areas

Image reprinted with permission from Intergraph.

Source: http://maps.intergraph.com/caltrans/default.asp.

properties of a spatial feature like a polygon and all properties of that polygon are contained in a single file, you could use simple link. When the user clicks on the polygon, the property information would be displayed. Here is an example:

```
<street xmlns:xlink=''http://www.w3.org/1999/xlink''
  xlink:type=''simple''
  xlink:href=''http://www.dgis.org/street/pics/
road.doc''
  xlink:role=''document''
  xlink:title=''Feature property document''
  xlink:show=''embedded''
  xlink:actuate=''onLoad'' width=''150''
height=''300''>
</street>
```

In this example, the first line declares the XLink namespace. The XLink namespace must be mapped to the http://www.w3.org/1999/xlink namespace URI. The second line indicates that this is a simple link. The third line, xlink: href=, indicates the remote resource of the link, the file you want to reference, and its location. If a value is not provided for the *href* attribute, the syntax is still OK, but it means that the link is untraversable. Such a link may still be useful, for example, to associate properties with the resource by means of XLink attributes.

The fourth line, xlink:role=, specifies the meaning of the connection between the resources. It indicates the remote resource is a document in this example. The fifth line indicates the description of this linked file, such as what a tooltip might show when the user moves the cursor over the link.

The sixth and seventh lines are optional link behavior attributes. The xlink: show attribute is an instruction to embed the ending resource in the current document at the location of the link element.

The xlink:actuate attribute value of "onLoad" shows the referenced ending resource should appear automatically when the link is loaded. The width="150" height="300" attributes indicate the width and height of the frame window when it is loaded in the XML document.

7.9.4 Extended Links

A simple link is more like a hyperlink in HTML, linking elements with target resources one at a time in one direction (outbound only). It cannot link elements with one or multiple target resources in bidirections. An extended link is created to handle linkages among multiple participating resources, remote or local. It is a link that associates an arbitrary number of resources in multidirections. An extended link is the only kind of link that is able to have inbound and third-party arcs.

Extended links are most appropriate in the following situations: (1) when one element links with multiple resources, such as one spatial element linking with multiple properties in different property files; (2) when the participating resources are read only; (3) where it is expensive to modify and update documents but inexpensive to modify and update a separate linking element; and (4) where the resources are in formats with no native support for embedded links (such as many multimedia formats).

For example, a city block is represented as a spatial polygon feature element. It is associated with remote resources such as property data, land use data, environmental data, census data, transportation data, and crime data, all of which are located in different locations in respective departments (Figure 7.2).

With a simple link, you cannot associate these remote resources. However, they can be associated and interconnected with the extended link, as shown in Figure 7.2. The lines emanating from the extended link represent the association it creates among the resources. But the lines or links do not have directions. This means that the resources are interconnected with no particular order; they could be accessed in either direction. To indicate directionality, you need to create traversal rules (use arc-type element).

An extended link in XML is represented by an xlink:type attribute with the value of "extended." A local resource is represented by a resource element

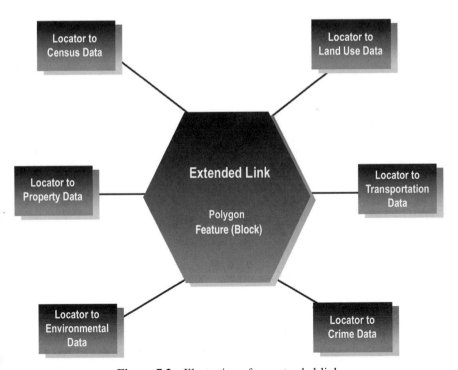

Figure 7.2   Illustration of an extended link.

with an attribute value pair of xlink:type=resource. A remote resource is represented by a locator element that has an attribute value pair of xlink: type=locator. An arc element is represented by an arc element that has an attribute value pair of xlink:type=arc.

In addition, the extended-type element may have the semantic attributes *role* and *title*. The *role* attribute indicates a property that the entire link has, and the *title* attribute indicates a human-readable description of the entire link.

A locator contains a URI for the location of resources and is typically represented by an XLink:href attribute, as shown in the following example, in which the street element contains two locator elements that identify the historical data of this street in the years 1990 and 1995:

```
<street XLink:type=''extended'' xlink:label=''current''
xlmns:xlink=''http//www.w3.org/1999/xlink''>
<streetName)Broadway</streetName>
<speedLimit>45</speedLimit>
<numberLanes>4</numberLanes>
<majorRoad/>
   <history year=''1990'' xlink:type=''locator''
xlink:href=''archive/1990.xml''
    xlink:label=''archive90''
    xlink:role=''http/www.dgis.org/''
    xlink:title=''street archive data 1990''/>
   <history year=''1995'' xlink:type=''locator''
xlink:href=''archive/1995.xml''
    xlink:label=''archive95''
    xlink:role=''http/www.dgis.org/''
    xlink:title=''street archive data 1995''/>
</street>
```

In this example, the street element is linked with two remote resources 1990.xml and 1995.xml that are located at "archive/." The xlink:label attribute indicates a reference point that will be used later to connect one point to the next. The xlink:role attribute is used to provide more information about the remote resource. The xlink:title attribute describes a title of the remote resource.

Besides linking with remote resources, the XLink can also connect the link element with elements in the same file or local resources. The xlink: type="resource" is used to indicate the linkage with local resources. A local resource element has the same attributes as a locator element, that is, xlink: href, xlink:label, xlink:role, and xlink:title. For example, the ⟨street⟩ element that has an xlink:type="resource" attribute becomes a local resource that can be connected to later:

```
<streetName xlink:type=''resource''>Broadway</
streetName>
```

Once you have the reference points or possible link points, whether they are remote or local resources, the next step is to connect them. This is done by using the arc element. Arcs are paths between resources. An arc element is defined by an xlink:type attribute with the value "arc." The arc element uses an xlink:from attribute to identify the link's source and an xlink:to attribute to identify the link's target. The values of the xlink:from and xlink:to attributes are not URIs: rather it is the value of the xlink:role attribute.

Based on the definition of the above example, we can make the following connections among the current data, the 1990 and 1995 archive data using the arc element:

```
<!-- arc elements make the connection -->
<nextyear xlink:type=''arc'' xlink:from=''current''
xlink:to=''archive90'' xlink:show=''replace'' xlink:
actuate=''onRequest'' /)
    <nextyear xlink:type=''arc'' xlink:from=''archive90''
    xlink:to=''archive95'' xlink:show=''replace'' xlink:
    actuate=''onRequest'' />
    <prevyear xlink:type=''arc'' xlink:from=''archive95''
    xlink:to=''archive90'' xlink:show=''replace'' xlink:
    actuate=''onRequest'' />
    <prevyear xlink:type=''arc'' xlink:from=''archive90''
    xlink:to=''current'' xlink:show=''replace'' xlink:
    actuate=''onRequest'' />
```

where "nextyear" and "prevyear" are connection elements that contain the connection information. This extended link can be represented as in Figure 7.3. Besides the links shown in Figure 7.3, other links are also possible. For example, if you want to link the current street information with all archive information in past years and let the user decide which year he or she wants to see, you could change the xlink:role attribute value to the same value:

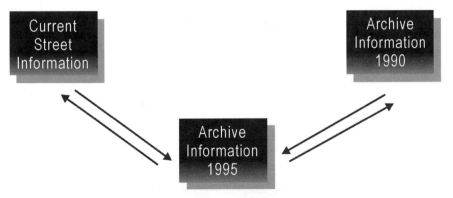

Figure 7.3   Extended link with three resources and four arcs.

```
<street XLink:type=''extended'' xlink:label=''current''
xlmns:xlink=''http//www.w3.org/1999/xlink''>
<streetName>Broadway</streetName>
<speedLimit>45</speedLimit>
<numberLanes>4</numberLanes>
<majorRoad/>
  <history year=''1990'' xlink:type=''locator''  xlink:
href=''archive/1990.xml''
    xlink:label=''archive''
    xlink:title=''street archive data 1990''/>
  <history year=''1995'' xlink:type=''locator''  xlink:
href=''archieve/1995.xml''
    xlink:label=''archive''
    xlink:title=''street archive data 1995''/>
</street>
  <!- arc elements make the connection ->
  <seeArchive xlink:type=''arc'' xlink:from=''current''
  xlink:to=''archive'' xlink:show=''replace'' xlink:
actuate=''onRequest'' />
```

This extended link can be represented as in Figure 7.4.

In addition to linking to a whole document, you could also link to a specific element or range of elements using XPointer. An XPointer is syntax to locate points or ranges in an XML document. It is usually attached to the end of the URI to indicate a particular part of an XML document. It is similar to the name attribute in the ⟨a⟩ element in HTML but without requiring insertion of a named anchor inside the document.

XPointer offers links based on the structure of the XML document. Unlike HTML, with which you have to declare a "Name" attribute explicitly in the ⟨a⟩ element, XPointer enables you to specify the link to a specific element or range of elements without requiring the target document to have predefined

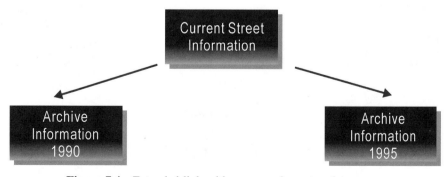

Figure 7.4 Extended link with one arc element and two arcs.

link anchor elements. For example, XPointer offers syntax that allows you link to two specific points anywhere within a target document as the beginning and end of a link resource by using Xpath.

An XPointer is often expressed as a location path in the XPath expression. The following are some valid XPointer examples:

- Xlink:href = "http/www.dgis.org/#streetad-dress[streetName = "Broadway"]" points to the streetaddress element and the specific content in the streetName element.
- Xlink:href = "http/www.dgis.org/archive/ #xpointer(streetaddress[streetName = "Broadway"]/rang-to (streetaddress[streetName = "Main"])" points to the range between streetName of "Broadway" and "Main."

7.9.5 Use XLink to Encode Feature Collections

A feature collection is a collection of one or more feature elements called feature members. A feature collection could also be a feature member of another feature collection. A feature collection can use the *featureMember* property to show the feature members it contains. Any feature collection element must derive from gml:AbstractFeatureCollectionType and declare that it can substitute for the (abstract) ⟨gml:_FeatureCollection⟩ element.

For example, a bus route contains bus *line segment* and bus *stop* feature members. The busRoute feature collection can be encoded in GML as follows:

```
<BusRoute fid=''route15''>
<dateCreated>May 2002</dateCreated>
<gml:featureMember>
 <LineSegment fid=''ls1045''>....</LineSegment>
</gml:featureMember>

<gml:featureMember>
 <stop fid=''stop2568''>....</Stop>
</gml:featureMember>

<gml:featureMember>
 <stop fid=''stop3812''>....</Stop>
</gml:featureMember>
</BusRoute>
```

Notice each GML features has a *fid* attribute of type ID inherited from gml:AbstractFeatureType. The *fid* must be unique in a GML document. The above GML document is associated with the following application schema fragments:

```
<element name=''BusRoute'' type=''ex:BusRouteType''
 substitutionGroup=''gml:_FeatureCollection''/>
<element name=''LineSegment'' type=''ex:
LineSegmentType'' substitutionGroup=''gml:_Feature''/>
<element name=''stop'' type=''ex:StopType''
substitutionGroup=''gml:_Feature''/>


<complexType name=''BusRouteType''>
 <complexContent>
  <extension base=''gml:
AbstractFeatureCollectionType''>
   <sequence>
    <element name=''dateCreated'' type=''month''/>
   </sequence>
  </extension>
 </complexContent>
</complexType>

<complexType name=''LineSegmentType''>
 <complexContent>
  <extension base=''gml:AbstractFeatureType''>
   <sequence>....</sequence>
  </extension>
 </complexContent>
</complexType>

<complexType name=''StopType''>
 <complexContent>
  <extension base=''gml:AbstractFeatureType''>
   <sequence>.....</sequence>
  </extension>
 </complexContent>
</complexType>
```

The unique *fid* attribute can be used in XLink to unambiguously reference specific features within a GML document or in a remote GML document. In the above example, the ⟨featureMember⟩ elements contain a set of features, but it could also point to remote features or features elsewhere in the same document using XLink. For example, if bus stop features are located at http://www.dgis.org/busstop.xml, the above XML document would become (no change to the application schema is necessary)

```
<BusRoute fid=''route15''>
 <dateCreated>May 2002</dateCreated>
 <gml:featureMember>
```

```
<LineSegment fid=''ls1045''>....</LineSegment>
</gml:featureMember>

<gml:featureMember xlink:type=''simple''
xlink:href=''http://www.dgis.org/
busstop.xml#stop2568''/>

<gml:featureMember xlink:type=''simple''
xlink:href=''http://www.dgis.org/
busstop.xml#stop3812''/>

</BusRoute>
```

7.9.6   Use XLink to Encode Feature Associations

There are generally two ways to describe feature associations: *containment* and *linking*. Containment is used to describe binary relationships (i.e., it connects pairs of features) only as shown in the above feature collection example, where external features were linked with feature collection using links. Linking is generally used to link multidirectional relationships among multiple features and relationships that cannot be described by containment, such as an adjacency relationship among multiple polygon features.

For example, assume we want to encode three adjacent land parcels, as shown in Figure 7.5. We could use the *adjacentTo* property and simple links to encode the land parcel "adjacency" association as follows:

```
<Parcel fid=''P4097''>
<Owner>Dave Smith</owner>
<gml:extentOf>...</extentOf>
<adjacentTo xlink:type=''simple'' xlink:
href=''#P4098''/>
```

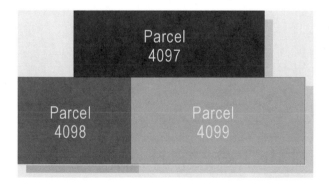

Figure 7.5   Illustration of three land parcels.

```
<adjacentTo xlink:type=''simple'' xlink:
href=''#P4099''/>
</Parcel>
....
<Parcel fid=''P4098''>
 <Owner>Peter Jones</Owner>
 <gml:extentOf>...</extentOf>
 <adjacentTo xlink:type=''simple'' xlink:
href=''#P4097''/>
 <adjacentTo xlink:type=''simple'' xlink:
href=''#P4099''/>
</Parcel>
....
<Parcel fid=''P4099''>
 <Owner>Sue Davis</Owner>
 <gml:extentOf>...</extentOf>
 <adjacentTo xlink:type=''simple'' xlink:
href=''#P4097''/>
 <adjacentTo xlink:type=''simple'' xlink:
href=''#P4098''/>
</Parcel>
```

The associated application schema is as follows:

```
<element name=''Parcel'' type=''ex:ParcelType''
 substitutionGroup=''gml: Feature''/>
<element name=''adjacentTo'' type=''ex:AdjacentToType''
 substitutionGroup=''gml:featureMember'' />

<complexType name=''ParcelType''>
 <complexContent>
  <extension base=''gml:AbstractFeatureType''>
   <sequence>
    <element name=''Owner'' type=''string''/>
    <element ref=''gml:extentOf''/>
    <element ref=''ex:adjacentTo'' minOccurs=''0''
    maxOccurs=''unbounded''/>
   </sequence>
  </extension>
 </complexContent>
</complexType>

<complexType name=''AdjacentToType''>
 <complexContent>
```

```
<restriction base=''gml:FeatureAssociationType''>
<sequence>
<element ref=''ex:Parcel''/>
</sequence>
</restriction>
</complexContent>
</complexType>
```

This is a simple way to identify the adjacency association. It is sometimes referred to as a "lightweight" relationship (OGC, 2001a). That is, the XLink is a "simple" link and the adjacency relationship is binary and bidirectional (it can be navigated in both directions). But the relationship itself has no identity in this encoding, and it is not possible to record properties on the relationship.

The same adjacency relationship can be encoded in more complex "extended" links in XLink, as shown below:

```
<! Describe Parcel 4097>

<Parcel fid=''P4097''
XLink:type=''extended'' xlink:label=''current parcel
4097'' xlmns: xlink=''http//www.w3.org/1999/xlink''>
<Owner>Dave Smith</owner>
<gml:extentOf>...</extentOf>
   <adjacetParcel fid=''p4098'' xlink:type=''locator''
xlink:href=''#p4098''
    xlink:label=''adjacent parcel''
    xlink:title=''adjacent parcel 4098''/>
   <adjacetParcel fid=''p4099'' xlink:type=''locator''
xlink:href=''#p4099''
      xlink:label=''adjacent parcel''
      xlink:title=''adjacent parcel 4099''/>
</Parcel>
   <!- arc elements make the connection ->
   <seeAdjacentParcel xlink:type=''arc'' xlink:
   from=''current parcel 4097'' xlink:to=''adjacent
   parcel'' xlink:show=''new'' xlink:
   actuate=''onRequest''  />
```

This GML encoding of parcel 4097 will link the current parcel (4097) with all adjacent parcels (4098, 4099, or any others) and let the user decide which adjacent parcel he or she wants to see. This is done by using the same label "adjacent parcel" to describe all adjacent parcels. When the user requests to see adjacent parcels of parcel 4097 by clicking on the parcel 4097, all adjacent parcels are displayed in a new window. This is just a simple example. More complex feature associations can be described using XLink, XPointer, and XPath.

7.10 MAKING USE OF GML DATA ON THE WEB

Now that we have all spatial features encoded in GML, how do we use them on the Web? There is nothing in GML that describes how the features should be displayed. How do we make vector maps based on GML data on the Web?

A GML document is simply a text. Like other text-based documents, it is viewed as text. But this text can be viewed in tables or graphics in spreadsheet; it can also be viewed as maps in a GIS program. Similarly, the original GML documents can be viewed in different forms of presentation. It can be viewed as text, as HTML form in a Web browser, as maps on the Web, or in WML form in wireless phones or PDAs. But in order to display GML in different formats as a presentation, GML must be transformed into a different format. The process of transforming XML documents into presentation formats that are different in presentation devices is called *styling* or *style transformation* (Hjelm and Start, 2002; Lake, 2001). For example, to make text-based features in GML into maps, GML must be styled into a graphical form such as a SVG, VML, or W3D. In order to display a GML document into wireless phones or PDAs, the GML document must be transformed into a WML format or XHTML format.

There are two basic ways to style or display XML documents on the Web, the XSL and the CSS. Both are W3C standards to be used with XML documents for displaying on the Web. CSS also works with HTML documents.

CSS is a non-XML syntax and does not need transformation. It is simple and straightforward, requires very little coding, and is thus an ideal starting point for formatting an XML or HTML document for Web publishing. A CSS defines the appearance of the XML document by applying rules to the elements in the XML documents, for example, rules for displaying ⟨LineString⟩ elements, rules for ⟨point⟩ elements, and rules to fill a ⟨polygon⟩ element. But CSS has its limitations. For example, CSS cannot reorder elements, which is important if you want to display only a subset of a large document.

The XSL, on the other hand, is an XML application and follows the rules governing any XML documents. Using XSL to display XML documents also requires transforming the XML document to a format that is appropriate for the display media, such as a Web browser or a wireless device. That is, XSL usually has to work with XSLT. XSL is used to define the appearance of an XML element such as fonts, text size, bolding, line spacing, and other aspects of a document's visual design. XSLT is a subset of XSL for transforming an XML document in one format to an XML document in another format, for example, from an XML document to a WML-formatted document.

XSL can do much more than CSS. It offers rules to format and display the layout of text and graphics on a page. It further offers a scripting language that allows the user to rearrange document content and to conditionally execute instructions based on evaluation of the document's data, structure, or other properties.

XSL applies formatting rules to each type of XML elements based on the structures in DTDs or schemata. Since all DTDs or schemata have the same structure, this makes it easy and more efficient to create stylesheets in conjunction with DTDs or XML schemata. Thus the same element in the XML document would have the same formatting rules.

For example, the appearance of an element may be based, among other criteria, on an element's position, an element's ancestors, and/or an element's attribute value. For example, assigning one style if an element's road attribute has a value of "expressway" and another if it equals "major road."

The XSL-generated stylesheet must specify formatting rules independent of viewing technology. The stylesheet processor generates an intermediate representation of the document as formatting objects. Then the XSLT transforms the XSL stylesheet into the syntax of different page layout languages.

XSLT provides the mechanism for translating the data from one format to another, such as from GML to WML. It shifts the focus of information exchange from defining common data formats for all applications to defining the transformations necessary to deliver data to each application in the format it desires.

7.10.1 Making Vector Maps on the Web with GML Data

Making a vector map on the Web with GML data is similar to making maps on the desktop using desktop GIS programs. It involves three basic steps:

1. *Feature Extraction* Extracting and interpreting geospatial features from the GML documents in a GML file server. This can be implemented in OGC's Web feature server.

2. *Map Styling* Styling the GML features into a graphic presentation using, for example, graphic symbols, line styles, and area patterns, and transforming the geometry of the GML data into the geometry of the graphic presentation (Lake, 2000) such as SVG, VML, and X3D.

3. *Graphic Rendering* Transforming the graphic presentation into a viewable image on the Web. Several graphic renderers are commercially available, such as the VML processor built into Internet Explorer 5.0 or above, Adobe SVG viewer plug-in, and Java applet SVG viewer from Ionic Software.

A simplified process of making vector maps on the Web is shown in Figure 7.6, where GML data are stored at the GML data store and GML features

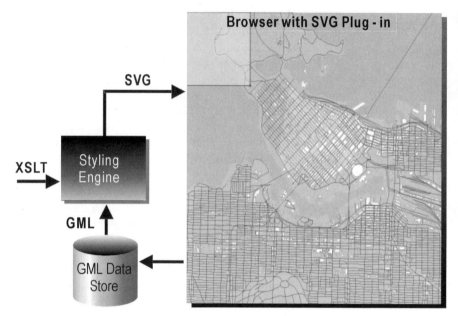

Figure 7.6 Process of making vector maps on the Web with GML. (Courtesy of Galdos Systems Inc. Figure reprinted with permission from Galdos Systems Inc.)

are extracted to feed into the styling engine. In the styling engine, XSLT styles GML features into a graphic presentation, such as an SVG. Here a map style sheet may be used to display specific graphic symbols, styles, and patterns as a means of portraying different spatial features. A styled SVG file is then sent to a Web browser for display. At the Web browser, an SVG plug-in needs to be installed to display the SVG file in the form of vector maps.

Figure 7.7 is an example of making vector maps on the Web with GML data. In this simple example there are only two road features to represent the line geometry feature, a lake feature to represent the polygon geometry feature, and a building feature to represent the point geometry feature. To make a vector map on the Web using this example data, first you need download an XSLT engine, such as Xalan or Saxon. To download Xalan, go to the Web site http://xml.apache.org/xalan-j. To download Saxon, go to the Web site http://users.iclway.co.uk/mhkay/saxon. In this example we use Saxon as an XSLT processor. With this XSLT processor we can transfer GML data into an SVG file.

The users can interact with the SVG map. For example, when the user wants to zoom or pan to a different area of the map, that request is then sent back to the GML data store, where the Web feature server will extract new GML features and send them the style engine for styling. The styled SVG will send to the Web for display again.

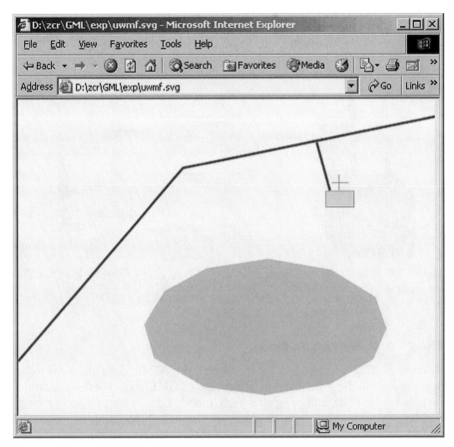

Figure 7.7   Example of making vector maps on the Web with GML data. (Provided by Chuanrong Zhang, Department of Geography, University of Wisconsin-Milwaukee.)

To facilitate the map-styling process, some programs have been developed. With these map-styling programs, styling the GML feature into a graphic presentation is as easy as making maps in desktop GIS programs. For example, Galdos Systems has developed a Freestyler as shown in Figure 7.8 (available at http://www.gmlcentral.com or http://mapstyles.galdosinc.com/ MapStyle.html). This is a simple, graphic, Internet-accessible application that you can use to create a GML XSLT map style sheet for SVG.

If you currently do not have GML data but have other data sources such as Shapefile, ArcInfo Coverage, or other formats, you need to have a conversion program to convert these file format to GML before you style them to create vector maps in the form of SVG on the Web. There are already some conversion programs that can convert your shapefile into GML. Oracle also has a GML conversion utility to convert Oracle data into XML. More are yet to come in the near future.

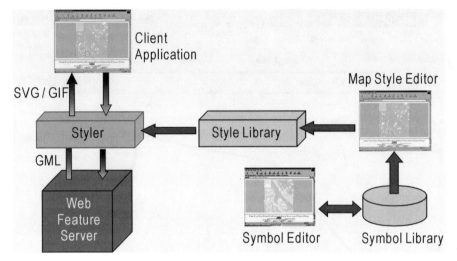

Figure 7.8   Illustration for Galdos Freestyler. (Courtesy of Galdos Systems Inc. Figure reprinted with permission from Galdos Systems Inc.)

7.10.2 GML Web Applications

Besides making vector maps on the Web using GML data, there are other applications on the Web that take advantage of GML data. These applications include Internet GIS and geospatial Web, mobile GIS, location-based services, building geospatial information infrastructure and information communities, and so on.

With the adoption of GML, all geospatial data could be interoperable and interconnected to reflect the true relationships of the real world. The capability of XLink and XPointer also allows the spatial relationship to be distributed. That is, data at one location in the earth can be dynamically related to data elements in another. This would essentially allow us to build a geospatial Web with one GML data server connecting with another data server, as shown in Figure 7.9.

When users request data about a particular area from the Internet, they can reach all the data that are available on the Web because all data are now interconnected. Furthermore, since GML is open and nonproprietary and no single vendor controls it, everyone can build components and software to capture, share, transform, and utilize these interconnected GML data Web. It also provides a mechanism for data providers to sell data to the end users in real time. Accessing geospatial data by Internet GIS programs anywhere on the Web in real time is made possible by using GML.

The interconnected, geospatial data Web would become part of the information infrastructure that, once built, could serve the whole information com-

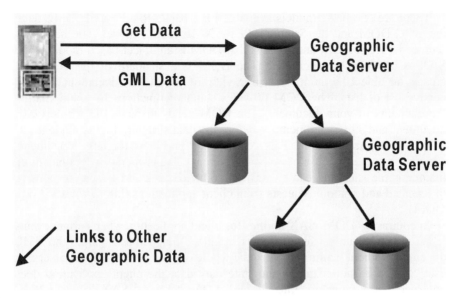

Figure 7.9   Illustration of geospatial data Web. (Courtesy of Galdos Systems Inc. Figure reprinted with permission from Galdos Systems Inc.)

munity that collects, manages, and utilizes geospatial information. Based on this integrated geospatial data Web, many applications and spatial services could be developed, such as location-based services, telecommunication planning and monitoring, utility planning and coordination, disaster management, accident investigation, and so on (Lake, 2001).

7.10.3   Extracting Features from GML Documents

GML has great implications for data sharing, exchange, and distribution over the Internet among different parties. But GML itself is nothing but a passive and structural format to encode geospatial data. A GML document is not too useful if it is not properly processed. For distributed GIS to take advantage of the structured GML data, an XML parser that is capable of accessing, interpreting, and extracting the underlying GML-coded geospatial data is needed.

The XML parser is used to read XML documents. The parser parses the contents of the XML document to the application piece by piece. The application that receives data from the parser may be a Web browser such as Netscape or Internet Explorer that displays the document to a reader, a desktop or Internet GIS program, a database server that stores XML data in a database, or a program written in Java, C, Visual Basic, or some other language.

There are two basic models available for a parser to access the GML data: the W3C DOM and the Simple API for XML (SAX) model. The basic difference between DOM and SAX is that DOM is an object-based model while SAX is an event-driven model. The DOM is a standard of the W3C that creates an object hierarchy or a tree view of your GML document as a recursive list of lists. The DOM provides standard functions for manipulating the elements in your document. The DOM needs to parse the whole GML document and store it in memory before the data structure tree is made available to client applications. This could create problems for large documents and requires intensive use of machine memory. The in-memory document tree maintains the state and can be further modified. So DOM is very responsive to repeated and random requests from client applications (http://www.w3.org/DOM).

In contrast to DOM, SAX notifies the client application when certain events (e.g., certain document features are recognized) happen as it parses the XML document without waiting to parse the whole document. The XML parser that uses SAX does not maintain the state; any data the client application does not specifically store are thus discarded. Furthermore, SAX is not a W3C standard. It was developed by an informal group of participants of the XML-DEV mailing list (http://www.xml.org/xml/resources_focus_sax.shtml).

Why or when would you use SAX or DOM? This partly depends on your XML document and partly on the client applications that receive the parsed XML data. If your document is very large and you only need a few elements in a large document, using SAX will save significant amounts of memory when compared to using DOM. On the other hand, the rich set of standard functions provided by DOM is not available when you use SAX. If the client application is responsive to real-time messaging from a GML document, or if you're parsing the XML document directly into a database, SAX is more appropriate. But if the client application uses an XML document as a simple database and needs to constantly access the data, DOM is more suitable. In addition, the DOM provides a W3C-standardized, complete, and editable view of the document's contents. It is thus more suitable if you intend to add codes or scripts within the document to explore and even change the document's contents (http://www.w3.org/DOM/faq.html).

For geographic features in the GML document, OGC has proposed an OGC Web feature server specification that developed interfaces for describing data manipulation operations such as querying, extracting, creating, deleting, and updating features that are stored in the GML documents or other data formats. The OGC Web feature server specification defines interfaces required to support query and transaction operations on geographic features stored in Web-accessible OGC simple feature datastores.

The basic architecture of the OGC Web feature server specification is shown in Figure 7.10. A client application such as a Web browser makes a feature request to the HTTP server, which forwards the request to the Web feature server. The Web feature server reads and processes the requests by

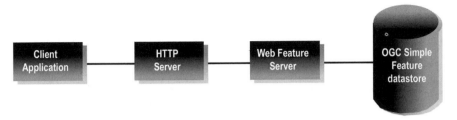

Figure 7.10   Architecture of OGC Web feature server.

getting the feature data from the OGC simple feature datastore. The datastore can be any type of system—for example, SQL database, flat file system, or GML documents. Features are stored in the datastore. A Web feature server request consists of a description of query or data transformation operations.

The Web feature server in Figure 7.10 is a program or module to support transaction or query requests. It has two major roles. First it translates client requests into the language of the target datastore and then passes the requests to the datastore engine for executing. Second, it sends any results back to the client.

Here is a very simple example of a client application requesting to retrieve feature instances from a datastore. In this example, all the properties of feature type "ParcelType" are fetched for an enumerated list of feature instances. The ⟨FeatureId⟩ element is used to reference each feature to be fetched:

```
<GetFeature>
 <Query typeName=''ParcelType''>
  <Filter>
   <FeatureId fid=''P4097''/>
   <FeatureId fid=''P4098''/>
   <FeatureId fid=''P4099''/>
  </Filter>
 </Query>
</GetFeature>
```

Besides ⟨GetFeature⟩, the OGC Web feature server specification also specifies other query and transaction functions, such as ⟨DescribeFeatureType⟩ to describe the structure of any feature type upon request; ⟨LockFeature⟩ to process a lock request on one or more instances of a feature type for the duration of a transaction; ⟨Transaction⟩ to service transaction requests such as create, update, and delete operations on features; and ⟨GetCapabilities⟩ to describe the capabilities of the Web feature server, such as which feature types it can service and what operations are supported on each. (For a detailed discussion, see the OGC Web feature server specification at http://www.opengis.org/ogcSpecs.htm.)

As you can see, the OGC simple feature datastore includes other data types besides the GML document, so users do not have to recode every existing geodatabase into GML. GML is not a database itself. You will not replace an Oracle or other geospatial data server with GML. But in order to publish your data as a valid GML document on the Web, you need to have software such as a Web feature server to retrieve data from a geospatial database in a GML format or to store GML document in a database. To store a GML document in the geospatial database on the server, software in the client side sends the XML document to the server using an established network protocol such as TCP/IP. Software on the server side receives the GML data, parses it, and stores it in the geospatial database. To retrieve a GML document from an existing geospatial database, you need a middleware product such as Enhydra that makes SQL queries against the database and formats the result set as XML before returning it to the client. Such an example is Oracle's XML SQL (XSQL) servlet.

7.11 SUMMARY: GML AND INTERNET GIS

GML provided a means of encoding the structure of spatial information, but it did not specify how the data might be presented to the Web or other mapping application programs. To view GML-encoded feature data on the Web, XSL or other stylesheets (e.g., Cascade Style Sheet) should be used for nonspatial data elements. To display spatial feature elements in the forms of maps, graphic Java applets or XML graphics technologies such as SVG or X3D renderer could be used to render the GML feature elements into maps. But GML itself does not specify the presentation elements. The separation of structure and presentation ensures that GML is not dependent on any particular XML graphical specification.

GML provides a mechanism to transport and store geospatial data in the Internet environment. The spatial features can be retrieved from the data server and transferred in GML format to the client applications for further process. The client applications then can manipulate and process the GML data for specific tasks or mapping functions. It is the client applications or server applications that make use of the data that enrich the functionality of Internet GIS.

One concern of the current GML is the simple spatial data model. The current version of GML (GML 2.1) only supports the simple feature geometry model. To be valid, all data models used by different GIS vendors have to be converted into the simple feature model. Some more complex features such as true curves and circles cannot be represented in the simple feature model. Furthermore, even within the goespatial information community, different data models are used to describe different spatial objects. For example, a data model used for transportation systems may not be the same as that for environmental systems. Even for the same transportation system, many data

Internet GIS Showcase: AxioMap

AxioMap—Application of XML for Interactive Online Mapping—is a Web map publishing kit and a customizable virtual map interface that allows for the display and manipulation of multiple point, line, and area layers, database query, mapping, hyperlinking, map labeling, and annotation. The major program developer of AxioMap is Ilya Zaslavsky at San Diego Super Computer Center, University of California, San Diego. This software is made possible by the advent of XML and XML applications for two-dimensional vector rendering such as VML and SVG. AxioMap generates VML shapes "on the fly" from XML-encoded geographic data that can physically reside on different servers. A thin-client-side solution, AxioMap provides for better interactivity than traditional map server–based approaches (Zaslavsky, 2000). A free version of the software can be downloaded from www.elzaresearch.com/landv.

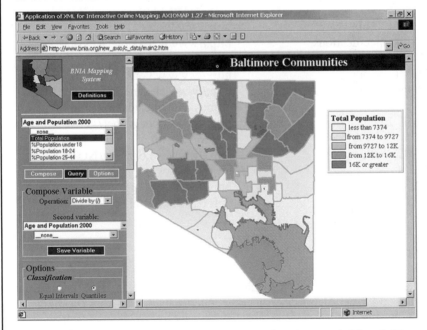

Application of XML for Interactive Online Mapping: AxioMap 1.27.

Image reprinted with permission from BNIA Website by Baltimore Neighborhood Indicator's Alliance. Software application used: AxioMap by ELZA Research.

models coexist. To be valid, different domain-based schemata (feature schemata, geometry schemata, and XLink schemata) have to be developed to support different data models. The dilemma is that if we create too many domain-based schemata, it will become another problem for data exchanges and data interoperability over the Internet.

This issue and others are being addressed in GML 3.0, which is addressing the convergence of GML and a geospatial encoding standard supported by ISO (ISO 19118). GML 3.0 will make important improvements in areas such as geometry and spatial reference systems and will offer much richer functionality than the current GML 2.1, while at the same time, GML 3.0 will ensure backward compatibility with GML 2.1.

WEB RESOURCES

Descriptions	URI
FGDC metadata standards	http://www.fgdc.gov/metadata/metadata.html
ISO/TC 211	http://www.isotc211.org/
XLink specifications	http://wwww.w3.org/TR/XLink
XML specifications	http://www.w3.org/XML/
SGML resource	http://www.w3.org/MarkUp/SGML/
Metadata model of ADL project	http://www.alexandria.ucsb.edu/docs/metadata/metadata001.html
GML 2.0	http://www.opengis.net/gml/01-029/GML2.html
VML, W3C	http://www.w3.org/TR/NOTE-VML
DOM	http://www.w3.org/DOM/
SAX	http://www.xml.org/xml/resources_focus_sax.shtml

REFERENCES

Baldonado, M., Chang, C. K., Gravano, L., and Paepcke, A. (1997). The Stanford Digital Library Metadata Architecture. *International Journal on Digital Libraries,* 1, pp. 108–121.

Castro, E. (2001). *XML for the World Wide Web.* Berkeley, California: Peachpit Press.

DuCharme, B. (1999). *XML: The Annotated Specification.* Upper Saddle River, New Jersey: Prentice-Hall.

Federal Geographic Data Committee (FGDC). (1995). *Content Standards for Digital Geospatial Metadata Workbook,* Version 1.0. Reston, Virginia: FGDC/USGS.

Federal Geographic Data Committee (FGDC). (1998). *Content Standards for Digital Geospatial Metadata.* FGDC-STD-001-1998. Reston, Virginia: FGDC/USGS.

Federal Geographic Data Committee (FGDC). (1999). *Content Standards for Digital Geospatial Metadata Part 1: Biological Data Profile.* FGDC-STD-001.1-1999. Reston, Virginia: FGDC/USGS.

Federal Geographic Data Committee (FGDC). (2000). *Content Standards for Digital Geospatial Metadata: Extensions for Remote Sensing Metadata (Public Review Draft).* Reston, Virginia: FGDC/USGS.

Fegeas, R. G., Cascio, J. L., and Lazar, R. A. (1992). An Overview of FIPS 173, The Spatial Data Transfer Standard. *Cartography and Geographic Information Systems,* 19(5), pp. 278–293.

Gardel, K. (1992). A (Meta-) Schema for Spatial Meta-data. In *Proceedings of Information Exchange Forum on Spatial Metadata.* Reston, Virginia: FGDC (Federal Geospatial Data Committee), pp. 83–98.

Gardner, S. R. (1997). The Quest to Standardize Metadata. *BYTE,* November, 22(11), pp. 47–48.

Harold, Elliotte Rusty, and Scott Means, W. (2001). *XML in a Nutshell: A Desktop Quick Reference.* Sebastopol, California: O'Reilly.

Hjelm, J., and Start, P. (2002). *XSLT: The Ultimate Guide to Transforming Web Data.* New York: Wiley.

ISO/TC 211/WG 1. (1998a). *Geographic Information—Part 1: Reference Model.* ISO/TC 211-N623, ISO/CD 15046-1.2.

ISO/TC 211/WG 3. (1998b). *Geographic Information—Part 15: Metadata.* ISO/TC 211-N538, ISO/TC 211-N538, ISO/CD 15046-15.

Kuhn, W. (1997). *Toward Implemented Geoprocessing Standards: Converging Standardization Tracks for ISO/TC 211 and OGC,* White Paper. ISO/TC 211-N418.

Lake, Ron. (1999). *Introduction to Geography Markup Language (GML).* URI: http://www.jlocationservices.com/company/galdos/articles/introduction_to_gml.htm.

Lake, R. (2000). *Making Maps for the Web with Geography Markup Language (GML).* URI: http://www.jlocationservices.com/company/galdos/articles/GMLMapMaking_gml.htm.

Lake, R. (2001). *GML 2.0—Enabling the Geo-spatial Web.* URI: http://www.jlocationservices.com/company/galdos/articles/GML3.htm.

Lanter, D. P., and Surbey, C. (1994). Metadata Analysis of GIS Data Processing: A Case Study. In *Proceedings of the 6th International Symposium on Spatial Data Handling,* Edinburgh, UK, T. Waugh and R. Healey (Eds.). London, U.K.: Taylor Francis, pp. 314–324.

Moellering, H. (1992). Opportunities for Use of the Spatial Data Transfer Standard at the State and Local Levels. *Cartography and Geographic Information Systems,* Special Issue, 19(5), pp. 332–334.

Open GIS Consortium (OGC). (1998). *OpenGIS Abstract Specification,* Version 3. Wayland, Massachusetts: Open GIS Consortium. URI: http://www.opengis.org/techno/specs.htm, May 11, 2000.

Open GIS Consortium (OGC). (2001a). *OpenGis Geography Markup Language (GML) Implementation Specification,* Version 2.0. Wayland, Massachusetts: Open Gis Consortium.

Open GIS Consortium (OGC). (2001b). *GML 2.0, Recommedation Paper,* February. Wayland, Massachusetts: Open GIS Consortium.

Ray, E. T. (2001). *Learning XML.* Sebastopol, California: O'Reilly.

Smith, T. R. (1996). The Meta-Information Environment of Digital Libraries. *D-Lib Magazine,* July/August.

Tsou, M. H., and Buttenfield, B. P. (1998). An Agent-Based, Global User Interface for Distributed Geographic Information Services. In *Proceedings of the 8th International Symposium on Spatial Data Handling,* Vancouver, Canada, pp. 603–612.

W3C. (2001). *XLink,* Version 1.0. URI: http://www.w3.org/TR/XLink.

Wu, C. V. (1993). Object-Based Queries of Spatial Metadata. Unpublished Ph.D. dissertation, State University of New York at Buffalo, Geography Department, Buffalo, New York.

Zaslavsky, I. (2000). A New Technology for Interactive Online Mapping with Vector Markup and XML. *Cartographic Perspectives,* 37, pp. 65–77.

CHAPTER 8

COMMERCIAL WEB
MAPPING PROGRAMS

One of the major errors made when selecting a Web mapping solution is in the implicit view by many that it's simply an extension of existing enterprise GIS/desktop mapping activities. This is not the case. Web mapping solutions are directed at a different audience than GIS/desktop mapping packages.
—W. Fredrick Limp (2001, p. 8)

8.1 INTRODUCTION

Previous chapters introduced the fundamental architecture of distributed Internet GIS and related software component technologies, such as Java, COM+/.NET, and CORBA, as well as a new way of coding geospatial information, the Geographic Markup Language (GML). This chapter will focus on the actual implementation of Web mapping services and the available commercial packages. You will see that some of the technologies we covered in the previous chapters have been implemented in the actual commercial Internet GIS software programs. In 2000, there were over 30 different Web mapping software packages or solutions provided by different GIS vendors. Thousands of Web mapping servers and image servers are running around the whole world. Many government institutions and private companies are all looking forward to adopt Web mapping tools for their own applications. However, different types of mapping tasks require different types of Web mapping servers. Choosing the right Web mapping products for specific GIS applications is truly a major challenge. From a manager's perspective, the major consideration will focus on the prices of Web map server packages, the training of technical staffs, and the maintenance of Web servers. From a GIS professional's perspective, the focus of Web mapping servers will be the functionality provided by the server, the implementation procedures, and GIS database connectivity. This chapter will provide some aspects from both views to help GIS professionals and managers to make informed decisions.

The following discussion will focus on four of the most popular Web Map Servers (WMSs), ESRI ArcIMS, Autodesk MapGuide, GeoMedia WebMap server, and MapInfo MapXtreme. Also, GE SmallWorld Internet application server and ER Mapper's image Web server will be illustrated separately in showcase windows. This introduction will focus on the architecture, major components, and communication mechanisms of these Web mapping programs. For a detailed user tutorial, refer to the user's manual for each software package.

8.2 ArcIMS FROM ESRI

ESRI is one of the major GIS vendors today. The early development of WMSs in ESRI includes ArcView IMS and MapObject IMS. ArcView IMS was built upon the popular GIS desktop software, ArcView, with an IMS extension to connect socket-based Web mapping server. MapObject IMS was created upon the ESRI programming tool, MapObject, with techniques from COM-based CGI functions and Java applets. Figure 8.1 shows an example of ArcView IMS applications that was developed by the University of Colorado at Boulder in 1998. The task of this Web mapping example was to create a virtual GIS class with on-line laboratory exercises for distance learning and education purpose (Buttenfield and Tsou, 1999).

Due to the limitation of network technologies and GIS software, the early development of ArcView IMS and MapObject only provided a "Band-Aid"-like solution which had very serious problems in the stability of map servers and the scalability of mapping applications. In 2000, ESRI developed a brand-new, stand-alone Web mapping package, called ArcIMS. The performance and stability of ArcIMS have been significantly improved compared to the early version of ArcView IMS and MapObject IMS. ArcIMS has its own mapping engine and three-tier architecture, which can generate images or features for different types of Web mapping services. The following discussion will introduce the software architecture, major components, and communication frameworks of ArcIMS based on the ArcIMS 4 architecture (ESRI, 2002).

8.2.1 Three-Tier Architecture Overview

ArcIMS has implemented a typical three-tier architecture that consists of presentation, business logic, and data storage tiers (Figure 8.2). The *presentation* tier provides an interface for users to access, process, and interact with the maps, tools, and geospatial data. It consists of two standard ArcIMS viewers, an HTML viewer and Java viewers. Customers can also build their own view-

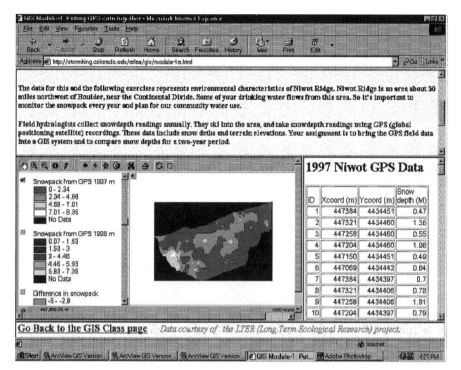

Figure 8.1   On-line GIS class using ESRI's ArcView IMS (Tsou and Buttenfield, 1999).

ers using ASPs and ActiveX controls to form an ActiveX viewer or Cold-Fusion to build a ColdFusion viewer. Furthermore, ArcIMS also supports a non-Web standard viewer called ArcExplorer (ESRI, 2002).

The *business logic* tier is used to receive and process user requests from the client viewers, produce maps, and manage the site. It has four major components, including the Web server, ArcIMS application server, ArcIMS application server connectors, and ArcIMS spatial server.

The *data storage* tier includes the sources of data and data server. The *typical* partition of this three-tier architecture is that the presentation tier is located at the Web browser and the business logic and data storage tiers are located in one or more computer servers.

When a user makes a request at the Web client, the request is sent to the Web server. Since the Web server cannot fulfill the request, it must send the request to a map (spatial) server that knows how to make maps and/or extract

Figure 8.2   Simplified ArcIMS three-tier architecture.

data from a data source. Once the map server generates the maps, it will send the maps back to the Web server, which forwards it to the Web client.

The actual process is much more complicated than this. First, in a distributed environment, there may be many spatial servers running at the same time to respond to thousands of user requests. A load balance service or an application server is needed to efficiently handle the request load and direct individual requests to a specific spatial server. This application server keeps track of which spatial server is working on which request, manages load, and assigns tasks to spatial servers.

Key Concepts

ArcXML *is the ArcIMS version of XML. The purpose of ArcXML is to convey the user request to the server through ArcXML code and send the information back to the user at the Web client.*

Second, the Web server does not communicate directly with the ArcIMS application server. It has to go through one of the application server connectors. The connectors provide a communication channel between a Web server and the ArcIMS application server. The role of the connectors is to establish a socket connection with the ArcIMS application server for each request. Once the communication channel is established, requests are converted into ArcXML format and are then sent to the application server. All connectors must reside on the same machine as a Web server. Now the simplified Figure 8.2 becomes Figure 8.3.

Communication between the tiers is handled through ArcXML, which is the ArcIMS version of XML. The purpose of ArcXML is to convey the user

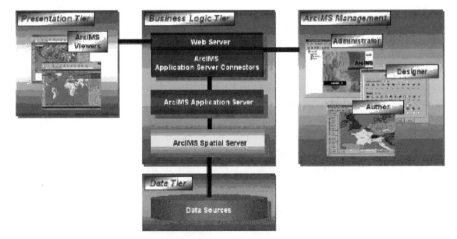

Figure 8.3 ArcIMS three-tier architecture (ESRI, 2002, p. 1). (Graphic Image provided courtesy of ESRI.)

request to the server through ArcXML code and send the information back to the user at the Web client. ArcXML is used to create map service configuration files about the area of the map and the layer of information the user is requesting.

- *Map service configuration files* describe how a map should be rendered, including such information as a list of layers used and symbols to be applied in each layer. This information is used by the feature servers and image servers to make maps or send out data to the client.
- *Metadata configuration files* provide information about the locations of the metadata and other associated information to support metadata services.
- *Requests* are user request information that is sent to map services, including request for maps, attribute data, and metadata. For example, a request could set a filter on an existing map service configuration file that specifies which part of a map and associated data will be acted upon.
- *Responses* send the information back to the client.
- *Administration* tasks such as adding, deleting, and stopping ArcIMS spatial servers, virtual servers, and services are handled using ArcXML.

The following is a sample of ArcXML. It sets the XML codes to request an image with the minimal coordinator of $(-180, -90)$ and the maximal coordinator of $(180, 90)$. Notice the elements used are different from the elements used in GML:

```
<ARCXML version=''1.0''>
    <REQUEST>
        <GET_IMAGE>
            <PROPERTIES>
                <ENVELOPE minx=''-180'' miny=''-90''
                maxx=''180'' maxy=''90''/>
            </PROPERTIES>
        </GET_IMAGE>
    </REQUEST>
</ARCXML>
```

8.2.2   ArcIMS Components in the Business Logic Tier

The ArcIMS business logic tier contains the components that are needed to process user requests and manage the site. These components include a Web server, the ArcIMS application server, the application server connectors, and the spatial server.

Figure 8.4 shows the structure of the ArcIMS business logic tier, which illustrates the relationship between the different components. When the

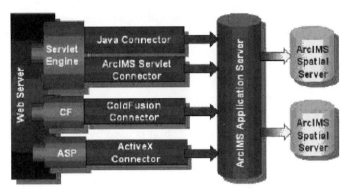

Figure 8.4 ArcIMS business logic tier (ESRI, 2002, p. 4). (Graphic Image provided courtesy of ESRI.)

ArcIMS user makes a request, the Web server first receives it and then passes through one of the connectors to the ArcIMS application server. The application server then hands the request to an ArcIMS spatial server.

Key Concepts

*The **Web server** in ArcIMS receives requests from the Web browser and responds with a document from the server to the Web browser. The **application server** handles the load distribution of incoming requests. It expects requests to be written in ArcXML. The **application connector** translates user requests into ArcXML format before handing the requests to the application server and translates ArcIMS spatial server responses into HTML before returning back to the Web server. The **ArcIMS monitor** is used to track the state of the ArcIMS spatial server.*

8.2.2.1 Web Servers The basic function of a Web server is to receive requests from the Web browser and respond with a document from the server to the Web browser. If a user requests anything other than an existing document, the Web server is unable to handle it. It thus has to pass on the request to other application servers, in this case the ArcIMS application server. The Web server is like a middleman or a messenger. Its role in ArcIMS is to get the user requests and pass them on to the application server. The Web server communicates with the Web client (browser) through HTTP.

8.2.2.2 ArcIMS Application Server The application server handles the load distribution of incoming requests. It also serves as a bookkeeper for keeping track of which map services (discussed later in this Chapter) are running on which ArcIMS spatial servers and allocates an incoming request to the appropriate spatial server. The application server can communicate with multiple Web servers, and thus it can reside on a different computer from the Web servers.

8.2.2.3 *ArcIMS Application Server Connectors*   Application server connectors are communication middleware betwen the Web server and the application server. ArcIMS offers four connectors:

- servlet connector,
- ColdFusion connector,
- ActiveX connector, and
- Java connector.

8.2.2.3.1 Servlet Connector   The servlet connector uses a Java servlet engine as the communication link between the Web server and the ArcIMS application server. Servlet engines are a Java platform technology for extending Web servers. It requires Java VM and servlet API. The servlet engine plugs into a Web server and provides the link between the Java VM and the Web server. Therefore, the servlet engine is used to communicate between the Web browser and Web servers. Because all servlets use the same API and the same protocol, ArcIMS can communicate with just about any Web server that can communicate with a Java servlet engine. The servlet connector is the default connector for ArcIMS. It works on Windows NT and UNIX platforms.

A new feature for ArcIMS 4 is compliance with OGC's WMS specification. The servlet connector supports a WMS connector so that any WMS requests from any clients on the Internet can be handled by the ArcIMS through the servlet connector. By complying with the WMS specification, the spatial server would become a global resource for any clients to access and utilize.

8.2.2.3.2 ColdFusion Connector   The ColdFusion connector uses a ColdFusion server as the communication link between the Web server and the ArcIMS application server. ColdFusion is an application server that extends the functionality of the Web server to handle complex user requests such as map rendering and data query. The ColdFusion application server receives requests from the Web server and passes them to the ColdFusion connector, where the request information is translated into ArcXML before being handed to the application server. After the request is processed by the spatial server, a response returns (through the same channels) to the ColdFusion application server, which generates an HTML page and sends it to the Web server. The ColdFusion connector works at both Windows NT and UNIX platform.

8.2.2.3.3 ActiveX Connector   The ActiveX connector is a COM DLL for use with COM applications such as Microsoft ASP. A linkage is established between the Web server and the ArcIMS application server by a connector object. A connector object can be created in ASP, Visual Basic, C++, Delphi, or another COM-compliant language. User requests can be made to handle map operations using the ActiveX connector object model API. The ActiveX connector translates these requests to ArcXML before handing them to the application server. Similarly, when it receives the response from the ArcIMS application server, it generates an HTML page and sends it to the Web server. The ActiveX connector is available only on the Windows platform.

8.2.2.3.4 Java Connector The Java connector is a set of Java beans (refer to Chapters 3 and 5 for detailed discussion of Java beans) that offer users the flexibility to create client and server applications, customer servlets, and Java Server Page (JSP) applications. It is available on all supported platforms.

8.2.2.4 Additional Processes on the Server In addition to the application server, two more background processes (Windows NT services/UNIX daemons) are used for supporting the spatial server.

- The *ArcIMS monitor* is used to track the state of the ArcIMS spatial server. The purpose of the monitor is to start new map services and spatial servers. Therefore, when the system is rebooted, the ArcIMS monitor automatically restores the spatial servers and map services.
- The *ArcIMS tasker* is used to remove output image files. These files, generated by the spatial server to support image map services, are removed at a user-defined time interval (e.g., every 5 minutes or 10 minutes).

8.2.2.5 ArcIMS Spatial Servers The backbone of ArcIMS is the ArcIMS spatial server. The spatial server can produce maps, access data, and bundle maps into an appropriate format based on the user requests. The spatial server contains several supporting components: Weblink, the XML parser, and the data access manager. Weblink is the communication gateway between the ArcIMS application server and the spatial server. The XML parser is used for parsing ArcXML requests. The data access manager provides a link between the spatial server and any data sources.

There are 7 kinds of component servers of an ArcIMS spatial server, including image server, feature server, geocode server, query server, extract server, metadata server, and ArcMap server (Figure 8.5).

Key Concepts

The **image server** *generates and sends maps to Web browsers as JPEG, PNG, or GIF images. The* **feature server** *sends or streams shapefiles and ArcSDE data sets from the server in a compressed binary format to a Java applet in the Web browser. The* **geocode server** *locates ad-*

Presentation Tier

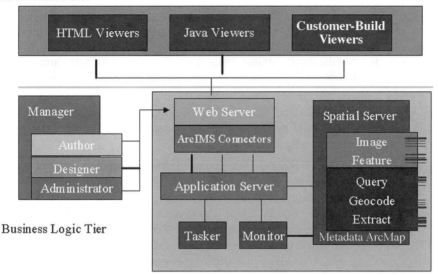

Figure 8.5 ArcIMS components (ESRI, 2001b). (Graphic Image provided courtesy of ESRI.)

dresses on maps based on address information in shapefiles and ArcSDE data sets. The **query** *server returns associated data for spatial and tabular queries. The* **metadata server** *is a repository for documents that contain information about maps, data, and services. The* **ArcMap server** *generates images using an ArcGIS ArcMap document as the input.*

- The *image server* generates and sends maps to Web browsers as JPEG, PNG, or GIF images. Cartographic images can be generated from shapefiles, ArcSDE data sets and other supported image formats such as ERDAS (Earth Resource Data Analysis System) GIS and LAN, ERDAS IMAGINE, GeoTIFF (Geographic Tagged Image File Format), GIF, MrSID (Multiresolution Seamless Image Database), Tagged Image File Format (TIFF), and GRID.

- The *feature server* sends or streams shapefiles and ArcSDE data sets from the server in a compressed binary format to a Java applet in the Web browser. Feature streaming is a temporary format that remains in the client computer's cache only as long as the Java applet is open. The Java applet receives instructions on how to assemble the received data. Feature streaming allows for more functional capabilities on the Web browser such as client-side labeling, changing the appearance of a map, and client-side spatial selection.

- The *geocode server* locates addresses on maps based on address information in shapefiles and ArcSDE data sets. The geocode server returns either an exact match or a list of candidate matches based on user inputs.
- The *query server* returns associated data for spatial and tabular queries. Queries can be built against shapefiles, ArcSDE data sets, and joined external tables.
- The *data server* extracts and returns data in shapefile format to the client. This process is different from feature streaming because data are actually sent to the client as a zipped shapefile rather than as a binary file as in feature server.
- The *metadata server* is a repository for documents that contain information about maps, data, and services.
- The *ArcMap server* generates images using an ArcGIS ArcMap document as the input. The behavior and types of requests are similar to the image server. The difference between the ArcMap server and the image server is the communication channel or information input. The image server accepts ArcXML while the ArcMap server uses an ArcMap document as input.

8.2.2.6 ArcIMS Spatial Server Instances Inside each spatial server, there are one or more instances to handle specific user requests: "A spatial server instance is a thread that can process one request at a time" (ESRI, 2002, p. 102). When a spatial server is initiated, it automatically opens two instances (the default value). But you can add or remove instances as they are needed. For example, if spatial server A has 6 instances and B has 4 instances, the virtual server has access to a total of 10 instances. An incoming request for a map service can be sent to any of the 10 instances running depending on the load condition of each instance.

8.2.2.7 Map Services A *map service* is a process that runs on one or more ArcIMS spatial server instances. It provides instructions to a spatial server on how to draw a map when a request is received (ESRI, 2002). There are four types of map services: image, ArcMap image, feature, and metadata.

An *image service* uses the image server component of the spatial server. It is used to generate a map image on the server upon receiving a user request and then send the response to the client as a JPEG, PNG, or GIF image format. Each time a client makes a request, a new map image is generated, even for a simple inquiry such as a zoom or pan. With image services, you could also have internal access to the query, geocode, and extract servers. That is, even if you choose the image server rather than the feature server, you could also use the functions in the query, geocode, and extract servers.

While the image service generates maps on the server and sends *map images* to the client, a *feature service* sends, or streams, the *data* from the server to the client in a binary format. Once the data are streamed to the client, they are temporarily cached at the client computer. The user then processes and interacts with the data in the Java applet on the client side. When the user wants to access more data than initially streamed, the Java client sends a request for additional data from the feature server. With feature services you could also have internal access to the geocode and extract servers for handling geocode and extract requests, respectively. The input to both the image service and the feature service is an ArcXML map service configuration file.

An *ArcMap image service* is very similar to an image service. That is, the ArcMap image service generates a map image and sends it back to the client as graphic formats such as JPEG, PNG, or GIF. New map images are generated each time a client makes a request. But the difference is that the *ArcMap image service* uses the ArcMap server while the image service uses the image server. Another difference is that the input to the service is an ArcMap document rather than ArcXML as in the image server. The third difference is that ArcMap servers have the equivalent of the query server already built in, but geocode and extract functionality is not available.

Metadata services allow users to search for a metadata repository for documents related to mapping, data, and services. Metadata services use the metadata server.

A map service is assigned to a specific virtual server rather than directly to an individual ArcIMS spatial server. The map service will start on all the spatial servers within the virtual server group. But what is a virtual server? We now turn to this concept.

8.2.2.8 *ArcIMS Virtual Servers* If there are too many incoming requests for the spatial servers to handle efficiently, more spatial servers need to be added. The added spatial server can be located in the same machine as the original ones; it could also be distributed across an ArcIMS site in different computers. That is, there could be one or more spatial servers that are physically located in the same machine or different machines. Each spatial server could have multiple instances running at the same time. To better manage these distributed spatial server instances and services, ArcIMS creates an ArcIMS virtual server. A virtual server is a logical grouping of one or more spatial server instances; it is not a physical entity.

Once the new spatial servers are added, they become part of the virtual server and can be assigned to run map services. Therefore, a virtual server is made up of multiple like instances, such as image server instances, of one or more spatial servers like image servers. There are seven types of virtual servers corresponding to seven component servers of an ArcIMS spatial server: image server, feature server, query server, geocode server, extract server, metadata server, and ArcMap server. Virtual servers can be added or deleted as needed.

The advantage of virtual servers is better service reliability and scalability. For example, if an ArcIMS spatial server goes down, the virtual server can still assign the work to another spatial server within the same virtual server. Furthermore, more spatial servers can be added to a virtual server to help meet increased loads.

8.2.2.9 Assigning User Requests to Spatial Server Instances When the application server receives a user request from the Web server, it starts a service. Once the service starts, it has to be assigned to a virtual server (an image, ArcMap, feature, or metadata virtual server) rather than directly to an individual ArcIMS spatial server. For example, the two image servers, A and B, and their respective instances in Figure 8.6 have been grouped into the same virtual server (e.g., image server). The image virtual server then assigns the service to all image servers (A and B in this case) and on all instances (e.g., instances A1–A6 and B1–B4 in Figure 8.6) within the virtual server group.

Once the image service starts, the query, geocode, and extract virtual servers can launch automatically depending on the functions requested. For example, a query server is always used to respond to attribute data requests. Therefore, the query virtual server assigns the service to the query server instances within that virtual server group. Similarly, if geocode or extract functions are used in the service, the geocode and extract virtual servers would assign the service to the geocode and extract server instances, respectively.

Likewise, when the feature service starts, a feature service is first assigned to a feature virtual server, which further assigns the service to all feature

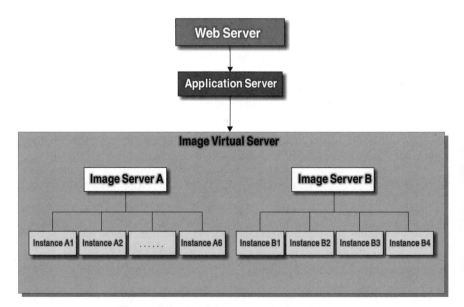

Figure 8.6   Spatial server and virtual server.

server instances within the virtual server group. If the service includes geocode and extract functions, the geocode and extract service would start on all instances within the feature virtual server automatically.

The ArcMap image and metadata services work the same way. The difference is that these two service types have no secondary servers, such as geocode, extract, and query servers, associated with them (ESRI, 2002)

Figure 8.7 illustrates that seven virtual servers are running on a spatial server called CEZANNE_1, where for the virtual feature server Feature-Server1, the extract virtual server ExtractServer1 is also running. For the image server ImageServer1, the geocode virtual server GeocodeServer1 and the query virtual server QueryServer1 are also running. The system administrator can use the administration tools provided by ArcIMS to create multiple virtual servers and configure the instances for each virtual server.

8.2.3 ArcIMS Components: Data Sources

ArcIMS supports different data formats for different services (e.g., image, feature, and ArcMap). Table 8.1 provides a summary of data formats available for image, feature, and ArcMap image services. Notice that ArcMap service supports the most data formats, while the feature service supports the least data formats. There is no direct support for other data formats from other GIS vendors.

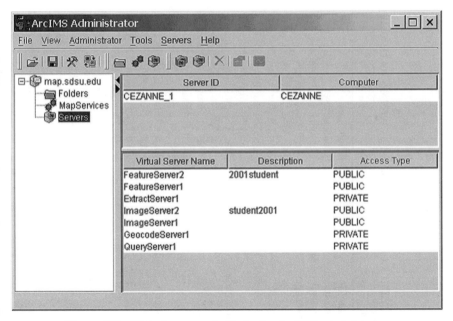

Figure 8.7   Virtual servers shown in the ArcIMS administrator.

TABLE 8.1 Formats Supported by ArcIMS by Data Type

Data Types	Data Format	Image	Feature	ArcMap
Shapefile	Shapefiles	Yes	Yes	Yes
Geodatabase	Geodatabases	No	No	Yes
Personal geodatabase	Personal geodatabases	No	No	Yes
Coverages	ArcInfo coverages	No	No	Yes
	PC ARC/INFO coverages	No	No	Yes
ArcSDE	ArcSDE for coverages	Yes	Yes	Yes
SQL Server, Informix,	ArcSDE features	Yes	Yes	Yes
	ArcSDE—versioned layers	No	No	Yes
DB2, Oracle	ArcSDE multiraster and 32-bit raster (Oracle)	Yes	No	Yes
	ArcSDE raster (SQL server, informix, DB2)	Yes	No	Yes
CAD	DWG (AutoCAD Drawing Format)	No	No	Yes
	DXF (Drawing Interchange Format)	No	No	Yes
	DGN (Director General de Normas)	No	No	Yes
Raster	Image catalog (raster catalog)	Yes	No	Yes
	ADRG (Arc Digitized Raster Graphics) image (.IMG)	Yes	No	Yes
	ADRG overview (.OVR)	Yes	No	Yes
	ADRG legend (.LGG)	Yes	No	Yes
	Band Interleaved by line (.BIL)	Yes	No	Yes
	Band interleaved by pixel (.BIP)	Yes	No	Yes
	Band sequential (.BSQ)	Yes	No	Yes
	Bitmap—Windows (.BMP)	Yes	No	Yes
	Controlled image base (.CIB)	Yes	No	Yes
	CADRG (Compressed Arc Digitized Raster Graphic, .CRG)	Yes	No	Yes
	DIGEST ARC Standardized Raster Product (ASRP)	Yes	No	No

Format			
DIGEST UTM/UPS (Uninterrupted Power Source)	Yes	No	No
Standardized Raster Product (USRP)	No	No	Yes
Digital Terrain Elevation Data (DTED) Level 1 and 2 (.DT1)	Yes	No	Yes
ERDAS Image (.IMG)	Yes	No	Yes
ERDAS 7.5 LAN (.LAN)	Yes	No	Yes
ERDAS 7.5 GIS (.GIS)	No	No	Yes
ERDAS raw (.RAW)	No	No	Yes
ER mapper (.ERS)	Yes	No	Yes
ESRI GRID	No	No	Yes
ESRI GRID stack	Yes	No	Yes
GIF (.GIF)	Yes	No	Yes
Impell bitmap (IMPELL)	Yes	No	Yes
JPEG (.JPG)	Yes	No	Yes
MrSID—LizardTech (.sid)	Yes	No	Yes
National Image Transfer Format (.NTF)	Yes	No	Yes
Portable Network Graphics (.PNG)	Yes	No	Yes
SunRaster file (SUN)	Yes	No	No
Tagged Image File Format (.TIF)	Yes	No	Yes
TIFF with Geo Header (.TIF)	Yes	No	Yes
Other			
Annotation layers	No	No	Yes
TIN	No	No	Yes
VPF	No	No	Yes

Source: ESRI, 2002, pp. 11–13.

Special care should be taken if ArcSDE is used as a data source. One ArcSDE connection should be available for each instance of an ArcIMS spatial server. For example, in Figure 8.5, spatial server A has six instances running. Therefore, the number of ArcSDE connections needed is also six— one connection for each instance of spatial server A.

But the number of ArcSDE connections required is not related to the number of Services running. Two or more of the same services can use the same group of ArcSDE connections if the same virtual server and the same ArcSDE instance are used.

Additional connections to ArcSDE are needed for the query, geocode, and extract servers. For example, one query server is required for each image server. Therefore, in the above example, if spatial server A is an image server, 6 ArcSDE connections would be needed for the image server and an additional 6 connections would be needed for the query server for a total of 12 connections. Additional ArcSDE connections are needed for the geocode server and extract server if called upon by a map service. The large number of ArcSDE connections required is not a problem because ArcIMS is a trusted client to ArcSDE, which means that an unlimited number of connections with ArcSDE are available. However, the large number of connections with ArcSDE takes more server resources. Connection pooling could be used to reduce the number of ArcSDE connections. In connection pooling, two or more instances of the same spatial server type, such as image server instances, can share one connection with ArcSDE.

8.2.4 ArcIMS Components: Client Viewers

ArcIMS includes ArcIMS viewers as clients for users to access, view, and manipulate geographic data. A client viewer usually includes a graphic presentation (e.g., map) and some methods or toolboxes for interacting with the map. There are two types of client viewers in ArcIMS, those standard ones that come with the ArcIMS software package that can be generated using ArcIMS Designer and those custom-built ones that can be built using ArcIMS application server connectors.

8.2.4.1 ArcIMS Standard Viewers ArcIMS provides three standard viewers: HTML viewers, a Java standard viewer, and a Java custom viewer. These three viewers have standard interfaces that are generated by the ArcIMS management tool–ArcIMS designer.

ArcIMS standard viewers communicate with an ArcIMS site through the ArcIMS servlet connector using ArcXML, as shown in Figure 8.8. When a user clicks on a map, the ArcIMS viewer generates a request that is translated in ArcXML and sent to the ArcIMS site through the Web server, servlet engine, servlet connector, and application server. The application server forwards the requests to the virtual server where the requested service resides and sends the request to one of the spatial server instances running on the

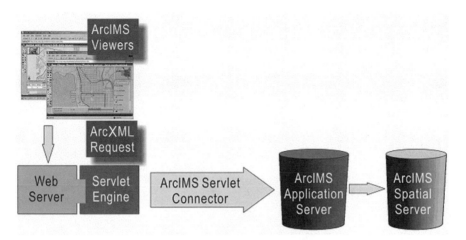

Figure 8.8 Communication process of ArcIMS standard viewers (ESRI, 2002, p. 20). (Graphic Image provided courtesy of ESRI.)

virtual server. The response follows the same path in reverse order. Notice that the response is in the form of ArcXML and is processed by the ArcIMS viewer at the client side. Therefore, the ArcIMS standard viewers generate requests and process responses in ArcXML on the client side. The ArcIMS standard viewers can be modified or customized or you can build a brand new viewer using the combination of HTML, DHTML, and JavaScript.

8.2.4.1.1 HTML Viewer The HTML viewer is written in HTML, DHTML, and JavaScript. This HTML viewer is a thin client that only supports map images on the Web browser and does not support feature data at the client side. Furthermore, only one image service or ArcMap image service can be displayed at a time. When a user clicks on a map or toolbox on the HTML viewer, a request is generated by the viewer using JavaScript. When the HTML viewer receives the response back from the spatial server in the form of ArcXML, the responses are then again parsed using JavaScript. The advantage of the ArcIMS HTML viewer is that the client-side browser will not require any plug-in preinstalled or running Java VM. Because of the use of JavaScript, the Web browser must be Internet Explorer or Netscape version 4.x or higher. Figure 8.9 is an example of the HTML viewer.

8.2.4.1.2 Java Viewers The Java viewers support image, ArcMap image, and feature services and are thus thicker clients than the HTML viewers. The Java viewers use a Java 2 applet for processing requests and displaying the information. The Java viewers support feature streaming. Data that are streamed to the Java viewers are temporarily cached on the client machine. The temporary cache is removed when the viewer is closed.

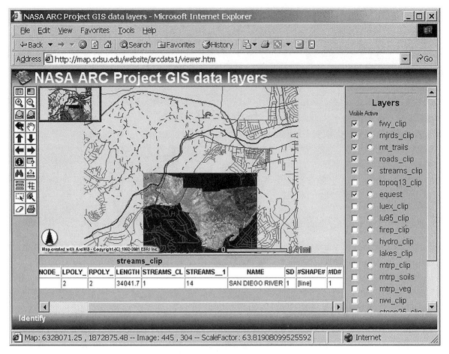

Figure 8.9   ArcIMS HTML viewer.

The Java viewers have more functions built into the client viewers so that the user can conduct more client-side processing than the HTML viewer. For example, the user can render maps, make queries, and do other processing inside the viewer. This is in contrast with the HTML viewer where all processing has to go through the spatial server. Therefore, most requests are processed on the client machine by the Java viewer. If the request requires data that are not currently in the cache, the client would send the request to the server to either retrieve more data or process data residing on the server. Furthermore, the Java viewer allows the user to combine local data and streamed feature data from the server to process requests in the same Java viewer.

ArcIMS includes two Java viewer clients: Java custom and Java standard. Both have the same functionality (Figures 8.10 and 8.11). Java standard viewer tools and functions are predefined and cannot be customized, while the Java custom viewer can be customized using methods in a viewer object model API. The Java custom viewer uses JavaScript to communicate with the applets, but the Java standard viewer does not use JavaScript. ArcIMS also includes ArcExplorer, a stand-alone viewer that does not require a browser.

8.2.4.2 ArcIMS Custom-Built Viewers   In addition to ArcIMS standard viewers, you could also design your own viewers using one of the ArcIMS

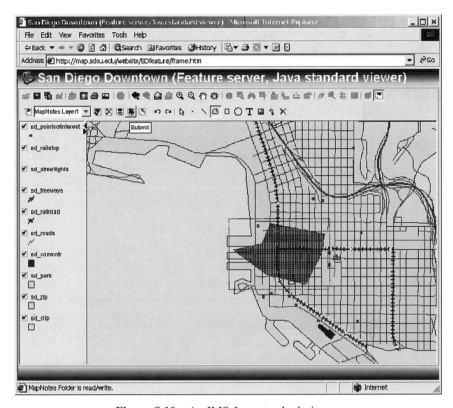

Figure 8.10   ArcIMS Java standard viewer.

application server connectors: the ActiveX connector, ColdFusion connector, and Java connector. The ArcIMS software program does not provide the standard interface, but it provides some sample custom-built codes.

8.2.4.2.1   Viewers Using ActiveX Connector   ActiveX viewers can be custom built using the ActiveX connector. Since the ActiveX connector is a COM DLL that is compatible with any COM application, ActiveX viewers use Microsoft ASP as a communication tool. That means that the communication between the ActiveX viewer and the Web server is through the ASP client and ASP server, as shown in Figure 8.12. The user makes a request from the Web browser using ASP, the request is received by the ASP server, which passes the request to the ActiveX connector. The ActiveX connector receives the request and translates it to ArcXML. Once the request is translated to ArcXML, the rest of the process is the same as the ArcIMS servlet connector (ESRI, 2002).

The ActiveX viewer looks identical to the HTML viewer to the end user and appears to have similar functionality. But the ActiveX viewer and HTML viewer are fundamentally different. First, the ActiveX viewer relies on ASP

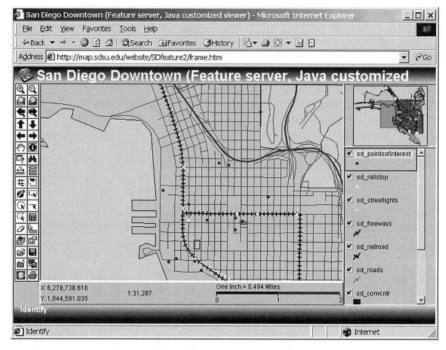

Figure 8.11 ArcIMS Java custom viewer.

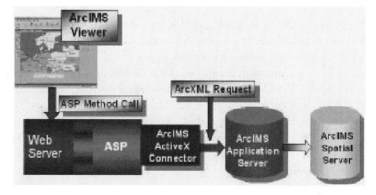

Figure 8.12 ArcIMS viewer using ActiveX connector (ESRI, 2002, p. 23). (Graphic Image provided courtesy of ESRI.)

for communication, while the HTML viewer relies on the servlet engine. Second, requests are not generated by the ActiveX viewer itself but, rather, are made through the ASP server, while the HTML viewer has to generate requests and convert the requests to ArcXML at the client. Third, the ActiveX viewer does not need to parse the response either, because the response is handled by ASP and an HTML page is generated on the fly, while the HTML viewer has to parse the ArcXML response at the client. Therefore, in the ActiveX viewer, all processing is handled on the server side and it is thus even thinner than the HTML viewer.

In addition, the ActiveX connector can also be used to build stand-alone client applications for the intranet using COM-based languages such as Visual Basic and C++.

8.2.4.2.2 Viewers Using ColdFusion Connectors ColdFusion viewers can be custom built using the ArcIMS ColdFusion connector. The ColdFusion viewer is very similar to the ActveX viewer. Both are HTML based and both process requests are at the server. Both look identical to the end user. The major difference is that the ColdFusion viewer uses the ColdFusion application server along with the Web server, while the ActiveX viewer relies on the ASP server working together with the Web server.

The process for making a request and receiving a response is illustrated in Figure 8.13. The user makes a request by triggering events in the client viewer; this executes ColdFusion tags on the ColdFusion application server. The ColdFusion application server then hands the tags to the ArcIMS Cold-Fusion connector, which translates the requests to ArcXML. Once the requests are translated to ArcXML, the remaining process is the same as the ArcIMS servlet connector and ActiveX connector.

The response follows the same path as the request in reverse order. Once the ArcIMS application server passes the ArcXML response to the Cold-

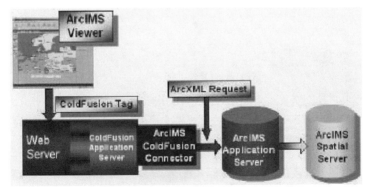

Figure 8.13 ColdFusion viewer using ColdFusion connector (ESRI, 2002, p. 24). (Graphic Image provided courtesy of ESRI.)

Fusion connector, an HTML page is generated on the fly by the ColdFusion application server. The Web server then sends the generated HTML page to the ColdFusion viewer on the Web browser. As you can see, just like ActiveX viewer, all processing is done on the server side. It is a very thin client.

8.2.4.2.3 Viewers Using Java Connector If the Java viewers that come with ArcIMS do not satisfy your specific needs, you could build a totally different viewer using the Java connector by adding JSP or a custom servlet. The process for making a request and additional components of the Java connector are shown in Figure 8.14. Notice how the servlet engine has two additional components: JSP and a custom servlet. The Java connector now consists of a group of Java beans and a JSP tag library. The tag library is a group of tags and attributes that provide an interface to the Java beans. The role of the tag library is similar to ArcXML.

Unlike the standard Java viewer, where the viewer generates user requests in ArcXML, the viewer using the Java connector generates user requests in the format of JSP. Once the request is received by the JSP server, it is passed to the Java bean through the JSP tag library. The Java connector then translates the request to ArcXML. Once the request is translated to ArcXML, the rest of the process is the same as other connectors. The response uses the same path as the request in reverse order. When a response is generated by the spatial server and passed to the ArcIMS application server, the ArcXML is processed by the Java connector. The Java connector then generates the HTML page on the fly, which is returned to the client by the Web server. In addition, custom stand-alone client applications for the intranet can also be built using Java connectors. But this is not an easy task and is not for the faint of the heart.

In addition to these client viewers, ArcIMS services can also be accessed from other stand-alone clients from the ESRI family programs, such as ArcInfo 8, ArcEditor 8, ArcView 8, ArcPad 5 or higher, or other hand-held or wireless devices.

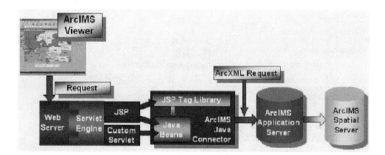

Figure 8.14 Viewers using Java connector (ESRI, 2002, p. 25). (Graphic Image provided courtesy of ESRI.)

8.2.5 ArcIMS Manager

The ArcIMS manager is a set of instructions for Internet GIS designers to set up and administer Internet services and to author, design, and publish map services on the server. It is a suite of Web pages that provides access to all ArcIMS server-side functions and tools. The ArcIMS manager consists of three stand-alone components:

- ArcIMS author—to author map service files;
- ArcIMS designer—to design Web pages;
- ArcIMS administrator—to publish and administer ArcIMS spatial servers, virtual servers, and map services.

The ArcIMS manager combines these applications—the ArcIMS author, ArcIMS designer, and ArcIMS administrator—into one wizard-driven framework, as shown in Figure 8.15.

8.2.5.1 ArcIMS Author The ArcIMS author is used to generate map configuration files. As we discussed before, the map configuration files are written in ArcXML and are the input to ArcIMS services. The ArcIMS author is also used to set up layers, define symbology, set scale dependencies, and define other mapping parameters. It can access shapefiles, ArcSDE data sets, and some image formats. Besides ArcIMS author, map configuration files can also be created and edited using an XML editor.

The second authoring tool is ArcMap. ArcMap is used to create map configuration files for ArcMap image services. The procedure for creating map configuration files for ArcMap image services is the same as for image or feature services. But the output file is in a binary format rather than ArcXML.

The third authoring tool is a text or XML editor to author metadata configuration files for metadata services. The metadata configuration file contains information about where in the ArcSDE those metadata files are located.

8.2.5.2 ArcIMS Designer The ArcIMS designer is used to design Web pages, including the Web page layout, available toolbox, and functionality of the Web page. The ArcIMS designer provides the user with a series of panels to select which services to use, which page style to use, and which operations and functions will be available in a client Web browser. The three standard client viewers—an HTML viewer, a customizable Java viewer, and a noncustomizable Java viewer—can be selected in the ArcIMS designer. The output from ArcIMS designer is a group of HTML pages.

8.2.5.3 ArcIMS Administrator The ArcIMS administrator is used to publish and administer services, including adding, starting, stopping, and deleting services. Since the input to a service is a configuration file, any service re-

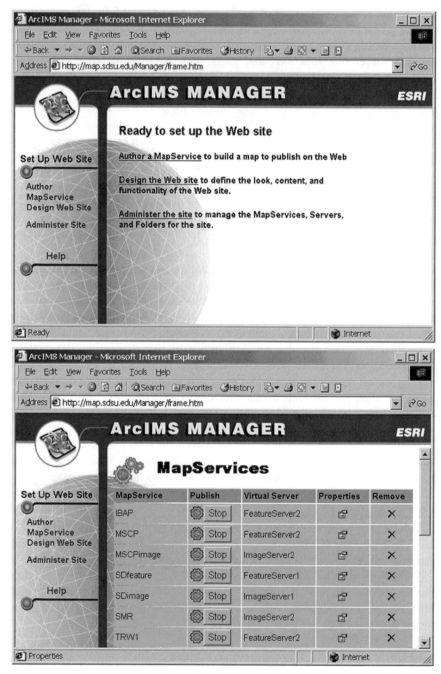

Figure 8.15 ArcIMS manager.

quests, whether starting, adding, stopping, or deleting, should translate into the configuration file in ArcIMS servlet connector or other connectors.

For example, when the start button is clicked in the ArcIMS administrator, an administrative request in ArcXML is sent from the administrator to the ArcIMS servlet connector, as shown in step 1 in Figure 8.16. In step 2, the ArcIMS servlet or other connector passes the request in ArcXML to the ArcIMS application server. The ArcIMS application server will determine to which virtual server the service should be assigned.

The ArcIMS administrator is also used to add or remove ArcIMS spatial and virtual servers. As new demands arise from the clients, additional spatial servers and instances can be started and added to either existing virtual servers or new virtual servers on the same server machine or different machines. The site configuration can be saved. This would allow the same configuration to be restarted if there is downtime on one or more machines (ESRI, 2002).

8.2.6 ArcIMS Summary

Like any other applications, ArcIMS has a three-tiered architecture:

- presentation tier—ArcIMS clients;
- business logic tier—Web server, ArcIMS application server connectors, ArcIMS application server, and ArcIMS spatial server; and
- data storage tier—data sources.

These tiers communicate through ArcXML. In addition to handling requests and responses, ArcXML is used for map service configuration files. When the user makes a request at the ArcIMS clients, it is sent to the Web server. The Web server receives the request and hands it to one of four ArcIMS application connectors:

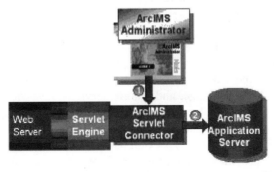

Figure 8.16 Process of sending an administrative request (ESRI, 2002, p. 18). (Graphic Image provided courtesy of ERSI.)

- ArcIMS servlet connector, including the WMS connector;
- ColdFusion connector;
- ActiveX connector; and
- ArcIMS Java connector.

These application connectors process the request and send the request in the ArcXML format to the application server. The ways these connectors handle and pass requests are different for each connector. For the servlet connector, the requests are processed and directly passed to the ArcIMS application server. But when ColdFusion and ActiveX are used, the incoming request is handled and processed by the ColdFusion or ActiveX application servers first before being sent to the application server through a connector.

When the request reaches the application server, the application server decides to which spatial server the request should be sent, based on the availability and the loads of each spatial server. The application server acts as a load balance service that handles load distribution and keeps track of which services are running on which spatial servers.

The spatial server is the backbone of ArcIMS. It provides seven services through its seven servers: image server for image rendering, feature server for feature data streaming, geocode server for geocoding, query server for attribute data inquiring, extract server for feature extraction, ArcMap server for ArcMap image rendering, and metadata server for serving metadata. These servers are accessed by four ArcIMS service types: image, ArcMap image, feature, and metadata services. Spatial servers are not accessed directly but rather through virtual servers. The Virtual server is a logical (or virtual) entity that consists of one or more spatial servers. It is a tool for managing multiple spatial servers. Virtual servers are designed to improve service reliability and scalability in a distributed environment so that when one spatial server fails, other spatial servers can take over and more servers can be added to handle the increasing amount of client requests. Virtual servers assign service to be processed in one of the spatial server instances within the virtual server group.

ArcIMS comes with three standard client viewers: HTML, Java standard, and java custom. The HTML viewer supports only image and ArcMap image services. Maps displayed at the HTML viewers are images, and only one image service can be viewed at a time. Java viewers use a Java 2 applet and support image, ArcMap image, and feature services. Java viewers have more client-side processing functions, support feature streaming, multiple services and local data within the same viewer. More client viewers can be built from the ground up using the ActiveX, ColdFusion, and Java connector.

ArcIMS also provides management tools to access the different components of ArcIMS, such as the ArcIMS author, designer, and administration tools.

8.3 GEOMEDIA WEBMAP PROFESSIONAL FROM INTERGRAPH*

Intergraph has two similar products for Internet GIS: GeoMedia WebMap and GeoMedia WebMap Professional (formerly GeoMedia Web Enterprise). At the time of this writing (June 2002), both products are shipped with GeoMedia product suite, version 5.0. GeoMedia WebMap is the product used to publish maps over the Web. GeoMedia WebMap Professional is similar but has much more functions than GeoMedia WebMap (Figure 8.17).

8.3.1 Architectural Overview of GeoMedia WebMap Professional

The basic architecture of GeoMedia WebMap and GeoMedia WebMap Professional follows the typical three-tier architecture shown in Figure 8.18. The client tier is a Web browser with a client viewer, an ActiveX control component for Internet Explorer, and a plug-in component for Netscape. This is where the end user interacts with GeoMedia WebMap or Web Enterprise.

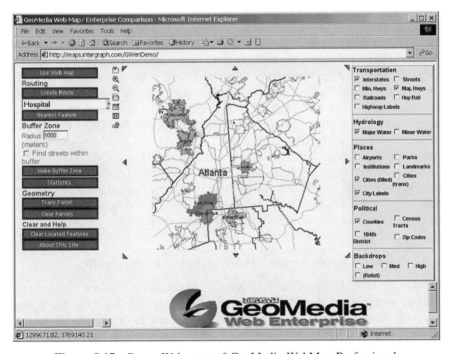

Figure 8.17 Demo Web page of GeoMedia WebMap Professional.

*The authors would like to thank Roger D. Harwell at Intergraph Corporation for reviewing this section and providing some important technical materials.

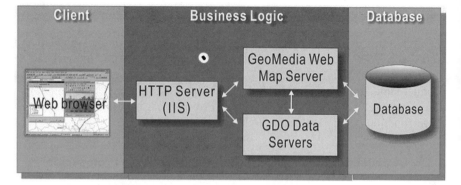

Figure 8.18 Three-tier architecture model of GeoMedia WebMap. (Figure reprinted with permission from Intergraph.)

The business logic tier includes an HTTP server, specifically Microsoft's IIS, a GeoMedia WebMap server, and a Geographic Data Objects (GDO) server. The purpose of this business logic tier is to process user requests and to make data connections. The role of IIS is to communicate with the Web browser, including receiving user requests from the browser and sending output to the browser client. It also acts as middleware to transport user requests to the GeoMedia WebMap server and GDO server. The GeoMedia WebMap or Web Enterprise server is a map server that processes user requests and produces maps. The GDO data servers are created to communicate with different data sources with different data formats. A specific GDO data server is developed for every data format so that GeoMedia WebMap can support and retrieve any data format without pretranslation (Integraph, 2001).

The database tier stores data with their original formats and projections in their sources, which can be at one or more locations. The data can be retrieved and combined using the GDO data servers. The following sections will discuss the functions of these three tiers in more detail.

8.3.2 Viewer Clients for GeoMedia WebMap Professional

GeoMedia WebMap Professional provides two client viewers, an ActiveX control viewer for Internet Explorer and a plug-in for Netscape Navigator. For any other operating systems (such as Macintosh or various flavors of UNIX), a JPEG image may be generated which is viewable by the Web browser on those systems without requiring a plug-in. The client-side viewers (the ActiveX control or the plug-in on a Windows operating system) used to view vector graphics, have built-in client-side functionality for users to perform GIS analysis on the client side. If JPEG images are published, dynamic HTML may be used to give more client-side functionality.

The ActiveX control viewer for Internet Explorer and the plug-in for Netscape Navigator are both delivered with GeoMedia WebMap Professional and are distributed to end users on demand. The ActiveX control viewer may be set up to be automatically downloaded as needed, while the Netscape plug-in must be manually downloaded and installed prior to viewing the ActiveCGM maps.

Users can request interactive maps by filling out query forms on their browsers or by connecting to hyperlinks that execute predetermined requests. GeoMedia WebMap offers the capability of searching for geographic information that matches user criteria and querying a map interactively. Spatial data search is conducted by using ASP, while interactive map querying is done by clicking on spatial features in the ActiveCGM map. When a user clicks on a map feature, GeoMedia WebMap provides associated information about that feature from the respective local or distributed databases.

8.3.3 Server Components of GeoMedia WebMap Professional

When the user makes a request from the client viewer, this request is transferred to the GeoMedia WebMap server and is then processed by an object in GeoMedia WebMap called MapServerManager. MapServerManager and its associated MapServer object have many functions and properties, as shown in the WebMap automation model of Figure 8.19.

Starting from the upper-left corner of the diagram in Figure 8.19, the following is a brief description of each object in the automation model:

- *MapServerManager* manages a pool of map server objects. It handles client requests by allocating tasks to specific map servers when the client

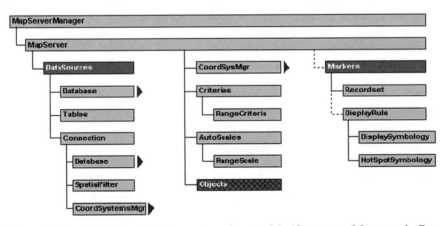

Figure 8.19 GeoMedia WebMap automation model. (Courtesy of Intergraph Corporation. Figure reprinted with permission from Intergraph.)

requests a map. It conducts load balancing and keeps track of tasks being performed by each map server.

- *MapServer* is the main GeoMedia WebMap object. It controls database connections, map queries, display symbology, and map generation.
- *DataSource* represents a database connection. DataSource must be created in order to open the database and prior to executing a query on the database. DataSource is created with the map server connect method.
- *Database* is the GDO database object. Each GDO data server is assigned a unique program ID.
- *Recordset* contains all database records that meet a specified criterion.
- *Criterias* is a collection of RangeCriteria objects.
- *RangeCriteria* objects set the map extents in which a feature set is visible.
- *AutoScales* is a collection of RangeScale objects.
- *RangeScale* objects modify the area of coverage and size limits of maps.
- *Objects* is a heterogeneous collection of objects.
- *Marker* contains the query input and the query results for a database query on the DataSource object.
- *DisplayRule* is used to assign the symbology for features and hotspots in the map, to set the tooltip text, to specify the action to be taken when a feature is selected, and to include/exclude the range criteria that will be displayed in the map. This object is created by the Marker object.
- *ActiveCGM* is an ActiveX control that renders the map feature set in the browser window.

GeoMedia WebMap Professional also provides additional analysis functions such as spatial query, buffering, address matching/geocoding, labeling, thematic display, and measurement through GeoMedia analysis objects. Furthermore, GeoMedia network analysis objects also provide for the WebMap Professional the functions to load network and create stops, find optimal paths, generate reports, and find service areas.

It can be seen from this model that most processing are handled by the MapServerManager. This approach allows the users to work on a very thin client such as a Web browser.

The current version of GeoMedia WebMap Professional 5.0 also supports the WMS specifications from the OGC. It provides a configurable application that demonstrates how to build an OGC WMS site to enable customers to create their own OGC-compliant site. It creates an Intergraph OGC WMS viewer (www.wmsviewer.com) to communicate with different OGC Web servers across the Internet and to share distributed resources and collaborate using OGC WMS-compatible data (Figure 8.20).

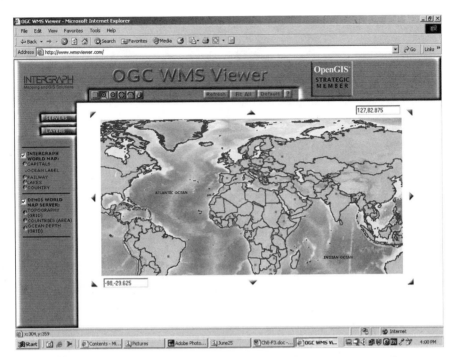

Figure 8.20 Intergraph OGC WMS viewer (snapshot from www.wmsviewer.com. Figure reprinted with permission from Intergraph.)

8.3.4 Database Management Components

A unique characteristic of GeoMedia WebMap Professional is its data server technology. Databases are stored in the data server in their native formats. There is no need to translate the source data into a new format for publishing. The data server transforms the coordinates of the user request into a common projection and retrieves the data from different data sources based on the coordinate. There is no need to make a copy of the databases either. The data server is shown in Figure 8.21.

The data server uses a set of interface components known as GDO, which is based on the fundamental concepts and emerging standards of the OGC. These GDO data servers allow data to be retrieved in a standardized way from virtually any GIS database without pretranslation and combined into a single, seamless map view. This data server technology enables GeoMedia WebMap's applications to connect to any data sources with multiple data formats, possibly stored in different map projections, to retrieve spatial and other data directly. This is a great improvement over having to convert data to a common database storage format, and it enables GeoMedia WebMap to

Figure 8.21 GeoMedia data servers (Courtesy of Intergraph Corporation. Figure reprinted with permission from Intergraph.)

combine all sorts of data into a single, seamless map containing direct connections to the source data.

A data connection between the Web browser and data server should be established for the user to access existing GIS and other databases. The database is also called a warehouse in the GeoMedia family of products. This is accomplished via functionality in one of the GDO servers. For example, if you need to connect to an MGE database, you will be using the MGE GDO server, while for an Access database, you will be using the Access GDO server.

Specifically, this is accomplished by creating a map server object and then by calling its connect method as shown below:

```
Dim MSM, MS
Set MSM = Server.CreateObject
(''GMWebMap.MapServerManager'')
Set MS = MSM.MapServer('''')
MS.Clear
MS.SetCoordinateSystem ''<path>:\usa.csf''
MS.Connect ''Access.Gdatabase'', _
            ''<path>:\ussampledata.mdb'', _
            '''', ''MyConnection''
```

In addition to vector data, GeoMedia WebMap data servers also support raster data such as aerial photographs, satellite images, and shaded relief images. These raster image files serve as backdrops of maps only, and users would not click on features on the images to obtain specific data.

8.3.5 Communication Process of GeoMedia WebMap Professional

Key Concepts

A **Map Definition File** (*MDF*) *is predefined by a map author who determines the information and parameters of a map to be displayed at the Web browser as well as the hyperlinks under each feature area of a map. An* **ActiveCGM** *(ACGM) file is a vector graphic file that is compliant with the ISO CGM standard. A* **marker** *or a record set is derived from the user-supplied query results that meet certain attribute and spatial filtering criteria.*

The communication process of GeoMedia WebMap and WebMap Professional is shown in Figure 8.22. GeoMedia WebMap and WebMap Professional adopt the Microsoft ASP technology in its data access and communications process. The client requests a map from a browser by kicking off an ASP or other scripts. The ASP may get the map parameters, including map layers, map boundaries, and others, from a MDF, or it may create all those dynamically at runtime. An MDF is predefined by a map author who determines the information and parameters of a map to be displayed at the Web browser as well as the hyperlinks under each feature area of a map. The ASP then loads a MDF, and based on the user requests, it can modify the MDF.

The ASP sends the MDF to the GeoMedia WebMap server, which transforms coordinates into common projection. The GeoMedia WebMap server extracts data from different data servers based on the requested data and creates an ACGM file. The ACGM file is a vector graphic file that is compliant with the ISO CGM standard (ISO 8632:1992). ActiveCGM was devel-

Internet GIS Showcase: Ascension Parish Tax Assessor's Office, Louisiana

This intranet site allows parish personnel to perform advanced property searches and view intelligent property maps on-line from their parish offices.

Ascension Parish personnel are able to search and display properties with the use of intelligent maps. The intelligent maps can be enhanced with digital raster images. The system also provides aerial photographs as a point of reference for researchers.

This system also allows parish personnel to perform advanced searches. Once results have been retrieved, users can export them to Excel, view property details, or view the intelligent property map.

Source: http://www.ats-us.com/solutions/solution.asp?Detail=7.

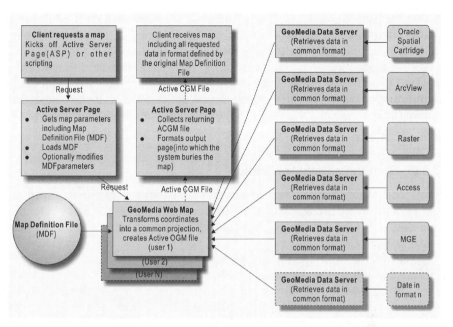

Figure 8.22   Communication process of GeoMedia WebMap. (Courtesy of Intergraph Corporation. Figure reprinted with permission from Intergraph.)

oped by Micrografix's wholly owned subsidiary, InterCAP, for creating, viewing, and adding intelligence to maps such as hyperlinks and animation.

ActiveCGM allows attribute information to be hyperlinked to graphics objects or groups of objects so that when a mouse moves over a polygon or a line on a viewer client, the associated attributes about the polygon or a line are displayed. Moreover, each feature in the vector map may be hyperlinked with other information such as reports, images, sounds, or other Web pages.

Once the ActiveCGM file is created at the server, it is returned by the ASP, which formats the ActiveCGM file and sends it back to the client viewer at the Web browser. The client viewer receives the map, including all requested data, in the format defined by the original MDFs. The user now can further render the map, query, and conduct other GIS analyses based on the retrieved ActiveCGM files.

Notice the ActiveCGM file is a vector graphic element format. The ActiveCGM file is transported to the client viewer if the ActiveX control or plug-in is installed in the Windows environment. However, if other operating systems such as Macintosh or UNIX are used, the ActiveCGM file will not be recognized by the client viewer. Rather, a map image in the form of JPEG may be generated by the map server to be displayed at the browser.

An MDF can be customized to include part of all layers of information from a vast amount of GIS data sources. For example, the developer can create MDFs to combine data from different sources on a single map, allowing end users to retrieve data from a transportation database, a utility database, and a land use database and display them together on a map. An MDF defines a very narrow slice of a map of some or all layers, because different users have different needs. For example, it can define a map to show streets, bridges, and traffic incidents for traffic operators and police officers. It can also define a map to show parcels, planning, and zoning information for planners and the city council.

8.3.6   Creating a Web Application

GeoMedia WebMap provides the administrator tool to "publish" a map on the Web by creating a compiled MDF, including defining data sources (warehouses) to be connected to, features to be displayed, and ways to display these features (Figure 8.23).

A GeoMedia Web application generator is used to enhance its visual authoring capability. This GeoMedia Web application generator runs as a command in GeoMedia that allows a developer or author to set up the contents of a map window (including, e.g., all connections, features classes, legend styles, and zoom position) and quickly produce a GeoMedia WebMap implementation (Figure 8.24).

A major task of creating a map for display is to create a "marker." A marker in GeoMedia WebMap is defined as a record set that is displayed as

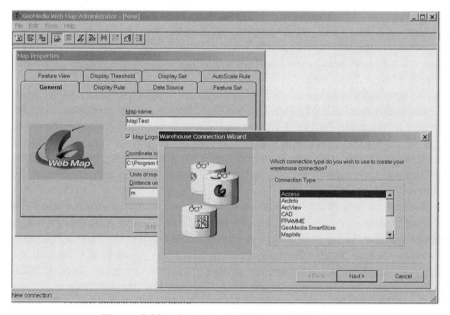

Figure 8.23   GeoMedia WebMap administrator.

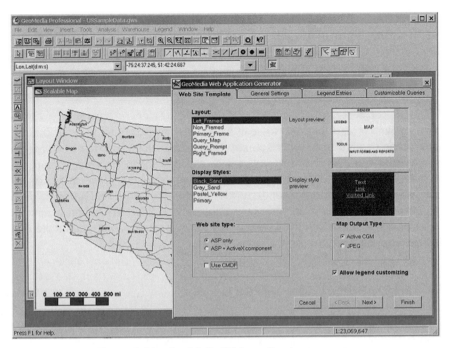

Figure 8.24 GeoMedia Web application generator.

a map. The purpose of creating a marker is to provide users with information about every map features so that the end user can retrieve information about any spatial features on the fly. This means that each feature on the map needs to have sufficient intelligence to know which database to search for its attribute information when clicked.

The marker or a record set is derived from the user-supplied query results that meet certain attribute and spatial filtering criteria. Based on the user requests, one or more markers can be created and displayed in different ways. Individual features within a marker may also be hyperlinked.

For example, a user may request a map that shows all parcels that have been sold within the previous six months and their zoning types. Based on this request, a marker is created to represent all parcels that meet this criterion from the assessor's database, color coded by zoning classification. Another marker may be created to represent all parcels that are within a flood plain. Make them red crosses. Then find all wetland within that region and make them blue.

All of these markers may be included in a single map, and each one will have the ability to query the database in a different way when a feature is clicked. So, for example, clicking on the flood plain parcel could report the risk factors of flooding; clicking on the wetland could report the kind of wetland habitat and the number of days without water, and so on; and clicking on parcels, color coded by zoning or simply displayed as a backdrop, could report ownership of those that may be affected. These items of information can be drawn from either the same or different data sources, depending on the specific data configurations.

8.3.7 Summary

Being able to roam the Web to search for data and seamlessly incorporate it into a map created from other sources is a unique characteristic of GeoMedia WebMap. The GDO data servers provide the first means to achieve data interoperability in the GIS software industry. Many spatial analysis and network analysis functions are provided by GeoMedia WebMap Professional. This demonstrates that the Web browser is not only a place to publish a map; it can also be used for sophisticated spatial analyses traditionally conducted on the desktop GIS programs.

8.4 MAPXTREME FROM MAPINFO

MapXtreme from MapInfo has two versions, MapXtreme for Windows (version 3.0) and MapXtreme Java (version 4.0) for any platform, including Windows. Since the Java version has more general applications, only the Java version—MapXtreme Java (version 4.0)—is discussed here, which is based on *MapInfo MapXtreme Java Edition Developer's Guide Version 4.0* (MapInfo, 2001).

8.4.1 MapXtreme Architecture

Like other applications, the MapXtreme Java application includes three general components: presentation, business logic, and database (Figure 8.25). The presentation tier is a client viewer in the forms of HTML and Java applet. Users interact with the HTML page and applet to make requests and view

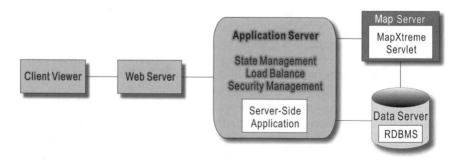

Figure 8.25 MapXtreme three-tier architecture.

results. The business logic tier components include the Web server, application server, and map server. The Web server (e.g., Sun's Java Web server, Apache's Web server, or Microsoft IIS) communicates with the client viewer by receiving user requests from the client and sending results to the client viewer. The application server (e.g., BEA's WebLogics) receives user request from the client via a Web server and passes the request to the map server— the MapXtreme servlet. Additional roles of the application server include state management, load balance, fault tolerance, and security management. The map server MapXtreme servlet processes the client request by drawing data from the data sources and making appropriate maps. Based on the setup, MapXtreme sends either a map image or stream vector data to the Web client via the application server and the Web server. The data tier is the database stored in the database server or RDBMS and is connected with the servlet or applet via JBDC.

8.4.2 MapXtreme Java Components

There are four major components to the MapXtreme Java edition: MapXtremeServlet, the MapJ object, renderers, and data providers. These components work together to complete the tasks of accepting user requests, responding to the requests, and providing a map or feature data to the user at the Web client.

Key Concepts

MapXtremeServlet *is the mapping server in MapXtreme that makes maps, generates map images, provides vector map data to clients and provides map metadata. The* **MapJ object** *maintains the state of a map by keeping information about a map's center and zoom extent, coordinate systems, distance units, and map layers. A* **renderer** *is a component that displays map data or renders the map.*

8.4.2.1 Map Server: MapXtremeServlet MapXtremeServlet is the mapping server in MapXtreme that provides three major functions:

· makes maps and generate map images,
· provides vector map data to clients,
· provides map metadata (e.g., the column names of a layer in a map).

The requests to MapXtremeServlet services are usually from MapJ objects (discussed below) through HTTP POST request methods. The communication between MapJ objects and MapXtremeServlet is through XML. Therefore, the client could be custom built using MapXtreme Java's XML enterprise protocol (discussed in Section 8.4.2.5).

MapXtremeServlet only works in a servlet container such as Sun's Java Web server and BEA's WebLogics application servers. Other servers that do not have the servlet container, such as Apache's web server or Microsoft's IIS, have to install additional servlet container plug-ins such as JRun or Tomcat in order for MapXtremeServlet to work. MapXtremeServlet relies on its parent servlet containers to handle such important tasks as load balancing, fault tolerance, and security management. This allows MapXtremeServlet to focus explicitly on fulfilling mapping tasks.

MapXtremeServlet is stateless. It relies on the MapJ object or other clients to keep track of the state of the client request. For example, if the client request is for map images, the state information would be kept in the MapJ request. This mapping request is then handled by a multithreaded "renderer server" within MapXtremeServlet. Similarly, if the request is for feature data, the state information would again be kept in the MapJ request, and the data request is then handled by a multithreaded "data provider server" within MapXtremeServlet. MapXtremeServlet is highly scalable because both the renderer server and data provider server are multithreaded, and the parent servlet container is capable of handling load balancing.

8.4.2.2 Map Server Client: MapJ Object The role of the MapJ object is to manage the state of a map. The MapJ object maintains the state of a map by keeping information about a map's center and zoom extent, coordinate systems, distance units, and map layers. In every request to MapXtremeServlet, MapJ includes this state information and sends it to a MapXtremeServlet instance. Once MapXtremeServlet receives these request information, it fulfills the request and sends the response (either map images or feature data) back to the MapJ object.

In most cases and in most deployments, MapJ is configured as a client of MapXtremeServlet, communicating directly with MapXtremeServlet. But in some cases, MapJ can also work stand alone to directly obtain map data and produce map images by working directly with different types of renderers and data Providers. This flexible configuration is one advantage of the MapJ object. The other advantage of MapJ is its small memory footprint, made possible by its limited maintenance of map state (MapInfo, 2001).

8.4.2.3 Map Rendering: Renderers If feature data are sent to the client from the server, a software component is needed to display it as the map. This software component is called a renderer in MapXtreme. As the name indicates, the role of renderers is to display map data or render the map. Renderers usually work together with the MapJ object. There are four types of renderers: LocalRenderer, MapXtremeImageRenderer, IntraServletContainerRenderer, and CompositeRenderer (MapInfo, 2001). LocalRender is called "local" because it is required to be in the same process as its associated MapJ object. It uses data providers (discussed below) to directly obtain map features for each layer in a map and then draws the features into its component's graphics object. In other words, LocalRenderer draws maps directly from feature data. It can be created from any Java Abstract Window Toolkit (AWT) component.

On the other hand, MapXtremeImageRenderer does not draw maps directly from feature data; rather it has to ask MapXtremeServlet to draw the map and receives a map image file from the servlet. When MapJ uses MapXtremeImageRenderer, it means that it wants to defer map rendering to an instance of MapXtremeServlet. The servlet then returns a map image to the MapJ client in one of the raster formats such as GIF, JPEG, and PNG that are supported by MapXtremeServlet. MapXtremeImageRenderer can be created from a URL reference to an instance of MapXtremeServlet.

It can be seen that the difference between LocalRenderer and ImageRenderer is that the former creates maps directly from feature data while the latter does not draw maps itself but rather requests MapXtremeServlet to draw the map. Therefore, inside MapXtremeServlet there is a "renderer server" that manages rendering requests by using instances of LocalRenderer and exporting images to the desired raster formats.

IntraServletContainerRenderer is an optional variation of MapXtremeImageRenderer. It is used in servlet forwarding. This renderer does not require socket connections between the renderer and MapXtremeServlet, as is necessary with MapXtremeImageRenderer. The benefit of this renderer is that the raster image can be sent directly to the client rather than have MapXtremeServlet write the image to the middle tier and then have the middle tier rewrite it back to the client. The limitation, however, is that the application must be deployed in the same container as MapXtremeServlet (MapInfo, 2001).

CompositeRenderer only draws layers with changed data. It is thus particularly useful for creating "animation" layers.

8.4.2.4 *Data Connection: Data Providers*   Data providers are data link components for the MapJ object to access data sources and return vector data. In other words, data providers are the data connection between the MapJ object and the map data. One unique feature of MapXtreme is that each layer object within MapJ has its own internal data provider. As discussed before, when MapJ uses LocalRenderer to render maps, it has to rely on data providers to access the data sources.

A MapJ object has two ways of accessing a data source: a direct approach and an indirect approach. The direct approach is to access the data directly from the data source. The indirect approach is to ask an instance of MapXtremeServlet to get the data from the data source. Once it receives the data request from the client MapJ object, MapXtremeServlet uses a data provider from its "data provider server" to access the data source. As MapXtremeServlet obtains data from the data source, it will stream the data back to the client MapJ object.

These two ways of accessing a data source are specified for each layer object through a "data provider reference." *LocalDataProviderRef* indicates a direct-data-access approach. It means that data access should occur "local to" or within the same process space as the MapJ object. *MapXtremeDataProviderRef* denotes an indirect-data-access approach. It suggests that a MapXtremeServlet instance will act as an intermediary in accessing the data source (MapInfo, 2001). When the MapJ object is located in the client side, the data streamed by MapXtremeServlet to the MapJ object are compressed.

MapXtreme has data providers for accessing the following data sources:

- MapInfo tables,
- Oracle 8i with spatial option,
- Informix universal server SpatialWare DataBlade,
- DB2 SpatialWare extender,
- JDBC-compatible tables containing longitude and latitude columns,
- ESRI shapefiles,
- raster files,
- data binding (where data from two sources are joined), and
- MapInfo grid.

8.4.2.5 MapXtreme Communications: MapInfo Enterprise XML Protocol

Key Concepts

The MapInfo enterprise XML protocol uses **document type definitions** (*DTDs*) *to specify a set of rules for the structure of an XML document and the syntax of each valid element.*

The MapXtreme Java version uses XML to move user requests from clients to server and responses from server to clients. MapInfo introduced and published a MapInfo enterprise XML protocol to define how to communicate requests and responses between client and server in MapInfo enterprise products, such as MapXtreme Java, MapMakerJ server, and RoutingJ server. The MapInfo enterprise XML protocol uses DTDs to specify a set of rules for the structure of an XML document and the syntax of each valid element. For example, there is a DTD to define requests and responses between client and server for map images; there is another DTD to define requests and responses between client and server for named resources. There are two DTDs to define OGC standards: the OGC Coordinate Reference System definition and the OGC simple geometry elements for points, polygons, polylines, and geometry collections as defined in GML.

The publication of the MapInfo enterprise XML protocol in the form of DTDs offers MapXtreme application developers some standards and guidance to write their own clients, stand alone or Web based, in any languages and any implementations.

8.4.3 MapXtreme Partitions

The partition for the MapXtreme Java version is flexible. The partition point can be located at the Web client, at the business logic, or even at the database. If the partition point is located at the Web client, MapXtreme becomes a thin-client application. If the partition point is located at the business logic or database, it is a thick-client application. If the partition point is located in between the Web client and business logic, it becomes a medium-client application. Based on the location of the partition point, MapXtreme Java can be constructed as two- or three-tier Web applications (see Figure 8.26). The difference among thin client, medium client, and thick client is in how much software and data are sent to the client, specifically where the MapJ component is located, on the server or client (MapInfo, 2001).

8.4.3.1 Thin Client: Three-Tier Configuration In a thin-client deployment, the presentation tier is the Web browser as a client. The user makes requests from the HTML pages on the Web browser, and all requests are processed by MapXtremeServlet through the Web server. The output map is typically a GIF or other image files embedded in the HTML. There is no Java applet or vector data to be downloaded on the Web client.

The thin-client deployment typically uses the three-tier configuration: a Web browser as the client, MapJ objects and MapXtremeServlet in the middle

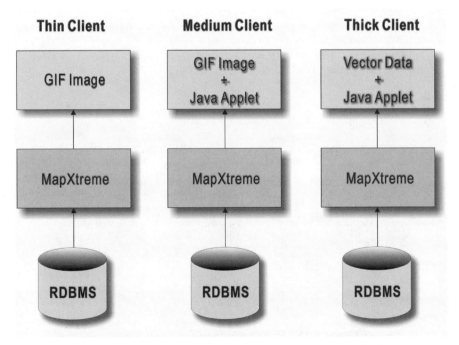

Figure 8.26   Three application options of MapXtreme Java.

tier located in the server, and the database in the database tier. The middle tier and database are located at one or more servers (Figure 8.27). The application resides at the same process as the Web server and may utilize any combination of servlets, JSPs, or enterprise Java beans in the middle tier (MapInfo, 2001).

In this three-tier configuration, when the client issues a request through the browser, the request is sent to the Web server that forwards the request to the application (e.g., servlets, JSPs, or enterprise Java beans). The application, in turn, sends the requests to the MapJ object and the requests may be used to update the state of the MapJ object. The MapJ object then serves as a client to make a mapping request to MapXtremeServlet. If the request is for a map image (not for attribute data), MapXtremeServlet will return an image of the map in a graphic format such as GIF or PNG to the application components. The application can then generate an HTML page and embed this map image within the HTML page by taking advantage of MapXtreme's servlet library of custom JSP tags and return the HTML page to the end user's Web browser (MapInfo, 2001).

This three-tier architecture has the following characteristics (MapInfo, 2001, p. 57):

- MapJ is deployed in the middle tier within custom application.
- MapXtremeServlet is deployed in the middle tier.
- Java is not required on the client. Client can send HTTP requests and can receive HTML pages as responses.
- Produces minimal network traffic: Applets are not required, so there is no applet to download. Vector data is not sent to the client, only HTML pages with embedded raster images. Raster formats such as GIF typically produce map images that are 15–25K in size.

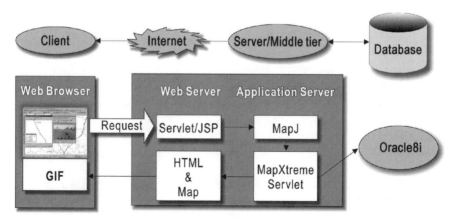

Figure 8.27 MapXtreme three-tier configuration.

Therefore, this three-tier configuration has the least demand on the user's client machine. There is no requirement or limitations on the client machine. It is thus most appropriate for users who do not have Java-enabled browsers and/or have low network bandwidth and slow network connections.

8.4.3.2 Thick Client: Two-Tier Configuration A thick client uses Java applets to facilitate users to interact with vector data on the Web client side. Java applets offer users more flexibility to interact with maps so that they can make maps and queries on the Web client without going back to the map server MapXtremeServlet. MapXtreme Java returns vector data instead of a raster image. Because part of the vector data are streamed to the Web client side, the partition point is located far into the database.

In the thick-client configuration, MapJ and other business logic components are deployed at the client side inside a Java applet within the browser, resulting in a two-tier configuration (Figure 8.28). That is, most of the business logic, map rendering, and data processing are located in the client side; feature data are also streamed to the Java applet at the Web browser. The only function left for MapXtremeServlet is to stream the data to the client.

In this two-tier deployment, a user at the Web browser needs to download an applet containing the Java beans from the Web server. Once downloaded, the applet will run on the Web browser automatically in runtime. The applet then establishes a communication with the data source and/or MapXtremeServlet. It no longer needs to communicate with the Web server. The applet can fetch the data directly from the data sources and do map rendering right from the client's machine.

The two-tier architecture has the following characteristics (MapInfo, 2001, p. 58):

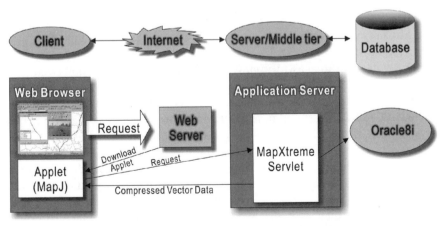

Figure 8.28   MapXtreme two-tier configuration.

- MapJ is deployed client side.
- MapXtremeServlet may or may not be deployed in the middle tier.
- Java required on client: The client browser must have support for a Java 2 Platform VM (or have a suitable plug-in).
- Heavier network traffic: The applet containing the JavaBeans must be downloaded. Vector data may be sent to the client and the size of the vector data is much more variable than the size of raster files.

The advantages of the two-tier architecture are threefold: First there is more user interactivity. The Java beans in the Java applet provide many visual map tools, toolbars, wizards, and map display components that offer users many more ways to interact with maps and vector data. Second, there is more rapid development of applications. Applications can be created much more rapidly working with MapXtreme's Java beans in a visual RAD (Rapid Application Development) environment than working at the lower level MapJ API, because Java beans provide many existing visual map tools, toolbars, wizards, and map display components ready for inclusion in the application. Third, with thick clients there is the potential for users to interact with local data on their local machines (MapInfo, 2001). The drawback of the thick-client configuration is that the Java applet and vector data download may initially take some time. A high network bandwidth may thus be required, and more powerful computers may also be necessary on the client since map rendering is done on the local machine. Therefore, this option may be best deployed on the intranet environment with a fast Internet connection.

8.4.3.3 *Medium Client: Two-Tier Hybrid Configuration*   The medium-client application is a partial implementation of the thick client. There are two meanings to this partial implementation. First, like the thick client, the medium client supports Java applet on the Web client but the applet is structured to use MapXtremeServlet to obtain map images rather than to obtain feature map data (Figure 8.26). The Java applet is used to allow users to interact with map images to improve user interactivity. The applet can give the user a more sophisticated user interface than straight HTML as well as additional map tools, such as a marquee selection tool (MapInfo, 2001).

Second, if data are stored in an RDBMS such as Oracle 8i, the Java applet cannot draw data directly from the data source. Rather, a JDBC driver is required to access an RDBMS. Thus, MapXtremeServlet may be needed to facilitate the data access using a JDBC driver, because JDBC drivers do not work well within an applet. In this case, the Java applet communicates with MapXtremeServlet, which access the data source through the JDBC driver. The Java applet then goes through a MapXtremeServlet instance to obtain some or all of its data.

Internet GIS Showcase: AT&T Growth and Global Markets

AT&T Corporation, one of the world's largest communications companies, has a division called Growth and Global Markets that sells in the lucrative but competitive medium-sized business market. As a means of increasing new revenues, Growth and Global Markets is deploying a business intelligence, data-mining, and spatial analysis solution called the "Attack" database that uses MapInfo's MapXtreme software.

This database system provides sales personnel with a vast amount of information about customer characteristics and AT&T services. Sales representatives use this information to make efficient and effective use of their time with potential customers. The system also allows the company to hook up services more quickly.

The Attack database is a multiserver Web-based application that combines MapInfo MapXtreme with one of the largest Microsoft SQL 2000 database deployments. The database brings together several types of customer and network data and allows field forces and sales management to create reports and maps that highlight sources of potential revenue.

Reports and query capabilities are designed such that sales representatives can manipulate and view data in several ways. For example, a sales representative can query a list of all customers within a selected postal code, street, wire center, or metropolitan service area that have international locations.

The implementation of the Attack database has had a measurable effect for AT&T, and the company plans to expand the user base to include the entire business services group, which numbers approximately 11,000 employees.

Source: http://www.mapinfo.com/community/free/library/
att_casestudy.pdf. Material reprinted with permission of MapInfo Co.

8.4.4 Enterprise Manager

The Enterprise Manager (EM) is developed by MapInfo to manage various aspects of MapInfo enterprise products, including MapXtreme Java map services, MapMarker J Server geocoding services, and Routing J Server routing services. Currently, only the MapXtreme Java services are enabled, although geocoding and routing services can also be added.

The EM itself is a Java client/server application. The client GUI can be run as a stand-alone application or as an applet in a browser. The communication between the server and the client is through XML. The EM uses prebuilt components to carry out specific mapping tasks. The developer can pick and choose which components to use in his or her applications. Currently, the EM has three major functions:

- Use the map definitions panel to manage and configure map layers; the buttons on the map definitions panel are Java beans that allow developers to add into his or her application.
- Use the named resources panel to manage named resources, including maps, layers, and renditions.
- Use the Web applications panel to rapidly build prototype Web applications. The Web applications use JSP technology. The EM also provides a set of JSP tags in the custom tag library for development uses.

8.4.5 Summary

The MapXtreme Java version is a pure Java application of Internet GIS. It takes advantage of the Java component technology to construct specific components with specialized functions. The MapXtemeServlet component is developed to handle user requests for feature data, map images, and metadata. The MapJ component is used to be a client of MapXtremeServlet and help keep track of the state of user requests. The map renderer component works together with MapJ to display and render maps. The data provider component establishes a data connection between the MapJ object with the data sources. The communication between these different components is through XML. Developers can also develop their own client components using XML based on the published MapInfo enterprise XML protocol. In addition to these four major components, the MapXtreme Java version also relies on the servlet container at the Web server or application server to handle such chore tasks as load balance, state management, fault tolerance, and security management.

The MapXtreme Java version could be configured and deployed as thin-, medium-, and thick-client applications. In a thin-client configuration, the user interacts with the HTML page on the Web browser and no Java applet is needed. The MapJ component is located at the middle tier at the server. All user requests are processed by the mapping server MapXtremeServlet and the response is generated map images. In the thick-client configuration, the user interacts with the Java applet on the Web browser. MapJ, along with the map renderer and data providers, is bundled in the Java applet. The feature data are streamed to the client from the data sources, or through MapXtremeServlet if JDBC is required, through data providers and displayed by the local map renderer. In the medium-client configuration, the Java applet is still used, but not to interact with the vector data streamed from the data sources; rather, it is used to interact with the map images generated by MapXtremeServlet on the server.

8.5 MAPGUIDE FROM AUTODESK

Autodesk's MapGuide is one of the earliest Web mapping programs. MapGuide is a stand-alone package for creating Web mapping applications. Users can combine the use of MapGuide with Autodesk's popular GIS software, AutoCAD and Map2000i, to create interactive query and display functions for both vector and raster GIS datasets (Figure 8.29). The following introduction will focus on the architecture, major components, and communication mechanisms of Autodesk MapGuide. The content is based on *Autodesk MapGuide Release 6 User's Guide* and *Autodesk MapGuide Release 6 Developer's Guide* (Autodesk, 2001a, b).

8.5.1 MapGuide Architecture Overview

Key Concepts

The **MapGuide author** *provides an interface to create maps by setting up* **Map Window Files** (*MWFs*) *by embedding the MWF or creating a*

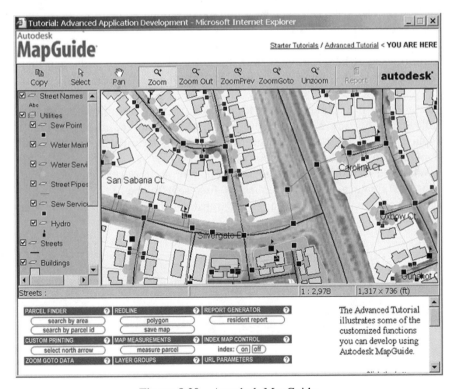

Figure 8.29 Autodesk MapGuide.

link to it in a Web page. The **MapGuide viewers** *display a map and some map-rendering tools in the user's Web browser. The* **Web server** *receives requests from the Web client and passes it to map agents and MapGuide map Servers, responds by sending published MWF files directly to the Web client, and also connects directly with the ASP or ColdFusion application server, which links with the database. The* **MapGuide server** *is a map server that serves map data in response to requests from: MapGuide author and viewers.*

MapGuide is another implementation of a three-tier architecture. The main components in MapGuide include a map viewer, a MapGuide server, and a database management server. In addition, MapGuide offers a map author to help Web designers to set up the MWF—a graphic element format to create maps for the map viewer. The Simplified MapGuide architecture is shown in Figure 8.30.

The Autodesk MapGuide author is used first to create maps by combining different resource data, such as spatial data (spatial data files and raster image files) and attribute data (from databases) in a MWF. The MWF contains the complete specifications of how the map will look and function. Once the MWF is created, you then publish it by copying it to a location where the Web server can access it. Web page authors can then embed the file in their Web pages or create links to it.

To view the map, users can install an Autodesk MapGuide viewer at the Web browser. A map viewer is designed to display maps, reports, and map-rendering tools. It is an interface for users to interact with the map to view, query, and select map elements. When the user opens a Web page that contains a MWF file or clicks a link to a MWF file, the Web browser automatically loads the Autodesk MapGuide viewer to display the map. The viewer displays the map according to the MWF settings specified in the Autodesk MapGuide author (Autodesk, 2001a).

When a user sends a request to view a map or use the Autodesk MapGuide author to create a map, the request is first sent to the Web server, which passes it to a map agent. A map agent translates that request to a MapGuide

Figure 8.30   Simplified MapGuide architecture.

server. The MapGuide server reads the request, extracts the data from the database server, and sends the data back to the map viewer on the Internet via map agents and the Web server (Autodesk, 2001a).

The partition point of the MapGuide architecture is at the middle of the business logic, as shown in Figure 8.31. That is, the presentation and some business logic (map-rendering tools) are at the client side (Web browsers), while some business logic (Web server and MapGuide server) and the database are located at the server side.

8.5.2 MapGuide Components: Map-Authoring Components

Before the user can view the map on a Web browser, a Web page has to be set up. Autodesk MapGuide provides a set of map-authoring tools to create maps and set up Web pages.

8.5.2.1 Autodesk MapGuide Author The MapGuide author provides an interface to create maps by setting up MWFs by embedding the MWF or creating a link to it in a Web page (Figure 8.32). When a user opens that Web page (or clicks its link in the Web browser), the Autodesk MapGuide viewer appears and displays the map.

MWF is not a vector data format; it is a graphic element format. In fact, MapGuide stores each individual map as an MWF, and the terms "map" and "MWF" are used interchangeably in MapGuide. One of the main functions of the MapGuide author is to create the MWFs.

Each MWF contains the specifications of the map window elements and references to underlying data *resources* on the server. Some specifications of map elements include the boundary of the map, the background color, and display configuration specifications such as menus and legends. Referenced

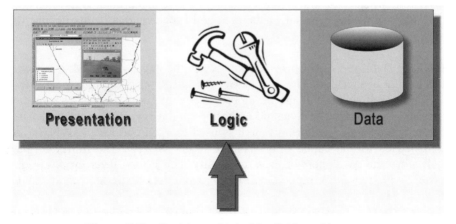

Figure 8.31 Partition point in MapGuide architecture.

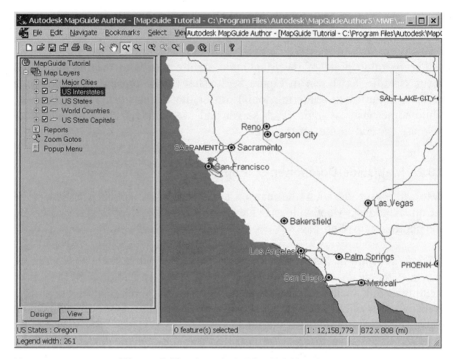

Figure 8.32   Autodesk MapGuide author,

underlying data resources include Spatial Data Files (SDFs) that define map objects, attribute data, raster data such as satellite images, and zoom go-to location (geocoding) data.

Typically, the MWF does not contain the map data but simply references the data files to use. When creating a layer, you specify the underlying data resources that contain the specific data files. You can also create static map layers, in which case the MWF contains a copy of the data. But when the end user views the map, all of the data and specifications of map elements appear simply as a single map.

8.5.2.2 Dynamic Authoring Toolkit   Once the MWF is created, it is fixed and cannot be changed. The map viewer will display it according to its specifications. But sometimes the developer and/or the user wish to change the feel and look of it. The dynamic authoring toolkit is designed to fulfill this wish. The dynamic authoring toolkit is used to open and save MWFs in an XML-based format called Map Window XML (MWX). The MWX format will allow the developers or users to modify map layers, layer groups, toolboxes (e.g., zooming), reports, and so on, using any text editor or an XML editor.

For example, users can submit a ColdFusion or ASP request from the Web browser to display only those spatial elements in a MWF that meet certain criteria. To accommodate this user request, the MapGuide author can create an application that invokes the dynamic authoring toolkit conversion component to convert the MWF to MWX and apply the user's changes in the MWX. The same dynamic authoring toolkit conversion component can then convert the MWX back to MWF and send the modified map to the user's viewer for display.

The kinds of dynamic changes in WMX are limited. In Autodesk MapGuide Release 6, only three changes can be made: (1) changing zooming extents and map center, (2) displaying only those themes that meet certain user-specified criteria, and (3) finding and modifying internal map settings at user requests (Autodesk, 2001a).

8.5.2.3 *SDF Loader*   The MapGuide server and viewer support different data formats. While the MapGuide viewer supports MWF, the MapGuide server supports its own native format—the SDF. MapGuide provides a command-line utility—the Autodesk MapGuide SDF loader to convert other spatial vector data formats into SDFs. The other data formats supported by the SDF loader include Autodesk Map DWG and DXF files, MapInfo Data/MapInfo Interchange File (MID/MIF), ESRI Arc/Info coverage files, Intergraph DGN (Director General de Normas) files, ArcView Shapefile files, and Atlas BNA (boundary) files, as well as ASCII comma-delimited CSV (Comma-Separated Variable) files. It should be noted that the SDF loader can convert the coordinates from their original coordinate system into latitude/longitude, but it cannot perform datum shifts. All data must be based on the same datum (Autodesk, 2001a).

8.5.2.4 *SDF Component Toolkit*   In addition to the SDF loader, MapGuide also provides an SDF component toolkit, a set of COM objects to read and write SDF, Spatial Index Files (SIFs), and Key Index Files (KIFs), the native spatial data file formats of Autodesk MapGuide products. The SDF component toolkit can be used to write applications to convert other spatial data types to SDFs or to create server-side applications that read and modify existing SDFs, and so on.

The SDF component toolkit is more powerful and flexible than the SDF loader. For example, The SDF component toolkit can be used to write applications to work with individual features within an SDF or convert individual features in the SDF rather than the whole file at once, which is impossible with the SDF loader.

Other MapGuide authoring tools include a raster workshop to optimize raster images to improve access performance of those image files, and a symbol manager to create and modify symbols.

8.5.3 Mapguide Components: Map-Viewing Components

The MapGuide viewers display a map and some map-rendering tools in the user's Web browser when a user opens a Web page that contains an embedded MWF or when the user clicks on a link to an MWF. The Autodesk MapGuide viewer is available in three versions: a plug-in, an ActiveX control, and a Java applet. In addition, there is LiteView, which does not require users to download anything. The viewers do not have to work with the Web browser; they can write a stand-alone C++, Visual Basic, or Java application that hosts the MapGuide viewer without a Web browser.

Different viewers are supported in different browser operating environment. Also, different viewers determine the programming or scripting languages to be used to develop the applications. Table 8.2 summarizes the different viewers supported by Autodesk MapGuide in different operating systems and Web browsers.

The plug-in version of the viewer is for use with Netscape Navigator on Windows systems. To create an application that is viewable by the plug-in viewer, the developers could use JavaScript and Java to access the plug-in viewer API objects in the Netscape LiveConnect technology.

An ActiveX control viewer works only with Microsoft Internet Explorer on Windows systems. To create an application that is viewable by the ActiveX control viewer, the developers could use VBScript, Jscript (Microsoft's implementation of JavaScript), and Java to access the ActiveX viewer API objects.

A Java edition was designed primarily for use with Netscape Navigator on Sun Solaris and Microsoft Internet Explorer on Apple Macintosh systems. It can also be used on Windows, but Autodesk recommends the use of the plug-in or ActiveX control on Windows. The Java viewer API is accessible from JScript, JavaScript, and Java.

These three MapGuide viewers are not portable. The Web developer has to develop an application for each version in order to reach the widest audience. If you develop an application *only* for the ActiveX control version of the viewer, users can access that application with Internet Explorer only; anyone with any other browsers will not be able to view your application. Similarly, if you develop an application *only* for the plug-in version of the viewer, users can access that application with Netscape only; anyone with any other browsers will not be able to view your application. Theoretically, the Java viewer has the potential to be used in both Internet Explorer and Netscape. But some new features have not been incorporated in the Java viewer yet. MapGuide provides three viewers free of charge.

If the user does not wish to download and install any of the three viewers discussed above, MapGuide LiteView can be used. LiteView is a Java program that runs on the server side as a servlet. It serves maps in a raster or graphic format by converting an MWF file into a PNG image and returns it as an HTTP response to a request. But LiteView has very limited interactivity.

8.5.4 MapGuide Components: Map-Serving Components

MapGuide map serving components process and serve data to the map, including the Web server, map agents, and the MapGuide server.

TABLE 8.2   MapGuide Viewers for Different Operating Systems and Browsers

Operating System	Browser	Viewer	Language
Windows	Internet Explorer	Autodesk MapGuide viewer, ActiveX control	HTML, VBScript, Jscript, JavaScript
		Autodesk MapGuide viewer, Java edition	HTML, JScript, JavaScript, Java
	Netscape Navigator	Autodesk MapGuide viewer, plug-in	HTML, JavaScript
		Autodesk MapGuide viewer, Java edition	HTML, JavaScript, Java
	Any browser that supports PNG file format	LiteView	ColdFusion, ASP, JSP, or Perl
	None (stand-alone application)	Autodesk MapGuide viewer, ActiveX control	Visual Basic
Mac	Internet Explorer	Autodesk MapGuide viewer, Java edition	HTML, Java
	Any browser that supports PNG file format	LiteView	ColdFusion, ASP, JSP, or Perl
	None (stand-alone application)	Autodesk MapGuide viewer, Java edition	Java
Solaris	Netscape Navigator	Autodesk MapGuide viewer, Java edition	HTML, JavaScript, Java
	Any browser that supports PNG file format	LiteView	ColdFusion, ASP, JSP, or Perl
	None (stand-alone application)	Autodesk MapGuide viewer, Java edition	Java

Source: Autodesk, 2001b, p. 19. *See* http://www.mapguide.com/help/ver6/PDF/en/MGUser-Guide.exe, pp. 64–65.

8.5.4.1 Web Server The MapGuide Web server has the same functions as Web servers in other Internet GIS programs we have discussed above, that is, ArcIMS, GeoMedia WebMap Professional, and MapXtreme. It has at least three functions in MapGuide:

1. It receives requests from the Web client and passes it to map agents and the MapGuide map server.
2. It responds by sending published MWF files directly to the Web client.
3. It connects directly with the ASP or ColdFusion application server, which links with the database. Some requests for data and report can be handled by the ASP or ColdFusion application server without going through the MapGuide server.

Key Concepts

The **Map agent** *is a communication interface between the Web server and the MapGuide server. The* **MapGuide server** *is a map server that serves map data in response to requests from MapGuide authors and viewers. The* **MapGuide server Administrator** *offers tools to manage and administer the Autodesk MapGuide server.*

8.5.4.2 Map Agents The map agent in Autodesk MapGuide is a client or agent to the MapGuide server. It is a communication interface between the Web server and the MapGuide server. The roles of map agent are to receive the user requests for map data from the MapGuide author or the MapGuide viewer via the Web server, process the requests, and then distribute them to the Autodesk MapGuide server. There are three types of map agents available: the CGI, the ISAPI, and the NSAPI.

CGI is the simplest map agent and works on any type of Web server. However, as we discussed in Chapter 4, CGI applications do not scale well and are slower than the ISAPI and NSAPI, because CGI does not run in the same process as the Web server. When there are many simultaneous requests from Web clients, the system will perform poorly since the opening and closing of the new process consume a lot of overhead on the server. CGI may become a bottleneck or even a failing point in the whole system.

The ISAPI map agent is an alternative to CGI on a Microsoft IIS. The ISAPI agent is in the form of a DLL. The ISAPI DLL stays resident on the computer and in the same process as the Web server. It eliminates the overhead of communicating between the Web server and a separate CGI executable, so it is faster than the CGI agent. ISAPI only works with Microsoft IIS.

The NSAPI map agent is similar to the ISAPI map agent except that it only works on the Netscape enterprise server. Like ISAPI, it is faster than CGI and thus performs better.

The ISAPI and NSAPI map agents are much better in load balancing than the CGI map agent, because the former can keep track of the number of

Internet GIS Showcase: Maricopa County (AZ) Assessor's Interactive Mapping Page

This system was designed to provide citizens and other users with easy access to parcel maps and property information. To achieve this goal, an interactive mapping application was created that allows users to search for parcels and subdivisions and to bring up a map containing parcels, streets, subdivision boundaries, and other land information.

Customers can search for a parcel by address, parcel number, or subdivision name entered into a search form. Once that property has been found, a click on the blue highlighted property confirms the choice and brings up the property characteristics. To see this property on the map, the user clicks the Zoom Selected button and the map zooms into an area around that parcel.

Other features provided by the Interactive Mapping Page are as follows:

- Maps can be printed or the map URL can be bookmarked for future access.
- Parcel area in feet and acres is provided.
- Neighboring parcels can be selected and displayed with parcel information using a selection radius.
- Aerial photographs are available in some places.

The Maricopa County (AZ) Assessor's Interactive Mapping Page used Autodesk's MapGuide as the developing software.

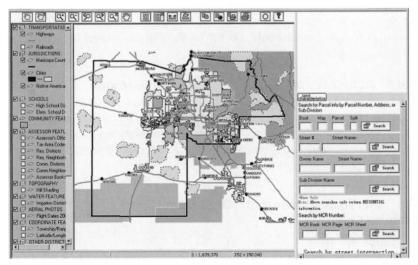

Source: http://www.maricopa.gov/assessor/gis/welcome.asp.

requests each Autodesk MapGuide server is currently handling while the latter cannot. When a new request comes in, the ISAPI and NSAPI map agents assign the new request to the Autodesk MapGuide server that is currently handling the fewest requests, while the CGI map agent distributes the requests randomly among the server because it has no persistent memory between requests.

The MapGuide map agent uses RPC to forward client requests to the MapGuide server. Because RPCs can be made across a network, the map agent does not need to reside on the same computer as the MapGuide server. This is important for enhanced security because it allows you to publish maps from the public server while keeping the data, MapGuide server, and communication between them secure on the private internal LAN.

8.5.4.3 Autodesk MapGuide Server   The Autodesk MapGuide server is a map server that serves map data in response to requests from MapGuide authors and viewers. When the map agent passes the request to the MapGuide server, the MapGuide server reads the request, determines which data to provide, and then sends the proper data from the database back to the client via the map agent and Web server. Additionally, the MapGuide server can also be used to control access to the data sources by checking for passwords, user IDs, and other optional security settings.

The MapGuide server is a multithreaded application, which means that it can process multiple requests at the same time (asynchronously), as opposed to one after another (synchronously). This is done by setting up the map layers to use data from different MapGuide servers. For example, layer 1 makes the request from server A, layer 2 makes the request from server B, and layer 3 make the request from server C. The requests to these servers go out asynchronously, so the server processes the requests simultaneously. Each server then sends its processed data back to the viewer. Therefore, if you have three servers, processing time could be up to three times faster than using one server.

8.5.4.4 Autodesk MapGuide Server Administrator   The Autodesk MapGuide server administrator offers tools to manage and administer the Autodesk MapGuide server. For example, with the MapGuide server administrator, you can manage server security, configure data source directories and database access, start and stop the server, track server requests, and generate log files as shown in Figure 8.33.

8.5.5 MapGuide Components: Data Sources and Database Connections

MapGuide directly supports five data types: a SDF, a raster image file or raster image catalog (RIC), a DWG file, attribute database, and a zoom go-to database. An SDF is a proprietary format that contains vector data, such

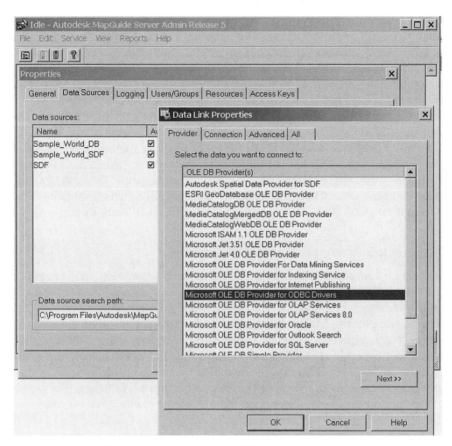

Figure 8.33 Data link in Autodesk MapGuide server administrator.

as points, polylines, and polygons. A raster image file contains a raster image, such as a satellite photo, and a RIC contains a set of raster image files, such as a set of satellite photos that make up the area of interest. A DWG file is an Autodesk drawing file format. An attribute database is a SQL table that contains the attributes of spatial objects. A zoom go-to database contains information for geocoding, including a street address database and special objects such as airports and landmarks.

Autodesk MapGuide also offers data providers to enable Autodesk MapGuide to access spatial and attribute data directly from a variety of traditional GIS, CAD, or relational spatial databases without converting the data to SDF format. In its Release 6, MapGuide Spatial Data Providers (SDPs) are provided (for purchase) for Autodesk GIS design server extension, Shape File (SHP), and Oracle spatial data.

If data providers are not used and you are using other data types that are not directly supported by Autodesk MapGuide, those data have to be converted into one of the supported formats. As we mentioned above, the MapGuide author provides several toolkits such as the MapGuide SDF loader or SDF component toolkit to convert other spatial data formats. The Map-Guide raster workshop is another utility to generate TIFF files from standard image formats, create RIC files, and manipulate the images that are referenced by the RIC.

For attribute data, the Autodesk MapGuide server can directly access an Autodesk DWG data source or any other data source through standard OLE DB data providers. These OLE DB data providers are included for Microsoft Access, SQL server, Oracle, or any ODBC data source (using the Microsoft OLE DB provider for ODBC drivers), and so on. Figure 8.33 illustrates possible database connection mechanisms for MapGuide, which include many different types of OLE DB connections and native SDF data format.

8.5.6 MapGuide Application Development

MapGuide can be used to create client-side and server-side applications or the combination of the two. A client-side application runs in the Web browser on the end user's computer. Users interact with the MapGuide viewer on data that are sent by the MapGuide server. A server-side application runs on the server and is typically used for things like generating custom map reports, generating dynamic HTML pages, and updating databases or SDFs. In many cases, MapGuide applications are a combination of client side and server side, and often the distinction between them can be fuzzy. Furthermore, a standalone application can be developed that does not require a Web browser to display maps in the MapGuide viewer.

A client-side application is developed to display MWFs on the Web browser by the MapGuide viewers. Users interact with the map (in the form of MWF) inside the MapGuide viewer, making queries, rendering maps, and selecting spatial objects. For simple operations like query and map rendering, users can get responses directly from the MapGuide viewer without going back to the server.

A server-side application, on the other hand, is developed to run on the server (or host). It processes the user request at the server and sends the process results to the client. With server-side application, the standard MapGuide viewer is not required; rather LiteView could be sufficient, because LiteView simply displays a raster map image generated by the MapGuide server. Besides the limited demand on the client machine, there are several other applications that are most suitable for MapGuide server-side applications or the combination of server- and client-side applications, including generating and serving reports or updating data on the data server by a user from his or her Web browser. A report in MapGuide applications is information about the selected map objects or about a point the user specified. This is usually generated at the server by a third-party tool such as ColdFusion or Microsoft ASP. In some cases, the database might be queried directly by these third-party application servers, bypassing the MWF completely. Another server-side application allows users to update the database from the browsers. This is also accomplished by a third-party tool such as ColdFusion and ASP.

8.5.7 How MapGuide Communicates

When a MapGuide client makes a request for a map or data, the request travels first to the Web server and then through the map agents to the MapGuide server. The MapGuide server contacts the database sources. The responses follow the same path in reverse order. The communication between the Web server and the MapGuide server use CGI, ISAPI, or the NSAPI map agent depending on the Web server used.

The communication process for the *client-side* application is described in the following steps:

1. Users at the client viewers send a request to a MapGuide site.
2. The Web server receives the request and passes it to the map agent.
3. The map agent translates the request for data to the MapGuide server.
4. The MapGuide server links the underlying data sources through the spatial data providers or OLE DB data providers and sends the response back to the client viewer through the reverse order of the initial request.
5. The MapGuide viewer receives the response and displays the maps (MWFs).
6. Further interaction between the MapGuide viewer and the maps can be done in the Web browser if the requested data are already sent to the MapGuide viewer.

The communication process for the *server-side* application is described in the following steps:

1. Users at the client viewers send a request to a MapGuide site.
2. The Web server receives the request and passes it to the map agent.
3. The map agent translates the request for a map and/or report to the MapGuide server.
4. The MapGuide server generates the map image or report and sends the response back to the client viewer through the reverse order of the initial request.
5. The viewer receives the response and displays the map image or report.

In some cases, the communication process for the server-side application can be simplified by using third-party tools such as ColdFusion or ASP as shown below:

1. Users at the client viewers send a request to a MapGuide site.
2. The Web server receives the request and passes it to the ColdFusion application server or ASP server.
3. The ColdFusion application server or ASP server goes directly to the data sources and generates the response based on the user requests.
4. An HTML page is automatically generated by the ColdFusion application server or ASP server and passed onward to the Web server and browser.
5. The viewer receives the response and displays the HTML page.

8.5.8 Summary

MapGuide has a three-tier architecture:

- presentation tier—MapGuide viewers (plug-in, ActiveX control, and Java applet) or Liteview.
- business logic tier—Web server, map agents, and MapGuide server; and
- data storage tier—data sources.

When the end user opens a Web page that contains an MWF or clicks on a link to an MWF, the Web browser automatically loads the MapGuide viewers to display the map. MapGuide has three viewers working with a Web browser: a plug-in for use with Netscape Navigator, an ActiveX control for use with Microsoft Internet Explorer in the Windows system, and a Java applet for use with Sun Solaris and Apple Macintosh system. The first-time user needs to download one of these viewers first. The user could choose to use LiteView without downloading, but this limits the functionality of LiteView because only the raster map images are displayed at the LiteViewer rather than the WMF maps.

The viewer displays the map according to the settings specified in the MapGuide author. The MapGuide author sets up MWFs to create the looks of maps. MWF is a graphic element format that contains the elements of the map and the references to the underlying data sources. Several toolkits are provided to convert other formats of spatial data into the SDFs, the native vector data format of MapGuide.

Typically, the request from the viewer is received by the Web server. The Web server then passes the request to the MapGuide server through map agents. Three MapAgents are available: CGI, ISAPI, and NSAPI, depending on the Web servers used. CGI is easier to implement but is less scalable and has no load-balancing ability. ISAPI and NSAPI are more scalable, have better performance, and can handle load balancing.

When the request is passed to the MapGuide server, the MapGuide server extracts the underlying data sources and sends the corresponding data back to the client viewer. MapGuide directly supports SDF and DWG files and also provides spatial data providers for some limited spatial data formats. Other spatial data formats that have no SDP support have to be converted to SDF through the SDF loader or SDF component toolkit. The MapGuide server links with the attribute database through numerous OLE DB data providers.

MapGuide is scalable and supports multiple servers by using distributed data or mirrored systems or a combination of the two. The distributed data model puts the MapGuide server and data in different servers so that the request for data can be made asynchronously to different servers that process the request simultaneously (Figure 8.34). The mirrored systems model has multiple servers with identical setup so that when one server fails other servers can take over. The distributed data system has faster response but is not fault tolerant, while the mirrored systems is fault tolerant but may not be as efficient as the distributed system. A carefully designed combination of the two may improve the system reliability and performance.

A MapGuide application is mostly client-side operation, that is, the map rendering, query, and spatial object selection can be done at the browser on the user's computer. But MapGuide applications can also be designed to be server-side systems, especially for generating customized report and data and allowing users to update spatial and attribute data on the server.

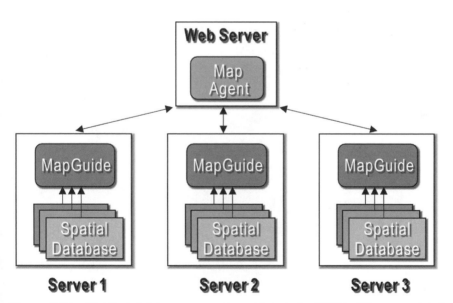

Figure 8.34 Distributed data access systems (Autodesk, 2001b). (Figure reprinted with permission from Autodesk.)

Internet GIS Showcase: GE Network Solutions—SmallWorld Internet Application Server

GE Network Solutions Smallworld Internet Application Server (SIAS) serves geospatial applications in a multitier, client–server architecture. It is built upon GE Network Solutions' popular object-oriented GIS package. SIAS can provide Web mapping application by connecting to standard relational database, such as Oracle 8i and Microsoft SQL server, or any ODBC compliant database, as well as to the Smallworld geospatial database, Smallworld Version Managed Data Store (VMDS).

One of the unique features of SIAS is that the functionality of Web mapping is much more comprehensive compared to other types of Web mapping software. For example, the network tracing function can perform sophisticated network operations such as trace-out, shortest path, and proximity searches on-line interactively. In addition, SIAS provides multiple types of query functions, including both the simple query box and complete SQL query functions against any integrated databases. Query results can be returned to the client in plain text, HTML, or XML.

Another unique feature is that SIAS includes a data dictionary to allow users to discover what data sets, tables, and field names are available in a Web map server. With the help of data dictionary, data providers can publish their Web mapping services more effectively and users can search for data they need more efficiently.

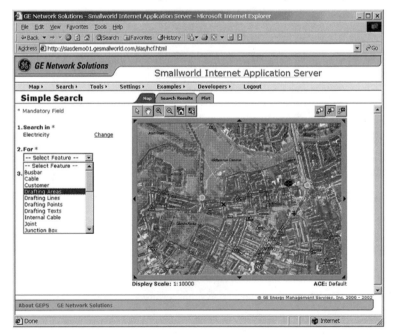

Image reprinted with permission from GE Network Solutions.

Internet GIS Showcase: GE Network Solutions—SmallWorld Internet Application Server *(Continued)*

The architecture of SIAS is a multitier framework that includes an HTTP Web server, a Smallworld dispatcher, multiple Internet application servers, and database connections.

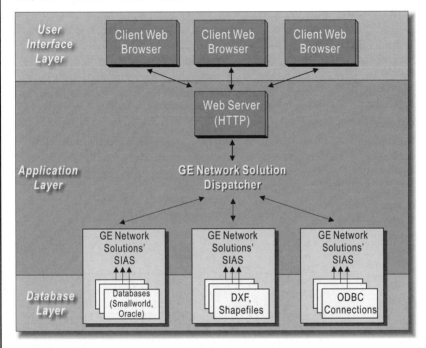

Architecture of GE SIAS. Figure reprinted with permission from GE Network Solutions.

SIAS services return information to clients in many standard formats, including HTML, XML, PNG, or JPEG. Where appropriate, services are also designed to conform to GML and WMS developed by OGC.

In general, SIAS is a flexible and scalable Web-enabled geospatial application server, which may be appropriate for the AM/FM industry, in which applications are required to perform complicated network analysis procedures.

Source: http://www.gepower.com/dhtml/network_solutions/en_us/index.jsp.

Internet GIS Showcase: ER Mapper's Image Web Server

Image Web Server is an Internet imaging services solution developed by Earth Resource Mapping. (ER Mapper). Image Web Server enables a Microsoft Web server to send images of unlimited size over the Internet directly into the user's web browser or application. The major advantage of Image Web Server is that the server utilizes the power of the patented Enhanced Compressed Wavelet (ECW) wavelet compressed image format (developed by ER Mapper) to stream terabyte-sized (1000GB+) images directly to any computer, with minimal server load. For example, an aerial photograph mosaic built from thousands of images can be compressed and served to a wide user base via the Internet. Image Web Server also uses the ECW protocol (ECWP) (under the HTTP level) to define a bidirectional communications channel from the desktop application (such as a Web browser) to the server. This channel allows an efficient, decoupled interaction between the client and server.

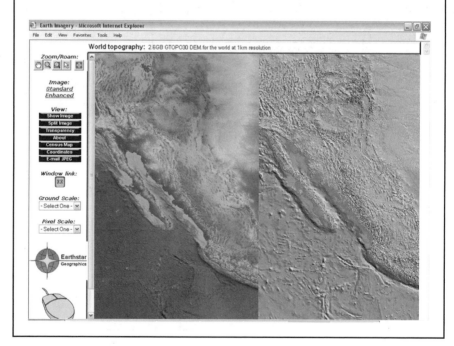

Internet GIS Showcase: ER Mapper's Image Web Server (*Continued*)

Traditional image-serving solutions attempt to serve terabyte-sized images via a complicated solution by storing thousands of image tiles on the server and then having to perform large amounts of server-side processing to extract out a suitable image view every time a user pans or zooms an image. But Image Web Server allows a client to pinpoint a particular geographic area of interest. This frees the server to do what it was designed for–serving data, not processing.

The most recent version of Image Web Server is 1.7, which can be integrated with many commercial Internet mapping applications, such as ESRI's ArcIMS and MapObjects IMS, MapInfo's MapXtreme, Autodesk's MapGuide, Intergraph's GeoMedia WebMap, UMN's Map-Server, and other GIS map servers.

Accessing Image Web Server from ArcIMS.

Image Web Server clients at the Web browser can dynamically integrate imagery from the server with multiple GIS layers from a range of data sources, including imagery, vector GIS layers, and database queries. The interactive, real-time roam and zoom capabilities over the imagery and GIS data can enhance the performance and perception of the Web site.

Source: http://www.earthetc.com. Image reprinted with permission from ER Mappers.

8.6 CONCLUSION

Internet GIS has a promising future for the GIS community. Many GIS professionals and vendors are developing a variety of packages and applications. By adopting commercial Web mapping packages or developing specialized servers and programs, GIS data providers can publish their own products online or allow users to preview their data sets or results. The general public can use Web mapping services to query the location of shopping malls, post offices, or bus stations or to plan a one-day vacation tour. The scope of Web mapping services will be much broader than traditional GIS applications. However, as Limp mentioned, "one of the most confusing aspects to selecting the best Web mapping solution is that everything is interlocked. For example, the speed of the Internet connection affects the amount of data transferred. The type of data also affects transfer volumes, which, in turn, also influences the speed, etc." (Limp, 2001, p. 9).

Even with the help of commercial packages, the establishment of Web mapping services is still difficult and complicated from both management and technical aspects. Many GIS vendors are now trying to simplify the implementation procedures of Web mapping servers. Besides the installation of Web mapping servers, project managers and GIS professionals need to consider the following aspects before the actual implementation.

First, data compatibility will be the major consideration in choosing vendor-based Web mapping products. Although many software vendors claim that their Web map engines can access multiple types of databases via an "adopter" or middleware, the extra procedures of converting data formats from one type to another may reduce the performance of Web mapping significantly. Some software packages may require additional components to access different GIS databases or require users to convert the data before online data serving. These extra procedures may become a significant problem during the actual implementation of Web mapping servers.

The second consideration is user profiles. It is important to have a better understanding of who may access the Web mapping servers before they are implemented. How many users will your Web mapping server be capable of serving? How many sessions can your Web server allow simultaneously? What kinds of functions will your map server provide—map display, query, buffering, or network analysis? What kinds of Web browsers do the users have? What kinds of network connections do they use—telephone modem or high-speed ADSL or Ethernet? Although it is very difficult to predict the actual usage of your Web mapping server and user profiles, the server configuration needs to account for both the capacity of the Web mapping servers and users' accessibility (bandwidth). The hardware and prices of enterprise Web servers are quite different compared to small Web servers for small work groups. If a Web server is expecting 10,000,000 hits per day, it may require high-end hardware, multiple CPUs, and huge amounts of memory. But if a Web mapping server is designed for in-house use only or departmental applications, a standard PC may be good enough for serving basic Web mapping functions. Also, different users may have different types of accessibility in terms of network bandwidth. The selection of client-side viewers in Web mapping applications will need to consider the bandwidth of user's networks. For example, if a user has only a regular 56K modem connection, the possible choice of client viewers will be limited to an HTML-based viewer instead of plug-ins or Java applets. Otherwise, the user may need to wait for 10 minutes to download large Java applets or plug-ins. Also, if different users will be given access to different levels of data, security protection and password access will need to be implemented. Such protection may not be available in some packages.

The third consideration is the capability to customize the client-side user interface and the availability of System Development Toolkits (SDK) for Web mapping servers. Some software packages only provide very limited functions and programming capability in order to protect their proprietary software techniques. If your Web mapping project requires a significant customization for both user interface and Web mapping functions, you may want to choose the software package which can provide the original source code on the client viewer or the SDK for customizing mapping functions. Currently, there are many toolkits or programming languages adopted for the development of Web mapping functions, such as Java, JavaScript, VB.NET, and ActiveX controls.

In general, the implementation of Internet GIS and software packages needs to consider the scalability, customizability, and usability of Web mapping servers. Although it is very challenging work, the reward of Web mapping services is very significant.

Besides Internet GIS, there are many other alternatives to distribute geographic information, such as PDF documents, CD-ROM, and mobile GIS. The next chapter will introduce the concept of mobile GIS and the mechanisms of wireless communication for mobile GIS applications.

WEB RESOURCES

Descriptions	URI
ESRI ArcIMS	http://www.esri.com/software/arcims/index.html
Integraph GeoMedia WebMap Professional	http://www.ingr.com/gis/gmwe/
MapInfo MapXtreme	http://dynamo.mapinfo.com/miproducts/Overview.cfm?productid=1162
Autodesk MapGuide	http://usa.autodesk.com/adsk/section/0,,939487-123112,00.html
GE SIAS	http://www.gepower.com/dhtml/networksolutions/en_us/index.jsp

ER Mapper Image Web Server   http://www.earthetc.com
On-line Internet mapping   http://map.sdsu.edu/geo596/lecture/
lecture (week 6)   week6.htm

REFERENCES

Autodesk. (2001a). *Autodesk MapGuide Release 6 User's Guide.* San Rafael, California: Autodesk.

Autodesk. (2001b). *Autodesk MapGuide Release 6 Developer's Guide.* San Rafael, California: Autodesk.

Buttenfield, B. P., and Tsou, M. H. (1999). Distributing an Internet-Based GIS to Remote College Classrooms. In *Proceedings, ESRI International User Conference,* July 1999, CD-ROM, Redlands, California: ESRI. Diego, California.

Environmental Systems Research Institute (ESRI). (2001a). *ArcIMS 3.1 Installation Guide.* Redlands, California: ESRI.

Environmental Systems Research Institute (ESRI). (2001b). *Using ArcIMS.* Redlands, California: ESRI.

Environmental Systems Research Institute (ESRI). (2002). *ArcIMS 4 Architecture and Functionality,* White paper. URI: http://www.esri.com/library/whitepapers/pdfs/arcims4_architecture.pdf.

Integraph. (2001). *GeoMedia WebMap Professional Online Documentation.* Huntsville, Alabama: Integraph.

Limp, W. F. (2001). User Needs Drive Web Mapping Product Selection. *GEOWorld,* Feburary, pp. 8–16.

MapInfo. (2001). *MapInfo MapXtreme Java Edition Developer's Guide Version 4.0.* Troy, New York: MapInfo.

Sileo, T. (2001). Smallworld Internet Application Server: When You Need More Than Just Map on the Internet. *Worldwired News,* 2(4), pp. 30–32.

CHAPTER 9

MOBILE GIS

The Internet is the infrastructure that will define the 21st century. In the future, 80% of Internet access will be wireless from mobile devices.
—Larry Smarr (Ainsworth, 2002, p. A)

So far we have discussed access to data and applications through desktop computers in the office or at home over the Internet or intranet. This type of access is limited because many people who work in the field need real-time access to data in the office. People on the move with cellular phones also need access to location-based information. How do we make data and applications available to those mobile users? Can mobile users access the data the same way as people with a desktop computer in the office? If not, how can we make the data available to those mobile users? These are the questions being addressed by mobile GIS and which we will discuss in this chapter.

Mobile GIS refers to the access and use of GIS data and functions through mobile and wireless devices such as mobile laptop computers, PDAs like Palm Pilots and pocket PC devices, and Web-accessible smart phones. With the improvement and convergence of GPS, Internet, and wireless communication technologies, mobile GIS has the potential to play an important role in field data acquisition and validation and in real-time access to spatial and other data by mobile users.

The major users of mobile GIS are the field workers and consumers of location-based services. For example, utility workers in the field can use GPS and mobile GIS to validate, add, and make changes to an existing database about the location of utility facilities. Hand-held mobile device users can use a location-based service to gain real-time traffic and routing information or location-specific information about where to eat, stay, or seek services.

This chapter discusses first the business case for mobile GIS followed by the general system architecture of mobile and wireless GIS. It then introduces several wireless GIS software products from major vendors. The chapter ends with some examples of wireless GIS applications.

9.1 BUSINESS CASE FOR MOBILE GIS

The use of hand-held devices is increasing dramatically. People on the go and in the field are hungry for real-time and location-specific information. What is the traffic ahead of me? Where is the cable that needs repair? Is there any gas line near the cable? Where is the nearest hospital? Where is my next service call and what is the shortest path to get there? To answer these questions, we need mobile GIS. Figure 9.1 illustrates an example of mobile GIS software developed by Microsoft called Pocket Streets. This software can display the local area maps with information about schools, restaurants, gas stations, and so on.

The next section will provide a user scenario to illustrate how mobile GIS can be applied in daily activity and help people to get the geospatial information they need. This scenario will be compared between traditional GIS solution and mobile GIS solution.

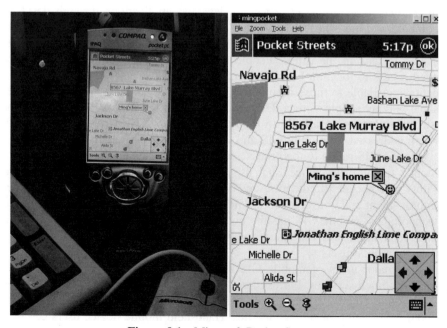

Figure 9.1 Microsoft Pocket Streets.

Key Concepts

Mobile GIS *refers to the access and use of GIS data and functions through mobile and wireless devices such as mobile laptop computers, PDAs such as Palm Pilots and pocket PC devices, and Web-accessible smart phones. The major users of mobile GIS are the fieldworkers and consumers of location-based services.*

9.1.1 User Scenario

Eva wants to visit her friend in Superior, Colorado. She will use her pocket PC with Colorado maps to connect with the GPS device in her car. She can use the pocket PC to locate her friend's home address, 1400 Begonia St., Superior, Colorado, and find a Chinese restaurant close to her friend's home for the dinner tonight.

In this scenario there are three actors (Figure 9.2). Eva, a GIS user, wants to use the GPS navigation system to find her friend's home and also locate a Chinese restaurant nearby. Ron, a GIS map data provider, provides location-based information services for GPS users. Dave, a Chinese restaurant owner, needs to update the restaurant address in the map server when he relocates his restaurant.

9.1.1.1 Traditional GIS Solution The first step in this scenario is acquiring the Colorado road data for Eva's pocket PC. Eva will need to uses her desktop computer and a [US roads 2001] CD-ROM to convert a map area which includes Boulder and Superior into a pocket PC readable map format. Then, Eva connects her pocket PC to her desktop computer using a Universal

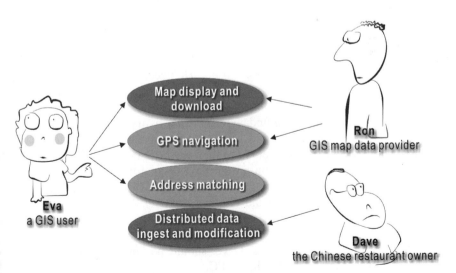

Figure 9.2 Mobile GIS scenario.

Serial Bus (USB) cables and uploads the road map into her pocket PC. Eva also uses this CD-ROM to find a Chinese restaurant located at 1100 Rock Creek Parkway, which is close to her friend's home.

Eva connects the GPS device on her car and links it to the pocket PC. Eva starts her car and types the destination address 1400 Begonia St., Superior, Colorado. However, the pocket PC shows the error message "Unable to match this address" on the screen. The problem is that Eva's friend lives in a brand-new community and Begonia Street was only built one year ago. Obviously, the [US roads 2001] CD-ROM has not updated the changes in this area for the last year. Eva has to call her friend to ask for detailed directions. She gives up on the idea of using the GPS navigation tool. After 30 minutes, Eva arrives at her friends' home. Two hours later, they decide to have dinner together. Then they drive to Rock Creek Parkway and try to find the Chinese restaurant, shown in the [US roads 2001] CD-ROM. Unfortunately, they cannot find the restaurant on Rock Creek Parkway because the restaurant has moved to a new location. Again, the CD-ROM did not update the restaurant address information. Eva and her friend have to cancel their dinner plans.

9.1.1.2 Mobile GIS Solution

In the mobile GIS environment, here are the actions that each player (Eva, Ron, Dave) would take.

EVA'S ACTIONS   Eva can use her pocket PC with a wireless modem connection to download the Boulder–Superior maps directly from the Internet via her cellular phone. Eva connects the pocket PC with the GPS in her car (Figure 9.3). Eva types in her friend's address as 1400 Begonia St., Superior, Colorado. The pocket PC shows the destination location on the screen and designs an appropriate route for Eva. On the way to Superior, Eva also selects a Chinese restaurant close to her friends' house and identities the Empress Restaurant at its current location on McCaslin Blvd. Eva visits her friends and they have a very nice dinner together.

RON'S ACTIONS   Ron's task is to provide the most updated GIS/location-based data for the multipurpose GPS users. He needs to maintain a database server 24 hours a day, 7 days a week, so all pocket GIS users may download geographic information, including roads and points of interest, to their devices. The data server must be updated frequently by database administrators. Ron also has a fieldwork team to check for the update of new streets and street changes. These fieldworkers have their own PDAs and update the street and other geographic information on the field by real-time connections with the data server. Once the street information is updated in the field, it is available immediately on the server and to any users over the Internet.

Furthermore, Ron also provides a mechanism for distributed data providers to update their own geospatial and attribute information. For example, the owner of the Empress Restaurant, Dave, should be able to update the new address of his restaurant via his home PC.

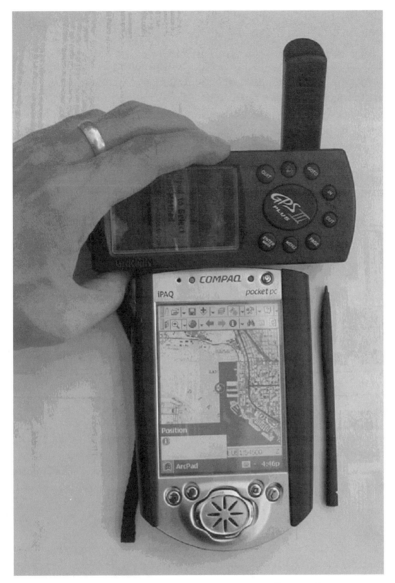

Figure 9.3 GPS connected with pocket PC GIS application.

DAVE'S ACTIONS   Dave's task is to make sure the new address of his restaurant is current in the GIS databases. He gets the permission from Ron to update the "restaurant" databases in the GIS database. Whenever the restaurant moves to a new location, Dave can use his home PC and modem to access Ron's GIS databases and update the change immediately.

Figure 9.4 illustrates the software architecture of this scenario. This scenario can be established dynamically on two GIS nodes (Eva and Ron) and one home PC (Dave).

One unique feature for location-based information is the dynamic change of information. The integration of mobile GIS and wireless Internet mapping will be able to provide the most updated geospatial information for multiple GIS applications, such as car navigation systems, facilities management, and emergency response systems. This scenario demonstrates the advantages of combining real-time information update and access with GPS and pocket PCs, which can provide a better solution for both fieldwork and location-based services.

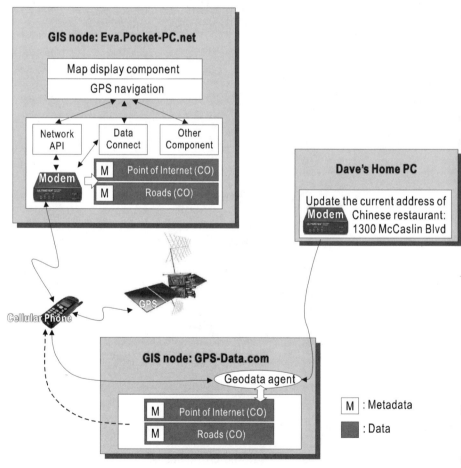

Figure 9.4   Dynamic architecture of mobile GIS for GPS navigation.

9.1.2 Mobile GIS for Field Work

In the above scenario, the data update delay of one year is not uncommon due to the long process of collecting field data, processing the data in the office, and validating the data again in the field. Data collected or edited in the field are usually uploaded to the data server after the data have been collected. There is no real-time connection. Once the data are validated and updated, there is also a delay in making the data available to the public. Therefore, data collected in the field cannot be made immediately available in the database at the data server for other applications. This time delay is not desirable for time-critical applications. With Mobile GIS, a fieldworker can update and validate the geospatial data right in the field. With wireless communication, the updated data could be saved in the data server at the office and made available for other applications and to the public as soon as the data are entered in the field.

Mobile GIS could also benefit time-critical fieldwork such as telecommunication facility and utility repairs. Traditionally, most data for fieldwork are predownloaded into fieldworkers' laptop computers or other mobile devices. Since the data are usually spatial or graphic in nature and require large amounts of storage space, CD-ROM or large hard-disk drives are needed in fieldwork. Field crews obtain information from the office by periodically downloading new data from the data server. Because of the delay of data updating, any changes in the database cannot be immediately reflected in the field and used by fieldworkers. Furthermore, data and software administration and management for field computers can be costly and complex. It is difficult for systems administration staff to maintain data and applications stored on field computers.

Therefore, field projects should have a solid wireless connectivity component. Mobile GIS, coupled with GPS, is needed to identify facility locations in the field and to facilitate information exchanges between the existing spatial and other databases and the fieldwork. Mobile GIS should allow full two-way communication between the field and the client's base of operation by wireless or modem connection.

Key Concepts

Field GIS workers can use mobile GIS devices for the data collection and validation process in the field, such as facility management and emergency rescue services. Mobile GIS is usually coupled with GPS and wireless communication to facilitate information exchanges between the existing spatial server and mobile devices.

9.1.3 Mobile GIS for Location-Based Services

The above scenario demonstrates an application of mobile GIS for location-based services. This will become widespread as telematics and in-car navigation systems evolve and gain popularity. There is potentially a huge demand for access to location-based services and geographic information from wireless hand-held devices and telematic devices inside vehicles.

People on the move are hungry for information, especially when they are in an emergency situation and/or in a new location. They are eager to find answers to such questions as "where am I," "what's surrounding me," "where can I find certain services," and "how do I get there?" The emerging technologies of cellular location systems that allow people to identify their location using a cellular phone would spur the growth of location-based services and the use of mobile GIS. This will create a great opportunity for wireless service providers and location-based information providers to offer location-based information and services, such as real-time traffic information, routing, nearest business and attraction locations, and the integration of location-based service and emergency response systems.

The time has come for moving location-based data and GIServices to mobile users and to serve even greater markets than desktop PC users. GIS is going mobile, there is no question about it. The next section will introduce the basics of mobile GIS, from the mobile devices and wireless networks to the architecture of mobile GIS. It will also introduce some mobile GIS programs and applications to serve fieldwork and location-based services.

Key Concepts

Location-based services *refer to applications that have geospatial data-handling functions and the integration of georeferenced information with other types of data. For example, car navigation systems, realtor services, and pizza delivery are some representative location-based services. Mobile GIS has become the perfect platform for the development of comprehensive location-based services.*

9.2 WIRELESS ENVIRONMENT FOR MOBILE GIS

Notwithstanding the business needs of GIS going mobile, mobile GIS is younger and much less developed than the wired Internet GIS due to the constrained wireless environment and the limitations of hand-held devices.

Mobile GIS relies on wireless communications for data exchanges and hand-held devices for display and data input. The communication environment—the wireless network—is more constrained than wired Internet connections. The hand-held device is also different from the desktop or laptop computer. It has limited display device and data input devices. Thus they pose major challenges to the construction of mobile GIS.

9.2.1 Hand-Held Mobile Devices

Mobile devices include three major categories: laptop computers, PDAs/ pocket PCs, and smart phones. Laptop computers are very similar to desktop computers in terms of operation systems, input devices, and output displays. Laptops can be connected with wireless data networks using a wireless modem such as a Cellular Digital Packet Data (CDPD) modem (we will discuss CDPD later in this chapter). Once connected, working with a laptop computer is not much different from working on desktop computers, except for a noticeable bandwidth or performance difference. However, PDAs and smart phones are drastically different from desktop computers.

PDAs include tablet PDAs and clamshell PDAs (Figure 9.5). A tablet PDA is one that does not have a keyboard and relies on a touch-sensitive screen for input. The Palm series PDAs are typical examples of tablet PDAs. Palm is to equip all its tablets with mobile data capability so that the user can access the Web. Palm also developed its own operating system (PalmOS) that competes with rivals Microsoft's Windows CE and and pocket PC as well as Symbian's EPOC. The other, now less popular PDA is the clamshell PDA. The clamshells resemble the laptop design, where the screen folds over the keyboard. The most successful company in this market is Psion, which developed its own operating system EPOC. Other clamshells typically use Microsoft's Windows CE as the operating system.

Smart phones are mobile phones with some extra computer-type functions (Figure 9.6), such as accessing the Web, data upload and download, information search and query, and interface with computers and PDAs. They have the features of both PDAs and cellular phones. These features may become standard in the next few years.

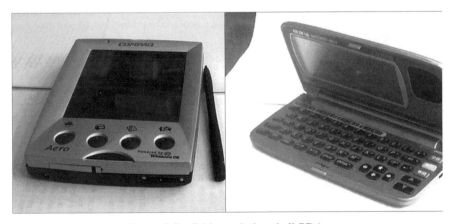

Figure 9.5   Tablet and clamshell PDAs.

Figure 9.6 Sendo Z100 Smartphone. (Photo provided courtesy of Sendo Limited. http://www.sendo.com.)

The difference between the PDA/pocket PC and the smart phone lies in their size and input interface. The PDA/pocket PC has a larger display screen and sometimes a QWERTY keyboard, while the smart phone has a much smaller display screen and more limited and restrained input interface.

Because of the limited screen size and the lack of a mouse, the user interface of a wireless handset is fundamentally different from that of a desktop computer. It has to use different input devices such as a phone keypad, handwriting recognition, and voice input. Since the display of maps requires a larger screen than the display of text, this poses another big challenge to the development of mobile GIS. Furthermore, the limited bandwidth of wireless networks poses yet another challenge to mobile GIS.

Key Concepts

The difference between the tablet PDAs, clamshell PDAs, pocket PC, and the smart phones lies in their size and user interface. The fundamental software and functionality are actually similar. For example, Microsoft's pocket PC operation systems can be applied on hand-held PDA, pocket PC, and smart phones.

9.2.2 Wireless Voice and Data Networks

The basic concept of a wireless network or cellular network system, whether voice or data network, is the network of *cells.* A cell is the coverage area of a single base station. The center of the cell is a powerful radio transceiver or base station (Figure 9.7). The size of the cell depends on the radio frequency and number of customers within each cell. Higher frequencies require smaller cells and more base stations (Dornan, 2002).

Cellular networks consist of many base stations. Base stations are connected with each other using fiber-optic cables or high point-to-point wireless links. The individual mobile devices generally communicate with one base station at a time. When the mobile user in motion (such as a cellular phone user inside a moving vehicle) moves from cell to cell, the cellular network has to be able to switch the user from one cell to another when the call is still in place. This process is called a *handoff* or a *handover.*

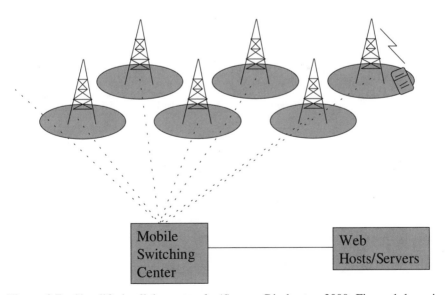

Figure 9.7   Simplified cellular network. (*Source:* Rischpater, 2000, Figure 1-1. p. 4. Figure reprinted with permission from Rocket Mobile Inc.)

The handoff is a very complex procedure and involves a calculation of when a user is crossing the cell boundary and the coordination between base stations. Base stations coordinate with each other through a *Mobile Switching Center* (MSC). The MSC is linked with external networks such as the Internet and the phone system through wires (Figure 9.7).

9.2.2.1 Evolution of Cellular Networks We have often heard terms like 2G and 3G networks. But what do they mean? Why do we care? Well, they mean the different stages in the evolution of cellular networks. We care about them because they have different information transmission speeds and thus affect the performance of our mobile GIS programs.

The *first-generation* (*1G*) cellular networks are *analog* networks, which means that the information transmission within the network is in the form of a continuously varying wave. The 1G cellular networks can only be used for transmitting voice, not for data. Even for voice, the call quality is poor and unstable. 1G networks are also very insecure; it is easy to eavesdrop the message being transmitted. The only 1G standard that is still in use today is the Advanced Mobile Phone System (AMPS) being used in North America (so the name AMPS changed to American Mobile Phone System). Mobile GIS cannot be built based on the 1G cellular network.

The *second-generation* (*2G*) cellular networks are the digital cellular networks, populated by the Personal Communications Service (PCS). Voice messages made in the 2G cellular networks are converted into digital codes, which are further encrypted before being transmitted for security reason. The call quality is also better in 2G networks than in the 1G analog networks. Data can be transmitted in the 2G digital networks, but the speed is painfully slow, with usually less than 10 kilobits per second (kbps) (Dornan, 2002). With such a slow speed, mobile GIS is almost impossible. The most popular 2G standard is the Global System for Mobile Communications (GSM), in addition to many others. Some data-only devices, such as Palm, also count as 2G mobile devices.

The *second-and-a-half-generation* (*2½G*) cellular networks are the interim solution to the more advanced third-generation networks (3G). Basically, telecommunication operators simply update their current 2G network into 2½G networks with a higher data transmission speed. The 2½G data speed is about the same as a modem. Although faster than the 2G networks, 2½G networks are still not fast enough for the wide deployment of mobile GIS.

The *third-generation* (*3G*) cellular networks offer much faster data speed than the 2G and 2½G networks, with data speed at up to 2 megabits per second (Mbps). This is similar to the speed of the current ISDN or T1 line. With this kind of network speed, more advanced mobile services could be offered, including videoconference, multimedia, and mobile GIS. Furthermore, 3G networks also offer the possibility of linking mobile devices with other devices, such as printers, TVs, microwaves, and other household appliances. Bluebooth is one example of 3G networks. The 3G networks are coming to play in the next few years and will cover the technology up to 2010.

The *fourth-generation* (*4G*) networks, still in research laboratories, will come to play by 2010. They will provide data transmission rate of up to 100 Mbps, which is equivalent to the current LAN speed. Mobile GIS would thrive in the 4G environment.

It can be seen that although the current 2G and 2½G networks provide big challenges for the development of Mobile GIS, the future development of fast digital cellular networks will dramatically improve the data transmission speed and will provide the foundation for the widespread use of mobile GIS. But we cannot afford to wait for faster wireless network before developing mobile GIS technology. At the same time, we have to live with the current cellular network environment and find the means to overcome the slow network speeds.

Key Concepts

First-generation (1G) cellular networks are analog networks. Second-generation (2G) cellular networks are the digital cellular networks, populated by the PCS. Third-generation (3G) cellular networks offer much faster data speeds than the 2G, with data speeds up to 2 Mbps, which can support more advanced mobile services, including videoconference, multimedia, and mobile GIS. Fourth-generation (4G) networks, still in research laboratories, will come to play by 2010. They will provide data transmission rates up to 100 Mbps and can provide more powerful mobile services.

9.2.2.2   Basics of Cellular Networks   Communication between a wireless terminal (i.e., a mobile device) and its base station is through radio frequencies. The frequencies used by cellphones all lie within the UHF microwave band, ranging from 400 to around 2000 MHz. Cellular networks based on higher frequencies require smaller cell size and more base stations, while cellular networks based on lower frequencies can serve large areas and require less base stations. But higher frequencies have higher capability than the lower frequencies. Mobile data transmission requires higher capacity than the voice communications; therefore, the 3G digital network systems usually use higher frequencies, around 2000 MHz (Dornan, 2002). The radio frequency spectrum is limited, resulting in many issues in governmental licenses.

9.2.2.2.1   Duplexing   Simple two-way, or *duplex,* communication between one wireless user and the base station requires two channels, or two frequency bands, one for receiving and the other for transmitting. The frequency at which the user receives information from the base station is called the *downlink,* and the frequency at which the mobile user transmits information to the base station is called the *uplink.* This paired spectrum is called Frequency Division Duplex (FDD). FDD is usually symmetric, meaning the amount of data or voice can be transmitted equally in both directions, downlink or uplink. This symmetric system is good for voice communication but not for many wireless Web applications, including mobile GIS applications. Mobile GIS in most cases is asymmetric: users download much more information than they upload. A symmetric connection to the Web wastes most of its bandwidth. However, for some mobile GIS applications that need constant uploading of field validation data to the server, a symmetric system is ideal. Most current cellular systems use FDD.

The other communication between the wireless user and the base station is the Time Division Duplex (TDD). TDD uses only one frequency channel but alternates between transmitting and listening. This is similar to two persons talking over the same channel, one talking and the other listening. When the other person wants to talk, he or she has to switch to the talking mode. The advantage of TDD is that bandwidth can be allocated dynamically between the uplink and downlink, allowing asymmetric data links. TDD is common in short-range wireless LAN systems (Dornan, 2002).

9.2.2.2.2 Multiplexing Duplexing deals with communication between one user and the base station. This is helpful but not very useful since there are many users using the same cell at the same time. Multiplexing deals with multiple users sharing the same available spectrum by using multiple-access technology. There are several technologies for multiplexing, including Frequency Division Multiple Access (FDMA) based on frequency division, Time Division Multiple Access (TDMA) based on time division, Code Division Multiple Access (CDMA) based on codes, and some other multiplexing technologies (Rischpater, 2002; Dornan, 2002).

FDMA is the simplest scheme to share frequency between multiple users by simply dividing the available spectrum into subbands and assigning one specific frequency to a user for the duration of a communication. During the course of a communication session, an assigned frequency is only used by its assignee. When the user ends a call, the assigned frequency is assigned to another caller. FDMA is generally used with analog systems or simple digital systems such as local-area or medium-area cordless phones, broadcast radio, and television. AMPS, the only analog cellular network still in widespread use in North America, uses FDMA.

The biggest problem with pure FDMA is that nearby frequencies interfere with each other, such as some radio stations. To reduce frequency interference, nearby frequencies need to be separated by a gap. This means some frequencies cannot be used, which is very wasteful of an already limited frequency spectrum.

TDMA divides a frequency band into several time slots and assigns each user a specific time slot over a unit of time called a *frame*. The mobile user transmits data in an assigned time slot; at other times, the user simply buffers the data and waits for his or her turn to transmit. Since one time slot is very short, the mobile user usually needs to transmit or receive on more than one time slot. TDMA is popular with packet-switched digital wireless data systems.

Most current digital wireless systems use TDMA, including the Digital Advanced Mobile Phone System (D-AMPS) or U.S. Digital Cellular (USDC) in the United States, the Global Systems for Mobile Communications (GSM) in Europe, and the Personal Digital Cellular (PDC) in Japan. Many advanced digital networks also use TDMA, including GPRS (General Packet Radio Service) and HSCSD (High-Speed Circuit Switched Data). Some others use the combination of FDMA and TDMA, such as the Cellular Digital Packet Data (CDPD) system.

While FDMA assigns the user the same frequency during the communication session, which is easy for eavesdropping, Frequency-Hopped Multiple Access (**FHMA**) uses many small chunks of frequency and transmits them in a pseudorandom fashion to the receivers. For the receiver to understand what was sent, the receiver and sender have to agree in advance on how to decode the seemingly random frequencies. Therefore, each user on the FHMA network is given a unique code so that a receiver can decode the sequence of the frequencies. The mobile device transmits on each frequency for only a very small amount of time, which makes a user's signal appear to "hop" from frequency to frequency, hence the name FHMA. FHMA is popular among local- or medium-area networks, such as those employed by wireless PCMCIA (Personal Computer Memory Card International Association) cards (Rischpater, 2002).

Similar to FHMA, *CDMA*, developed by Qualcomm, is another application of *spread-spectrum technology* developed during World War II. The spread spectrum was developed to make it difficult for an eavesdropper even to know that the communication was taking place because the communication is through a broad-spectrum wave in a similarly random fashion. Only the recipient who has the secret code would be able to decode the signal. This makes it difficult for the enemy to listen to or jam the signal with more powerful transmission.

CDMA sends every signal at once but encodes each one differently. The receivers receive these individual signals and assemble them together based on some secret codes. CDMA covers a very wide range of frequencies, transmitting on all frequencies at once. In other words, unlike TDMA, in CDMA every mobile terminal can broadcast at the same time, but each using a different code. But this means that CDMA requires a very high bandwidth, usually of the order of megahertz rather than kilohertz. The extra bandwidth is used to send extra copies of the transmitted signal to reduce the signal interference (Dornan, 2002).

CDMA allows every mobile user the same frequency at the same time. This means all available spectrum in the cell can be fully utilized, in contrast with other systems that can only use a third of the spectrum due to nearby frequency interference. CDMA supports both packet data and voice transmission.

Another advantage of CDMA is its soft handoff feature. This means that in the boundaries of cells one caller communicates with two base stations simultaneously. This allows *soft handoff* or seamless handoff from one cell to the other. CDMA also allows for pinpointing of the mobile device's exact location by triangulation. This is very useful for mobile GIS to provide location-based services.

9.2.2.3 Data Transmission in PCSs (2G and 2½G) PCSs were originally used by the U.S. Federal Communications Commission (FCC) to describe a particular group of licenses in the 1900-MHz band but now refer to any digital system that provides high-quality voice and narrowband data. It is the second generation of wireless technology. The PCS also includes the 2½G systems that provide higher data rates through relatively simple upgrades to the 2G networks (Dornan, 2002).

There are three main categories of PCS, including digital cell phones that provide both voice and data communications, data-only PCS, and the non-cellular or private mobile ratio system. Each has different features and maximal data transmission speeds, as shown in Figure 9.8. Most existing PCSs are based on TDMA because FDMA would waste too much bandwidth and CDMA was not available at the time PCS is standardized.

9.2.2.3.1 GSM GSM has the best coverage in the world (more than half of the world mobile users use GSM) because of its early adoption and standardization by European governments. It has a relatively high quality voice and a more reliable connection than the other TDMA systems. GSM uses separate paired frequency channels to send and receive voice and data. GSM divides the channel into eight time slots. Each phone only transmits and re-

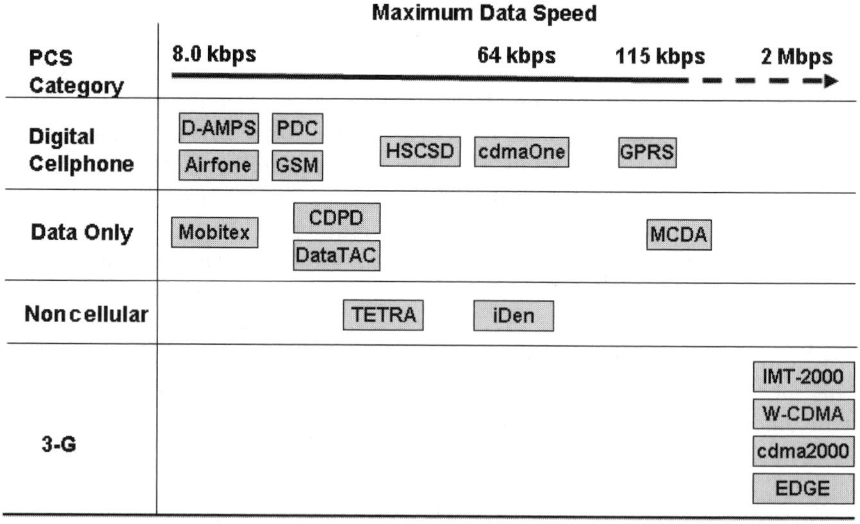

	Maximum Data Speed			
PCS Category	8.0 kbps	64 kbps	115 kbps	2 Mbps
Digital Cellphone	D-AMPS PDC / Airfone GSM	HSCSD cdmaOne	GPRS	
Data Only	Mobitex / CDPD / DataTAC		MCDA	
Noncellular	TETRA iDen			
3-G				IMT-2000 / W-CDMA / cdma2000 / EDGE

Figure 9.8 Maximal data speed for different PCSs and 3-G.

ceives for one-eighth of the time. Therefore, for each user the effective data rate is much lower than the maximal data rate. But GSM is upgradable to higher data rates.

9.2.2.3.2 HSCSD One upgrade to GSM is HSCSD, which gives each user more than one time slot in the multiplex. Each time slot has a rate of 14.4 kbps. The maximal time slot for each user is four. Therefore, the maximal rate for data transmission in HSCSD networks in 57.6 kbps. It allows users to choose only two or three time slots and to form intermediate steps of 28.8 or 43.2 kbps.

9.2.2.3.3 GPRS Another upgrade to GSM is GPRS. It is designed for data communication for the wireless Internet and is the most popular among telecommunication operators. Unlike GSM and HSCSD, which use circuit switching, GRPS uses packet switching. As you recall from our discussion in Chapter 2, in *circuit-switched* networks, the mobile device has to establish a direct point-to-point link with the base station for the duration of the communication, while in *packet-switched* networks, messages travel in small pieces, or *packets*. At the sending end, the message is broken down into many packets, which may take different routes to get to the same destination. At the receiving end, the receiver reassembles all packets to form the same message sent by the originator. It means that there is no need for the mobile device to keep a continuous linkage with the base station and the Web server.

The differences between circuit switching and packet switching lead to the major difference in operation between the HSCSD and GPRS. Under the GMS or HSCSD, each mobile user has to keep open a full circuit of 9.6 kbps or more for the duration of the time on-line, even though most of time the user does not request any information transmission from the base station and Web server. This means that users have to pay for the whole duration of the time on-line even though they only make data communication occasionally and the operator has to commit valuable spectrum to the users regardless of whether they use it or not. This is wasteful for both consumers and operators. You can see clearly that this is the limitation of circuit switching.

Packet switching, on the other hand, uses bandwidth only when there is a need for data packet transmission. When the user does not request data transmission, even if the connection is still on, the bandwidth is free to serve other users. Therefore, GPRS can use the single 14.4-kbps time slot to serve hundreds of users on-line as long as those users do not request data transmission at the same time. Each user is still continuously connected but with a very low rate of 0.1 kbps or less, but that rate can burst to higher speed whenever the user requests a transmission of data, such as clicking on a hyperlink or downloading data from the server (Dornan, 2002). GPRS has a clear advantage for both mobile users and service providers.

GPRS is being deployed by telecommunication operators. Operators have to build an entire new backbone network in order to support GPRS. The first generation of GPRS can only get a fraction of its full potential data transmission rate. But it offers a great promise for 3G wireless technology.

9.2.2.3.4 D-AMPS D-AMPS is a digital version of the older analog AMPS technology that is still widely used in the United States. D-AMPS uses the same paired spectrum and structure as regular AMPS, but it divides each of the 30-MHz frequency channels into three time TDMA slots (recall GSM has eight time slots), meaning each channel now can be shared by three simultaneous users. Each channel has a data rate of 28.8 kbps, and each user occupies the channel for one-third of the time, giving each individual user a 9.6-kbps data transmission rate.

9.2.2.3.5 PDC/JDC Similar to the D-AMPS in the United States, Japan has developed PDC, sometimes called Japanese Digital Cellular (JDC). Unlike D-AMPS, PDC/JDC is designed to be backward compatible with its own J-TACS analog system. PDC has the same data transmission rate of 9.6 kbps. It is the second most popular mobile standard (behind GSM) due to the successful I-mode, a mobile Internet system developed by NTT DoCoMo.

9.2.2.3.6 cdmaOne The only conventional digital cellphone system that uses CDMA is cdmaOne. CDMA is very efficient in terms of spectrum because the same channel can be used by several different phones and by every base station. This means that all cells can use the same frequencies, which is impossible for FDMA and TDMA. Therefore, one operator's entire frequency allocation can be used by each cell.

cdmaOne uses *Walsh* codes, a set of 64 numbers, each 64 bits long, to code every bit of data. Each user has a different Walsh code. Theoretically, since there are 64 codes, up to 64 users could share each channel. But in practice, only 10–20 users typically fit into one channel, which is still more efficient than the FDMA and TDMA. In the future, more upgrades to include more users in one channel are possible.

A big disadvantage of the CDMA system is the power consumption. CDMA transmits 64 duplicates of the same information to ensure that at least one gets through. This causes a consumption of battery of 64 times faster, and users get 64 times as much microwave radiation from the phone (Dornan, 2002).

One unique feature of CDMA in general and cdmaOne in particular is that the location of every base station is accurately measured using GPS. Therefore, the location of each phone can be estimated based on the time it takes for a radio signal to travel to and from various base stations, known as the *triangulation* method. This is a very useful feature for Mobile GIS.

9.2.2.3.7 CDPD and Other Packet Data Systems CDPD is a simple packet-switched data-only system that works on the AMPS and D-AMPS cellular networks. It uses a single channel in each cell for data transmission.

Its maximal data transmission rate is 19.2 kbps for downlink and 9.6 kbps for uplink. But since every user shares this single channel, the data rate for individual users could be much slower than this maximal rate depending on the number of users at a time. But it is usually cheaper than the circuit-switched D-AMPS because CDPD is packet switched. CDPD is always on, but users pay only when they require data downloading or uploading. Depending on the operators, CDPD customers usually pay a flat monthly fee or are billed per kilobyte of data transmission, not the airtime of connection.

CDPD is widely available, in virtually all major cities in the United States and Canada, and is the only way of sending data through an analog AMPS network. A CDPD modem is required inside PC cards designed for laptop computers to communicate with the CDPD networks. CDPD modems are increasingly often built into AMPS or D-AMPS phones for data communication as well.

CDPD is a good choice if you have a laptop computer and would like to download and upload data in the field from the office server. But it is very slow and you need a lot of patience. Tacoma Public Utilities in the state of Washington has successfully used CDPD to link its fieldwork force with a database in the server at the office (Teo, 2001).

The other two packet data systems include Mobitex by Ericsson and DataTAC or Ardis by Motorola and IBM. Both have very low data transmission rate, which has applicability for Mobile GIS applications.

In addition, there are also Private Mobile Radio (PMR) systems such as TETRA in Europe and the integrated Digital Enhanced Network (iDEN) in the United States. Both offer higher data transmission rate (see Fig. 9.8). TETRA was designed for police uses and thus offers extreme security. iDEN is a popular proprietary system in the United States, developed by Motorola and operated by several other companies.

It can be seen that, compared with wired networks, wireless data networks have a significant disadvantage in data transmission speed or bandwidth. In the current stage, the typical data network has a maximal data transmission speed of between 9.6 and 115.2 kbps, which compares poorly to the wired T1 line speed of 1.544 Mbps and Ethernet speed of 10–100 Mbps. This would be a serious limitation of deploying mobile GIS.

Other major limitations in wireless networks are power consumption and battery life. A limited supply of power is the most significant obstacle in mobile GIS development because the battery life of a mobile device is very limited. As bandwidth increases, the handset's power consumption also increases. Therefore, even as wireless networks capitalize on higher bandwidth in the future, the power of a handset will always be limited by battery capacity and size. Thus a major challenge is the amount of data throughput.

9.2.2.4 *Third-Generation Systems*   The 3G system was originally defined as a standard that provided mobile users with the performance of ISDN or better, with the absolute minimum rate of 144 kbps and ideal rates of from

384 kbps to 2 Mbps. The 3G system was later expanded to include two additional requirements, that is, they should support Internet protocols and be based on a packet-switched network backbone.

The ITU named the 3G standard as IMT-2000. IMT stands for International Mobile Telecommunications and 2000 stands for three meanings: the year 2000, when the 3G systems will be made available; data transmission rates of 2000 kbps; and the frequency range of 2000 MHz. Obviously, the first aspiration has not been fulfilled; even the third one is questionable in fulfillment, but the name IMT-2000 is often interchangeable with 3G (Dornan, 2002; Pandya, 2000).

How do the 3G systems increase the data transmission speed? Generally speaking, the increase of performance in 3G systems can be achieved by the combination of three methods. The first and the obvious one is by using extra spectrum. The second method is to use new modulation techniques that squeeze higher data rates from a given waveband, for example, by using the octal system rather than the binary system. The octal system allows every symbol to have eight values instead of only two. The third method is to use CDMA rather than TDMA to take advantage of the better ability of CDMA to cope with new users (Dornan, 2002). However, not all 3G systems are based on CDMA.

There are three main 3G systems in the IMT-2000 standard, Wideband CDMA (W-CDMA), cdma2000, and Enhanced Data Rates for GSM Evolution (EDGE). All will offer packet-switched data at rates exceeding 384 kbps and up to 2 Mbps.

W-CDMA, also known as Universal Mobile Telecommunications System (UMTS) in Europe, is designed to be backward compatible with GSM and requires new spectrum. The higher data transmission rate in the W-CDMA is the use of wide bandwidth. W-CDMA uses the channel bandwidth of 5 MHz, which is four times that of cdmaOne and 25 times that of GSM. The wider bandwidth allows for higher data rate. In addition, W-CDMA does not send every bit of information 64 times, as does cdmaOne. It sends every bit of information between 4 and 128 times. Therefore, when the signal is strong, W-CDMA sends less duplicated information (called gains) and thus has more bandwidth available. The maximal data rate for W-CDMA is 4 Mbps. W-CDMA is the first 3G system being deployed in Japan (Dornan, 2002; Pandya, 2000).

cdma2000 is a narrowband CDMA standard. It is an upgrade to cdmaOne and is backward compatible with existing IS-95 systems. For current cdmaOne networks, the update is simple, just update existing networks with new software or modulation rather than by building new radio systems. But there are several standards within cdma2000, including, but not limited to, cdma/1XRTT (an implementation of the multicarrier cdma2000 standard that uses one 1.25-MHz carrier) and cdma/3XRTT (an implementation of the multicarrier cdma2000 standard that uses three 1.25-MHz carriers).

EDGE is a straightforward upgrade to GSM and is a 3G system that is not based on CDMA. EDGE is also compatible with other TDMA systems, such as D-AMPS and PDC. In 2000, the ITU accepted EDGE as a model of IMT-2000 and made it a standard Universal Wireless Communications (UWC)-136.

The deployment of 3G systems or IMT-2000 will allow many more applications to be introduced to a worldwide base of users. The new 3G networks will address the growing demand of mobile and Internet applications for new capacity in the overcrowded mobile communications sky. The 3G networks increase transmission speed to 2 Mbps and could establish a global roaming standard. With 2 Mbps, mobile GIS would exceed the performance of current Internet GIS using the T1 network.

In summary, mobile GIS works in a much more challenging wireless network environment than the wired Internet environments, at least for now. Wireless data networks tend to have the following:

- Less bandwidth: Bandwidth is the amount of content that can be passed down the network at a given time.
- More latency: Latency refers to the length of time it takes the content to flow a certain distance. It is a measure of the amount of time it takes for a request to make a round trip from the client and back.
- Less connection stability: The mobile user in motion has to hand off from cell to cell, which could cause information drops during the handoff process.
- Less predictable availability.

9.2.3 Wireless Web

As we have discussed so far, mobile GIS has two big challenges: the low bandwidth of the wireless network and the small screen and user input devices of hand-held devices. Most existing wireless networks only allow data transmission speeds of between 9.6 and 14.4 kbps. The cellphone screen is usually big enough to show a few lines of text messages. PDAs have larger screen, some even have colors, but they are still significantly smaller than desktop computer displays. Most cellphones and some PDAs do not have a keyboard for user input. Therefore, fitting Web contents that are rich in graphics and animations into a cellphone or PDA screen is simply not possible. This requires us to rethink the design of wireless Web in general and mobile GIS in particular. Figure 9.9 illustrates a comparison of using small-screen pocket PC and desktop PC to display the same Web pages.

To "mobilize" the Web to mobile devices, several different open standards and proprietary systems have developed for the wireless Web, including Compact HTML (C-HTML), XHTML, Web clipping, Handheld Device Market

Figure 9.9 Comparison of full-size HTML Web page displayed in desktop PC (Left) and pocket PC (Right).

Language (HDML), WML, Mobile Execution Environment (MExE), and WAP. All use a *microbrowser,* which keeps only the core elements of HTML and restricts the file sizes. All are independent of the underlying wireless networks. However, all of them are not mutually compatible; a browser for WAP cannot be used to read sites designed for C-HTML.

9.2.3.1 C-HTML C-HTML is a simplified version of HTML that was published by the W3C in 1998. It keeps the core element of textual display of HTML and trims off many additions to standard HTML, such as fonts, frames, tables, and style sheets. It still supports graphics but not animations or Java applications. The specifications of C-HTML are quite similar to the original version of HTML. Therefore, any browser can interpret C-HTML-coded Web pages. The disadvantage is that this lowest common denominator approach may not work efficiently for many devices. For example, tables and cascading style sheets that are ideal for laying out text and graphics for laptop computers are missing in C-HTML. This does not bode well for mobile GIS.

C-HTML has been adopted by the I-mode service by NTT DoCoMo in Japan. All content provider sites usually redesign separate pages for the wireless Web contents using C-HTML rather than strip the noncompact features from their regular sites. In other words, each site has two versions, one for standard HTML and the other for C-HTML. Besides C-HTML, the success of I-mode is also a result of packet switching in its wireless network, allowing phones to remain connected to data services at all times.

9.2.3.2 Web Clipping Web clipping was developed by Palm's parent company 3COM to be used in the Palm VII organizer that runs over the Mobitex network. It is similar to C-HTML and uses a subset of standard HTML. Web clipping does not support such features as tables and frames. Palm VII cannot directly interpret HTML; it has to rely on a special program called a PQA (Palm Query Application), which tells Palm which bits of which pages on the site it should download. A site must install PQA before it can be accessed by Palm VII users. The need for PQAs severely limits a Palm VII user's surfing options. In later versions of Palm devices, other systems such as WAP are used.

9.2.3.3 MExE MExE is another proposed wireless Web standard from ETSI, a standards body in Europe. MExE provides specifications for running programs on a mobile phone. It requires a Java interpreter built into the phone so that applications written in Java can be run just like Java applications running on a PC-based Web browser.

Unlike PCs, current mobile phones have very limited processing power. Running Java applications on a mobile phone is quite challenging, at least for now. Sun has thus produced a scaled-down version of the Java language– J2ME. MExE also has different levels of specifications to support different levels of computing power in mobile phones. Many vendors and operators plan to support MExE, including Nokia, Sony, and NTT DoCoMo.

9.2.3.4 HDML HDML is drastically different from HTML. It created two new text layout metaphors: *cards* and *decks*. A card is defined as "a single user interaction," such as a menu or a page in HTML. A single HDML file called a deck could include many cards. This is different from HTML, where each file is one single page.

There are three simple cards in HDML: display, entry, and choice cards. A *display* card presents data to the user. With this card, the user can only move on to the next card. An *entry* card allows the user to input, such as type a phone number or an address. A *choice* card has menus, similar to the hyperlink in HTML. Users can navigate between cards by selecting different items from choice cards.

The advantage of HDML is that when a deck is downloaded, users can select items from a menu and see them instantaneously, rather than waiting for each single card request going through the wireless networks to the server and then back. This is very important for small wireless devices because the small screen limits the amount of information that can be displayed in one card. HDML has been adopted by AT&T and has become the foundation of the WAP and WML standards.

9.2.3.5 WML WML is a special subset, or *schema,* of XML based on HDML. WML offers more flexibility than HTML to fit the specific needs of mobile devices.

WML requires very little bandwidth resources compared to HTML. This is very important for the currently low bandwidth wireless environment.

WML also requires less processing strength to render on the microbrowser. Less processor power means longer lasting batteries at the mobile devices. Furthermore, WML is designed to fit the limited screen size of mobile devices.

Microsoft Mobile Explorer (MME) supports both HTML and WML. The MME devices are really two completely separate devices in one; one supports HTML and the other supports WML. But WML 2 is no longer a prime authoring tool in the Wireless Application Protocol version 2.0 (WAP 2); it is intended only to support conversions of WML 1 documents for the purpose of backward compatibility with WML 1 in WAP 1 (WAP Forum, 2002).

9.2.3.6 XHTML Mobile Profile Markup Language (XHTMLMP) XHTMLMP is defined by WAP as a markup language in WAP-enabled mobile devices. It extends the BASIC profile of XHTML as defined by the W3C by defining additional markup features for enhanced functionality in wireless devices. It uses the XHTML modularization approach, so that additional language elements can be added as needed. It is compatible with the core XHTML BASIC language, meaning documents written in the core XHTML BASIC language are operable on the XHTMLMP browser (WAP Forum, 2002).

9.2.3.7 WAP WAP was designed by the wireless industry, including more than 80 companies, such as Openwave Systems (used to be Phone.com), Nokia, and Ericsson, to serve Internet contents and Internet services to wireless clients with WAP-enabled mobile devices such as mobile phones. WAP is not a proprietary system but an open system. It is not simply a markup language either. Rather, it is a complete new stack of protocols designed to overcome some of the wireless network's specific problems, such as low bandwidth and high latency. It is becoming an important standard for wireless access to the Internet resources and the Web. For more information about WAP, visit http://www.wapforum.org.

9.2.4 Operating Systems for Mobile GIS Applications

Currently, there are two major operating systems for mobile devices: Microsoft WinCE and Palm OS. The following discussion will focus on their programming capability for mobile GIS applications.

9.2.4.1 Palm OS Palm OS is the earliest example of mobile operating systems (Figure 9.10). Many hand-held devices and PDAs adopt Palm OS, such as Palm pilot, Handspring Visor, and Sony Communication, Link, Information, and Entertainment (CLIE). Palm is a pioneer in the field of mobile and wireless Internet solutions and a leading provider of hand-held computers. Basically, the design strategy of Palm OS is simplicity, light weight, and small

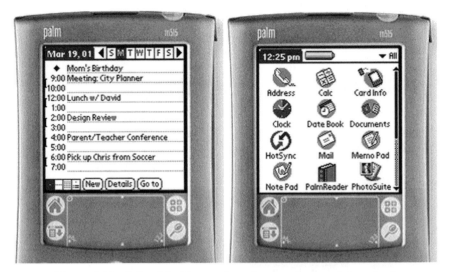

Figure 9.10 Palm OS. (*Source:* http://www.palm.com/education.)

size. There are three major types of platform applications in Palm OS: hand held, conduit, and Web clippings.

Hand-held GUI applications that run on Palm OS are generally single-threaded, event-driven programs. Applications run one at a time. Users do not quit or exit an application; they simply choose to run a different application.

As mentioned before, a Web clipping application is a set of HTML pages compressed into a special format called PQA. Web clipping programs can be downloaded onto the device. Users can fill out the HTML forms and send their request to the web server via wireless communications.

A *conduit* is a plug-in to HotSync technology that synchronizes data between the application on the desktop and the application on the hand-held device. Conduits are usually written using Visual C++, Visual Basic, or Java along with the Palm Conduit Development Kit (CDK).

One advantage of Palm OS is the close connection to Java technology. Java developers can develop a PDA profile, which is targeted to provide a standard set of Java APIs for small, resource-limited hand-held devices. The PDA profile is based on the CLDC and J2ME (see http://www.palmos.com/dev/start). Currently, MapInfo's MapXtreme application is running on the Palm OS.

9.2.4.2 WinCE Microsoft WinCE was first released by Microsoft in 1996, called the hand-held PC version 1.0 (Figure 9.11). The current version of WinCE is called Pocket PC 2002, and the platform has been integrated with the Microsoft .NET framework (Microsoft, 2002). Different from Palm OS, the design strategy of WinCE is to provide comprehensive desktop PC func-

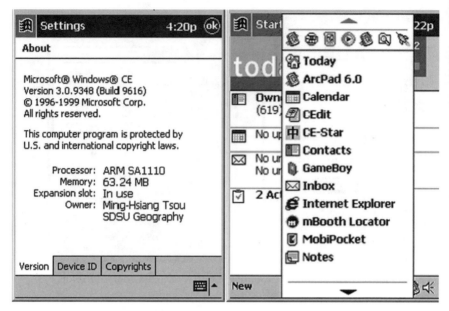

Figure 9.11 WinCE platform.

tions for mobile devices. There are many WinCE applications similar to desktop PC, such as Media Player, Internet Explorer, Pocket Word, and Excel.

ESRI's ArcPAD is a mobile GIS software adopting the WinCE platform. The user interface of WinCE is very similar to the regular Windows environments, such as Windows 2000 and XP. Also, Microsoft provides several free development tools for mobile and embedded applications under its .NET framework, such as the Microsoft Mobile Internet Toolkit (MMIT). MMIT adopted WML and C-HTML, which can be applied for the development of mobile GIS applications.

The most recent release of the Microsoft WinCE family is the pocket PC 2002/phone edition, which has the additional phone functions besides the regular pocket PC functions.

9.3 GENERAL SYSTEM ARCHITECTURE OF MOBILE GIS

As data communication via wireless networks evolves and standards are developed, mobile GIS is emerging and more applications are being developed. To understand mobile GIS, we now turn to the basic system architecture and the major components of Mobile GIS.

9.3.1 Major Components of Mobile GIS

The unique features of mobile GIS are the mobile clients (e.g., laptop computers, pocket PCs, PDAs, and cellular phones), nonstationary users (moving objects), and wireless networks. So we need technologies to determine the locations of mobile clients and process the location information. As the mobile clients move from one location to other locations (or from one cell to the other), we need to track them so that when the mobile clients request location-based information at a specific location we know what data to send and from which distributed data server. We also need technologies to receive and process the client requests and to send the information to the mobile clients through the wireless networks in the right format so that the information can be displayed properly at the clients. Therefore, the following major components are needed to construct a Mobile GIS:

a. Position-determining components
   1. Position-determining equipment
   2. Position-processing technology—mobile positioning center
b. Location Information components
   1. Geographic information (e.g., streets and boundaries)
   2. Location-specific information (e.g., yellow pages)
c. Information processing service components
   Process user requests and provide location services, such as geocoding, routing, finding nearest services, and map generation
d. Gateway service components
   Connect the wired Internet and the wireless networks and mobile devices.
e. Wireless network components
   Different wireless networks
f. Internet-enabled mobile devices

9.3.1.1 Position-Determining Components Position-determining components refer to components to determine the locations of the mobile devices in real time. They include hardware and software to identify and track the locations of mobile devices. Two major components are needed to determine the locations of mobile devices in real time.

1. *Position-Determining Equipment* Position-Determining Equipment (PDE) identifies the location of the mobile device (MapInfo, 2002). There are generally two ways to determine the location of a mobile device: a network-based system that relies on triangulation of the cellular informationas in CDPD or a handset-based system that relies on GPS. A network-based system

requires PDE to be placed at the switch centers. A handset-based system requires GPS to be placed in the mobile handset.

2. *Position-Processing Technology* Position-processing technology is a software technology (usually a server) to process, track, and manage the location information sent from the PDE. It is usually located at the Mobile Positioning Center (MPC) and is thus also synonymous with MPC (MapInfo, 2002). The role of the position-processing technology is to help other applications to query and retrieve the location information from the PDE. In some cases, the location information is only used for real-time information; in other cases, the location information is stored at the server. There are several position-processing technologies available, such as Alcatel's iMLS, Nokia's mPlatform, Ericsson's MPS, and Nortel's eMLC.

9.3.1.2 Location Information Components

Location information refers to content information about a geographic location. It includes reference information to a geographic location, such as a street file or a boundary file, and specific information about a particular location, such as Yellow Pages information. Mobile GIS needs this information to determine the location of a mobile device and to send location-specific information to the mobile user.

1. *Geographic Information* Geographic information is usually referred to as base map information. It consists of street networks and basic boundary files (e.g., zip codes, cities, and counties) that could be used for geocoding and navigation purposes. That is, the geographic information must contain street address and routing information (e.g., permissible turns, one-way streets). Accurate and up-to-date geographic information is essential for mobile GIS and location-based applications.

2. *Location-Specific Information* Location-specific information refers to attribute information about a particular location. This includes such information as landmarks, businesses, schools, hospitals, government agencies, telecommunication and utility facilities, or any information needed by the mobile user, such as the menu of a restaurant. It could also include such dynamic information as real-time traffic information and weather information.

9.3.1.3 Information Processing Service Components

Information processing service components concern the processing or services of geographic information, that is, how to provide the location information in a usable way to mobile users. They are some of the traditional GIS functions. Some examples of these location information services are as follows:

(a) location-specific information query and display services—providing wireless access and use of location-specific information such as Yellow Pages directories;

(b) geocoding services—finding the location of a specific address;

Internet GIS Showcase: Dutchtone Customer Relations Department

Dutchtone N.V. is a telecommunication company with its own national network covering mobile telephony based on GSM technology. The company offers an extensive range of services to both residential and business customers. In an effort to achieve greater operational efficiency and customer satisfaction, Dutchtone turned to Internet GIS.

Dutchtone's on-line system is designed to quickly display the current coverage of the Dutchtone network and reveal any faults in the coverage. This information allows Dutchtone's customer relations department to respond to customer questions more quickly and more accurately.

Coverage Assistant, the centerpiece of Dutchtone's system, uses Autodesk's MapGuide application to assess coverage and locate faults. Dolmen, a GIS Autodesk systems center, used existing Dutchtone data sources to make up-to-the-minute location and time-sensitive information available through Coverage Assistant.

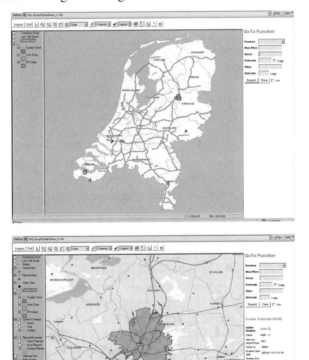

Image reprinted with permission from Autodesk.

Source: http://www.autodeskgovernment.com/News/
CaseStudiesfromAutodesk/Dutchtone.pdf.

(c) spatial search services—searching location-specific information surrounding a geographic position, such as finding businesses or services that are within a quarter mile of a position;

(d) routing services—calculating the routing information from point A to point B; and

(e) map display and rendering services—displaying location information in the form of maps, images, and geospatial features.

9.3.1.4 Gateway Service Components Gateway services are middleware between the information processing components and the mobile devices. The purpose of the gateway service is to make the services provided by the server (e.g., application server and Web server) compatible with the mobile devices. So the roles of the gateway service include (1) getting the current position of the mobile device from the mobile position center, (2) processing the requests from the mobile devices and sending the user requests to the Web server portal or other applications, and (3) converting the responses from the information processing services into a format that can be displayed in a mobile device, regardless of the operating services and platforms being used by different mobile clients. Some examples of gateway services are Lucent's MIG, Oracle's 9iAS Wireless, and Openwave's UP.Link (MapInfo, 2002).

9.3.1.5 Wireless Network Components Mobile GIS relies on the wireless network to transport information. Therefore, its performance is affected by the kinds of wireless network architectures because different wireless networks (e.g., GSM, GPRS, UMTS, W-CDPD) have different data transmission rates. The wireless network components are beyond the control of mobile GIS, but a basic understanding of the different wireless networks, including networking, routing, and core communication capabilities, is important to understand mobile GIS.

9.3.1.6 Internet-Enabled Mobile Devices Mobile GIS requires Internet-enabled mobile devices. This means that the mobile devices have to be able to access and display Internet services. That is, they have to be able to interpret and display correct images and styles. There are many platforms for the mobile devices, such as Sun's J2ME and WAP. The design of mobile GIS should ensure that the application can support a variety of mobile devices with different operating systems and platforms. Today, many of the most popular mobile phones and PDAs support J2ME, WAP, and XML or WML.

Those are the main components that are required to construct a mobile GIS. We will now look at how they are put together—the generic architecture of mobile GIS.

9.3.2 Generic Architecture of Mobile GIS

The architecture of mobile GIS is very similar to the wireline-based Internet GIS. It has three major components: the client, the server, and the network services. The traditional client in the Internet is a personal computer. As client devices get smaller, they become mobile. A full range of client devices can be used for mobile GIS, from mainstream laptop computers with all the computing power of a PC to PDAs or pocket PCs with smaller screens to cellular telephones with even smaller displays, simpler input devices, and limited processing capability. Effective mobile GIS solutions should deliver data and applications to the entire spectrum of client devices.

Servers are system components that determine the position of the mobile devices and make data and applications available to the mobile client devices. Server applications should be flexible enough to serve all client devices regardless of whether the client device is wireless or wired. Therefore, they have to support a variety of protocols and APIs.

Network services refer to the communication infrastructure that allows clients and servers to connect with one another. Network services can be broken down into three sections: the so-called first mile, the last mile, and remaining miles. The first mile of the communication path is the portion starting with the server. It involves a LAN that may be operated by a given service carrier. The last mile of the communication path is the portion starting from the client device. It could be wired for an Internet connection or wireless using one of many available radio frequency air interfaces. The balance of the network, or the remaining miles, is based on traditional telecommunications and WAN technology. It could be

- a voice network carrying modulated data to a modem where it enters a digital network,
- both a voice and a data network, and
- exclusively a data network.

Key Concepts

Wireless network services can be broken down into three sections: the so-called first mile, last mile, and remaining miles. The **first mile** *of the communication path is the portion starting with the server. It involves a LAN that may be operated by a given service carrier. The* **last mile** *of the communication path is the portion starting from the client device. It could be wired for an Internet connection or wireless using one of many available radio frequency air interfaces. The balance of the network, or the* **remaining miles,** *is based on traditional telecommunications and WAN technology. Most ISPs and cable companies, such as AOLTime Warner, are interested in the services of the last mile.*

It can be seen from the above discussion that the differences between a wired Internet GIS and a mobile GIS are the client devices and the last mile of the communication path. The server and the remaining networks are the same or very similar. Mobile GIS extends the reach and scope of server-based Internet GIS applications. Therefore, server-based applications should be developed with both wired and wireless clients in mind, not just for wired Internet users or wireless users.

The conceptual general architecture of mobile GIS is shown in Figure 9.12. The presentation element is the user interface on an Internet-enabled handheld device. The business logic elements include a base transceiver station, a mobile switching center or mobile position center, a gateway service, a Web server, a GIS server (or information processor), and other server applications. The data element includes a location content server (data server) and database sources.

The user with an Internet-enabled mobile device in a wireless network (GSM, W-CDPD, GPRS, etc.) makes a request for location-based information or other information from a URI. The mobile device first checks if it already has an open connection with the telecommunication service provider; if not, it dials up the modem attached to a dial-in server (RAS, or Remote Access Service) in the base transceiver station. The PDE in the base transceiver station determines the location of the mobile device. The position information is processed and maintained by the mobile switching center. This dial-in server gives the mobile device access to the protocols it needs and assigns it an IP address. The request for the URI is sent to the gateway service (e.g., a

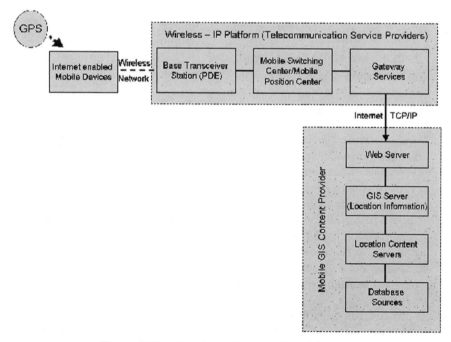

Figure 9.12   Generic architecture of mobile GIS.

Web gateway service or WAP gateway service). The gateway service then requests the URI with a normal HTTP request from the Web server. The gateway service is a piece of middleware that links the wireless and the Web server, basically giving the Internet-enabled mobile device access to the Web servers and other content provider servers. The presence of the gateway service is a major difference between the wired Internet GIS and the mobile GIS. The gateway service acts as a middleware or translator between the mobile device and the Web server. For example, it translates the requests from the mobile device to HTTP requests for the Web server. It also converts the response from the content provider server and Web server to a format (e.g., WML) that is readable by the mobile devices.

On the content service provider side, there is a normal Web server which links with the GIS server (or information process server), location content server, and databases. Sometimes, the Web server and the gateway service are combined as a "gateway server." It should be noted that the normal Web browser could also access the Web server. The Web server, depending on which type of browser the requests come from (microbrowser inside the mobile device or Web browser from a PC), sends out contents in different formats.

When the Web server returns the contents provided by the GIS server and/or content provider server, the gateway service has to compile or repurpose them into the format that can be understood by the mobile device and can minimize the bandwidth usage. The transformed content is then passed back to the mobile device for display.

Finally, the built-in microbrowser inside the Internet-enabled mobile device receives and reads the file and renders and displays the contents on the mobile device for the user. This is how the majority of mobile devices are connected to the Web servers and GIS servers on the Internet.

This conceptual mobile GIS involves three parties (mobile device users, telecommunication service providers, and mobile GIS content providers) and two networks (wireless networks and wired Internet networks). The mobile device users utilize the services provided by the telecommunication service providers and mobile GIS content providers. The telecommunication service providers offer a wireless network linkage, facility, and equipment to provide air interfaces, networking, routing, and other communication services (called wireless IP platforms) for mobile device users. The content providers offer location information contents and information processing services. The network linkage between the mobile device users and the wireless IP platforms is the wireless network, while the linkage between the content providers and the telecommunication service providers is the wired Internet connection.

It can be seen that mobile GIS can offer services and applications similar to thin-client applications on the Internet GIS. Information that is sent to the mobile devices is merely the repackage of applications on the server in a different format to meet the needs of the mobile device. It is what is inside the applications that matters, not necessarily how it is packaged. Most contents and process results on the existing Web and GIS server can be repackaged and sent to the mobile devices via gateway services. However, the limitation of the mobile devices can offer little or no client-side processing. Thus mobile devices have only very limited display and rendering capabilities. In addition, the current wireless network offers very limited bandwidth for data exchanges. These limitations of the mobile devices and the wireless network pose a big challenge for the development of mobile GIS. Nevertheless, several mobile GIS programs have been developed by GIS software vendors. We now turn to the discussion of some samples of mobile GIS programs.

Key Concepts

The mobile device users utilize the services provided by the telecommunication service providers and mobile GIS content providers. The **telecommunication service providers** *offer a wireless network linkage, facility, and equipment to provide air interfaces, networking, routing, and other communication services (called wireless IP platforms) for mobile device users. The* **content providers** *offer location information contents and information processing services.*

9.4   SAMPLES OF MOBILE GIS PROGRAMS

There are several mobile GIS programs available from major GIS vendors. They offer different functions and features and serve different applications. This section provides an overview of these products. They are by no means a complete list. They are chosen because of their popularity in the applications and the authors' own familiarity with them.

9.4.1   MapXtend from MapInfo

MapXtend from MapInfo is a development environment for creating and delivering enterprise location-based applications to a variety of mobile devices, including laptops, PDAs, and cellular phones. In fact, MapXtend supports any device based on J2EE and J2ME specifications, including Handspring Visor, Palm (IIIc and V), HP Jornada and Compaq iPAQ pocket PC, Motorola iDEN and *NTT DoCoMo,* browser based, and WML platforms (MapInfo, 2001). It is targeted to provide location-based services but could be used in many other applications, such as service calls, infrastructure repairs, and 911 emergency response.

The architecture of MapXtend (Figure 9.13) is very similar to the generic mobile GIS architecture. MapXtend is a very thin client application with most of the processing done at the server. It has three major components: client, server, and data. In addition, MapXtend also provides an administration ser-

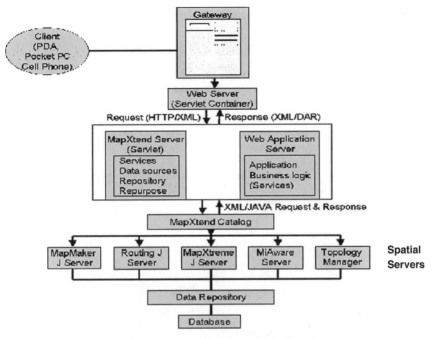

Figure 9.13 Architecture of MapXtend.

vice (MapXtend administrator) to configure and monitor the server operation and performance.

9.4.1.1 MapXtend Client MapXtend's client is very thin due to the limited resources of the mobile devices. Only the user interface and some local storage modules are put at the client on the mobile device. The client implementation of MapXtend is based on J2ME. This means that the client applications, although thin, can still offer rich user interfaces and support persistent data on the mobile device for off-line use.

The MapXtend client (Figure 9.14) has two-way communication with the MapXtend server via wireless communication networks. It allows the user to dynamically download and manage raster, geospatial locations, GML geometries, and textual data from the server. The user can select, view, query, update, and store the data into the local database on the PDAs. The user can use MapXtend to capture new data in the field and upload it to the server using XML stream. MapXtend also supports GML geometries which can be displayed, picked, captured, and edited at the client. Furthermore, MapXtend also offers the client an API to customize map renderers and symbology at the client. The MapXtend client communicates with the MapXtend server via XML and HTTP protocol. The client may use other protocols when accessing servers other than the MapXtend server (MapInfo, 2001).

Figure 9.14   MapXtend client in Palm IIIc.

9.4.1.2 *MapXtend Server*   The server part of MapXtend actually includes four components, as shown in Figure 9.13: Web server, MapXtend server, MapXtend catalog, and spatial servers.

The major role of the *Web server* is to host a servlet container to hold the servlet, because the MapXtend server is a servlet. As discussed in Chapter 4, a servlet is a Java-based Web server extension application that is used to respond to client requests. A servlet requires a Web server with servlet container enabled such as Sun's JavaWebServer and BEA's WebLogics application servers or servlet container plug-ins such as JRun or Tomcat.

The *MapXtend server* is a service manager or dispatcher to allocate client requests to appropriate services. It is a servlet that receives the HTTP requests from the mobile devices and sends back responses. The MapXtend servlet is initialized by the *Web application server*. The requests and responses between the MapXtend server and the Web server/clients are XML documents that follow the specific syntax and semantics in the corresponding DTDs.

The MapXtend server is managed by an XML-based server configuration file, which contains four main components: server, repository, data source, and services. Each component is separately managed by a designated MapXtend manager; therefore, it can be defined as a separate individual file. Each component also has a separate DTD file (MapInfo, 2001):

- The *server* component defines the server information and settings, as well as the administrator configuration. It is instantiated by the Server Manager.
- The *data source* component is designated to be a central placeholder for any database connection information, data descriptors, or external servers properties the MapXtend services may access.
- The *Repository* component is a collection of themes. The repository configuration allows the users to organize and specify the data sets to be cached. A theme is a collection of layers and represents a logical data structure. The themes offer a convenient way to categorize data.
- The *services* component is one of the most important for the server configuration. It defines the available server services. Each service must have at least a name and a class specified. Whenever a class implementing the service is stored across the network (jar files or class directories), a URI address can be provided for remote class loading.

The server and services components must be specified; the other two are optional. The configuration file can be changed and customized, either locally or remotely across the network.

In addition, MapXtend contains a *repurposing* engine (Xalan for Apache) that allows responses in XML to be converted into other markup languages such as HDML, WML, and HTML to fit individual mobile devices.

The MapXtend *catalog* is a collection of registered services or classes that allow the MapXtend server to dispatch client requests to different servers that provide certain services. These MapInfo servers include MapXtreme Java, MapMarker, Routing J server, Topology Manager Enterprise, and Data Manager. The MapXtend server provides interfaces with these servers so that it can take advantage of all the functions of these servers. For example, the MapInfo MapMarker J server provides geocoding functions to convert street address information into a latitude/longitude location that can be used by any application. The MapInfo Routing J server provides the routing function to determine the best way to get from one point to another and delivers back step-by-step directions. The MapInfo MapXtreme Java edition provides two critical functions: proximity queries (*such as nearest and within "x" distance*) and delivering content information and location information on a map to a mobile device.

The MapInfo MiAware server provides a flexible and scalable XML environment for creating new services. MiAware is comprised of core MapInfo servers, XML APIs, and a series of location-services-specific functions such as position acquisition, find-the-nearest, reverse geocode (transform latitude/ longitude into usable location information), map generation, and Yellow Pages management and traffic condition information.

9.4.1.3 Databases Supported   MapXtend connects MapXtend services and spatial servers with data repositories and databases. The data repositories are implemented in the MapXtend server using XML-based JAR to improve performance and portability. The raster or spatial data as well as specific predefined views of data that need to be accessed often can be cached on the server.

Spatial data access is provided via DataProvider or JDBC to a number of data sources, including

- Oracle 8i,
- MapInfo SpatialWare for Oracle,
- MapInfo tables (TAB),
- Informix universal server,
- IBM DB2, and
- GML data

MapXtend provides an adaptor to convert different geometry encoding to and from GML. In addition, it also provides raster and spatial data adaptors specific to different devices (e.g., browsers, Palm, WML, and iDEN). The data transfer between the client and the server uses the standard Directed Asymmetric RowSet (DAR).

9.4.1.4 Information Flow inside MapXtend   A MapXtend client first submits a service request from a mobile device. This request is transmitted to the Web server and MapXtend server in the XML document as an HTTP/POST. The request is in the form of a URI, which must specify the "service" parameter in the name of registered service as part of the query string. For example, the following request specifies the registered service as "FindNearest":

```
http://129.79.82.199/servlet/
MapXtend?service=FindNearest
```

When the MapXtend receives the request, it dispatches the request to a corresponding service. The service will send off the request to a spatial server or third-party server to complete the task. The incoming request also needs to specify the output format the mobile device supported. For example,

```
http://129.79.82.199/servlet/
MapXtend?service=FindNearest&device=XML
```

The appropriate service is then instantiated by the server. Each incoming request has its own designated service instance. A service implements the application business logic, and it can call other services from within the server or remotely from other distributed servers such as the MapXtreme Java server (Mapinfo, 2001). The response is returned to the client in the reverse order.

9.4.1.5 MapXtend Administrator One of the unique characteristics of MapXtend is its administration function. The MapXtend administrator can interactively configure the MapXtend server configuration and its associated services on the Web without interrupting the server. It can also track server status and display parameters as well as provide on-line linkage with MapXtend documentation. The user can access the administrator anywhere over the Internet because the user interface of the administrator at the client side is Web based using simple HTML/DHTML and JavaScript. There is no Java applet required.

9.4.2 IntelliWhere LocationServer from Intergraph

IntelliWhere LocationServer is a server-based application or software platform that provides spatial information and spatial processing services to any applications or devices that need spatial information or spatial operation. These other applications are called the client's "main application system" in IntelliWhere LocationServer (Crisp, 2002).* The client's main application system can be any application that needs to perform some sort of spatial operation such as WAP applications to provide location-based information to users of mobile devices, Web applications to add maps to Web pages and perform basic spatial analysis, or tracking applications to display real-time position information.

The basic function of IntelliWhere LocationServer is to accept a spatial query from the calling application, process it, and send the result of that query back to the caller in the required format. The result can be incorporated into the calling application for further processing or delivered via the Internet to Web browsers or via wireless portals to a wide range of wireless devices. The communication between the client application and IntelliWhere Location-Server is XML.

9.4.2.1 Functions and Features of IntelliWhere LocationServer

9.4.2.1.1 Spatial Query Functions The most important function of IntelliWhere LocationServer is its spatial query capabilities. It is developing a range of query capabilities such as geocoding ("Where am I?" or "where is the service delivery point?") and network search ("How many households need to be serviced in my service area?" "How far away to my destination?"

---

*The discussion of IntelliWhere LocationServer in this section is based on Crisp (2002).

or "Where's the nearest hospital?"). For example, the built-in OrderBy-Distance query can return a list of addresses or features surrounding a specific address in order of increasing travel distance from that address. A map may also be returned showing the locations of those features along with specific text information at each point, such as a customer name or telephone number.

IntelliWhere LocationServer receives the queries as XML strings from the calling application. It processes the query using the location, map, feature, and event data and returns the result in a format that is appropriate or optimal for the calling device or application using the IntelliWhere LocationServer Device Behavior Metadata System.

9.4.2.1.2 Routing Functions IntelliWhere LocationServer comes with built-in routing capabilities and a geocoding capability called the Annotated Centerline Engine (ACE). ACE is specifically designed to locate addresses and generate points from a client-supplied address (typically containing a street number, street name and type, and locality information such as town name or postal code). Alternative geocoding software engines can also be used if the user so desires for particular jurisdictions or applications.

IntelliWhere LocationServer also provides routing information from points to points. It even can provide driving directions through a voice generator and relayed as voice instructions for drivers.

9.4.2.1.3 Device and Application Independence IntelliWhere Location-Server can support applications that work with any mobile devices that operate with Short Message Services (SMS), WAP, J2ME, iMode, or an HTML/XML browser, including EPOC, Palm, and pocket-PC-based devices. Similarly, it can also support many third-party applications. The calling application simply sends an XML request string to LocationServer, and the results are sent back in the form requested by the calling application.

In addition, IntelliWhere LocationServer relies on the GeoMedia data server for data access. Therefore, it can access many common proprietary GIS databases in real time, without data conversion.

9.4.2.1.4 Supported Communication Protocols Using the appropriate gateway, IntelliWhere LocationServer supports any data communication medium, including CDMA, CDPD, Time Division Multiple Access (TDMA), GPRS, UMTS, GSM, and W-CDMA.

9.4.2.2 Architecture and Main Components of IntelliWhere *LocationServer* The architecture of IntelliWhere LocationServer is shown in Figure 9.15. The *main application system* refers to any client applications that request services from IntelliWhere LocationServer. It handles the user interface and some of the query formatting and result display functions. It must have an application (e.g., a URI/XML formatter) to send the user request in URI in the required format to IntelliWhere LocationServer and an-

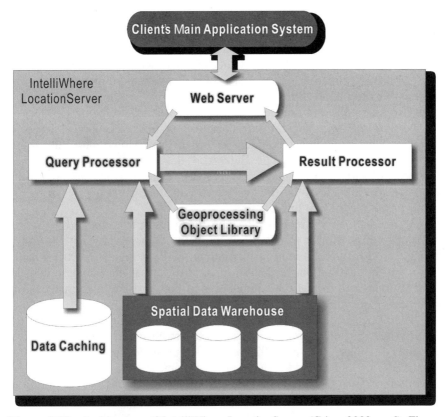

Figure 9.15   Architecture of IntelliWhere LocationServer (Crisp, 2002, p. 6). Figure reprinted with permission from Intergraph Queensland Branch.

other application (e.g., XML parser) to read the resulting XML string from IntelliWhere LocationServer. The XML string contains the results of the query in the required format. For example, the text results are returned as text strings, and generated maps are returned as a URI to the map.

The IntelliWhere LocationServer *Web server* is used to communicate with the main application system. The Web server receives URIs or XML strings from the main application system and passes them to the query processor. It also delivers the results as XML strings back to the client's Main Application System. The IntelliWhere LocationServer supports Microsoft IIS.

The *query processor* receives the requests from the Web server, parses the query, and then processes the query using information from the geoprocessing object library, spatial data warehouse, and data caching database. When the query has been processed, the result is passed on to the result processor for formatting into the required format for the receiving main application system. The query processor uses a URI/XML parser to read and parse the query and a set of ASPs for processing those queries. In its current release, the only query function supported is OrderByDistance.

OrderByDistance orders a list of spatial features by distance from a specific target location (e.g., my address). Other optional outputs include a map and a list of driving directions. This query could generate more interesting results such as "finding the feature that is the closest to me" or "finding the shortest route from one place to another." More spatial queries are being developed.

The *geoprocessing object library* contains a number of geoprocessing software modules, including GeoMedia WebMap Professional and other third-party vendor products such as Enterprise Resource Planning (ERP) and Customer Relationship Management (CRM) vendors. These geoprocessing modules are called upon by the query processor for spatial queries or by the result processor for building maps.

The *result processor* receives the processing result from the query processor and formats it prior to passing it on to the Web server. There are three formatters to format different results into an XML string: The text formatter to format the text output, the map formatter to format map images, and the XML formatter to format the combination of both text and map images. The result processor also formats various status messages into an XML string.

"The *Data Caching database* contains information held as a cache by the IntelliWhere LocationServer to enhance performance and reduce the size of the query strings" (Crisp, 2002, p. 8). For example, a database of place identifiers and street addresses within a namespace can be cached by LocationServer so that the main application system can pass a short string of identifiers instead of a much longer string of street addresses.

The *spatial data warehouse* is where the actual spatial data are stored. Other components, especially the query processor and result processor, rely on the data for processing spatial queries and for building maps. The data access could use the GeoMedia data servers to connect with a variety of sources managed by a metadata system for easier access and updates.

IntelliWhere LocationServer is not yet an off-the-shelf program; rather, it is a server-side application that needs to work with other applications. IntelliWhere does provide an off-the-shelf program, OnDemand, to assist field workers to obtain location-specific information from the server in real time when working in the field and use their hand-held device to update and upload the data. Field workers can use an occasionally connected PDA to access maps and can zoom in and out, run basic searches and queries, or bring up full details of an asset.

9.4.3 ArcPad from ESRI

ArcPad (version 6) is ESRI's mobile mapping and GIS software which runs on portable computers such as a pocket PC or hand-held or tablet PC with the Microsoft Windows CE operating system. It is designed for field data collection and validation. It can be integrated with the GPS or differential GPS (DGPS) to locate its positions on the field.

ArcPad acts as a client to ArcIMS. It supports such functions as map display, query, editing, and data capture. It also has the optional GPS plug-in capability (Figure 9.16). The mapping interface has ArcView-like buttons. ESRI also provides an ArcPad tools extension in its ArcGIS to prepare data for ArcPad and to postprocess data collected in ArcPad.

ArcPad supports a number of GPS receivers. It can display and automatically save GPS information in ArcPad data files. ArcPad indicates its position in the field by a moving crosshair on a map. GPS data can be used for navigation or to capture point, line, or polygon features into ESRI shapefiles.

ArcPad handles multilayer maps (Figure 9.17) and displays vector or raster images, satellite imagery, and aerial photos. ArcPad uses vector data in ESRI's shapefile format as well as several raster image formats such as JPEG, MrSID (compressed images), Windows bitmap, and CADRG raster maps. Users can combine vector and raster data to the limit of the speed and memory capacity of the hardware in use. ArcPad supports these vector and raster data formats directly without the need to convert to unique portable formats (ESRI, 2000).

Data are downloaded from the ArcIMS server to ArcPad using a TCP/IP connection, such as a wireless LAN, cellular phone, or a wireless modem.

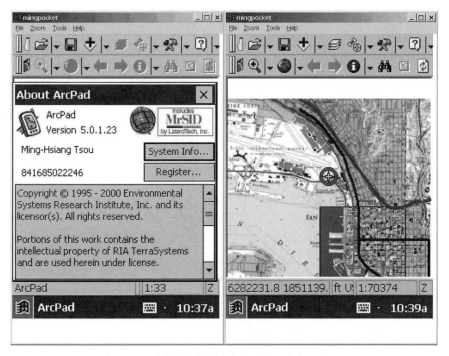

Figure 9.16 ArcPad 5.01 and GPS function.

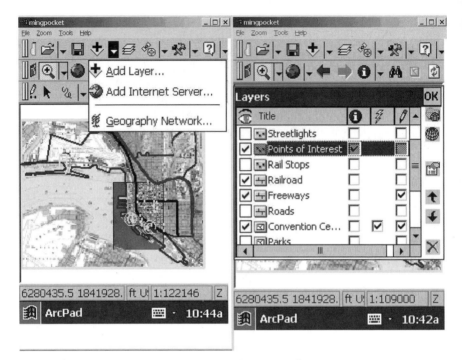

Figure 9.17 Adding new GIS layers in ArcPad from Internet map server.

Changes and additions in the field can be uploaded into the master database (ESRI, 2002).

ArcPad's potential applications include property damage assessment, habitat studies, military fieldwork, street sign inventory, road pavement management, power pole maintenance, and meter reading. It can also be used to provide location-based services to wireless devices, including mobile phones.

9.4.4 OnSite from Autodesk

Autodesk OnSite version 2 (Figure 9.18) is a software application that delivers digital maps and design information to a mobile workforce. It operates on all Microsoft WinCE-based hand-held and tablet-computing devices. OnSite offers workers in the field the ability to access, view, mark up (redline), and measure design drawings and geographic information at the point of work. Autodesk OnSite consists of two components: Autodesk OnSite View and Autodesk OnSite Enterprise.

Autodesk OnSite View is a stand-alone mobile design program or a client component to OnSite Enterprise that is installed on hand-held PCs to enable users to access digital drawings and maps in the field. It has three major functions in three modes of operation: *view* files or maps on a "select" mode,

Figure 9.18 Autodesk Onsite. (Reprinted from Autodesk Web site with permission.)

mark up or redline drawings or maps in a "markup" mode, and *measure* distances and areas in a "measure" mode. The view function is used to select and edit markups or to view tips and perform hyperlink jumps from the source map or drawing. The measure function is used to measure lengths, areas, angles, and perimeters and to find point coordinates. The markup function is used to redline on top of a source map or drawing file (Autodesk, 2001). OnSite Viewer's markup tools are XML-based, which also include audio notes and user-definable markup symbols. OnSite Viewer also provides a COM API for building customized mobile applications.

OnSite View communicates with the desktop or server computer via Microsoft ActiveSync technology. There is no real-time wireless connection yet. That means the field workers have to download the drawing files and map files from the desktop or server computer before going to the field using the ActiveSync software. At the end of the day, the field workers have to upload the files from OnSite View to the desktop or server computer. The OnSite interface is pen based–it mimics the traditional pen-and-paper mode of drawing.

Autodesk OnSite View version 2 only supports its own OnSite Drawing (OSD) format. When you use Microsoft ActiveSync to move DWG and DXF files from your desktop computer to your mobile device, the files are automatically converted to the Autodesk OnSite View file format, OSD. After creating markups on your mobile device, you can move the markup files back to the desktop computer in yet another conversion process from OnSite Markup (OSM) format to the Redline Markup Language (RML) format.

This file conversion process can also be done at the server site using Autodesk OnSite Enterprise, an OnSite server component which can convert native AutoCAD (DWG and DXF) and Autodesk MapGuide (MWF) files to the Autodesk OSD format. OnSite Enterprise can also be used with third-party synchronization software, such as Synchrologic, Aether System's

ScoutSync, or AvantGo, to download the converted files from the server to the mobile device and upload Autodesk OSM files to the server, which Autodesk OnSite Enterprise then converts to RML. The third-party synchronization software ensures the files on the desktop or server are in sync, meaning the latest version of the data files are always copied to either the server or the mobile devices. The communication between the third-party synchronization software and the mobile devices is through HTTP. Autodesk OnSite View is now the viewing client of OnSite Enterprise. With the OnSite Enterprise server, users can remotely access design and mapping data on an organization's central servers, review it, mark it up digitally, and send it back to the server. In addition, Autodesk OnSite Enterprise also provides a COM-compliant API for building customized enterprise-level applications (Autodesk, 2001).

It should be noted that Autodesk OnSite is not a client of MapGuide. Autodesk OnSite Enterprise can be integrated with MapGuide to convert MapGuide's MWF file into the Autodesk OSD format. But the functions of MapGuide cannot be used by OnSite. This is drastically different from its early version of the OnSite program where the OnSite application is built as a client to Autodesk's MapGuide software and Oracle 8i.

Autodesk OnSite focuses mainly on clients in the telecommunications and utilities industries as well as architecture designs. It allows workers in the field to review and revise design and redline changes in the maps on the field, and the revision is automatically moved to the server or desktop when the mobile device is synchronized with the desktop or server.

9.5   APPLICATIONS OF MOBILE GIS

Some applications of mobile GIS are emerging; many more are yet to come. We can categorize different applications of mobile GIS in terms of *field workers* and *consumers*. Mobile GIS can be implemented in organizations with a large field force, such as telecommunication and service providers who have a large contingent of field workers responsible for equipment service and repair, or a mobile sales force that needs access to real-time corporate data. It is important for field workers to constantly keep in touch with their corporation and have up-to-date information. Potential applications include field data collection and validation, retrieving historical data, updating work orders, responding to customer service requests, and so on.

Similarly, for consumers, it is important to have real-time information about where they are and what is around and how to get from point A to point B. Potential applications in these location-based services include service location search, driving directions, and tracking services, such as finding the nearest ATM, hotel, and restaurant using a WAP-enabled mobile phone.

9.5.1 Mobile GIS Applications for Fieldwork

Mobile GIS, coupled with GPS, has made headway in fieldwork applications, including field data collections and validations, incident investigation and site analysis, real-time work order management and dispatch, and so on.

9.5.1.1 Field Data Collections and Validations   By integrating with GPS, mobile GIS is becoming an important tool in collecting and validating field data. These applications include property damage assessment, habitat studies, street sign inventory, road pavement management, and meter reading. Users can collect field data and input them directly on the spatial database or make changes to existing data based on the field observations.

For example, the Natural Resources Conservation Service (NRCS) of Iowa uses mobile GIS and GPS in a PDA to conduct such fieldwork as inventory natural resources and survey flood damages (*Geospatial Solutions,* September 2000, p. 26). Iowa NRCS engineers used the GPS receiver to record the location of flood damage and measure the boundaries of the damaged areas. These data, alone and with other observation and digital imagery of damaged sites, are transmitted back to the operation center on a daily basis and are stored in an Emergency Watershed Protection (EWP) database. The data can then be evaluated quickly for type of damages, repair and assistance needs, and funding requirements. Having current data available improves the agency's ability to provide assistance more quickly.

Another application is to collect time-series data such as soil and crop condition from the same sampling point. With the GPS and the mobile-GIS-enabled PDA, the field worker can accurately locate the previous sampling points and enter data for the current time period.

The common practice at this time of low wireless bandwidth is downloading the data in the morning from the corporate office into a laptop computer and taking the laptop to the field. The field crew then works on editing and validating the data in the field. At the end of the day, the updated data are uploaded to the corporate data server. In the future, as the bandwidth of wireless networks improves, real-time communication between fieldwork and the office data server is inevitable. In fact, some experiments have been made to have real-time connection with the office data server, as demonstrated in the Tacoma Water Utility (Teo, 2001).

9.5.1.2 Incident Investigation and Site Analysis   By accessing real-time information, field workers can have more information in incident investigation and site analysis. For example, by being able to access and see information about the severity and status of network alarms on the telecommunication network in a hand-held computer or PDA, the field technician will have a clearer understanding of the extent of a problem and its possible causes and can therefore prioritize the repair tasks.

Mobile GIS can also be used in real-time site analysis and accident report. A field worker can retrieve historical data about a particular site from a hand-held device. For example, if police found a code violation of a property, they can immediately retrieve the information about the property, such as the property owner and previous code violation history. The police can even write an electronic ticket to the property owner. Similarly, with Mobile GIS and GPS, the police can write a more accurate accident report (Figure 9.19). Likewise, an architect in the field can retrieve design information about a particular site from a mobile device and make changes based on the field survey and validation. The changes can be made available to other architects in the office.

9.5.1.3 Real-Time Work Order Management and Dispatch Work orders such as repair requests and service calls can change at any time, some added, some altered, and some cancelled. It is extremely important for the field crew to have updated work order information. Mobile GIS can help the field workers have current data and information for the next work order, even for the work order that is added after the field worker leaves the office. The use of mobile GIS can eliminate unnecessary travel time to and from service depots for maps and other job-related information. It can eliminate redundant data input activities after jobs were completed. The field worker can also rely

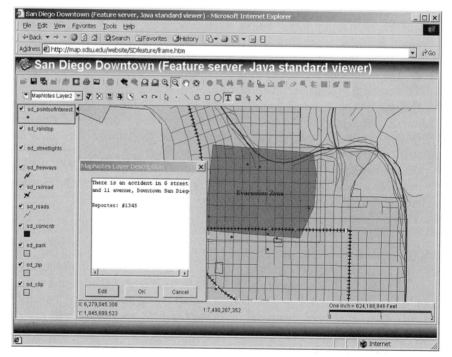

Figure 9.19 Police can submit an accident report by using mobile GIS.

on mobile GIS on the hand-held device to find the routing information to get to the next site, reducing travel time and response time. This is especially important for multiple service requests.

Mobile GIS can be used to identify and visualize the location of repairs, equipment, or property in the field. It can further update corporate databases on trouble tickets, repairs performed, or service calls completed in the field. By integrating with GPS, mobile GIS is a perfect tool for "call-before-you-dig" services, that is, to access, visualize, locate, and update corporate data on underground infrastructure to avoid costly cuts in lines.

For example, utility companies realize that many of their service disconnects made for nonpayment had to be reconnected the same day because the customer had made the payment after the field crew had been scheduled. By providing real-time access to that payment information, the utility could eliminate the return visits, with a substantial saving (Tadpole Technologies Inc., 2000).

Being able to access real-time information, mobile GIS can significantly improve dispatch efficiencies and overall productivity. It allows utility and telecommunication companies to consolidate their dispatch centers and reduce dispatch staffing.

9.5.1.4 Real-Time Responses to Customer Service Requests

Responding to emergency service requests for workers already in the field is difficult. For example, a breakdown of a telephone line requires the field workers to have information on the locations of the telephone cables and surrounding environment. The field workers also need to know the best way to get to that location. With mobile GIS, this information is easier to obtain. The field crew needs only to use a hand-held device to connect with the database on the corporate server and can easily retrieve the information.

9.5.2 Location-Based Services for Consumers

The FCC mandated the capability of reporting the location of a cellular phone. Phase 2 of the FCC mandate (also called the e-911 mandate) requires PDE and that the mobile device be located within 125 meters. This e-911 mandate has unleashed tremendous potential for mobile GIS. Since wireless carriers have to comply with this law, mobile GIS can play an important role in providing location-based services. Besides sending the location information to the appropriate emergency call center, hand-held devices will become a very important tool in travel-based location services and other location-based content.

The potential applications for location-based services could be very large. The following are some possible examples.

9.5.2.1 Location-Based Services for Vehicle Drivers

The first and probably the most important application of mobile GIS is in transportation, particularly the emergency response systems, in-car navigation systems, real-time traffic information, and real-time information locators.

The recently emerged Mayday system has gained popularity among new car buyers. The Mayday system is an in-car electronic equipment that can identify the location of a car and automatically notify the closest emergency response centers if the car is involved with an incident. The emergency response center would immediately send out a response team to help with the vehicle. The Mayday system has a GPS unit installed in the car to automatically locate the location of the car in real time. That information is automatically sent to the computer in the emergency response center if an incident occurs. Major car manufacturers have been supporting this Mayday system, including GM's OnStar system and Ford's RESCU systems. Along with the emergency response functions, these systems also provide turn-by-turn in-car navigation systems.

Through the deployment of ITSs and the traffic-monitoring systems, real-time traffic information such as traffic speed and traffic condition and accident information have been made available on the Internet. But the information is not specifically tailored toward the routes that a traveler is going to take. Furthermore, wireline-based traffic information is only good before the trip. Once you are on the road, information on the Internet is of little use if you do not have mobile access.

Mobile GIS, coupled with wireless Internet technology that allows drivers to retrieve information inside the car and/or over the cellular phone, could provide much more useful and timely information, even without a Mayday system installed in the car. With the PDE-equipped cellular phone, the server knows where you are and where you are going. Therefore, it can send you the traffic information that is specific to your particular location and travel direction in real time. An advanced traffic information system can even advise you of the alternative routes and the traffic conditions on those routes so that you can make a more informed decision.

In addition, travelers on the move are eager to know where things are and how to get there in the shortest amount of time possible. Mobile GIS can offer great help to obtain real-time location-based information. One example is a 411-like concierge service. A location-based concierge service that provide Yellow Pages–like service along with other similar services could offer great services to travelers and thus there may be potentially great demand from wireless users.

9.5.2.2 Tracking Another application of location-based services is tracking. The applications could cover a wide range of objects, from fleet tracking of trucks or taxicabs to tracking packages, children, pets, prisoners, and Alzheimer's patients. A PDE-equipped phone could make the tracking more affordable than the conventional GPS equipment. However, this application may meet opposition from privacy advocates. Nevertheless, used properly, tracking using PDE-equipped cellular phones can become very useful and could be deployed on a large-scale basis.

For example, PepsiCo uses GPS and mobile GIS to track its truck delivery (*Geospatial Solutions,* September 2000, p. 22). PepsiCo equips each rig with a fleeting tracking system that is comprised of a GPS receiver, a mall microprocessor, and a wireless CDPD modem. The internal GPS receiver updates the vehicle's position every 10 seconds and stores it onboard. The onboard unit transmits the Position/Velocity/Time (PVT) data to a fleet tracking service provider's network. The service provider immediately posts the PVT data on a secure Web site, which displays each truck's position on a street map.

A fleet manager can log onto the Web site and can monitor the location and the schedule of the trucks at any time. Managers at the bottling plants and merchandisers can also access the Web site and know the exact location of the truck and determine the arrival time. The tracking techniques help PepsiCo address fleet management and marketing issues as well as delivery. By tracking trucks' movements in real time, the fleet manager can dispatch trucks more efficiently and help reduce the amount of time the trucks are kept at any one stop. With just-in-time delivery, GPS and mobile GIS offer fleet managers the capability to know the locations of the fleet at all times.

9.5.2.3 Traffic Monitoring
PDE-equipped cell phones can be used to obtain and measure real-time traffic flow data. By tracking the movement of cell phones while motorists are talking on their cell phones, it is possible to measure the speed of their movement and thus the traffic conditions on the road they are traveling. This will create a lot of traffic data at very low cost.

A pilot program has been planned on a 15-mile stretch of the Capital beltway south of Washington, DC, between U.S. 5 in Maryland and the interchange of Interstates 95, 495, and 395 in Springfield, Virginia, for the Spring 2001. This experiment relies on signal triangulation to determine locations, that is, tracking on radio signals emitted from cell phones used by motorists by cellular towers. Computers will pinpoint the caller's location and calculate the speed of the phone user's automobile based on how long the call lasted and how far the automobile traveled.

By monitoring when cars change speed, traffic operation officials might be able to predict backups—up to an hour before they happen—and post messages in variable message signs or even send information back to travelers to alert the changes of traffic conditions and encourage the use of alternate routes.

9.5.2.4 Real-Time Bus Location Information Systems
Imagine you are waiting at the bus stop for a bus and the bus does not come on time. You do not know whether you have just missed the bus or if the bus is late and, most importantly, when the next bus will show up. What you need is to be able to push a button in your cellular phone and retrieve the bus location information in real time. This is where mobile GIS can help.

Mobile GIS can be used to display real-time locations of different modes of public transportation (buses and trains) on mobile devices. For example, when you need to find out when the next bus will come to the stop, you can request the current location of the next bus. With the location-enabled cellular phone, your location is automatically identified. So the system can automatically send the location information of the next bus coming to the bus stop. Even without the PDE, you can still request the information by using a stop identifier number located at each bus stop. You can access the public transportation site through your mobile device and key in the stop identifier number and the bus route number. The bus location application in the Web server then knows exactly where you are and what bus you are waiting for and can display the location of the next coming bus based on the onboard Automatic Vehicle Location (AVL) System installed on each bus. In fact, the Infopolis project in Europe has successfully transmitted bus location information to cellular phone users (http://www.ul.ie/~infopolis).

In summary, mobile GIS is still in its infancy due to the very limited mobile devices and low bandwidth of the current wireless network. But the demand for mobile GIS and location-based services is high, from field workers to mobile consumers, and the bandwidth of the wireless is improving as the wireless industry is getting ready for 3G cellular networks. The future of mobile GIS holds great promise for developers, location service providers, GIS professionals, and consumers.

WEB RESOURCES

Descriptions	URI
Microsoft mobile devices	http://www.microsoft.com/mobile
Palm OS	http://www.palmos.com
WAP	www.wapforum.org.
WML	http://www.oasis-open.org/cover/wap-wml.html
Java J2ME wireless toolbit	http://java.sun.com/products/j2mewtoolkit
Autodesk OnSite	http://usa.autodesk.com/adsk/section/0,,702580-123112,00.html
ESRI ArcPAD	http://www.esri.com/software/arcpad/index.html
Integraph GeoMedia WebMap Professional	http://www.intergraph.com/gis/gmwe
MapInfo MapXend	http://dynamo.mapinfo.com/miproducts/Overview.cfm?productid=1065
Infopolis project	http://www.ul.ie/~infopolis
IntelliWhere	http://www.intelliwhere.com

REFERENCES

Ainsworth, B. (2002). *Technology Trailblazers, San Diego Union-Tribune,* May 5, pp. A1, A10.

Autodesk. (2001). *Autodesk OnSite Viewer 2 User's Guide*. San Rafael, California: Autodesk.

Crisp, N. (2002). *Introduction to IntelliWhere LocationServer,* White Paper, Milton Brisbane, Australia: IntelliWhere.

Dornan, A. (2002). *The Essential Guide to Wireless Communications Applications: From Cellular Systems to Wi-Fi*. Upper Saddle River, New Jersey: Prentice-Hall.

ESRI. (2000). *ArcPad™: Mobile Mapping and GIS,* White Paper, September 2000. Redland, California: ESRI.

ESRI. (2002). *What's New at ArcPad™ 6,* White Paper, April. Redland, California: ESRI.

Ferber, J. (1999). *Multi-Agent Systems: An Introduction to Distributed Artificial Intelligence,* English Edition. Harlow, England: Addison-Wesley.

MapInfo. (2001). *MapInfo MapXtend User's Guide*. MapInfo. Troy, New York: MapInfo.

MapInfo. (2002). *Mobile Location Services*. URL. http:www.mapinto.com/wireless/img/new%20mobile%20area.doc

Microsoft. (2002). *Development Tools for Mobile and Embedded Applications,* White Paper. Bellevue, Washington: Microsoft.

Open GIS Consortium. (2000). *A Request for Technology in Support of an Open Location Services (OpenLS™) Testbed*. Wayland, Massachusetts: Open GIS Consortium.

Pandya, R. (2000). *Mobile and Personal Communication Systems and Services*. New York: IEEE Press.

Rischpater, R. (2002). *Wireless Web Development,* 3rd Edition. Berkeley, California: APress L. P.

Tadpole Technologies Inc. (2000). *The Field Case for Field Force Automation*. Englewood, Colorado: Sharper Insight, Inc.

Teo, F. K. S. (2001). For Tacoma Water GIS Is a Wireless Wireless Wireless World. *Geospatial Solutions,* February, pp. 40–45.

WAP Forum. (2002). *Wireless Application Protocol (WAP) 2.0 Technical White Paper,* www.wapforum.org.

CHAPTER 10

QUALITY OF SERVICE AND SECURITY ISSUES IN DISTRIBUTED GIS

Good, fast, cheap. Pick any two.
—Geoff Huston (2000, p. 275)

10.1 INTRODUCTION

Over the last decade, many important and interesting applications have utilized the Internet as the media to disseminate information and services, such as Internet phone, video-on-demand, and Virtual Private Network (VPN). Internet GIS is one of newest Internet applications and will become more and more popular and important in the future.

However, these new Internet services and applications have quite different requirements from those for which the Internet was originally designed and implemented 30 years ago. Originally, the Internet was only designed for disseminating text-based information and document transfer functions, such as telnet, e-mail, gopher, and FTP applications. The nature of the Internet is a distributed-network model which may have unreliable packet delivery because each packet can be delivered to the same IP address from different routing paths. Under the current Internet technology and TCP, multiple services (such as voice, video, GIS mapping) are very limited because of the unreliable quality of network communication. Internet GIS applications may be inaccessible if client users have severe network congestion or connect with very low network bandwidth.

In order to solve these problems, the issue of quality of services (QOS) has recently become the major research focus, especially for the next generation of Internet frameworks. The computer science community proposed several possible solutions for QOS, such as integrated services, resource reservation setup protocol (RSVP), and differentiated service framework (Huston, 2000; Wang, 2001). This chapter will discuss the QOS-related issues for Internet GIS, including performance, scalability, functionality, portability, and security issues. One note is that the QOS topics discussed in this chapter are broader than the original QOS proposed by the computer science community.

Key Concepts

Quality of Services (QOS) *is a new research direction for the next generation of the Internet and focuses on the new technologies and standards to provide resource assurance and service differentiation for various Internet applications. There are two major technologies under the QOS framework, integrated services and differentiated services (Wang, 2001).*

From a system architecture perspective, two distinct approaches coexist in the current development of Internet GIS: the server-side approach and the client-side approach. Both approaches represent different ways to access and process data over the Internet. They are different in terms of performance, interactivity, portability, and security.

10.2 PERFORMANCE

Performance is one of the most important issues in the development of Internet GIS, because Internet users are usually impatient waiting for the system to respond. Performance is especially important for Internet GIS because GIS data, including vector and raster data, are large in volume.

10.2.1 What Is Performance?

When we talk about performance, we usually refer to the extent of system responsiveness. Actually, there are two major criteria for measuring performance: throughput and response time (Shan and Earle, 1998). *Throughput* is a server-oriented measurement that measures the amount of work done in a unit of time. For example, the throughput of an Internet mapping, Web site, such as MapQuest, can be evaluated by how many images (maps) are generated by their GIS engines per minute. Some use the number of transactions processed per second or per minute. But the transaction is difficult to measure because it is application and operation dependent. For example, buffering and identifying are two transactions, but it takes much longer to process buffering than to identify a spatial object.

Key Concepts

Performance *refers to the efficiency and effectiveness of computer systems and software applications. Similar to the term* productivity, *there are several objective measurement approaches, such as throughput and response time, to measure performance for specific computing tasks or software applications. The results of performance tests are usually called benchmarks.*

Response time is the amount of *user-perceived* time between sending a request and receiving the response. It includes client processing time, network time, and server processing time (Shan and Earle, 1998). The response time of each component, client, server, and network also depends on the amount of workload in each component and the speed of each component as well as the user's perception.

Throughput and response time are often related but not necessarily directly associated. A fast throughput is usually associated with a better response time and vice versa. However, a high throughput does not guarantee a fast response time. For example, assuming two Internet GIS programs have the same exact throughput but one displays partial maps or the outline of a map while the full data are loading and the other waits till all data are transmitted before displaying the full map, the former system would give the user the perception that it has a faster response time than the second one.

Here is another example of a server-side Internet GIS operation. There are two Internet GIS servers. System A has a more powerful server with a throughput of 5 units of work per second while system B has a less powerful server with a throughput of 4 units of work per second. Suppose two users, Bob and Susan, submit requests at the same time. Bob requests a "buffer" operation, which takes about 100 units of work to process, while Susan requests an "identify" operation, which takes about 1 unit of work for the server to process. Now we look at how the sequence of the processing can affect the response time.

For system A, Bob's request is processed first. This task will take $100/5 = 20$ seconds, and Bob's perceived response time is 20 seconds. Susan's request, submitted at the same time as Bob's, is processed next. It takes $1/5 = 0.2$ second. However, the response time for Susan's request is $20 + 0.2 = 20.2$ seconds. This means that Susan has to wait for 20.2 seconds to get the result back even though her request only takes 0.2 second for the computer to process. The average response time of the system as a whole is $(20 + 20.2)/2 = 20.1$ seconds.

For system B, Susan's request is processed first. It is completed in $1/4 = 0.25$ second, and the response time for Susan is 0.25 second. Bob's request is processed next. It takes $100/4 = 25$ seconds. The response time for Bob is $0.25 + 25 = 25.25$ seconds. The average response time of system B as a whole is then $(0.25 + 25.25)/2 = 12.75$ seconds.

This example shows that although system A has a more powerful server and runs 20% faster than system B, its average response time is even longer than system B with a slower server and lower throughput. By changing the order of processing user requests, the system with lower throughput gives the impression of better performance. This illustrates that increasing performance is not simply buying faster servers, network cards, and other hardware components. System design and smart use of system resources are also very important. Software engineers sometimes use sophisticated mathematical models based on queuing theory to analyze and optimize system throughput and response time (Shan and Earle, 1998). It should be noted that this example does not apply for multithread systems that can handle multiple requests at the same time.

Whether we are talking about throughput or response time, the issue of Internet GIS performance or any other client/server system can be addressed in the performance of different components of the whole system: client, server, and networking.

10.2.2 Client Performance

Component performance, whether it is client, server, network, or any other components, depends on two major factors: the amount of workload and the speed of the components. Client performance is also influenced by the user-perceived response time.

The client workload is determined by the amount of processing in the client side. A thin client such as server-side Internet GIS has a very light load on the client side. It thus has a better client performance. A thick client such as plug-ins, Java applets, and ActiveX controls has a heavier load on the client side and thus the client performance is not as good. But a thin client does not take full advantage of the client-side system resources and tends to increase the load on the server and the network.

The client speed depends on the hardware and the size of the client-side executable. A faster CPU, more RAM and disk capacity, and a faster graphics card can obviously increase the speed of the client machine. The size and the design of client-side applications also affect the client speed. A large number of plug-ins, applets, and ActiveX controls can slow down program startup time and increase the time needed for virtual memory swapping between the hard drive and RAM. Concurrency design, that is, sending out multiple requests concurrently rather than sequentially, would make the client-side program more efficient (Shan and Earle, 1998).

The server-side Internet GIS obviously has the best client performance. As far as client-side applications are concerned, GIS plug-ins usually suffer from a longer startup time than Java applets and ActiveX controls. The Java compilers built into Web browsers can make Java applets run almost (but not quite) as fast as native executables. Compared with Java applets, GIS ActiveX controls have a unique performance advantage because they can access the full platform functionality, for example, local files, memory, and hardware and software system controls, which are unavailable to a GIS Java applet.

On the other hand, GIS plug-ins and helper programs, as well as ActiveX controls, have to be installed in the hard disk of the client machine. Thus, the numerous plug-ins and ActiveX controls downloaded would consume disk space in the users' local machine and thus affect the client performance. This will be especially problematic if these programs become more complex and grow in size and number. Java applets, on the other hand, are installed at runtime on the Web browser when the user initiates the request and can be automatically uninstalled when the user quits the application. Therefore, the Java applet does not take any permanent disk space on the client's local machine.

Most client-side Internet GIS components cache the data from the server on the local machine to enhance performance. For example, when the user is working on a large map, the data have been loaded to the client from the server. If the user wants to zoom in to a smaller area of the map instead of discarding the data outside the zoom area, the client can cache them to avoid trips across the network to the server when the user decides to zoom out. This is an effective means to increase client-side performance. Many existing Internet GIS programs use the cache mechanism to enhance client performance.

To increase the user-perceived response time, several techniques can be used, such as rendering partial results, displaying status information, and allowing the user to do other things while waiting for the result. But the actual response time will not change; it just changes the users' perception. For example, when a user requests a large detailed map, rather than wait until all of the data are loaded before displaying the map, a partial map or the map outlines can be rendered first. Detailed contents of the map can be loaded progressively. Another example is to display the status or percentage of the data or map images that has been retrieved. This is especially important for the initial download of the plug-ins, Java applets, or ActiveX controls because these client-side applications tend to be large and need longer downloading time for slow network connections.

10.2.3   Server Performance

Similar to the client performance, server performance is affected by the server load and server speed. But, unlike the client performance, the user-perceived response time is no longer relevant because the user does not interact directly with the server. The server-side Internet GIS has the heaviest load on the server because all processing is done on the server side. The client-side Internet GIS shifts some processing to the client side so the server load is reduced.

Server speed is determined by the server hardware as well as by the configuration and design of server-side software. Faster and more powerful server hardware can obviously enhance the server performance. But this is not the only way. A properly configured server system and using multithread to allow the server to process multiple clients' requests concurrently can help better utilize system resources and improve throughput and response time (Shan and Earle, 1998). Many experience server administrators can tune performance by optimizing the server configuration. Much of the time spent in the Internet GIS server is for initialization of GIS server programs. Therefore, reducing or minimizing the initialization time can increase server performance significantly. Some programs such as ArcView Internet Map Server require the map server to be constantly running in the server to reduce initialization time.

For the server-side Internet GIS, the CGI script or DLL program handles all input from the Web browser and interprets all output for the GIS server. Because the CGI has to be reinitialized for every request, it becomes very difficult for the CGI script to handle a large amount of requests from users, especially simultaneous requests. For any frequently used site, this incurs a considerable load on the server. The CGI script could become a failing point or bottleneck. That is, when the CGI script or the GIS server fails to work properly, the whole system will fail to function. DLL has a better performance because it resides as the same process as the Web server and is always on, without the overhead cost of reinitialization for each request.

Furthermore, most prior stand-alone GIS programs and many current Internet GIS servers cannot handle multithread tasks; that is, they can only handle one request at a time sequentially and cannot run multiple requests simultaneously. Therefore, multiple copies or licenses of GIS servers are needed to handle multiple simultaneous requests. The capability of handling a large number of requests is limited by the number of copies or licenses of the GIS server. Even creating a separate server instance takes some program initialization time.

10.2.4 Networking Performance

The networking or communication performance is the most important aspect of the overall system performance of Internet GIS. It is usually the bottleneck of the entire system because sending messages across the WAN is often much slower than sending messages in the LAN, as discussed in Chapter 3. But measuring network performance is not an easy task because network performance varies from time to time and is affected by the amount of traffic on the network and other factors.

Networking performance can be measured on *throughput* and *latency*. The throughput of networking is the number of bits it can transfer per unit of time. Message latency is the user-perceived delay of information transporting on the network; it is similar to the *response time* on the client performance (Shan and Earle, 1998). Similarly, the networking performance in Internet GIS is generally affected by the speed of the network connection, the efficiency of communication or glue software between the client and the server, and the workload of the network.

A faster network clearly has better network performance. A T1 line connection is obviously better than the 56K modem. The speed of the Internet connection is improving with faster modem and faster communication connections. At higher education institutions, the next generation of the Internet, Internet 2, has increased the Internet speed dramatically (see Chapter 2). In addition, fiber-optic networks, DSNs and cable modems, are becoming available from commercial telecommunication providers, which will speed up Internet data communication and thus network performance.

In addition to buying a faster network, there are still other ways to get better communication performance by designing efficient networking or middleware software on both ends of the network. The number of trips across the network is usually the most important factor in network performance. Therefore, cutting down the number of information exchanges crossing the network has the most impact on network performance (Shan and Earle, 1998).

For example, Figure 10.1 shows an Internet GIS program that has five major components: map display, map render, display element generator, filter, and data source. The information exchange between the map display and map render is the most frequent. The first design (Fig. 10.1) put the map render component in the server, resulting in heavy network traffic between the client and the server. But, moving map render component to the client side, as shown in Figure 10.2, would significantly reduce the cross-network traffic between the client and the server. Therefore, in the design of Internet GIS, properly partitioning the application between the client and the server to minimize network traffic can significantly increase network performance.

Server-side Internet GIS that relies mainly on the server for processing has many more cross-network trips and thus has much more workload on the network than the client-side Internet GIS. Since CGI and the current HTTP are stateless, an HTTP Web server does not remember callers between requests. If a user requests a map and he or she wants to zoom in to the same map, the whole routine from the Web browser to the Web server and initial-

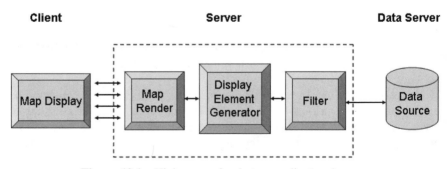

Figure 10.1 High messaging between client and server.

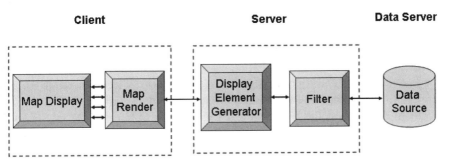

Figure 10.2 Low messaging between client and server.

izing the GIS server have to be repeated. Every single request has to be sent to the GIS server, and every result has to be transmitted to the Web client. Therefore, it generates heavy traffic on the network and has the lowest networking performance.

However, because of the large size of client-side applications such as plugins, ActiveX controls, and Java applets, the initial download of these client-side applications takes much longer time than server-side-generated graphic files such as GIF or JPEG. But once they are downloaded to the client side, the cross-network traffic is much less. Therefore, there is a trade-off between the size of the client-side applications and the cross-network trips of the server-side Internet GIS.

An efficient design of Internet GIS programs can accommodate a slow speed connection to some extent. For example, different Java applets can be deployed based on different client network speed connections. For slow network connection such as the 56K modem, a Java applet with server-side processing can be deployed, while for fast network connection, a Java applet with client-side processing can be deployed. Some existing Internet GIS programs such as MapXtreme allow the user to have such options.

10.2.5 Overall System Performance

The overall system performance of Internet GIS is dependent on the combination of client, server, and network performance, not the individual components alone. More specifically, the overall system performance depends on the bottleneck caused by the slowest component.

Therefore, the first step to enhance system performance is to identify the weakest component or the system bottleneck(s). Once the bottleneck is identified, you can then focus on performance tuning of the slowest element. This is the most effective means of enhancing performance. For example, Bar a in Figure 10.3 shows the unit of time (20 units in the server, 10 units on the

Internet GIS Showcase: Hong Kong Slope Safety

Hong Kong has a history of tragic landslides; in the 50 years after the end of World War II more than 470 people lost their lives. Landslide risk to the community has been substantially reduced since the Government established the Geotechnical Engineering Office (GEO) in 1977; however, Hong Kong's unique combination of dense urban development on steep hillsides and torrential summer rainfall ensures that there will always be some risk.

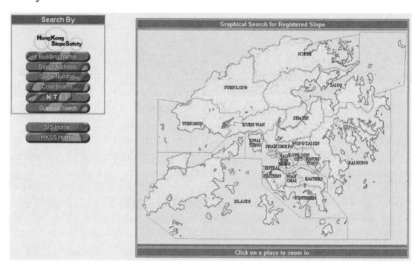

In 1998, the GEO completed an intensive four-year project to catalogue all sizable man-made slopes and retaining walls in the city. A total of 54,000 slopes were catalogued, including all cut slopes and retaining walls 3 m or higher, all fill slopes 5 m or higher (including any associated retaining walls), and those which are less than 5 m high but would pose a direct risk to life in the event of failure.

As part of this project, the GEO developed a computerized Slope Information System (SIS) containing data on all of these slopes and retaining walls. The SIS is a GIS which integrates different types of slope information with geographic and textual databases. The SIS is a valuable source of information to geotechnical engineers, property owners, management companies, and the general public.

In order to better serve the community, the SIS has been adapted to the Internet using GeoMedia WebMap to provide free and convenient public access to slope information from homes and offices.

Source: http://hkss.ced.gov.hk/hkss/eng/slopeinfo/index.htm.

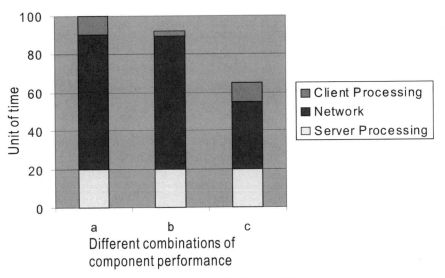

Figure 10.3 Performance optimization in different components.

client, and 70 units on the network) an Internet GIS program spends in client, network, and server processing from making a request to receiving the result. The system bottleneck is a slow network with 70% of the overall processing time spent on transferring the requests and results across the network between the client and the server.

Bar b in Figure 10.3 shows that if we focus on increasing the client performance by buying a powerful machine that runs 10 times faster, the unit of time spent on the client is then reduced to from 10 units to 1 unit. The total time spent in the whole system is reduced from 100 units to 20 + 1 + 70 = 91 units. Therefore, although the client processing is now 10 times faster, the overall system performance improves by only 9%. This is because the client processing is not a bottleneck. Bar c in Figure 10.3 shows that if our focus is on enhancing the performance of the network by a combination of a faster network, and reduced the data size and cross-network trips, the unit or time spent on the network is reduced by a half (35 units) and the total system processing time is reduced to 20 + 10 + 35 = 65 units, a 35% improvement of overall system performance.

Therefore, the key to obtaining the best system performance is to find the weakest component and to strike a balance among client performance, server performance, and cross-network traffic. The server-side Internet GIS has superb client performance but limited server performance and heavy network traffic. Client-side Internet GIS has more client processing and takes longer to initialize and download the data and client-side applications from the server, but it has much better server and network performance. A combination of server-side and client-side processing that properly partitions the applications between the server and the client has a potential to have the best overall performance.

10.3 SERVICE RELIABILITY

In addition to performance, reliability is another important QOS indicator for Internet GIS applications. A reliable service means the Web site is available when needed. When we say the system is not reliable, we mean the system fails to function from time to time. When the server is often "down," it is not a reliable service. There are several reasons that the system is not reliable. It could be due to human errors, component faults, or product defects (Shan and Earle, 1998). *Human error* refers to human errors in handling system configuration, administration, and operation. A *fault* refers to a function failure of a system component. A *defect* refers to component failure or malfunction due to system design, implementation, and documentation. A defect certainly causes faults, but not all faults are caused by defects. Faults can happen in any element of the system, from hardware (e.g., client and server machines, network devices) to system software (e.g., operating system) to application software (e.g., any Internet GIS programs).

10.3.1 A Measure of Reliability: Availability

When a system has a fault, the system may not be available. When the system is not available from time to time, we would say the system is not reliable. Therefore, the system's availability is an effective measure of reliability. But how do we measure availability? The availability (*A*) of any component is a function of the Mean Time Between Failure (MTBF) and Mean Time to Restore Services (MTTR). The MTTR includes time to respond, isolate, correct, and verify faults. Thus the availability of one component can be measured as (Shan and Earle, 1998)

$$A = \frac{\text{MTBF}}{\text{MTBF} + \text{MTTR}}$$

For example, if a component's MTBF is 20 days and the MTTR is 6 hours (0.25 days) (these data can be obtained from the component manufacturers or derived from operation statistics), the predicted availability of the component is

$$A = \frac{20}{20 + 0.25} = 98.77\%$$

This would mean that the component could be out of service for roughly 4.5 days in every year.

If the availability of individual components is known, we can derive the system availability. Because the system availability requires every individual component be available, the availability of the system is the product of the availability of individual components. For example, assume the system consists of three general components—network, server, and data storage—and the availabilities of the individual components are

$$A(\text{network}) = 92\%$$

$$A(\text{server}) = 82\%$$

$$A(\text{data storage}) = 99\%$$

Assume that failures on individual components are independent (this is an important assumption that must be met for this probability calculation), the expected overall system availability is

$$A(\text{overall system}) = A(\text{network}) \times A(\text{server}) \times A(\text{data storage}) = 74.69\%$$

It can be seen that the overall availability of the entire system is always worse than or equal to the availability of the weakest component. Therefore, identifying and working on the weakest component make up the most important first step to enhance the availability of the system. In this case, the weakest component is the server. If the system administrator decides to install a mirrored server as a backup, the availability of the server with the backup becomes

$$A(\text{servers}) = 1 - [1 - A(\text{server})][1 - A(\text{server})]$$
$$= 1 - (1 - 82\%)(1 - 82\%)$$
$$= 96.76\%$$

The overall system availability is now

$$A(\text{overall system}) = 92\% \times 96.76\% \times 99\% = 88.13\%$$

With the installation of the mirrored server, the overall system availability has increased from 74.69 to 88.13%, which is about 18% improvement over the single-server approach.

However, if the system administrator does not begin with the weakest component—rather he or she decides to build a mirrored data storage—the availability of data storage would increase from 99 to 99.99%. The overall system availability would still be

$$A(\text{overall system}) = 92\% \times 82\% \times 99.99\% = 75.43\%$$

This example shows the importance of identifying the weakest component and the effectiveness of working through the weakest component to increase the overall system availability. If the component is not the weakest component, upgrading it may not make much difference to improve the system reliability.

The same principle can be applied to the availability of components of the Internet GIS programs (e.g., client components, server components, and network communication components) as well as the overall system availability of an Internet GIS program. The expected availability of a whole Internet GIS program is the product of the availability of each component.

10.3.2   Increasing System Availability

A major approach to increasing system availability, besides a robust Internet GIS software design, is to create redundancy so that if some system components fail, there is a backup. The use of redundant components, such as a backup network and/or a mirrored server, can greatly improve system availability. This redundancy enables a component to switch, or "fail over," to its backup, a process commonly called *failover*.

Failover support can be stateless or can preserve application states (Shan and Earle, 1998). Stateless failover means that when the system fails over to its backup component, such as another server, it loses all the information and computation done in the original server. This may not be a big issue for noncritical and nontransactional Internet GIS operations because the user simply starts over. But the stateless failover could cause major problems for on-line transactions. For example, a user tries to buy a ticket on-line, and halfway in the process the server fails; the client will automatically fail over to another server, and the ticket may be purchased twice.

Failover mechanisms that preserve states can prevent the kind of situation just described by carrying over the states from the failing component to the next backup component. However, this is much more difficult to implement.

Some Internet GIS programs, such as MapGuide, have instituted the option of installing a mirrored servers. If possible, a mirrored server should be installed to enhance system reliability. In addition, Internet GIS software could differ greatly in terms of robustness, depending on how the software was designed and coded, how rigorously it has been tested under different exceptional conditions, and so on. A careful examination of the system should be made before committing to a software package.

10.4   USER INTERACTIVITY AND FUNCTIONALITY SUPPORT

Different types of Internet GIS have different levels of user interactivity and available GIS functions. For example, server-side Internet GIS uses HTML or DHTML as a form of user interface, which offers limited user interactivity. These shortcomings are overcome by the browser add-ons of Web client applications of plug-ins, applets, and ActiveX controls, which are used to extend the browser capabilities to interact directly with GIS vector data.

For example, Figure 10.4 indicates the differences between HTML viewers and Java applet viewers in the design of ESRI ArcIMS applications. Java applet viewers have more GIS functions, such as drawing and editing, but require a Java plug-in. An HTML viewer is lightweight and less functional but more accessible for lower bandwidth users.

The server-side Internet GIS that uses existing GIS programs as a GIS server can take advantage of many of the functionalities of existing GIS software that has been developed. Thus, the analysis tools and software programs developed in the last two decades can be utilized. Furthermore, an advantage of server-side Internet GIS is that it can handle very large data files to serve customer queries. The back-end GIS applications can be further linked with other relational database management applications to handle data queries more efficiently.

Some GIS analysis functions are prohibited by the limitations of HTML. The current version of HTML cannot allow the drawing of a rectangle, a

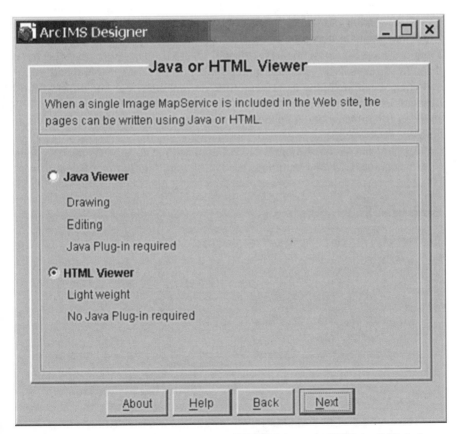

Figure 10.4 Differences between Java and HTML viewer in ESRI ArcIMS designer.

circle, or an irregular polygon on a map image to select an area feature. The user cannot select a line or a point feature on the map either. It is thus very difficult, if not impossible, to perform functions such as spatial data overlay and spatial object buffering. To overcome the limitation of HTML, some programs create a map viewer (such as MapCafe at ArcView IMS) or use DHTML at the client side to help users interact with maps. Moreover, the final result of all the work from the GIS server to the user at the Web browser is still static map images in the GIF or JPEG format.

ESRI's ArcIMS HTML viewer is one of the server-side Internet GIS examples which combine the use of static map images (GIF or JPEG) and JavaScript for dynamic interactions (Figure 10.5). However, the available function of the HTML viewer is much less than the ArcIMS Java viewer due to the limitation of HTML documents. Figure 10.5 illustrates that the HTML viewer cannot provide "layer" and "project" functions due to the limitation of HTML.

In contrast, the client-side Internet GIS uses GIS plug-ins, applets, and ActiveX controls to seamlessly support GIS vector data. Some GIS functionalities such as zoom, pan, and query can be built into those client-side applications and performed locally. GIS data are provided as a stream transferring from the server through the network.

Java applets are very flexible in creating and displaying graphics and maps. They are more dynamic than HTML and DHTML. Therefore, the user interface can be extended to include more complex client-side mapping, querying, and analysis capabilities. However, at the current stage, the functions of most Internet GIS programs are still limited to GIS mapping and viewing. GIS analysis functions such as spatial analysis, buffering, and overlay are yet to be fully developed.

GIS plug-ins, ActiveX controls, Java applets, and Java beans are all client-side executables. More functionalities can be added to the interfaces of these

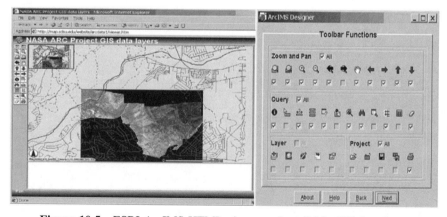

Figure 10.5 ESRI ArcIMS HTML viewer and available GIS functions.

small applications. Because of the ability of direct support of vector data, client-side Internet GIS offers more potential to support GIS analysis. But adding more functions would make these programs larger, which would have negative impact on performance at the client machine and the network.

Furthermore, it would be difficult to develop all GIS analysis functions using a single plug-in, applet, or ActiveX control. Since there are so many GIS analysis functions, the single plug-in, applet, or ActiveX control either has to be very large or has to be able to call a large number of other programs.

Java offers the most potential to create modularized GIS analysis functions with each analysis being a separate bean. Because Java applets and beans are downloaded at runtime, the partition of the application between client and server becomes more dynamic and flexible. For example, if the network connection is slow, the server can send analysis applets to the client for local processing, while if the network connection is fast but the CPU is less powerful, an applet can be used that relies on the server to do most of the processing. ActiveX control is a very competitive contender as well because of its strong support from Microsoft and its integration with the .NET technology. ActiveX has a better performance advantage due to its access to local resources.

Figure 10.6 illustrates one example of ArcIMS Java viewer. The Java viewer can provide more GIS functions, such as rearrange layers and change map symbols and colors.

To empower the analysis function of Internet GIS in the future, the trend will be to create standard individual function modules or components, as discussed in Chapters 3 and 5. These standard function components enable developers to produce "plug-and-play" GIS analysis tools. These function components are assembled on the client's Web browser in runtime when they are needed. This componentized or modularized approach is consistent with the direction of the world of computing, which is moving toward component-

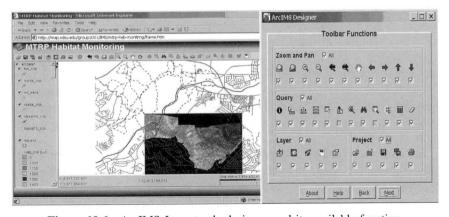

Figure 10.6   ArcIMS Java standard viewer and its available function.

ware and network-based computing. This approach demands a standard specification to ensure interoperability (Buehler and McKee, 1996).

10.5 PORTABILITY

Portability refers to the ability of Internet GIS components to adapt to different computer platforms. Server-side Internet GIS, HTML viewers, and viewers with Java applets have excellent portability. They are not platform dependent, which gives them the widest accessibility. The access to the server-side Internet GIS and HTML viewers is ubiquitous. Anybody who has Internet access can access and use it regardless of what operating systems and platforms are used. The drawback of Java applets, however, is that they have to be run in a Java-enabled browser with a built-in Java VM.

GIS viewers using plug-ins and ActiveX controls are platform dependent and thus have limited portability. Plug-ins are native to a specific platform on which the Web browser runs. The specifications for the development of Netscape/Microsoft plug-ins vary significantly by platform, so they are not platform neutral. Therefore, plug-ins and helper program developers must develop various plug-ins specifically for each platform. That is, individual plug-ins would need to be created for UNIX, Windows, and Macintosh operating systems.

Furthermore, viewers with ActiveX controls can work very well in Microsoft Internet Explorer. But Netscape Navigator does not directly support ActiveX controls on its own and will ignore any controls embedded in a Web page. A plug-in has to be installed in order for Netscape Navigator to support ActiveX controls.

Viewers on the client with plug-ins, ActiveX controls and Java applets are not compatible and can only communicate with the server from which they originated. Current viewers are not interoperable. For example, a viewer with a plug-in can only talk with its own server; it cannot talk with any other servers. Web users are required to download various viewers from different servers in order to display and process different GIS data from different servers. For example, the user has to download a plug-in from MapGuide, an ActiveX viewer from GeoMedia WebMap, a Java applet from ArcIMS, and another Java applet from MapXtreme to display information delivered from different GIS servers. There are no uniform viewers that can communicate with different servers. Similarly, viewers cannot talk with each other. For example, an ActiveX viewer cannot talk with a Java applet viewer on the same client machine.

For example, Figure 10.7 illustrates the common compatibility problem of accessing Internet GIS applications on a mobile device. This example shows that users cannot access an ArcIMS HTML viewer from a pocket PC device because the pocket version of Internet Explorer cannot support advanced JavaScript functions. Since the operating systems on the mobile devices, such

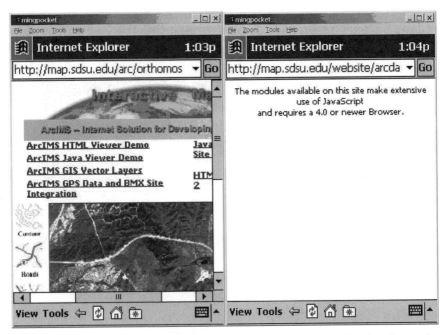

Figure 10.7 Error message indicating incompatible JavaScript function for pocket PC Internet Explorer.

as pocket PC, Palm OS, and smart phone, are quite different from the regular desktop PC or notebook computer, it is a challenge to develop a universal Internet GIS viewer which can be accessed from all types of mobile devices. This noninteroperability issue is being addressed by OGC, which is developing a Web mapping specification so that a standard viewer or at least the interoperable viewer can be developed to access the geospatial data and services from different servers that meet the specification.

10.6 SECURITY

Internet security is one of the most serious problems in the design of Internet GIS applications. In general, security is central to all kinds of computer systems and information services, whether they are stand-alone workstations, network computers, or systems that specialize in geographic information or financial transactions. As long as people share information and exchange data via computers and networks, the security problem will always be a major concern for system administration, implementation, and users. There is added risk for Internet applications because they are exposed to broader users and larger networks and thus present many possible points of entry to the system.

The computer science community has already proposed several types of security frameworks to protect Internet applications. The OSI security architecture, one of the most popular and acceptable models, provides a general framework of security services and countermeasure mechanisms in the OSI-RM document (ITU-T Rec. X.800, 1991). There are five types of security services (Oppliger, 2001):

1. authentication services,
2. access control services,
3. data confidentiality,
4. data integrity, and
5. nonrepudiation services.

The security threat usually comes from viruses, hackers, fraudulent users, or incompetent employees. Although Internet GIS shares the same security problems with other types of Internet applications, Internet GIS applications will require special considerations because of their specialized purpose and possible mobile functions. In general, there are three types of security issues in the design of Internet GIS: disclosure of information (interception), denial of service (DOS), and corruption of information (Jansen and Karygiannis, 1999).

Key Concepts

Internet security *is one of the most serious problems in the design of Internet GIS applications. The security threat usually comes from viruses, hackers, fraudulent users, or incompetent employees. There are three major security threats: disclosure of information, denial of service (DOS), and corruption of information. There are many antivirus, firewall, and network security software solutions available for various Internet servers. However, system administrators should always consider the balance of security protection and system performance and usability. If you apply too many security packages, the server will become very slow and few users will be able to access your services.*

The first type of security problem is disclosure of information. Since Internet GIS usually carries important information, such as sensitive infrastructure or critical environmental data sets, the information items carried or encapsulated by Internet GIS may be intercepted or retrieved from other programs or on-line users. The information can be disclosed in many situations, such as the unintentional release of a password, the exposure of sensitive data, or unauthorized access by other users or programs. One example of this type of security problem is when an unauthorized user claims the identity of an authorized user in order to gain access to services and resources. Another example is that valuable data and sensitive communication could be eavesdropped by hackers.

The second type of security problem in Internet GIS is DOS, where anonymous network computers and programs can launch attacks against the Internet map servers or systems by consuming an excessive amount of the server platform's computing resources. DOS is currently one of the most serious security threats to on-line information services because it can prevent other users, Internet or mobile GIS, from connecting or accessing the servers. The mechanism of DOS is to attack target GIS servers by repeatedly sending millions of messages in a very short time period. In February 2000, the first mass-distributed DOS attack was launched against many commercial Web servers, including Yahoo, E*TRADE, eBay, and CNN.com (Scambray et al., 2001). This attack took down these Web servers for several days and caused significant financial loss for these companies. Although the targets of this DOS attack example are commercial Web servers, Internet GIS servers are also vulnerable to such attacks.

The third type of security problems in Internet GIS applications is the corruption of information or GIS processing functions. The corruption of information can occur in both data (e.g., embedded sensitive data can be damaged or corrupted) and GIS programs. A computer virus is frequently the cause of the corruption. Usually, if a GIS program was affected by a computer virus, the affected program is no longer the original program and may generate wrong information or damage the GIS platform or runtime environment. In general, the corruption of information may reduce the accuracy of geospatial data, cause Internet GIS servers to crash, or perform malicious GIS operations.

For Internet GIS applications, there may be security risks to every components of the system, including the server, the client machine, and the network connection. Someone may break into the server to steal confidential data or information or even contaminate the original data. When users download client-side programs, the downloaded program may ruin the local systems. Someone may intercept sensitive information such as your password or credit card number when you try to send the information to the server via the Internet network.

10.6.1 Security Issues in the Server

There are many security risks to the server, but two are most dangerous to the Internet GIS server. The first is system break-in by unauthorized users to steal and change confidential documents and data or modify, contaminate, and even incapacitate the server machine system. The second risk is DOS attack by an outside intruder, which consumes most of the server resources and renders the server machine too busy to respond to legitimate user requests.

The DOS attacks could be launched by any malicious outside intruder. There is not much the system administrator can do to prevent this, at least not yet. But system break-in by unauthorized users could be caused by bugs or misconfiguration problems in the Web server, and the system administrator and Internet GIS developer could have some control.

To the Web server and GIS server, the main point of entry is through user requests. On the server side, usually CGI and other server-side CGI alternatives such as ASP and Servlet are used to accept and process user requests. These server-side programs could pose risks to the server and database, particularly the use of CGI scripts. The CGI protocol was not intrinsically designed to have the safeguard against system break-ins by any outside users.

CGI scripts exposed to the HTTP request line on the user's browser may contain information about the host system that hackers can use to break into the server system. If the CGI scripts can read or write files to the server or interact with other programs such as a GIS server or data server on the host system, there are even bigger security risks. A hacker can trick the CGI scripts to execute commands that cause damage to the server systems.

For example, if CGI scripts can read files on the server, that would allow hackers to gain access to the system information and resources of the server system. The hacker would use this system information to break into the server. If the CGI scripts can write files to the server, the hacker could trick the CGI scripts with vicious commands to modify or damage documents and data or even introduce Trojan horses to the server system to bring the server down. If the CGI programs can interact with or invoke other programs, as most server-side Internet GIS programs do, the hacker could potentially gain control of the other programs in the server or even the whole server machine.

In addition to CGI scripts, other server-side CGI alternatives such as DLL, ASP, and Servlet may also have potential security holes. Even caching the server database data on the client side (a common practice to improve performance) can bring in security risk because this would create another entry point for hackers to gain control of the data. As long as there is a separation of the server and the client and there are information exchanges between the client and the server, there is potential security risk to the server.

Controlling access to the system is a common mechanism to enhance server security. There are several ways to control access, including the use of an authorized user with passwords, the creation of an Intranet, and the use of a firewall. For Internet GIS applications that target the general public, restricting access by using passwords would not be convenient for the user. Using the intranet would defy the purpose. Therefore, firewall is a common mechanism to control access and to protect the server and database.

There are a number of ways to use a firewall for security purpose. You can use it to control all access from outside users, creating an intranet. Within an intranet, only computers within the LAN are able to access your server. Many intranets use a firewall to block all outside request to the Web server, as shown in Figure 10.8.

The use of a firewall in this case is the most secure way and serves well for internal users within an organization. However, if you want to serve Internet GIS to outside users on the Internet, this configuration would not work. If you want to serve Internet users, you will need to place the Web server outside the firewall, as shown in Figure 10.9, and place the GIS server and data server inside the firewall.

Internet GIS Showcase: U.S. Military Traffic Management Control

The U.S. military has commissioned a system to make troop movements between forts and strategic ports as rapid and efficient as possible. GeoDecisions (an Intergraph Team GeoMedia partner) is currently developing a Web-based GIS Rail and Highway Status Report System to provide decision makers with critical information necessary for rapid deployments. This information will include bridge locations and associated attributes, vertical clearance imaging, highway interchange imaging, construction locations, congestion, detours, weather predictions, and other related movement information. Certain vital information such as weather and congestion will be provided "real time" on the system.

The system is designed for use by unit movement planners and command center authorities for the expert management of deployment options.

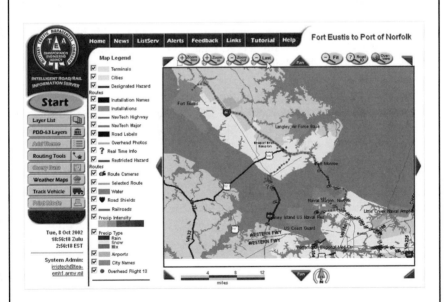

Image reprinted with permission of Gannett Fleming.

Source: http://www.geodecisions.com/irris/.

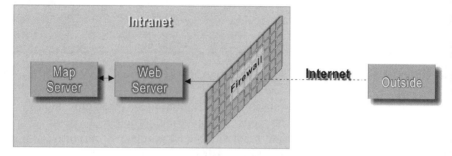

Figure 10.8 Use of firewall to construct an intranet.

Ideally, a firewall should be placed in its own machine. It is never a good idea to place the Web server on the same computer as the firewall machine, because any problems in the Web server will undermine the protection of the firewall. A proxy server, however, could be placed in the firewall machine (Figure 10.9). A proxy server is a shield between the Web server and the GIS server and the data server. It intercepts the incoming user requests from the Web server, inspects them, and verifies their validity. If the request is authentic, it then forwards the request to the map server and data server for processing.

In this configuration, the Web server is vulnerable of outside hacker attack, but the firewall protects the GIS server, data, and other resources in the inner network. This is only the basic configuration; there could be many variations of this configuration. For example, you could put some public information on the Web server and only the confidential information inside the firewall (ESRI offers several options to install ArcIMS and firewalls in its white paper "Security and ArcIMS," available at http://www.esri.com/library/whitepapers/pdfs/SecurityArcIMS.pdf).

10.6.2 Security Issues in the Client

Security risks in the client machine involve the downloading of client-side programs such as plug-ins, Java applets, and ActiveX controls, often from

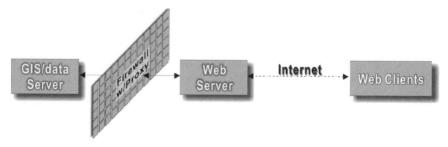

Figure 10.9 Use of a firewall to protect GIS server and data to serve Internet users.

machines over which the user has no control. Such application codes could contain fatal flaws or viruses that could potentially crash the browser, damage the user's system, breach the user's privacy, or merely create an annoyance. The other risk is the misuse of personal information knowingly or unknowingly provided by the end user.

10.6.2.1 Security Issues in Plug-Ins and ActiveX Controls

Plug-ins, helper programs, and ActiveX controls are binary codes executing directly on the local machine's hardware. Therefore, all of the dangers of running unknown software from the Internet apply to them. Unknown plug-ins and ActiveX controls downloaded from the Internet could potentially corrupt the client computer. How do you know that a downloaded plug-in or ActiveX control will not erase your hard drive (Chappell, 1996)?

Because viewers with plug-ins and ActiveX controls have full access to platform services, which place no restrictions on what a plug-in and ActiveX control can do, they involve greater risk to the local system. For example, they can be used to delete important local files and cause damage to the local operating system. To address this concern, Microsoft adopts a verification mechanism in its Windows Trust Verifications Services (Chappell, 1996) to verify that a control comes from a trusted source (Shan and Earle, 1998). Microsoft's Internet Explorer Web browser supports authenticated code-signing technology and Netscape uses the Object Signing protocol. This enables vendors of ActiveX controls and developers of plug-ins and other software components to digitally sign these components. The digital signatures are then certified by a trusted "certifying authority" such as VeriSign. When a digital certificate is granted, the software developer pledges that the software is free from viruses and other malicious components. When they are downloaded and the digital signature is recognized, a code signature certificate is displayed on the screen. This certificate ensures that the software component is coming from a trusted source and has not been tampered with, but it cannot guarantee that the software component is safe to use (Chappell, 1996). The user has to decide whether to install the applications on his or her local machine.

This security model places the security responsibility on the user's hand. For example, if an ActiveX control was not signed at all or was signed and certified by an unknown certifying authority, the browser presents a dialog box warning the user that this action may not be safe. The user has to make the decision whether to abort the transfer or continue the transfer and take his or her chances. To be safe, the Internet user would essentially have to reject all ActiveX controls not signed by a known authority, which may not be conducive to full use of Internet GIS. Users who have changed Internet Explorer's security level to "low security" or who agree to download and execute the controls despite the warnings have high risks of being attacked by malicious ActiveX controls. Even when an ActiveX control comes from a trusted source, there is no assurance that executing the control will not cause damage (Shan and Earle, 1998). The ActiveX certification process only allows you to know whom to blame if you download a signed ActiveX control after it crashes your machine. But the damages have been done. Even there, the task of identifying the ActiveX control responsible for damaging your system is not trivial.

10.6.2.2 *Security Issues in Java Applets*

The Java security model is considerably different from the ActiveX security model. All applets loaded over the network are assumed to be potentially hostile and are treated with appropriate caution. Therefore, Java has established the Java security framework to establish an intelligent, fail-safe stance for the execution of Java programs (Weber et al., 1996). The Java security framework consists of three major layers that create the Java execution environment, from the Java language to the Java compiler and verifier to security managers in the runtime. When Java byte code is loaded, it has to be verified by the Java verifier. If the verifier cannot confirm that the code being loaded was produced by the Java compiler, the code will not be loaded and executed.

On the browser side, a downloaded Java applet is wrapped in a secure cocoon (Java VM) during execution on the user's local device. Java achieves security by restricting the behavior of applets to a set of safe actions by a "security manager" object. The security manager does not ordinarily allow applets to execute arbitrary system commands, to load system libraries, or to open up system device drivers such as disk drives. Java scripts are generally limited to reading and writing to files in a user-designated directory only. For example, Java applets are not allowed to create, modify, or delete files or directories not authorized by the security manager on the local system. This restriction could even be stricter depending on the implementation. For example, the HotJava browser allows you to set this directory, while Netscape disallows all file manipulation. Thus Java applets have very limited access to local system resources such as local file systems. Each applet is playing in its own safe "sandbox"; thus this method of providing security is called *sandboxing* (Chappell, 1996).

Applets are also limited in the network connections they can make: An applet is only allowed to make a network connection back to the server from which it was downloaded. This is set up to prevent the applet from collecting local information on any machines on the LAN from behind a firewall and transferring them to any other host. It is also designed to prevent commands to be executed remotely.

Finally, the security manager allows Java applets to either read and write to the network or read and write to the local disk, but not both. Otherwise, an applet could spy on the client machine's private documents and transmit the information back to the server.

Therefore, the download and execution of Java applets are relatively more secure than executing ActiveX controls and plug-ins. However, Java applets are not error free either. Several flaws have been found in the Java security system, and there are undoubtedly still bugs lurking. Furthermore, this very restrictive security system imposes some restrictions on what can be done using Java applets in the development of Internet GIS. For example, Java applets are prevented from accessing other machines on the network beyond that from which the applets were loaded. This imposes serious problems in developing the Internet GIS as a distributed system. Furthermore, because applets are not allowed to write on the local system in some implementation such as Netscape, data cannot be written and saved to the user's local machine (Weber et al., 1996). This may not meet the needs of some Internet GIS users.

In addition, poorly or maliciously designed Java applets can take system resources such as memory and CPU time such that no other tasks can be performed. Applets running under the same browser could interfere with each other, so that a vendor could design an applet to deliberately make a competitor's applet appear to behave erratically.

10.6.2.3 *Privacy and Security Issues in Cookies*   A cookie is a small piece of information that the HTTP server sends to the browser when the browser connects for the first time. Cookies remain on the browser's local machine. Every time the browser connects with the same server, the browser returns a copy of the cookie to the server. The server then uses the information on the cookie to remember the user or to maintain the state.

There are two purposes for cookies: state preservation and authentication. A "cookie" is used to make up for the stateless nature of the HTTP protocol as mentioned before. For example, without cookies, it would be difficult to put all items in your shopping cart when you do shopping on-line. Therefore, cookies could be very helpful to preserve the state during the steps of a multipart transaction, such as an on-line ordering system or a bank transaction system. In terms of Internet GIS, cookies can also be used to help the server maintain the state with the client. But it is not part of the standard HTTP specification, so some Web browsers may not support it.

Cookies are also used as an authentication tool to allow users to automatically access certain Web sites without asking the server to look at authorized users at the database. The user's log-in name and password are stored in the cookie so that the user can access a subscribed Web site automatically each time the user clicks on the Web page.

The use of cookies has privacy concerns because cookies contain information about the URI of the Web page you accessed, the name and IP address of your computer, the brand of the browser you are using, and the operating system you are running. Other types of information in the cookies include those that you have voluntarily provided, such as user name and password for specific sites as well as other private information. This information can be legitimately used to establish a state in your Web browsing. But the information inside cookies could be shared by advertising agencies such as DoubleClick Network, a system created by the DoubleClick Corporation to create profiles of individuals browsing the Web and to present them with advertising banners customized to their interests. It is this kind of unauthorized use of cookies that intrudes upon users' privacy (URI: http://privacy.net/cookies).

Cookies can be categorized as transient cookies and persistent cookies. Transient cookies are active only during a browsing session and will disappear when you leave the browser. Persistent cookies are those that store user identification information over an extended period. These cookies raise security concerns because sensitive information such as user name and password can be stolen from the user's cookie database file. Furthermore, cookies passing from the browser to the server could be eavesdropped during the transmission process. The hacker then can use the stolen user name and password to enter the site that you subscribed to or even to make unauthorized transactions (URI: http://www.cookiecentral.com).

There are ways to minimize the risks of cookies:

For the User

- It is a good idea to frequently clean up the cookie files or use a proxy server that removes cookies and other identifying information from your URI requests.
- You can also allow only transient cookies but forbid persistent ones.
- You can forbid all cookies altogether. However, you may forego the usage of some Web sites, because some Web sites have to have cookies to keep the state information.

For Developers

- Create cookies to contain as little private information as possible.
- Never allow cookies to contain plaintext user names and passwords.
- Allow cookies to expire at a certain time. A cookie with an expiration date and time would limit the time it could be used by a hacker.
- Use a message authenticity check code to ensure that none of the fields of the cookie have been tampered with.

10.6.3 Security Issues in the Network: Eavesdropping

The major security risk in the network communication process is eavesdropping, which means interception of network data sent from browser to server or vice versa. Eavesdroppers can operate from any point on the pathway between browser and server, including

- the network on the browser's side of the connection,
- the network on the server's side of the connection (including intranets),
- the end user's ISP,
- the server's ISP, and
- either ISPs' regional access provider.

Security measures should be taken at both browsers and servers to protect confidential information against network eavesdropping. Without system security on both the browser and server sides, confidential documents are vulnerable to interception.

Encryption is one approach to protecting private information against network eavesdropping. Encryption works by encoding the text of a message with a key by using certain encryption algorithms. To decode an encrypted message, one has to have both an encryption key and the encryption algorithm. If a hacker only knows the encryption and decryption algorithms, he or she cannot crack the message without the key. Encryption keys are classified as private key and public key (Shan and Earle, 1998).

10.6.3.1 *Private Key Encryption*

Private key encryption requires that both the sender and the receiver have the same key and use it as a cipher to encode and decode the message. It is also called the symmetric encryption system. The key is typically a large numeric number and varies based on the encryption algorithm used. The Data Encryption Standard (DES) is the most common private key method. It works like this: A sender encrypts the message with a key and an encryption function. The encrypted message is sent through the network to the recipient. The recipient then extracts the message using the same key and a corresponding decryption function (Shan and Earle, 1998).

The weakness of the private key encryption is twofold. First, every pair of principles in the communication process has to have a unique private key. As the number of users in the network expands, the number of keys required can become very large. The second weakness is the transportation of the key itself. If the sender and the receiver agree on the key, they cannot communicate through the network because the key could be intercepted or eavesdropped by others. To solve this problem, one could use the Kerberos security system, in which all parties are required to communicate their keys through a trusted third-party server. Furthermore, the current DES is a 56-bit cipher because at the time the DES was created the U.S. spy agency NSA insisted on the use of the 56-bit key rather than the 128-bit key. This was because it did not want an algorithm that it could not break. But with the advance of computer technology, the 56-bit key could be cracked by hackers within a reasonable time. Therefore, a triple-DES that scales up to the 256-bit key will be adopted by NIST (Oppliger, 2001).

10.6.3.2 *Public Key Encryption*

In the private key encryption system, the same key was used for both encoding and decoding. In the public key, or asymmetric, encryption systems, keys come in pairs: One key is used for encoding and another for decoding. In this system everyone owns two keys: a public key that is widely distributed and available to everyone and a private key that is a secret key known to no one else. The public key is used to encode messages while the private key is used to decrypt incoming messages. The private key is mathematically related to the public key. Under this system, the sender who wants to send a secure message to the recipient can encrypt the message with the recipient's public key and send the message to the recipient through the network. Upon receiving the encoded message, the recipient decrypts the message using the private key. If someone intercepts the encrypted message, it is impossible to decrypt the message without knowing the recipient's private key. The best-known public key system is RSA, named after its inventors (Rivest, Shamir, and Adleman). With RSA, there is no need for a "trusted third party" such as Kerberos as long as you know your own private key. But the delivery of the private key requires something like Kerberos or Certificate Authority (CA) that authenticates the public key. This system can also be used to create unforgeable digital signatures.

However, the public key encryption method is not without problems. It is typically much slower (about 5–20 times) than the private key encryption (DES) (Shan and Earle, 1998). Furthermore, a public key database may be faked by a hacker, who then could masquerade as the trusted database and give out the wrong public keys. If the sender uses the wrong public key to encrypt a message, only the hacker can decrypt the message, not the intended receiver.

Given the strengths and the weaknesses of both private key and public key encryption schemes, most practical implementations of secure Internet encryption actually combine the two. For example, a sender can use the recipient's public key to encrypt a private DES key. After the recipient receives the private DES key, both parties can now use the private DES key to encrypt and decrypt the actual data to increase performance. Since commercial ventures have a critical need for secure transmission on the Web, there is very active interest in developing schemes for encrypting the data that pass between browser and server. Encryption will become better and better over time.

10.6.4 Implementation of Countermeasures

Internet security problems may jeopardize the use of Internet GIS and cause serious loss of critical information or resources. Fortunately, several countermeasures are available with modern computer technologies. These technologies can be applied in the platform of Internet GIS to prevent potential security problems.

The first general type of countermeasure is the adoption of encrypted information transmission, such as the public key or private key solutions mentioned above. This approach is the easiest and most effective way to prevent the disclosure of information and related security problems. Many software platforms and Web servers are equipped with this feature already. It has been used in many sensitive information transmissions, such as credit card numbers, user passwords, or financial transactions. The encryption of information can prevent the potential security problem when transmitting sensitive information via public networks that could be intercepted by others programs for unauthorized access (Scrambray et al., 2001).

The second approach is to have a recovery plan (backup procedures) for mission-critical database and software systems. Any security problems or accidental mistakes of a system administration may cause the crash of systems or the damage of Internet GIS servers. A comprehensive recovery plan is the most important procedure in protecting Internet GIS applications and their users. The plan may include creating a mirror site for Internet map servers and regular backup procedures for both on-line and off-line media.

The third approach is to design a sandboxing model for the runtime environment of client-side GIS applications. Although the isolation of client-side GIS applications from the computer platform will limit the capabilities of Internet GIS, the sandboxing model can prevent malicious client-side Internet GIS programs from directly accessing critical elements, such as memories, operating systems, and local hard disks. The sandboxing model will prevent software-based faults or potential computer memory leaking problems. Currently, many distributed software component frameworks are adopting this approach, such as Java's VM with Java applets.

The fourth type of countermeasure is the digital signature, where each client-side GIS application can carry a signed document to confirm its authenticity and integrity. The digital signature can be assigned by a server and the authorized GIS programs will carry the document to access specified systems or networks. When the client-side Internet GIS programs move to the specified computer or network environment, the client-side machine will send a request to the server to verify the signature and then grant access to the client-side Internet GIS programs. The adoption of the digital signature will provide more flexible access controls of Internet GIS applications because different GIS applications can carry different signatures (documents) and access different types of client-side machines. Also, a single mobile GIS program can carry multiple signatures assigned by different servers for accessing different types of networks. The mechanism of a digital signature mentioned here is similar to the visa function in our real world. For example, different countries can issue different types of visas for travelers to enter or visit their countries.

The final approach is the implementation of mobile program travel logs, where mobile component software will keep an authentic record of travel histories and events. The travel histories will indicate the possible security problems and maintain the integrity of Internet GIS applications. For example, if portable Internet GIS applications only travel around the intranet or LANs, it is safer than traveling to the Internet or WANs. By analyzing the travel logs of Internet GIS applications, Internet GIS servers or remote-access machines will be able to detect potential problems or security threats carried by malicious GIS programs or applets.

Currently, only a few approaches, such as encrypted information transmission and sandboxing, have been applied in the actual Internet GIS implementation. Other types of countermeasures for Internet GIS security are still under development. In the GIS community, software programmers and GIS professionals need to understand the security problems in distributed GIServices. With appropriate implementation of these countermeasures, the GIS community will enjoy flexible and comprehensive GIServices and will not worry that Internet security may jeopardize the runtime environment of GISystems, the accuracy of geographic information, or the proper procedures of GIS analysis.

Another important issue in the implementation of security countermeasures is the awareness of security problems for GIS users and system developers: "Technology by itself cannot solve security problems. Technology for security must be complemented by an awareness of security issues and disciplined application of the techniques" (Harrision et al., 1997, p. 238). Without proper training and education plans, curious GIS users and incompetent system operators may become the major source of security threats. The awareness of Internet security is the key in preventing the misuse of Internet GIS and in securing mission-critical information and resources.

"Security can be expensive. It is not only expensive in terms of the cost of hardware and software, but also in performance, administration, and user inconvenience" (Shan and Earle, 1998, p. 316). The more restrictive and more secure systems will cause more inconvenience to the user and system administrator. The performance of the system may also suffer. Therefore, there should be a balance between security benefits and the costs of implementing the security countermeasures. In fact, the problem of Internet security is similar to the transportation of goods. The more valuable transported goods will require more expensive and more secure transporting measures. But if the transported goods are not that valuable, the requirement of security will become much lower and the cost of transportation will be reduced.

10.7 CONCLUSION

This chapter has reviewed the major QOS issues for Internet GIS application, including system performance, service reliability, user interactivity, functionality, portability, and security. Table 10.1 shows a summary of QOS characteristics based on four different technologies: CGI, plug-in, ActiveX controls, and Java applets.

In general, the success of Internet GIS applications will rely on the high quality of Internet services and reliable network communications. The IT industry is developing several possible QOS frameworks and hopefully GIS applications can adopt these frameworks in the future. However, the role of the GIS community and GIS professional should not only be as end users of QOS but also as participants in the development of QOS specifications. Because Internet GIS applications are very unique and require additional security and communication mechanisms compared to other Internet applications, the participation of QOS standards will ensure that future Internet frameworks are suitable for various Internet GIS applications.

TABLE 10.1 Assessment of Internet GIS Development Approaches

		CGI Based	Plug-Ins	Java Applets	ActiveX Controls
Performance	Client	Excellent	Good	Good	Excellent
	Server	Poor to good	Good	Excellent	Excellent
	Networking	Poor	Good	Good	Good
	Overall	Fair	Good	Good	Excellent
Interactivity	User interface	Poor	Good	Excellent	Excellent
	Function support	Fair	Good	Excellent	Excellent
	Local data support	No	Yes	No	Yes
Portability		Excellent	Poor	Good	Fair
Security		Poor	Fair	Good	Fair

Besides the consideration of QOS, there are several other issues related to the design of portable Internet GIS applications:

- *Cross-Platform Implementation* The implementation procedure across different platforms is a major problem of Internet GIS applications. Because some Internet GIS programs need to travel between different systems and frameworks, choosing appropriate software development frameworks becomes a major challenge for the implementation of Internet GIS applications.

- *Size and Functions* Another concern of Internet GIS is the size of portable GIS programs versus the functions of GIS. In theory, the size of portable GIS components should be minimized because smaller GIS components will travel faster across networks. However, Internet GIS developers want to design more complicated portable GIS components with more functions, which will make the size of portable GIS components bigger. Many current Internet GIS research projects are facing the dilemma of the size and functionality of portable GIS components. One possible solution is to utilize on-time assembling to call and assemble many components at the client machine on demand and to use more lightweight programming technologies such as scripting languages (JavaScript or VBscript) to reduce the actual program size of each component during the runtime.

- *Protocol Development* The design of appropriate communication protocols is another challenge for Internet GIS applications. The requirement of Internet GIS communication protocol is different from the traditional network communication protocol, such as TCP/IP or HTTP. The Internet GIS communication protocol will be a high-level, application-oriented protocol, which needs to focus, for example, on the exchange of mapping characteristics, geometric features, attribute databases, and user commands. The protocol must be accepted by the GIS community and be customizable by different Internet GIS applications.

To sum up, the QOS of Internet GIS is essential for the future development of Internet GIS applications. Although several GIS companies have already begun to focus on these issues in the last few years, the current QOS is still far from the expectations of most Internet GIS users. The GIS community must focus on these key issues and work together to improve the quality of Internet GIS applications in the long run. Hopefully, the next generation of Internet GIS will be able to provide fast, secure, high-quality, and inexpensive information services for the public and GIS users.

WEB RESOURCES

Descriptions	URI
eSECURITY Technologies Rolf Oppliger	http://www.esecurity.ch
QOS	http://www.qos.net
Internet Performance Measurement and Analysis (IPMA) project	http://www.merit.edu/ipma
RFC 2828—Internet Security Glossary, R. Shirey.	http://rfc.sunsite.dk/rfc/rfc2828.html
ITU-T Rec. X.800, March 1991	http://fag.grm.hia.no/IKT7000/litteratur/paper/x800.pdf
Security architecture for OpenGIS interconnection for Commité Consultatif International de Telegraphique et Telephonique (CCITT) application, 1991	http://privacy.net/cookies http://www.cookiecentral.com

REFERENCES

Buehler, K., and McKee, L. (Eds.). (1996). *The OpenGIS® Guide: Introduction to Interoperable Geoprocessing.* Wayland, Massachusetts: Open GIS Consortium.

Chappell, D. (1996). *Understanding ActiveX and OLE.* Redmond, Washington: Microsoft Press.

Harrision, C., Caglayan, A., and Harrision, C. G. (1997). *Agent Sourcebook: A Complete Guide to Desktop, Internet, and Intranet Agents.* New York: Wiley.

Huston, G. (2000). *Internet Performance Survival Guide: QoS Strategies for Multiservice Networks.* New York: Wiley.

ITU-T Rec. X.800. (1991). *Security Architecture for Open Systems Interconnection for CCITT Application.* URI: http://fag.grm.hia.no/IKT7000/litteratur/paper/x800.pdf.

Jansen, W., and Karygiannis T. (1999). *Mobile Agent Security.* Special Publication 800-19, Gaithersburg, Maryland: National Institute of Standards and Technology, August.

Oppliger, R. (2001). *Internet and Intranet Security,* 2nd ed. London: Artech House.

Scambray, J., McClure, S., and Kurtz, G. (2001). *Hacking Exposed: Network Security Secrets and Solutions,* 3rd ed. New York: McGraw-Hill/Osborne.

Shan, Y.-P., and Earle, R. H. (1998). *Enterprise Computing with Objects: From Client/ Server Environments to the Internet.* Reading, Massachusetts: Addison-Wesley.

Wang, Z. (2001). *Internet QoS: Architecture and Mechanisms for Quality of Service.* San Francisco: Morgan Kaufmann.

Weber, J. et al. (1996). *Special Edition Using Java,* 2nd ed. Indianapolis, Indiana: Que Corporation.

CHAPTER 11

DISTRIBUTED GIS IN DATA WAREHOUSING AND DATA SHARING

Geographic data offer what is arguably one of the most compelling areas of application of the WWW.
—Michael F. Goodchild (1999, p. 133)

11.1 INTRODUCTION

The previous chapters illustrated the fundamental concepts and technologies of Internet GIS. Starting with this chapter, the discussion will shift focus to the actual application of Internet GIS in different areas, such as data sharing and data warehousing, transportation, planning, and resource management. This chapter will focus on the use of Internet GIS in data sharing and data warehousing.

The widespread adoption of GIS technology in the past decade has accumulated a large amount of geospatial data. Many digital GIS databases have been developed over the years among government agencies, private companies, academic institutions, and other GIS users. Much of those data can be reused by other projects and users. Some of these data providers are willing to share their data with or without charges, and most users are willing to reuse these data with a fee if these data are sharable and easily accessible.

But how do we share it? As a *data provider,* one way to share it is to put it on the Web for whoever wants to use it and download it, with or without a fee. The question is, how will the user find it? In the explosion of information on the Web, it is often difficult for users to find the right information. How do you ensure that your data will be found? As a *data user,* the question is, again, where do you go to find the appropriate data? Once you find the data, how do you know if the data have the quality, spatial reference systems, and format that you wanted?

This chapter attempts to develop some answers to these questions. The purposes of this chapter are to demonstrate ways of data sharing and access on the Internet, to help data providers better understand data sharing and provision on the Internet, and to help data users find the data on the Internet. It starts with the basic elements of sharing, discovering, and accessing geospatial data on the Internet. Followed by a discussion of two major geospatial data portals: the Geospatial Data Clearinghouse Activity (GDCA) at the U.S. Federal Geographic Data Committee and the Geography Network and G.Net managed by ESRI. It then presents a few examples of major geospatial data providers.

11.2 ACCESSING GEOSPATIAL DATA OVER THE INTERNET: BROWSING OR SEARCHING

GIS data are abundant and readily available on the Internet for either on-line viewing or direct download. But despite this availability, much of the data are underutilized because users do not know where to find the information.

One way to find data on the Internet is to browse through sites and follow links to other sites as they are revealed. The browse function, which allows users to travel through and across sites, page by page via HyperText links, is one of the revolutionary features of the Web. However, as a means of research, browsing is not a practical or productive method. It is like going to the library and searching for a book shelf by shelf without using the card catalog.

The alternative to browsing is an active search with the help of a search engine such as Google or Yahoo. Search engines can seek out and display links to spatial data throughout the Web. The problem with this method is that it typically returns many useless links and fails to return important links. To locate geospatial data about the city of Milwaukee, for example, users need to know what terms will lead them to the data they want. A Google search using *Milwaukee GIS* as the search terms generates a decent list of sites, but the user has no way of knowing if this list represents all available data or even if the best available data are on the list. Data providers need to ensure that their data will be revealed by a simple, straightforward search such as "Milwaukee GIS." Otherwise, they have wasted their money and effort in making the data available.

To genuinely improve the accessibility of on-line data resources, the GIS community needs to develop some sort of standardized system for efficiently sharing and accessing geospatial data on the Internet. The critical elements include a set of geospatial data portals that index data sources, a standard for metadata that describes the elements in each source, and an intelligent, spatial-aware search engine. A standardized data format and method of data transfer will also improve the ability of GIS users to use the data they find.

One improvement that has already begun is the development of GIS-specific search engines. Figure 11.1 is an example of such a site. While more likely to return useful data than a general-purpose search engine, this example is still not an adequate solution because it is a text-based system. Search engines such as Google or Lycos or our Internet GIS example (Figure 11.1) are all designed to search the Web for matching instances of text. Geospatial data, however, are not text based. Their spatial and temporal elements such as bounding coordinates and time spans are difficult to index as literal strings that a search engine can find.

The solution is not to replace text-based search engines, which are efficient and intuitive to use. Rather, the solution is to improve the text-based metadata that describe geospatial data. The GIS community needs to establish standards that will homogenize the content semantics, data elements, and formats of metadata. (Such standards have been discussed in Chapter 6.) Also, to enhance the efficiency of spatial data discovery on the Internet, it is necessary to construct a spatial data catalog or search key, a competent search and retrieval protocol, and an efficient search mechanism.

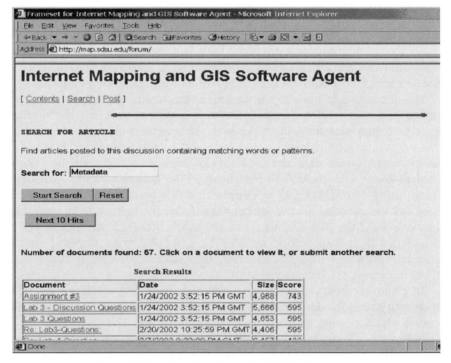

Figure 11.1   Text-based search of Web documents.

11.2.1 Index and Catalog Geospatial Information

If users are to find spatial data on the Internet, the data must somehow be searchable. This means that they must be indexed and cataloged in some way. The most important searchable key is location, followed by some combination of scale, date, theme, and possibly several other characteristics. The location key can be expressed either through a coordinate system, by finding the area of interest on a map, or through a gazetteer by finding the appropriate place name. The scale key can be expressed in at least four distinct ways: as the metric scale of the document used to create the digital data set, as a pixel size, as a spatial resolution, or as a measure of position accuracy (Goodchild, 1995). Themes are often organized and combined in complex ways and are therefore very difficult to catalog.

Keys to spatial and thematic data are generally based on metadata elements. Metadata, or data about data, describe the content, quality, condition, and other characteristics of data. Standards for metadata were first established in 1994 by the FGDC's Content Standards for Digital Geospatial Metadata. The standards provide a common set of terminology and definitions for documenting geospatial data. More specifically, the standards establish the names of data elements and groups of data elements to be used, the definitions of these data elements and groups, and information about the values that are to be provided for the data elements.

Despite these standards, valuable data sets exist in specialized data libraries that are incompatible with one another. There are a variety of data libraries or warehouses that focus on specific data types such as land information, housing and urban development, transportation, and environmental data. Each library has its own catalog, index, and metadata. This makes the data more difficult for general-purpose GIS search engines to locate. The solution to this problem is a larger, general spatial data library that uses consistent metadata semantics to catalog its data.

11.2.2 Metadata Catalog Services

One of the best ways to make the wide array of geospatial data easily accessible on the Internet is through metadata catalog services. Similar to how a card catalog organizes library books, a metadata catalog describes and provides links to available metadata servers. Metadata and data providers should register with catalog services to advertise their data. Ideally, they should also adhere to consistent metadata standards and services.

Metadata can organize data stored in a central location or distributed data stored in multiple locations. A centralized data storage system collects all geospatial data and puts them in a central location, such as a digital spatial data library or a GIS data warehouse/repository (Oswald, 1996). A distributed or federated data system, on the other hand, allows direct data providers to store the data at geographically diverse locations. The data remain accessible from one location, but different portions of the "library" will be stored at and retrieved from different servers.

The centralized approach improves ease of access, data quality, and format consistency. However, it also increases the difficulty of data maintenance and updates for data providers, who need to make changes through a third party—the library or a data clearinghouse. A distributed data system, on the other hand, is easier to keep current because the data are updated locally at the remote sites. No time delay is wasted for data delivery to a central storage center. Locally managed information tends to be more current and typically requires fewer system resources. The downside of distributed data systems is less consistency of quality and format.

The two approaches to data storage each have merits and drawbacks. Taken to extremes, neither approach makes sense. A single, central storage facility for all available geospatial data would be unwieldy and difficult to keep updated. At the opposite end of the spectrum, if all data items were stored on separate servers, consistency would be a major problem and small data providers would be overwhelmed by technical support needs.

The best solution is a mix of centralized and distributed systems appropriate to the type of data and resources of the provider. Agencies and organizations with the resources to host and maintain their own data should do so. Small municipalities and individual departments that lack the resources to host and maintain their data should be able to rely on a centralized county, state, or federal system. Currently, both approaches are used, but the majority of data provision systems use the distributed data system. Some specialized digital spatial data libraries and GIS data warehouse/repositories have adopted the centralized data management approach, but most have adopted the distributed data system. Examples include the ADL, the U.S. Geospatial Data Clearinghouse, the Geography Network, and G.net.

11.2.3 Efficient Search Mechanism

An essential component of any Internet-based GIS data service is a competent search-and-retrieval protocol.

There are two major information search protocols: the WAIS protocol and the Z39.50 protocol. The WAIS protocol, also referred to as Z39.50-1988, is the early version of the standard search-and-retrieve protocol provided by ANSI in 1988. WAIS is a protocol that serves queries against indexed text files. It allows for the indexing and search of full-text documents or fields within structured text documents on the Internet. It is well-suited for document retrieval based on free-text search and is widely used by the WAIS, freeWAIS, and related software products, such as many of the Internet search engines (e.g., Lycos and Yahoo).

The newer version of the WAIS protocol is the Z39.50 protocol, which is now the standard preferred by the ANSI/ISO for network information search and retrieval. The unique feature of the Z39.50 service protocol is that it allows one client (or gateway process) to access multiple servers at the same time, even servers with different software. The Z39.50 service protocol supports field-level searching. That is, in addition to searching for file names, it can also search for contents down to the field level of a data file. It is capable of indexing and discovering numeric fields used to define dates and coordinates. These functions are ideal for serving spatial metadata that have spatial and temporal contents at the field level.

In addition to the standard WAIS and Z39.50 search protocols, there are many proprietary search mechanisms developed for use with individual GIS programs.

11.2.4 Geospatial Data Transfer and Distributions

Once the appropriate data are located, the next step is retrieval. An important aspect of the retrieval step is the format of the data. The diversity of GIS software currently in use demands an equally diverse set of format options for users who are requesting data. Format requirements can be met in two ways: special-purpose translators or the use of a common or neutral format such as SDTS that is recognizable by most systems. Translators can make multiple formats available at all times or can reformat data "on the fly" per user requests.

It is also desirable to have a proper exchange format for metadata that allows metadata entries and collections to be transportable without loss of context or information. For this purpose it has been suggested that metadata providers consider making metadata entries available as SGML, an international standard for data markup. SGML can be produced for a metadata entry through use of the USGS metadata parser (mp) software (Nebert, 1996).

The actual transfer of the data can occur in two basic ways: on-line (electronically) or via more traditional physical means, such as magnetic tapes, floppy diskettes, or CD-ROMs. Figure 11.2 illustrates an on-line download in progress. While instant, electronic access is the ultimate goal of many data storage systems, some impediments to universal on-line access remain. The most important stumbling blocks are file sizes and data ownership issues. Some files are just too big to be transferred reliably over the Internet, or at all. Others are not made available for download because the owners are concerned about unauthorized access to the information.

11.2.5 Data Discovery Client

A client element is needed for users to make requests and retrieve the data from the data sources. A client could be a simple HTML Web page or a Java applet with interaction functions. In either case, a client should provide all the search elements, including map extent or name of the place, spatial data themes or data layers, time stamp of the data, and desired data format. These

Internet GIS Showcase: VanMap

VanMap is a Web-based map system that serves the city of Vancouver. VanMap users can choose from a variety of information layers when viewing maps, including street names, intersections, addresses, property lines, zoning, sewer mains, water mains, and public places.

VanMap allows users to research information about a specific building, to locate schools or parks, or to determine the best route to a destination. The site also provides links to community Web pages with more detailed information about specific amenities.

Through a preferences dialog box, VanMap users can customize the appearance of maps and the status bar. They can also use the preferences dialog box to specify the selection mode. The preferences specified are stored with the user's copy of VanMap, so all maps will use these settings—there is no need to set preferences for every new map.

VanMap was developed with Autodesk's MapGuide.

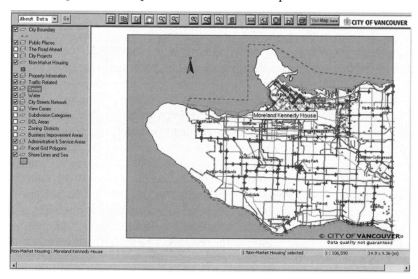

Image reprinted with permission from the city of Vancouver.

Source: http://www.city.vancouver.bc.ca/vanmap/gettingstarted/get_started.shtml.

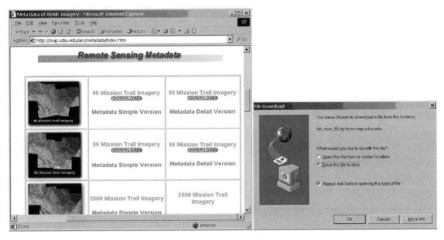

Figure 11.2 On-line geospatial data download function.

elements could be put on the client in the form of HTML forms or pull-down menus or graphic icons in a Java applet.

These are basic elements of searching, discovering, and accessing geospatial data from different data servers on the Internet. The actual implementation of these elements in different software and data portals may vary. The following sections will discuss the actual implementation of on-line data portals such as FGDC's National Spatial Data Infrastructure, Geography Network, and G.net by ESRI as well as some public and data providers.

11.3 NATIONAL AND GLOBAL SPATIAL DATA INFRASTRUCTURE

The concept of the U.S. National Spatial Data Infrastructure (NSDI) was created in 1994 by U.S. Federal Executive Order 12906. The basic role of the NSDI is to provide the technologies, policies, and people necessary to promote sharing of geospatial data among federal agencies; state, local, and tribal governments; the private and nonprofit sectors; and the academic community. The major purposes of the NSDI are to reduce costs and duplication of data collections among governmental agencies, to improve data quality, and to make geographic data more available and accessible to the public. There are four goals for the development of NSDI (FGDC, 1997):

1. Increase awareness and understanding of the vision, concepts, and benefits of NSDI through outreach and education.
2. Develop common solutions for discovery, access, and use of geospatial data in response to the needs of diverse communities.

3. Use community-based approaches to develop and maintain common collections of geospatial data for sound decision making.

4. Build relationships among organizations to support the continuing development of NSDI.

Similarly, in 1997, the concept of Global Spatial Data Infrastructure (GSDI) was created at the second GSDI conference. The idea of GSDI is similar to the U.S. NSDI but on a global scale. Its mission is to promote geospatial data sharing and the use of geographic information systems among countries and regions around the globe.

11.3.1 Geospatial Data Clearinghouse Activity

The FGDC supports the NSDI by providing the technology, policies, criteria, standards, and people as well as a structure of practices and relationships among data producers and users to facilitate geospatial data sharing and use. One of the implementations by the FGDC is its GDCA. The initial motivation of the GDCA was to collect the digital spatial data that were previously acquired by different federal agencies, share them with each other to minimize duplication of effort in spatial data collection, and promote cooperative digital data collection activities. The GDCA was later expanded beyond federal agencies to include state and local governmental agencies and nongovernmental organizations such as nonprofit organizations, private sector organizations, and academic institutions. The GDCA has now become a major source of geospatial data and a primary data dissemination mechanism to traditional and nontraditional spatial data users.

The GDCA provides a framework for individual governmental agencies, consortia, the private sector, and academic institutions to advertise, share, and access each other's available data. The GDCA framework is a federated system, meaning that the data are located in the servers of the data providers but the users can search and access all the data servers from a single user interface on the Web. The different data servers are called nodes.

To make the federated data system work effectively, the GDCA needs numerous data providers and users participating across the network. Therefore, each state and other nonfederal sector is encouraged to develop a node to form part of the GDCA. Many efforts have been made by the FGDC to develop and implement the GDCA since 1995. The FGDC established the NSDI Competitive Cooperative Agreements Program to help form partnerships with the nonfederal sector that will assist in the evolution of the NSDI. These cooperative agreement programs are used to support the development of the GDCA nodes for finding and accessing geospatial data; the development and promulgation of FGDC-endorsed standards in data collection, documentation, transfer, search, and query; and the development of software tools

Internet GIS Showcase: Denver International Airport

At 53 square miles, Denver International Airport (DIA) is one of the largest airports in the world and has a complex infrastructure to maintain. The airport's caretakers monitor its many-layered infrastructure through a series of on-line models and GIS maps. While the hundreds of thousands of images in the airport's Oracle facilities database are different file types—DWG, raster, vector—by the time they reach DIA's end users, most have been converted to AutoCAD Map files. The airport standardized its data on AutoDesk software because AutoCAD was already the front end for its facilities management system.

Users that need fast GIS information about the airport include engineers and facilities people, contractors, consultants, property managers, and administrators preparing reports. DIA needs Web-ready mapping software because many of those end users, particularly facilities engineers, need to be able to pull maps on their own from a browser-based interface. Some general users also read maps on the airport's Web-based intranet, which serves 1000 people. AutoCAD Map and MapGuide can quickly serve maps in Web formats because they embrace several kinds of map data. This allows them to function more efficiently than many less flexible GIS programs in the marketplace, which can take hours to convert files.

The engineers who use maps of DIA's underground utility locations use MapGuide, a browser-based interface that quickly delivers map data from several different GIS and CAD file formats. MapGuide also allows an office to generate statistics from maps. Its Internet platform, which comes packaged with ColdFusion Web development tools, is designed to enable on-line collaboration. The success of the GIS has led some "DIA engineers to envision future systems that once seemed outlandish, such as crews with hand-held devices displaying GIS data they download from on-site chips."

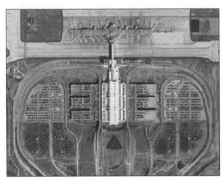

Image reprinted with permission from Directions Magazine.

Source: Directions Magazine, http://www.directionsmag.com/article.php?article_id=121.

or techniques to aid in the evaluation of geospatial metadata or data through the GDCA.

11.3.2 Framework of the GDCA

The GDCA is not a centralized or indexed repository of geospatial data; rather, it is an open federation of distributed spatial data servers. To make this open federation work, every data server needs to follow the same framework, that is, to use the same vocabulary and support the same search protocol. If the various data servers use different vocabularies (or metadata) to describe their data and use different search protocols, users will have difficulty finding and transferring the data.

11.3.2.1 *Metadata Requirements in GDCA* The search for geospatial data in the GDCA is the search for elements in the metadata. Therefore, metadata are a fundamental element in the FGDC's GDCA. If the actual data are available on a data server but the data elements are not included in the metadata, these data items may not be found. On the other hand, if metadata are available while the actual digital data are not directly linked with the metadata, that data will also not be found. The data provider could provide information about how to obtain the data, either through purchase or through CD distribution. The metadata would link with an order form in lieu of an actual data set. This metadata search provides a mechanism for providers of spatial data, both noncommercial and commercial, to advertise the available data to potential customers via the Internet by participating in GDCA activity.

In order for the metadata to be searchable, the metadata from different sources in multiple participating GDCA nodes have to meet a standard in terms of required metadata elements and format. The metadata in the participating GDCA nodes should be in the format of SGML.

The GDCA recommended three characteristics of metadata to be used by server and client software (Nebert, 1995):

1. definition and numeric tags for each attribute, down to the field level—for example, use the string text "East_Bounding_Coordinate" in the metadata to associate with the field "eastbc" in a relational database;
2. operators associated with attributes—for example, "Greater than" and "Falls-within"; and
3. presentation format information—for example, shapefile and GeoTIFF.

The metadata are closely associated with the digital data to keep the metadata current. When the digital data change, the associated metadata would also change.

11.3.2.2 GDCA Search Engine: Z39.50 The GDCA uses the ANSI standard Z39.50 for the query, search, and presentation of search results to the Web client. The Z39.50 Standard has both an indexer for indexing metadata and text information and a search protocol to search for specific elements in the metadata. It has a client and a server, where the client submits queries and passes queries to the server while the server uses the search-and-retrieval protocol to process the queries and passes the results back to the client.

The current version of Z39.50 decouples the indexer and the search server. That means that Z39.50 can use one or more search methods such as relational database search, text search engine, and spatial search (GIS) in its search server. The advantage of this decoupling is that Z39.50 could allow the user to query the spatial database directly in addition to querying the metadata repository. This is advancement from the early versions of WAIS and freeWAIS, where the indexer and the server were the same software, so they can only be used to search for metadata reports. Therefore, Z39.50 allows a user of the GDCA to search for data based on a query that can contain spatial, temporal, and text fields as well as full-text search.

The Clearinghouse for Networked Information Discovery and Retrieval (CNIDR) at Research Triangle Park, North Carolina, has developed Z39.50 server software, known as Isite. The Isite package (available at http://www.cnidr.org/Isite.html) includes a command line client (Z39.50 client), a Z39.50 compliant server (Zserver), an HTTP to Z39.50 gateway, search API, and a text search engine called ISearch (Figure 11.3).

The http to the Z39.50 gateway includes a Web server and a CGI software component, which acts as a Z39.50 client to pass user requests from the Web server to the Z39.50 server. The user could also directly interact with the Z39.50 client component, in which case no Z39.50 gateway is necessary.

The core of the Isite software is the Z39.50 communications server, the Zserver, which is designed to accept a request from a Z39.50 client or from the Z39.50 gateway, translate the search request through the search API to local or remote data servers, and return a list of search results. The search API supports free-text indexing and search of text documents. It also supports a command-line-based search protocol (script) that allows one to define a search script or pass along query terms and perform retrieval from a database or metadata. The ISearch routines are designed to index structured text files in which all fields are placed in indices. The Isite software is being extended to include numeric indexing and search to support coordinate and temporal search (Nebert and Fullton, 1995).

11.3.2.3 Architecture of GDCA The architecture of the GDCA is a distributed system. It consists of the client component, the Web server with a Z39.50 gateway, and GDCA nodes (with metadata repository and data servers). The client component could be a simple HTML-based Web browser as shown in Figure 11.4 or a Java applet as shown in Figure 11.5.

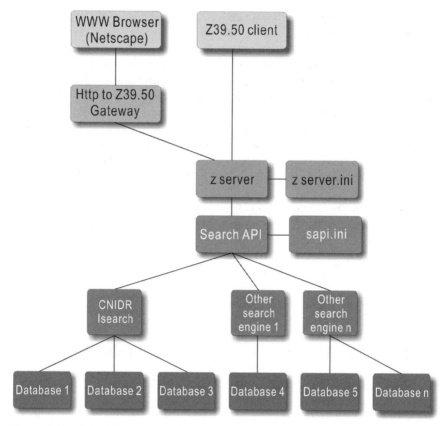

Figure 11.3 Isite information system architecture. (*Source:* Isite information systems http://www.cnidr.org/isite.html.)

The gateway is an important component in the GDCA. Besides acting as a Z39.50 client to the Z39.50 server, it also maintains a list of participating data servers from all participating GDCA nodes. This list of participating servers is a registry maintained by the FGDC (Nebert and Doyle, 1998). The gateway facilitates a distributed search. That is, when a user makes a request, that request could be sent simultaneously to one or more data servers. There are no central repositories of metadata. The advantage of this distributed index and search is very scalable. That is, the service is less likely to be down due to a failure of that centralized repository and more data servers or GDCA nodes can be flexibly added. The gateway software could be located at the Web server as a WWW-to-Z39.50 protocol gateway or at the Java applet that could be downloaded to the client.

Data servers or GDCA nodes are installed at the data sources or the lo-cations of participating data providers. Metadata are maintained by data pro-

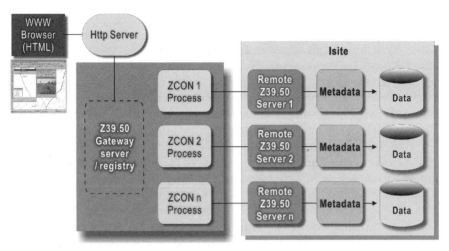

Figure 11.4   Architecture of GDCA with HTML client.

viders at the same server as the data. This setup encourages (but does not guarantee) the management, update, and maintenance of the metadata.

When the user makes a request, the user's query is passed on to the GDCA servers via the Web server and the gateway software (Zgate). These values are passed to the gateway as variables to a pair of scripts which reformat and pass the arguments to multiple Z39.50 servers at different GDCA nodes. Each query creates a separate search process or instance (Zcon in Figure 11.4 and

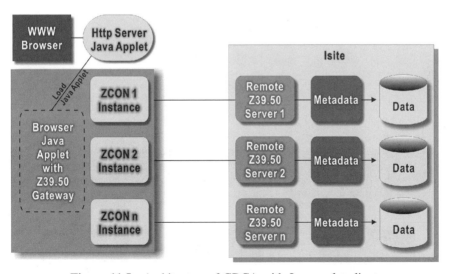

Figure 11.5   Architecture of GDCA with Java applet client.

Figure 11.5) which is maintained until the internal inactivity timer closes the session after a preconfigured period of time with no user input (Isite Information Systems at http://www.cnidr.org/isite.html). Zcon then communicates with the remote Z39.50 server. The results are passed back from the Z39.50 server to Zcon, back to Zgate, and back to the Web browser via the HTTP server. What the user gets from the search is a set of metadata entries in the form of HTML. Some metadata elements may have direct linkages with the data itself, but others may only have the order information linked underneath.

Figure 11.6 illustrates one example of a query mechanism in FGDC's GDCA node. First of all, one metadata query (search for "San Diego") was initiated from a user's Web browser. The Web browser then accessed an FGDC entry point node which has a Web server with Z39.50 gateway functions. The entry point server then distributed the user's query to multiple GDCA nodes simultaneously, including San Diego State University's clearinghouse node, ESRI's Geography Network, and NOAA's clearinghouse node. The Isite software in each local clearinghouse node has already indexed its metadata records on a regular basis. When each GDCA node received the request from the FGDC's Z-gateway server, its local Isite software would use the ISearch program to search its metadata index records and then send the results back to the FGDC's entry point. The FGDC's entry point then combined the query results and displayed them on the user's Web browser.

One major function of the NSDI data GDCA approach is the capability of querying multiple metadata repositories at the same time via the Z39.50 pro-

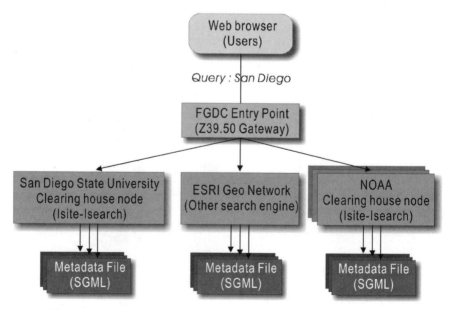

Figure 11.6 Mechanism of querying multiple FGDC's GDCA nodes.

tocol. Distributed data clearinghouse nodes can provide this unique function by installing a software package Isite. Figure 11.6 illustrates the mechanism of querying multiple metadata repositories from an NSDI GDCA node.

If you want to share and advertise your data with GIS users, an effective way is to become a GDCA node, which will ensure your data are searched whenever users access the FGDC's GDCA Web site (http://www.fgdc.gov). To become a GDCA, you need to develop your metadata according to the FGDC's recommended metadata standard, download and install the Isite software (available free at http://www.cnidr.org/isite.html) in your local server, and register your site with the FGDC's GDCA registry.

11.3.2.4 User Interface of GDCA   There are two user interfaces for the FGDC GDCA—an HTML-based Web browser and a Java applet. The HTML-based Web client sends and receives all information via HTML using HTTP. A Web client first downloads a form from the gateway host, selects one or more data sources, composes a field-based or free-text query, and specifies a geographic region using space name or bounding latitude and longitude coordinates. The gateway script reformats the results into HTML and passes them back to the waiting Web client (Figure 11.7). The Web browser acts as a user interface for the end user to request GIS data and for the server to present the data results to the users.

The Java client enables users to pick geographic areas of interest from an interactive map rather than entering coordinates. A map-based Java query tool is designed to enable the user to enter search points or regions graphically against an orthographic map of the world, and the software pastes the coor-

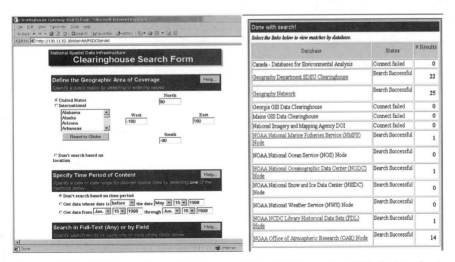

Figure 11.7 NSDI search results from multiple GDCA nodes (HTML interface). (Figure reprinted with permission from FGDC.)

dinates string into the text query window. In this case the Z39.50 gateway is also located at the Java applet at the client side.

With the Java client, the user can interact with an interactive map to search spatial data by clipping a map window or selecting polygons or other geographic features, rather than by specifying a place name or coordinates (Figure 11.8).

11.3.2.5 Limitations of GDCA The FGDC's GDCA was designed for sharing and accessing geospatial data. It assumes that the users are GIS professionals and would be able to take care of the data format, data quality, and data usage issues. So the GDCA makes no attempt to integrate and transform the original data.

But the issues of data consistency and data integration do not go away, even for GIS professionals. These issues are becoming more serious as casual users with no experience or training with the data or the software are beginning to use the geospatial data.

To deal with this lack of consistency or integration, the Office of Management and Budget (OMB) of the United States started a Geospatial One-Stop, an E-Gov initiative (http://egov.gov) in 2001. Geospatial One-Stop is a col-

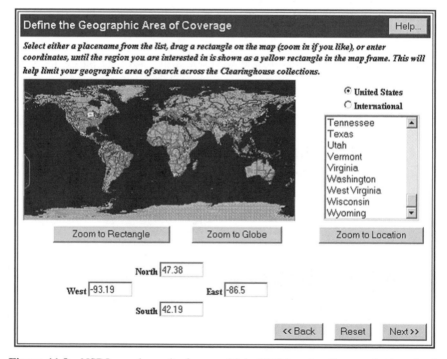

Figure 11.8 NSDI search results from multiple GDCA nodes (Java applet interface). (Figure reprinted with permission from FGDC.)

laborative effort to establish a comprehensive Web portal of geographic data to enhance the NSDI. The purpose of Geospatial One-Stop is to provide a common, consistent source of geospatial data for all spatial data customers, including federal, state, local, and other agencies; academia; and the private sector and the general public. To accomplish this, efforts are being made to develop a core data content standard for each of the seven NSDI themes. These content standards will become the core data standards needed for the exchange of information and will become national standards that can be implemented with existing software. Geospatial One-Stop is a cooperative effort between federal, state, and local governments with the assistance of the private sector and academia and under the support from the OMB.

Notice that the Geospatial One-Stop initiative is different from the FGDC's GDCA. The GDCA provides searches for existing data (regardless of data quality, content, and formats) based on the metadata, whereas Geospatial One-Stop could eventually provide consistent geospatial data that meet the core data content standard.

11.4 GEOGRAPHY NETWORK AND g.net

Geography Network is a collaborative and multiparticipant Web-based GIS portal for publishing, sharing, and using digital geospatial information. The Web site, GeographyNetwork.com, was introduced by ESRI in June 2000. The development of Geography Network is originally derived from ESRI's ArcData Online program and the development of ArcIMS software for online mapping functions (ESRI, 2001). Through Geography Network, GIS users can search and access geospatial data, publish their own data, or provide Internet mapping services by using ArcIMS software or OGC's WMS protocols (Figure 11.9).

In general, the role of Geography Network is very similar to the FGDC's NSDI Clearinghouse, but Geography Network focuses more on the proprietary technology implementation, such as ESRI's ArcIMS software and ArcSDE. Also, NSDI's clearinghouse is a public domain Web service, but ESRI's Geography Network is a private Web site where some GIS data are not free and users may need to pay a fee to access these data.

g.net is a new architecture proposed by ESRI in 2001 for sharing and accessing geographic information in distributed-network environments. The term g.net is borrowed from Microsoft's .NET framework, which has been mentioned in previous chapters. From a technology perspective, g.net is a GIS version of the .NET framework which will be able to provide multiple geospatial Web services for different users via distributed-network environments.

The g.net architecture (Figure 11.10) was originally developed from ESRI's Geography Network and ArcIMS software. In 2002, ESRI began to revise the

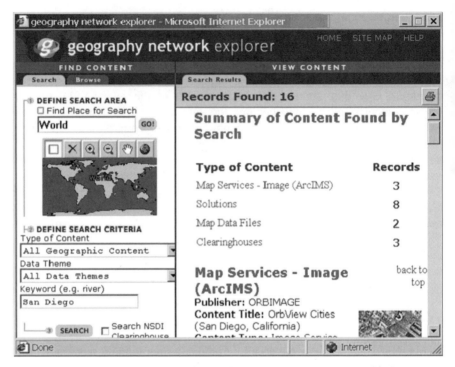

Figure 11.9 Geography Network (search keyword: *San Diego*). (Graphic Image provided courtesy of ESRI.)

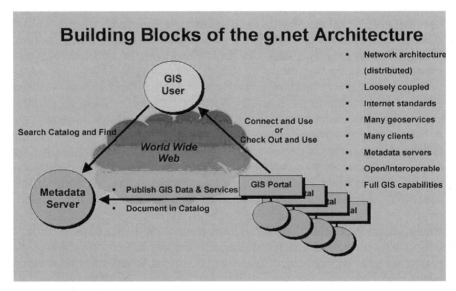

Figure 11.10 ESRI's g.net architecture. (*Source:* ESRI, 2002. Graphic Image provided courtesy of ESRI.)

g.net architecture and added a few new components to g.net, such as metadata catalog services and client-side mobile GIS applications.

The major technology behind the g.net architecture is ArcIMS, ArcSDE, and ArcGIS. ArcIMS is ESRI's Internet map server and can provide on-line map display and metadata catalog search. ArcSDE is the geospatial database management software which can allow ArcIMS to access multiple types of geospatial data sets. ArcGIS is the major GIS engine which provides comprehensive spatial analysis functions. These software components will be integrated in the g.net architecture and will provide an efficient mechanism for accessing and sharing geospatial information, spatial analysis functions, and Web services in the future.

11.5 SAMPLES OF OTHER GEOSPATIAL DATA PROVIDERS

The above geospatial data portals provide data users with an important starting point to discover and access geospatial data and data providers an important channel to advertise and publish their data. In addition, there are data providers that have their own Web sites that provide, with or without a fee, geospatial data on the Internet. Here are some examples.

11.5.1 U.S. National Imagery and Mapping Agency (NIMA)

BACKGROUND   The U.S. National Imagery and Mapping Agency (NIMA) was established in 1996 as a consolidation of the previously separate imagery and mapping agencies. Its primary role is to support the DOD, as well as other federal policymakers and government agencies, with imagery and geospatial intelligence.

NIMA is a major provider and distributor of imagery and geospatial data. Though originally only a provider for the U.S. military and its allies, the agency now also serves academics, commercial organizations, and non-DOD government customers. NIMA has shifted from traditional product delivery to on-line access as a means of serving its growing and diverse customer base.

PURPOSES OF NIMA DATA SERVICES   NIMA's new on-line services serve two basic purposes: They provide for the defense and intelligence communities' complex assortment of geospatial requirements and they allow public use of the geospatial information. More specifically, the data service Web site is designed to

- support national security with current, relevant, and accurate geospatial intelligence;
- allow all stakeholders easy access to geospatial intelligence databases;

Internet GIS Showcase: Georgia 2000

Georgia 2000 is a Web-based system for Georgia and the southeast that provides Internet access to data from the 2000 Census, county vital statistics, economic development agencies, school districts, and many more sources. This information, in combination with interactive maps and menus, makes Georgia 2000 a powerful decision support tool geared for business owners, consultants, community planners, educators, and researchers. Maps and statistical reports are free to registered users.

Because it is a GIS, Georgia 2000 can also be used to locate a specific area such as a senate district or to select geographic features with a buffer based on their relation to another geographic feature. The Georgia 2000 Information System makes interagency and public data access and data sharing more efficient processes. The extensive on-line data processing resources eliminate the need for users to have their own processing software. It can also eliminate duplicate data processing efforts.

The Georgia 2000 Information System is under development by the University of Georgia Information Technology Outreach Services. It is intended to function as a decision support tool for students, professionals, entrepreneurs, legislators, and public administrators.

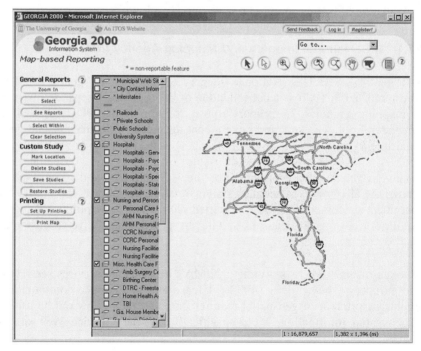

Image reprinted with permission from the University of Georgia information technology outreach service.

Source: http://ga2000.itos.uga.edu/map_interface/top.asp.

- create custom-designed geospatial intelligence, analytic services, and solutions; and
- ensure a solid knowledge foundation for planning, decisions, and action through geospatial intelligence in all its forms and from many sources—imagery, imagery intelligence, and geospatial data and information.

DATA PROVIDED   The geospatial data made available by NIMA comes in both vector and raster formats, including CADRG, Digital Orthorectified Imagery (DOI), DTED, and Vector Product Format (VPF). Much of the data are available to the public over the Internet. More sensitive data are available only to the U.S. military and its allies via extranet nodes.

The sensitive limited-distribution data are stored in Mil-Spec format on a storage array and are replicated across two geographically separate extranet nodes (Veritas File Replicator). Access to these nodes is controlled using Public Key Infrastructure (PKI) technology (see Chapter 10), which validates the identity and credentials of its users. This system allows for access to the limited-distribution data using standard Internet connections, a vast improvement over the traditional closed-network approach.

The less sensitive public data are hosted at http://geoengine.nima.mil, a site with identical architecture to the extranet geospatial servers but without the security measures. Access to the data is free.

In the data server, there are data "converters" to convert selected NIMA data to ESRI compatible data sets, which are stored in ESRI's Spatial Data Engine (SDE). There are also "translators" at each local server to populate the Oracle database with metadata and to generate coverage footprints. Reference map data are stored in SDE.

SYSTEM FUNCTIONS   NIMA data services utilize a variety of products and formats to improve the accessibility of the data. Among them are products from ESRI and Oracle, several standards-based technologies including OpenGIS Web mapping and Web feature servers, NSDI GDCA, and ESRI's Geography Network. The system uses custom servlets to design the Web sites (http://geoengine.nima.mil, https://gis.extranet.nima.mil), uses Metamanager to serve up NSDI-formatted metadata, and uses Internet mapping technologies from ESRI to display geospatial data and to serve up image services, feature services, and extract services (Bordick and Soltyka, 2001).

NIMA's geospatial engine provides access to imagery of Earth, maps, and other geospatial information. There are many functions in its geospatial engine Web site for accessing the public data:

- The *Map tab* on the Web site displays vector-based data in the map frame over the selected map extent.
- The *Find tab* allows the user to make the map extent recenter around a specified place (a city, a mountain, a stream, etc.) or to recenter to a

specific latitude/longitude using Degrees, Minutes, Seconds (DMS) or Decimal Degrees (DD).

- The *Coverage tab* allows the user to view a list of digital products.
- The *Metadata tab* allows the user to view a table of detailed or high-level metadata about the NIMA products.
- The *View tab* allows the user to generate an image of a product that is available over the map center.
- The *Download/Order tab* allows the user to download or order MIL-Spec-compliant NIMA data within the specified map extent or area of interest.
- The *Print tab* allows the user to do some customizing of a map to be printed out.
- The *Tool tab* lists software tools that can view and/or analyze NIMA data.

Users can also obtain software from NIMA to facilitate the viewing and processing of NIMA data. The two kinds they offer are EDGE Viewer, which provides for flexible data visualization in both two and three dimensions, and NIMAMUSE, a self-contained set of computer programs and computer utilities designed to work with NIMA Mapping, Charting, and Geodesy (MC&G) data and information. NIMAMUSE is being phased out in favor of commercial applications.

OUTCOME   A unique characteristic of NIMA's data services is its simplicity. The system uses only HTML geospatial engines without plug-ins, JavaScript, or applets that could slow down site access and downloading. Though streamlined, the site is user friendly and professional, making it useful to users not specifically trained in GIS and intelligence analysis.

11.5.2   NASA's Terra Sensors Data Services

BACKGROUND   The long-term global effects of human activity on Earth are just beginning to be understood. NASA has launched a comprehensive system to monitor the planet's oceans, lands, and atmosphere. NASA's Earth Observing System (EOS) uses space-based "remote sensors" to collect information that can be used to understand current changes and predict future trends. The flagship of this system is a satellite called Terra, which has been collecting global change data since February 24, 2000. Data from five types of instruments on Terra have been made available to the public via an online data services site (http://terra.nasa.gov): ASTER (Advanced Spaceborne Thermal Emission and Reflection Radiometer), CERES (Clouds and Earth's Radiant Energy System), MISR (Multiangle Imaging Spectroradiometer), MODIS (Moderate-Resolution Imaging Spectroradiometer), and MOPITT (Measurements of Pollution in the Troposphere) data.

THE SYSTEM AND ITS FUNCTIONS The Terra Web site allows users to view images on a program called Visible Earth. Visible Earth can be used to browse and search through the large database of available images. The browse function allows users to search by subject, satellite sensor, or location. The search function utilizes two search methods: basic and advanced. Both methods are text-based searches of metadata. Basic search can search for keywords in the title, description, free-text keywords, and location fields of a record. Advanced search can select any combination of terms or keywords to execute a search. For example, users can use Boolean operators—AND, OR—in the keyword field and can combine sensor, location, and parameters to do advanced searches.

OUTCOME Users are able to access some of the information directly on-line and without a charge. Data can also be obtained by mail—this service is offered for a fee and is provided by whichever data center receives the request.

11.5.3 Data Services at NASA's Distributed Active Archive Centers (DAACs)

BACKGROUND DAACs are the "treasuries" of NASA's Earth Observing System Data and Information System (EOSDIS). The DAACs process, archive, document, and distribute earth science data in support of earth science research. The central purpose of the archives is to help users find, order, and use earth science data.

DATA PROVIDED An abundant variety of data can be obtained from the DAACs Web site (http://nasadaacs.eos.nasa.gov), including Terra Sensors Data; NASA Landsat Data Collection; National Oceanic and Atmospheric Administration (NOAA)/NASA Pathfinder Advanced Very High Resolution Radiometer (AVHRR) Land Data; Goddard Earth Sciences (GES) DAAC Climatology Interdisciplinary Data Collection; SSM/I Pathfinder Atmospheric Moisture Products from DMSP-F8; AVHRR Oceans Pathfinder Monthly Global Best Sea Surface Temperature (SST) CD-ROM (NOAA, NASA), Product 067; Upper Atmosphere Research Satellite (UARS) Data; Near-Real-Time Ice and Snow Extent (NISE) data, and so on.

SYSTEM The DAACs provide data access through a unified system known as the EOS Data Gateway (EDG). Each EDG search engine supports three search methods: primary data search, data set lookup, and geographic region search.
 In addition to the EDG search engine, each DAAC also maintains a data catalog and some have developed separate search tools for their individual data collections. NASA data are accessible through the Global Change Master Directory (GCMD), which is a comprehensive directory of data sets useful

for global change research. This tool allows researchers to search the earth science data sets by parameters, location, instruments, platforms, and projects. The GCMD search engine does not allow any map-based searches, but it does provide free-text searches that include basic search, Boolean search, wild card search, and advanced search.

OUTCOME   Data transfers from the DAACs can occur in several ways. Users can download data from a staged FTP server, from an anonymous FTP server, or from a WWW server. Alternatively, users can use an on-line form to request hard copies of the data on CD-ROM or tape. While much of the data are provided at no charge, occasionally there are fees.

11.5.4 USGS Earth Resources Observation Systems (EROS) Data Center

BACKGROUND   The EROS Data Center (EDC) is part of the USGS National Mapping Division. The EDC handles data management and systems development for the largest data archive in the United States. Originally created to handle data from NASA Landsat satellites, the EDC handles a much wider array of information, including millions of satellite images and aerial photographs and a variety of USGS digital cartographic data products.

SYSTEM FUNCTIONS   Data searching and ordering are made possible on the EDC Web site (http://edc.usgs.gov) by a variety of information management systems, including Earth Explorer, Photo Finder, Map Finder, and EOS Data Gateway:

> *Earth Explorer* (see Figure 11.11) replaces the Global Land Information System (GLIS). New functionality includes a secure credit card/e-commerce system and cross-inventory search capabilities. Earth Explorer uses session (transient) cookies and Java applets.
>
> *Photo Finder* is a Web-based, quick-search tool for locating and ordering photographs from the USGS' National Aerial Photography Program.
>
> *Map Finder* is a Web-based, quick-search tool for locating and ordering 7.5-minute (1:24,000-scale) USGS maps of the continental United States.
>
> *EOS Data Gateway* is a Web-based query system provided by NASA. This information management system supports a variety of functions, including documentation, image browsing, and a data-ordering service.

11.5.5 TerraServer

BACKGROUND   Microsoft's   TerraServer   (http://terraserver.homeadvisor. msn.com/default.asp) was originally created as a demonstration project. It

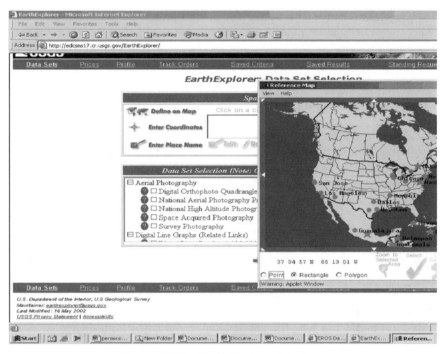

Figure 11.11 Earth Explorer interface.

showcased the capability of Microsoft's SQL server to handle a large database and thousands of user requests from around the world. The system is designed to work with commonly available computer systems and Web browsers over slow-speed communications links. The TerraServer Web site is a popular source for free satellite imagery and is claimed to be one of the world's largest on-line databases.

DATA   TerraServer's satellite images and maps of the United States were provided by the USGS. The images are in the public domain and are freely available for users to download, use, and redistribute. TerraServer uses two types of products to manage the data: Digital Orthophoto Quadrangles (DOQs) and Digital Raster Graphics (DRGs). These products divide the images into small, manageable tiles that can be easily retrieved and copied into JPEG format.

SYSTEM FUNCTIONS   Users can search for imagery in TerraServer in four ways: (1) enter a specific place name, (2) use TerraServer's "advanced find" search feature, (3) locate an area by clicking on the world coverage map, or (4) locate the place of interest by using TerraServer's list of famous places. There are also several advanced search methods, including search by U.S.

addresses, search by the type of places, and search by geographic location using longitude and latitude. A new Web service called TerraServer.NET enables Web developers to easily integrate TerraServer data into their own applications.

11.5.6 Space Imaging

BACKGROUND   Space Imaging launched IKONOS, the world's first 1-meter-resolution, commercial Earth imaging satellite in 1999. It then created an on-line data service Web site to advertise and disseminate IKONOS imagery products and Space Imaging's other satellite and aerial image-based products (http://www.spaceimaging.com/level2/level2products.htm).

DATA PROVIDED   Space Imaging's on-line data services offer a wealth of CARTERRA products created from IKONOS imagery and from a constellation of other remote sensing satellites that Space Imaging owns or has marketing rights. These other satellites include the U.S. Landsat satellites, the Indian Remote Sensing (IRS) satellites, Canada's RADARSAT, and the European Space Agency's Radar Satellite. Space Imaging also sells submeter-resolution aerial imagery for applications requiring a high degree of detail and accuracy.

SYSTEM FUNCTIONS   Space Imaging offers a set of proprietary tools for accessing and analyzing data. CARTERRA Online is a data access tool that assists in the searching and purchase of data at the Space Imaging Web site (carterraonline.spaceimaging.com). CARTERRA Analyst integrates GIS, remote sensing, imagery analysis, photogrammetry, and cartography tools into a single workstation. It enables customers to search, browse, and retrieve images and data from a database, perform multispectral analysis, create and combine image files, generate reports and products, and create GIS databases of reports, data layers, and images.

OUTCOME   Users can purchase and download data directly from the site, either as compressed, self-extracting zip files or as a TIFF images. The images available are more appropriate for use as visual art or for informational display; they are not suitable for performing remote sensing and spatial analysis.

11.5.7 American FactFinder: U.S. Census Bureau

BACKGROUND   The U.S. Census conducts the most comprehensive "survey" of the U.S. population of any other population survey. The term *survey* is not really accurate when making reference to the Census because the Census does not sample the population; it counts the whole population (or tries to—there are limitations to counting all persons that are beyond the scope of this discussion). The Census is conducted once every 10 years to obtain demographic data such as persons per household, age, sex, race, and other population characteristics. The U.S. Census Web site (http://www.census.gov) offers users the opportunity to download data and create maps.

PURPOSE   American FactFinder was designed for public use on the Internet so that Census and other demographic data could be readily available to a large audience. The site allows users to download population, housing, industry, and economic data and to illustrate such data in the form of tables and maps.

COMPONENTS   FactFinder is a service offered by the U.S. Census at http://www.census.gov/geo/www/maps that allows users to create tables and maps with selected data. FactFinder is comprised of three major components:

- Basic facts allow the user to create tables and maps from a small set of popular demographic data sets.
- Data sets allow the user to access supplementary surveys from the Decennial Census, American Community Surveys, and Economic Census as well as the complete set of the Decennial Census.
- Reference and thematic maps allow the user to visualize demographic data by country (United States), state, county, metropolitan area, zip code, census tract, and census block. The thematic mapping utility is more versatile in that the user can map a greater variety of data. The user can change the look of the maps by changing legend properties "Classes," "Boundaries," "Features," and "Title."

DATA SOURCES   Map boundaries and some other linear features are derived from the Census Bureau's TIGER (Topologically Integrated Geographic Encoding and Referencing) file, while other linear features such as major roads are derived from the Department of Transportation's National Transportation Atlas. The Digital Chart of the World from the National Imagery and Mapping Agency provide such features as streams and water bodies.

SYSTEM OPERATIONS   The basic facts component of American FactFinder processes a user's query and displays the resultant map with JavaScript. As for the data sets component, JavaScript is also responsible for querying the data, but here maps are generated with a Java applet.

In general, American FactFinder allows users to query geographic areas by

- list, which allows the user to select an area from a list of geographic areas;

- name search, which allows the user to search for the geographic area by name;
- map, which allows the user to select an area by clicking on the area of interest on a map (not available when querying a reference map); and
- address search, which allows the user to enter a specific address to which FactFinder responds with a list of geographic areas (e.g., state, county) that contain the address (only available for searching Census 2000 data).

American FactFinder also allows users to recenter a map by entering either an address or latitude and longitude coordinates. Figure 11.12 shows a thematic map of renter-occupied housing units in percent.

11.6   CONCLUSIONS AND SUMMARY

Internet access to geospatial data is a valuable tool for people doing GIS analysis. As on-line resources expand, efficient data search tools are becoming increasingly important. The basic data-finding function is *browsing*. Given the large volume of data available, *search* engines are necessary. Browse and search capabilities complement one another—neither is sufficient by itself as

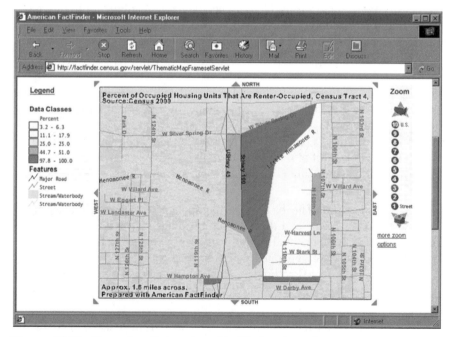

Figure 11.12   Example of American FactFinder thematic map. (*Source:* http://www.census.gov/geo/www/maps.)

a full Internet-based geospatial data service. However, the ability to search information directly is a more critical measure of interoperability. The efficiency of data searching depends on the data structure, metadata indexing and cataloging, a competent search protocol, and efficient information search architecture.

On-line data networks can be organized in two ways: centralized and distributed data systems. The appropriate organization depends upon the amount of data being stored and the capabilities of the organizations providing the data. Distributed data systems are preferred for large organizations, such as the federal government and state governments, because data updates are easier to perform and the system members are typically large enough to handle data hosting. Centralized data systems are more suitable for smaller organizations and agencies whose individual members do not have the resources to host their own data.

Spatial data searches are often difficult because spatial boundaries and time span information are fundamentally different from text information. To allow for effective searches, metadata must be developed that are searchable by text-based search engines. Compatible and concurrent metadata elements need to be established according to a metadata standard.

A distributed search that allows users to pass a short query string to many servers simultaneously provides an efficient means to search for GIS information on the Internet. When the specifications of OpenGIS's Web Map Server and Web Feature Server are implemented by all the GIS data portals and data providers, it will be possible for any user on the Internet with a spatial discovery client to search for any geospatial data and to do so efficiently and effectively. In the meantime, the best way to share GIS data is through a portal such as FGDC's GDCA or ESRI's Geography Network and G.Net.

WEB RESOURCES

Descriptions	URI
FGDC GDCA	http://www.fgdc.gov/clearinghouse/clearing house.html
Isite	http://www.cnidr.org/isite.html
Z39.50	http://www.loc.gov/z3950/agency
NASA TERRA	http://terra.nasa.gov/
NASA DAACS	http://nasadaacs.eos.nasa,gov/
USGSEDC	http://edc.usgs.gov/
Microsoft TerraServer	http://terraserver.homeadvisor.msm.com/
Space Imaging	http://www.spaceimaging.com/level2products.htm
BTS	http://www.bts.gov/gis
Census	http://www.census.gov/
HUD	http://www.huduser.org/datasets/pdrdatas.html

REFERENCES

Bishr, Y. A., Molenaar, M., and Radwan, M. M. (1996). Spatial Heterogeneity of Federated GIS in a Client/Server Architecture. In *Proceedings of GIS '96, Tenth Annual Conference on Geographic Information Systems,* Vancouver, British Columbia, Canada, March 1996.

Bordick, L., and Soltyka, N. (2001). Architecting Solutions for Distributing NIMA Geospatial Data and Services to All Variety of Users. URI: http://www.esri.com/library/userconf/proc01/professional/papers/pap1142/p1142.htm, June 14, 2002.

Dangermond, J. (2001). *g.net—A New GIS Architecture for Geographic Information Services.* Spring 2001 ArcNews online. URI: http://www.esri.com/news/arcnews/spring01articles/gnet.html.

Dueker, K. J., and Vrana, R. (1994). Sharable Digital Road Map Database: Functional, Representational, Institutional Issues. In *Proceedings of the 1994 Korea–US Symposium on IVHS and GIS-T.* Seoul: Korea Transportation Research Society, pp. 131–148.

ESRI. (2001). *An Overview of the Geography Network,* White Paper, ESRI. URI: http://www.esri.com/library/whitepapers/pdfs/gn_overview.pdf.

ESRI. (2002). *What is g.net?* White Paper. URI: http://www.esri.com/library/whitepapers/pdfs/what_is_gnet.pdf.

FGDC. (1997). *A Strategy for the NSDI.* URI: http://www.fgdc.gov/nsdi/strategy/strategy.html, June 11, 2002.

Goodchild, M. F. (1995). Future Directions for Geographic Information Science. *Geographic Information Science,* 1(1), pp. 1–7.

Goodchild, M. F. (1999). Introduction. In M. F. Goodchild, M. Egenhofer, and R. Fegeas (Eds.), *Interoperating Geographic Information Systems.* Dordrecht: Kluwer Academic. Chapter 11, pp. 133–134.

Li, B. (1996). Strategies for Developing Network Oriented GIS Software. In *GIS & Remote Sensing: Research, Development & Applications—Proceedings of Geoinformatics '96,* April 1996, West Palm Beach, Florida.

Nebert, D. D. (1995). *What Does It Mean to be an FGDC Clearinghouse Node?—Draft Position Paper for FGDC Clearinghouse Working Group.* Reston, Virginia: U.S. Geological Survey.

Nebert, D. D. (1996). Information Architecture of a Clearinghouse. Paper presented at the WWW Access to Earth Observation/Geo-Referenced Data workshop, World Wide Web Conference 5, Paris, France, May 6–10, 1996.

Nebert, D. D., and Doyle, A. (1998). *Discovery and Viewing of Distributed Spatial Data: The OpenMap Testbed.* URI: http://www.fgdc.gov/publications/documents/clearinghouse/eogeo98/eogeoddn.html.

Nebert, D. D., and Fullton, J. (1995). *Use of Z39.50 to Search and Retrieve Geospatial Data.* Reston, Virginia: U.S. Geological Survey.

Oswald, R. (1996). Shared Vision: Manitoba Partners Build Land Information Data Warehouse. *Geo Info Systems,* January.

Peng, Z. (1997). Integrating Internet with GIS: The Development of Internet GIS. Department of Urban Planning, University of Wisconsin-Milwaukee.

Peng, Z., Groff, J. N., and Dueker, K. J. (1995). An Enterprise GIS Database Design for Transit Applications. In *Journal of the Urban and Regional Information Systems Association.* Vol. 10(2), pp. 46–55, 1998.

CHAPTER 12

INTERNET GIS APPLICATIONS IN INTELLIGENT TRANSPORTATION SYSTEMS

> *Transportation applications of GIS have become increasingly popular in recent years, so much so that they are now routinely referred to by the acronym GIS-T.*
> —N. M. Water (1999, p. 827)

12.1 INTRODUCTION

Transportation was one of the first fields to benefit from the development of Internet GIS, popularized by on-line trip-planning programs such as Map-Quest (http://www.mapquest.com) or Yahoo! Maps (http://maps.yahoo.com). This chapter describes some applications of Internet GIS in transportation, especially transportation information dissemination and ITS. It also discusses a way of designing a Web-based transit information system that allows transit users to plan a trip itinerary and to query service-related information, such as schedules and routes, using Internet GIS technologies.

The Internet and the WWW are revolutionizing the process of information dissemination, communications, and transactions, and they have brought some important changes to the traditional functions of transportation services. For example, historically, transit agencies have relied on printed brochures to provide customers with information about transit routing and schedules. Transit users have had to select proper routes and transfer points based on the information printed on the brochure. Such brochures are often complex and can be confusing for many people. In addition, since schedules are infrequently updated, many service changes cannot be reflected in the brochure in a timely manner. For road constructions and traffic conditions, users have to rely on radio, phone, and even the newspaper for information.

With Internet GIS, schedule, traffic, and construction information can be provided on the Internet in a graphic manner. Most traditional customer service functions (e.g., schedules, routing, itinerary planning, and real-time traffic information provision) in transportation agencies can be enhanced or even replaced by Internet GIS or other Web-based information systems. The beauty of Web-based information is that it can be presented in a more intelligible manner than traditional brochures allow.

Many transit agencies are now in the process of creating and upgrading their transit information on the Web. With the rapid development of Internet technology and the proliferation of on-line information, the number and use of transit information Web sites are increasing rapidly. For instance, the U.K. Public Transport Information Web site had 1000 visits a month at the end of 1996; by July of 1998 it was receiving over 13,000 per month (http://www.ul.ie/~infopolis).

There are many on-line transportation information systems. These range from simple, static transit schedule displays to more sophisticated real-time systems that show bus locations or traffic conditions. This chapter provides a taxonomy to review existing transportation information systems on the Web to form a framework for developing future transportation information systems.

12.2   REVIEW OF TRANSPORTATION INFORMATION ON THE INTERNET

The purpose of transportation information on the Web, like any other on-line information systems, is evolving from information dissemination to interactive communications and on-line transactions. Transportation information dissemination serves the purpose of transportation information announcement and display, such as bus and train schedules and routing, service changes, and road constructions and closures. Users simply view the posted information. Interactive communication provides user interactivity and feedback channels. Users can actively manipulate and search for specific information based on their own needs and give feedback to the system providers. Transactions, as for ticketing and reservations, offer instant interactions between system providers and users.

Based on these evolving purposes, on-line transportation information systems can vary significantly in terms of content and function. Table 12.1 provides a framework for on-line transportation information systems and categorizes the quality of service provided on different Web sites based on the information content and functionality.

The rows in Table 12.1 represent the contents of information provided on an on-line transportation information site. The content information may range from static information to dynamic real-time information. At the most basic level, a transportation Web site can simply provide the basic information about the agency and its services (level A). While somewhat useful, it is not very informative, and the site may not attract users.

TABLE 12.1   Taxonomy of On-Line Transit Information Systems

		Functions and Interface		
Contents	Content Level	Information Dissemination (Function Level 1)	On-Line Decision Support (Function Level 2)	On-line Transactions (Function Level 3)
General information	A	A1	A2	A3
Static information	B	B1	B2	B3
Trip itinerary planning	C	C1	C2	C3
Real-time information	D	D1	D2	D3

The next level of content (level B) may include static information about transit routes, networks, service schedules, bus fares, or road construction and closure information. This content is the minimum that is needed for users to access basic transportation service information. At this level, the content may also include multimodal information, such as rail, air, and highway traffic information. This would help the travelers make a more informed decision on mode choices. Multimodal travel information would be especially valuable for Internet-enabled kiosks in airports or other trip generators. These can be used by travelers to make a more informed decision on whether to ride a bus, take a taxi, or rent a car.

The third level of content (level C) may include trip-planning or trip itinerary information. The trip-planning program can provide the user the optimal path between the user's trip origin and destination as well as the time of travel based on transit schedule information and/or traffic conditions.

The fourth level of content (level D) may include real-time information about traffic, bus locations and expected arrival times, possible delays, as well as incident or weather information. This real-time information is particularly useful for travelers before a trip departure.

The columns in Table 12.1 represent the level of function that the Web interface supports and the rows represent the content level. The entries in each cell represent their row and column locations. The function of a Web interface ranges from information dissemination to decision support and, finally, to on-line transactions. Web sites with information dissemination functions only disseminate transportation-related information such as traffic condition, bus schedules, and road closure information. Web sites with decision support functions provide not only transportation-related information but also functions and tools to assist users to make intelligent decisions on what to do. Furthermore, the on-line transaction function not only can provide decision support but also provides tools to get the job done once the users make a decision. The on-line transactions include on-line ticketing, on-line license applications, and so on.

The first level of function is information dissemination. It simply provides information. The ways to disseminate information include simple HTML, PDF, and static image maps. It could also provide map-based information query, search, and geocoding functions. For example, if the user enters an address or points to a specific location on the map, the system can find all the bus routes and stops within walking distance of that location. The user can render the transit network and street network maps by zooming in and out, panning, or conducting a spatial search.

The next level of function support is capable of providing decision support systems. This includes routing and departure recommendations given current traffic information. It further provides information customization and information delivery. For example, the user can customize the common travel origins and destinations for each day. The traffic or bus information will be delivered to the customer's e-mail, pagers, or cellular phone to advise on travel routing (Peng and Jan, 1999). Wireless information delivery is clearly a future trend of on-line information systems. Customization facilitates the timely delivery of personalized information. Use of real-time information is especially important for customized services.

Finally, at the highest level of function support (level 3), on-line transactions can be provided, such as on-line ticketing, on-line reservation in the case of para-transit, on-line matching for car poolers, or on-line license applications. Based on the level of content, there is also a different level of function support for this on-line transaction support. For example, the difference between B3 and D3 is that B3 is a conventional physical ticketing and D3 is an electronic ticketing. At level B3, transit tickets and passes can be purchased on-line using a credit card, but real tickets and passes have to be delivered via mail. At level D3, no physical ticket delivery is necessary. Customers can pay for the trip on the Web. The user does not even need to print out a ticket; the machine on the bus can automatically validate the electronic ticket or the payment from a smart card. In terms of on-line registration, such as on-line ride-share matching and on-line para-transit scheduling at the B3 level, users can enter personal information about home and work addresses, travel time to work, and so on, but they cannot obtain matching information on the fly, whereas at the D3 level matching information can immediately be made available on the Web. Furthermore, at the D3 level, registration for para-transit services can be done in real or near-real time based on actual bus location.

These levels of services reflect the current and foreseeable future development of Internet technology in intelligent transportation systems. It may evolve as Internet technology advances, but the basic concept may remain relatively stable over the near term. The level of service of a transportation information Web site ranges from the lowest at the upper left corner (A1) to the highest at the lower right corner (D3) of Table 12.1.

Most current Internet-based transit information services are at levels A1, B1, and C1. For example, the Toronto Transit Commission (Canada) Web site (http://www.city.toronto.on.ca/ttc/schedules/index.htm) is at level B1, where both schedules and route maps are in PDF form (Figure 12.1). The Metropolitan Atlanta Rapid Transit Authority's Web site (http://www.itsmarta.com) and the SunTran System of the Tucson Metropolitan area, Arizona (http://www.suntran.com/index.htm), are also at level B1, with schedules in HTML or PDF and route sketch maps (unscaled transit route illustrations) in PDF or GIF format. Washington Metro in Washington, DC, has implemented a clickable map for its rail system (http://www.wmata.com). Each rail station in a map image is linked with a description about the station and its related schedule and bus transfer information, a B1 service. It also has an itinerary planning system for its rail system (http://www.wmata.com), a C1 service. The Denver, Colorado, Regional Transportation District (RDT) (http://www.rtd-denver.com/index.html) has implemented a text-based searchable schedule database at the B1 level.

The San Francisco Bay Area Transit Information System (Figure 12.2) (http://www.transitinfo.org) is at level B1, where transit schedules can be directly queried from the transit maps and interactive mapping functions are provided.

The trip-planning system at the Road Management System for Europe (ROMANSE) in Southampton, England (http://www.romanse.org.uk) is at level C1, which provides pull-down list input boxes and text output. Another trip-planning system at the Los Angeles County Metropolitan Transportation Authority (California) (http://www.mta.net) is also at level C1. A map is shown to illustrate the trip origins, destinations, and routes, but there is no map-rendering function. Transit schedules are presented as separate PDF doc-

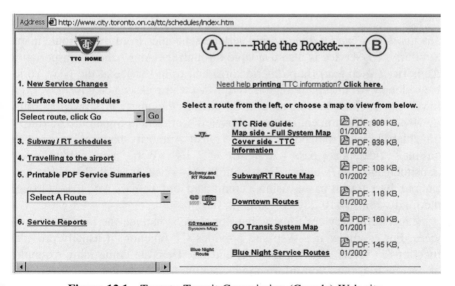

Figure 12.1 Toronto Transit Commission (Canada) Web site.

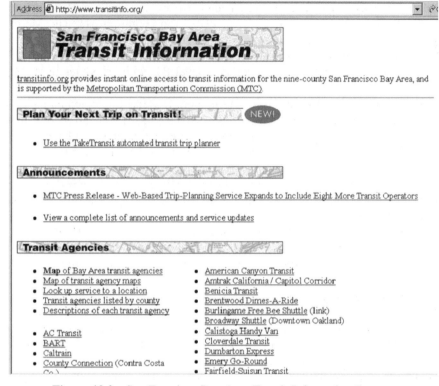

Figure 12.2   San Francisco Bay Area Transit Information System.

uments. Users cannot directly find service schedules from transit route maps. Similarly, C1 service is also available through the Ann Arbor Transportation Authority (Michigan) (http://theride.org) and train service at the New York–New Jersey–Connecticut region (http://www.itravel.scag.ca.gov/itravel). A few Web sites have service at level D1 such as the busview_X program (http://www.its.washington.edu/projects/busviewX.html) and Superoute 66 (http://travel.labs.bt.com/route66). Chicago Transit Authority has implemented an on-line ticketing pass system at its Web site (http://www.transitchicago.com/welcome/index.html), a service at level B3. Customers can purchase transit passes using a credit card and the pass will arrive through the mail within 7–10 days (Figure 12.3).

Higher level services usually, but not always, include the lower level services. For example, if a Web site provides C2 function, it usually provides functions at levels A1, A2, B1, B2, and C1. Here are some more examples of on-line transportation information systems.

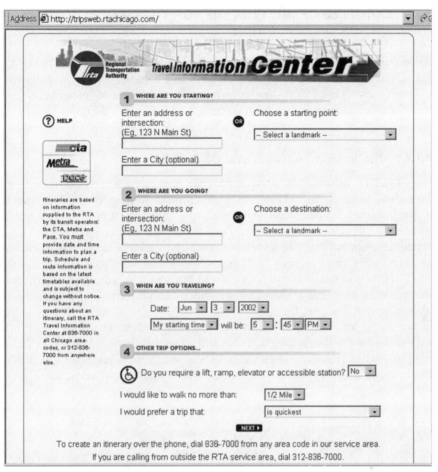

Figure 12.3   Chicago Regional Transit Authority (RTA)'s Web Information Services.

Key Concepts

The purpose of transportation information on the Web is evolving from information dissemination to interactive communications to on-line transactions. Transportation **information dissemination** *tools provide and display, for example, bus and train schedules, routing, service changes, and road constructions for customers.* **Interactive communication** *provides user interactivity and feedback channels for specific information based on a user's own needs.* **On-line transactions** *can offer instant interactions between system providers and users, such as ticketing and reservation.*

12.3   EXAMPLES OF INTERNET GIS FOR TRANSPORTATION INFORMATION DISSEMINATION

Most current on-line transportation information systems are at the levels A1, B1, C1, and D1. Some use Internet GIS, and some do not. The following examples are chosen to demonstrate the different levels of information contents and functions as illustrated in Table 12.1.

12.3.1   Static Map Service (Level B1 Service)

This kind of information service provides static, transportation-related information in a simple HTML or PDF format. A map is used simply as a display image. There is no map-rendering function. It can be well exemplified by the Houston Metro services (http://www.ridemetro.org/services/schedulesandmaps.asp), in which a general bus route map is posted on the Web in PDF format (Figure 12.4) and schedules of bus routes are also stored in PDF files and are linked to a long list of bus routes (Figure 12.5). Individual bus route maps (Figure 12.6) and schedules can be retrieved by clicking on a specific bus route on the list.

Browsing information in static Web pages is convenient. People with little experience in browsing the Web can follow text links to get relevant information easily. However, drawbacks of the static Web page are ostensible: Static maps are very easily overloaded with symbols and texts, there is no

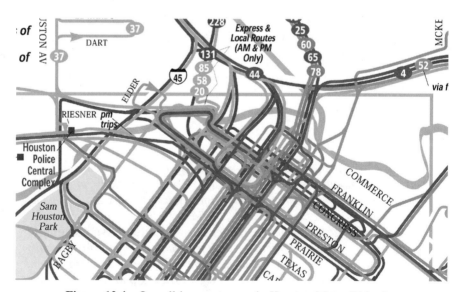

Figure 12.4   Overall bus route map in Houston Metro Web site.

20 Canal/Long Point Ltd.	102 Bush IAH Express
23 Crosstimbers Crosstown	131 Memorial Express
25 Northline/Richmond	132 Harwin Express
26/27 Outer/Inner Loop	137 Northshore Express
Crosstown	163 Fondren Express
29 TSU/UH Hirsch Crosstown	170 Missouri City Express
	201 North Shepherd Park & Ride
30 Clinton/Cullen	202 Kuykendahl Park & Ride
33 Post Oak Crosstown	204 Spring Park & Ride
34 Montrose Crosstown	205 Kingwood Park & Ride
35 Fairview/Leeland	206 Eastex Park & Ride
36 Kempwood/Lawndale	210 Katy/West Belt Park & Ride
37 El Sol Crosstown	212 Seton Lake Park & Ride
40 Pecore/Telephone	214 Northwest Station Park & Ride
41 Gulf Meadows Circulator	216 W. Little York/Pinemont Park & Ride
42 Holman Crosstown	221 Kingsland Park & Ride
43 Pinemont Plaza	228 Addicks Park & Ride
44 Acres Homes Limited	236 Maxey Road Park & Ride
45 Tidwell Crosstown	244/246/247 Bay Area Park & Ride
46 Gessner Crosstown	257 Townsen Park & Ride
47 Hillcroft Crosstown	261 West Loop Park & Ride
48 Navigation/West Dallas	262 Alief/Westwood Park & Ride
49 Chimney Rock Crosstown	265 West Bellfort Park & Ride
50 Harrisburg/Heights	273 Gessner Park & Ride
52 Hirsch/Scott	283 Kuykendahl/Greenway Plaza/Post Oak
53 Westheimer Limited	284 Kingwood/Townsen-Greenway/Uptown
54 Aldine/Hollyvale Circulator	Park & Ride
56 Airline Limited	285 Kingsland/Addicks-Uptown/Greenway
58 Hammerly	Park & Ride
60 South MacGregor/Hardy	291 Kuykendahl/North Shepherd/TMC Park

Figure 12.5   List of bus routes in Houston Metro Web site: schedules can be retrieved by clicking on a specific bus route on the list.

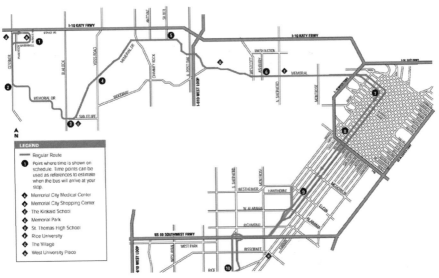

Figure 12.6   Individual bus route map in Houston Metro Web site.

way to find specific information that is not printed in the map, and it is the user's task to put different information (e.g., schedule, route) together in order to make a decision. This is basically a convenient way to distribute bus route schedules and maps on-line. But there is no additional value added beyond distributing exactly the same information as the printed bus schedules.

12.3.2 Location Maps and Driving Directions (Level C1 Service)

The most popular tool for accessing location and trip direction information available on-line might be MapQuest (http://www.mapquest.com/) and Yahoo! maps (http://maps.yahoo.com). A location map can be easily retrieved by entering an address or airport code at either Web site. Driving directions can also easily be obtained by giving an origin address and a destination address. The results of a driving direction include a path map and a textual direction (Figure 12.7). Certain map interaction capabilities are provided such as zoom in, zoom out, and pan the map. Moreover, map information can be displayed with different details at different zoom levels.

12.3.3 Real-Time Transportation Information Systems (Level D1 Service)

One of the real-time transportation information systems available is the Milwaukee MONITOR provided by Milwaukee Freeway Traffic Management

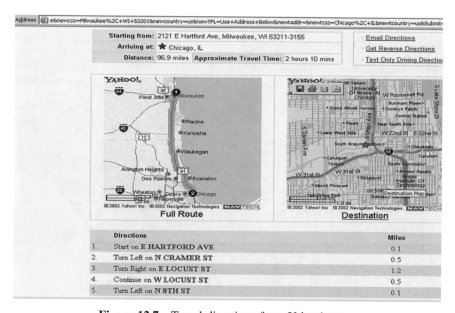

Figure 12.7 Travel directions from Yahoo! maps.

Center (http://www.travelinfo.org/milwaukee.html). MONITOR is a computerized system that collects data through the use of electronic detectors and closed-circuit television cameras. Based on the traffic information collected by field detectors, a congestion map is generated dynamically and put on the Web by the system (Figure 12.8). The congestion map is served by two viewers: an HTML viewer and a Java viewer. The former posts an image map on a Web page and updates it every half minute, and the latter streams data to the user PC and draws a map in the user Web browser using a Java applet. In addition to congestion, the map also shows other traffic information, including construction sites, closed ramps, message signs, and video camera images which are hyperlinked with the main congestion map. Since professional GIS programs may not be used in the application, the map quality is poor.

The California Real-Time Freeway Speed Information (http://www. dot.ca.gov/traffic) is a similar but more advanced system. The map viewer is a Java applet that offers better map manipulation capabilities, including zoom in, zoom out, display/hide map layers, and identify information (Figure 12.9). Freeways, streets, and real-time driving speed are displayed. Real-time data

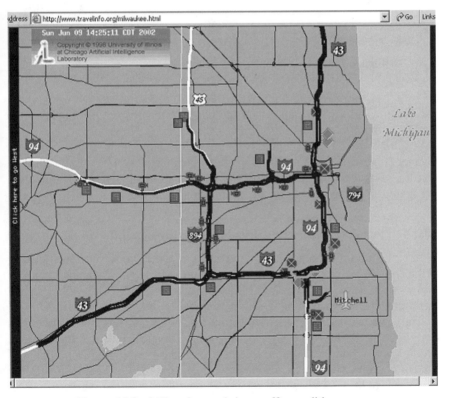

Figure 12.8 Milwaukee real-time traffic conditions map.

Figure 12.9 California Real-Time Freeway Speed Information.

are streamed to the Java applet on the client Web browser for display. Map quality is much better than the Milwaukee MONITOR system.

A third example of providing real-time information is the Cape Cod travel planner (Figure 12.10) (http://www.e-transit.org/cape_cod/map.asp). It offers real-time bus location information and basic map manipulation tools, including zoom in, zoom out, pan, and information identification.

12.3.4 Decision-Supporting Trip-Planning Systems (Level D2 Service)

With real-time traffic conditions, one could make decisions by simply looking at the congestion maps. A step further is to provide users a "what-if" analysis tool to make decisions. The TrafficView program in Seattle Sidewalk designed by Microsoft has such high-level functionality.

TrafficView in Seattle Sidewalk (http://trafficview.seattle.sidewalk1.com) is designed by Microsoft as an Advanced Traveler Information System. TrafficView obtains real-time traffic data from the Washington Department of Transportation (which also hosts its own real-time map of Seattle area traffic conditions at http://www.wsdot.wa.gov/PudgetSoundTraffic/cameras/), reformats the information into a user-friendly format, and presents it to commuters.

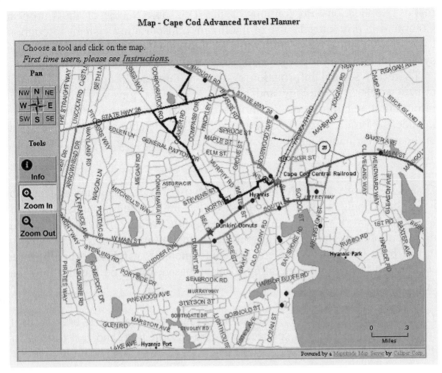

Figure 12.10 Cape Cod travel planner.

TrafficView provides a wide variety of traffic-related information, including real-time traffic maps, video cameras, and incident information.

A unique feature of TrafficView is that the user can get customizable information. For example, the user can specify a freeway entrance and exit location; the on-line trip calculator can estimate travel time based on real-time traffic conditions. It can also provide information about distance and driving time (based on real traffic conditions) of alternate routes. Moreover, the users can choose to receive real-time traffic information by electronic mail at a preselected time. These decision support features give commuters some options of which route to use given the current traffic conditions.

12.3.5 Traveler Planning with Infopolis 2 (Level D2 Service)

Infopolis 2 is phase 2* of a European-wide project initiated by the European Commission to look at how travelers use different modes of transportation (e.g., walking, bus, train, and park-and-ride) within the same trip. The Tele-

---

*The Infopolis 2 phase of the project was conducted from 1988 to 2000.

matics Applications Programme of the Transport Sector of the European Commission promotes the use of intelligent transportation systems throughout Europe. Currently, there are over 170 intelligent transporation information systems and over 600 public transportation Web sites in operation.

There are several information systems within the Infopolis network that utilize wireless technology and/or GIS to provide travelers with real-time information regarding various modes of transit. These information sources include the Public Interactive Terminal (PIT), the electronic bus stop display, on-board information, at-home/office information, the enquiry office terminal, and the hand-held terminal.

12.3.5.1 *Purpose of Internet GIS*

Because traveler planning is geographically based and the general purpose of the information systems within the Infopolis network was to provide travelers with real-time information regarding their chosen mode of transportation, the information systems need to utilize both GIS and the Internet—not to mention wireless technology—to accomplish the task of generating location-based, real-time traveler information.

12.3.5.2 *Data Sources*

Data sources include street networks, transit-specific schedules, transit connections, fare information, and real-time information as derived from AVL systems.

12.3.5.3 *System Operations*

All of the information systems utilized by the Infopolis project access a geographic database for spatial data. For the most part, they all also function in a Web environment. The at-home/office information system is the only system that provides alternatives to a Web environment* for obtaining information. The functions of each of the six information systems are as follows.

12.3.5.3.1 *Public Interactive Terminal*

PITs or information kiosks are terminals located at transit station and bus stop locations where travelers can generate route maps and itinerary information as well as general information about activities going on in the city. Most of these terminals operate within a single system; in other words, they operate off of a local database consisting of public transportation information; however, there are terminals that are connected to other systems so as to access information regarding other modes of transportation. (For example, SIAM in Spain provides information on public transportation and automobile traffic and TRIPlanner in the United Kingdom and FITs in Italy provide information on public transportation, au-

---

*Even though some PITs may operate autonomously from off of a local database, many PITs may access information from that same database and are "networked" together, sharing location information.

tomobile traffic, and air quality.) Most of these terminals connect to the control center via telephone lines and modems, although some systems utilize GPS (i.e., Genoa, Spain, with its Automatic Vehicle Monitoring (AVM) DGPS system for "service disturbance" information) to integrate real-time information into their databases.

12.3.5.3.2 Electronic Bus Stop Display These displays offer real-time information to travelers regarding the location of a bus. Buses are monitored with AVL systems, which determine the approximate locations of the buses along a route. The location information is transferred to the control center where the approximate time of arrival of any given bus to a particular bus stop is calculated and reported back to the traveler at the bus stop display.

12.3.5.3.3 On-Board Information This information is also generated through the use of AVL systems. Video screens on the buses themselves provide travelers with location information such as the bus's destination and where the bus is located along the route with respect to the next bus stop and up-and-coming connections. The on-board information is invaluable to passengers accessing buses in networks that change routes frequently throughout the day.

12.3.5.3.4 At-Home/Office Information This information is made available over the telephone, radio, television, or Internet. The purpose of this system is to provide pretrip information such as best route, fares, schedules, and connections prior to a traveler entering the transit system. In some cases, when information centers are connected to control centers of transportation companies, real-time information as to the location of transit vehicles is also made available through the information system.

12.3.5.3.5 Enquiry Office Terminal This type of terminal is used by an information provider at a transportation company for the purpose of providing travelers with the information they requested (i.e., best route, fares).

12.3.5.3.6 Hand-Held Terminal The hand-held terminals fall within one of two types of mobile devises: communicating or noncommunicating. The communicating hand-held terminals provide either one-way or two-way communication between the user and the service center. Both one- and two-way hand-held terminals provide real-time travel information, such as the arrival times for buses, but some two-way hand-held terminals allow the user to access services from the service center. One-way hand-held terminals work much like pagers, where the user can only receive a message. Noncommunicating hand-held terminals only provide the user with prerecorded information.

These examples illustrate some of the functions and content information provided on some Web sites that provide transportation information. Just looking at them, however, may not show how they were developed. The following sections will provide a transit trip-planning example to illustrate the development process and related issues.

12.4   ARCHITECTURE OF DISTRIBUTED TRANSIT INFORMATION SYSTEM

The purpose of designing an Internet transit trip-planning system is to allow users to plan itinerary trips on the Internet based on the travel origins, destinations, times of travel, and transit schedules. When a user specifies travel origin, destination, and time of travel, the system can automatically find the optimal travel route, including information about when and where to get on a specific bus, transfer points (if any), and the time and place to get off the bus. The system is map based. That is, the search and the output of the system are a map. In addition, the user can render the map by browsing through what is available around a particular landmark or a bus stop, for example.

The aim of the Internet trip-planning system is to provide service at levels C2 and possibly D2, as indicated in Table 12.1. The system uses the three-tier architecture (Figure 12.11) that is composed of the Web browser (client tier), Web server (server tier), and one or more server applications (application tier). The Web browser is a user interface used to gather user input. The Web server acts as middleware to handle users' requests and transfer the requests to an application server. The server application is used to process user requests. It is composed of three components: a map server, a network analysis server, and a database server. The map server is designed for map rendering and spatial analysis, the network analysis server is used to provide network analysis functions, and the database server is used to handle data management via database management systems.

The architecture shown in Figure 12.11 is a server-based information system. That is, users make queries at the Web browser but the process is con-

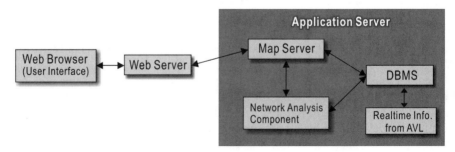

Figure 12.11   Three-tier architecture for Internet transit information systems. [Reprinted from Peng and Huang (2000, Fig. 1) with permission from Elsevier Science.]

ducted at the server applications. User queries from a Web browser are transferred to the Web server, which sends the user's request to the map server. Based on the user's request, the map server either processes the query itself or sends the task to a network analysis component and/or a database management system for processing. The output is then delivered to the Web server and ultimately to the user at the Web browser.*

12.4.1 System Components and Implementation

Although this three-tier architecture can be used in all Internet GIS programs in general and on-line transit information systems in particular, different implementations may result in totally different systems. This study implemented the system inside MapObjects by ESRI using Visual Basic. We used MapObjects to display maps, acquire input (e.g., trip origins and destinations, time of travel), and display graphic output to take advantage of its capability of feature rendering and map manipulation. The network analysis component is written in Visual Basic based on a modified search tree algorithm (Huang and Peng, 2002b). The MapObjects IMS is used to serve the application on the Internet.

12.4.2 User Interface Design

The system implementation process is shown in Figure 12.12. It starts with user input on trip origin, destination, and travel date and time. An interface needs to be developed where users interact with the application. Flexibility and ease of use are critical for this application (Howard and MacEachren, 1996). The first user interface is very simple, as shown in Figure 12.13. It offers the user several ways to enter travel origin, destination, and travel time. He or she can enter the exact street address or select from a pull-down list of landmarks. When the result is presented, the user will have the option of seeing the map or travel directions in text (Figures 12.14 and 12.15).

To enhance the interactivity between the user and the map, Web client-side applications such as plug-ins, ActiveX controls, and Java applets could be developed. But these client-side applications are perceived to be too technical for transit users by the transit agency for which this system was designed. Therefore, a server-side approach is used to build the interface to interact with users. This is the thin-client approach—the simplest, yet the most user friendly. It has no limitation on a user's computer platform and local resources. Basic map-rendering functions such as zoom, query, and search are provided in HTML form. Users are able to select features directly from a map image. They are not able to draw a box or a circle directly from

*Part of the materials presented here is from Peng and Huang (2000) and Huang and Peng (2002a,b).

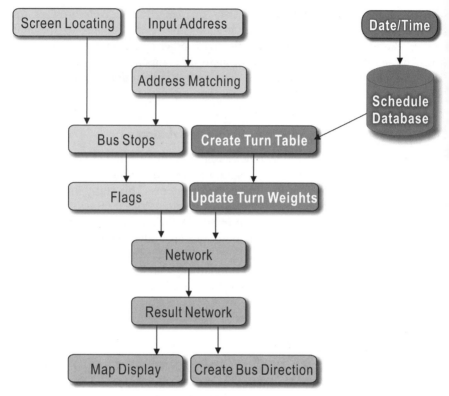

Figure 12.12 System implementation process. [Reprinted from Peng and Huang (2000, Fig. 2) with permission from Elsevier Science.]

a map image because of the limitations of HTML, but XML can be used for further improvement.

12.4.3 Map Server Function Design

The system is intended to offer users interactivity with the map by allowing users to browse service and other information directly on the map. Therefore, the system has to be able to offer map-rendering and address-matching capabilities. This is handled by the map server (e.g., MapObjects and MapObjects IMS by ESRI).

Trip origins and destinations are not necessarily on the transit network. Consequently, the system needs to find all bus stops that are within walking distance (a quarter mile, or 0.4 kilometer) of the user's trip origin and destination. The reason that all stops within walking distance are searched rather than just the one that is the closest is that the closest transit stop may not be on the shortest route path (Peng, 1997). Sometimes a little longer walking

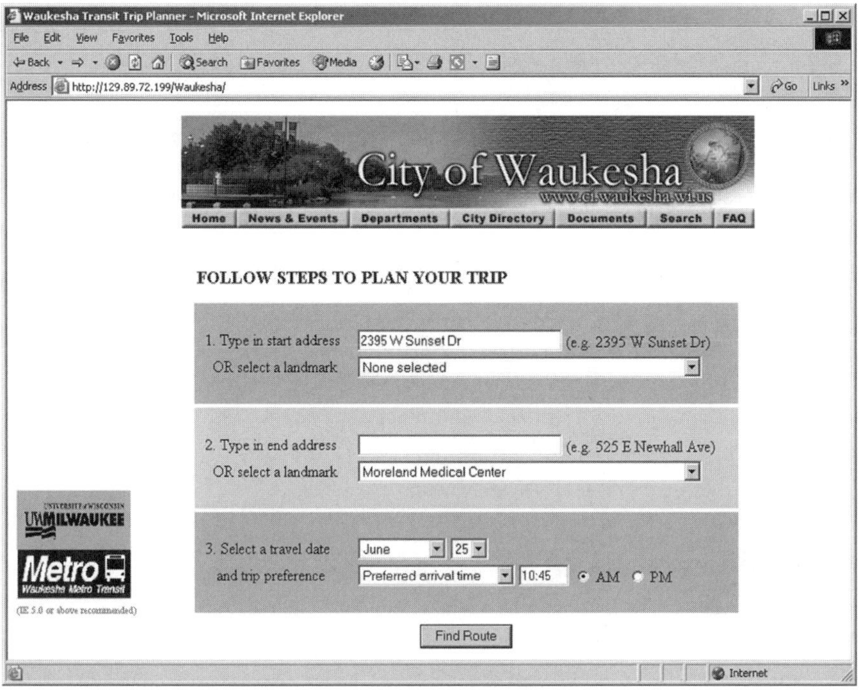

Figure 12.13   User input interface.

time may result in a shorter overall travel time. If there is no stop within walking distance of trip origin and destination, a longer walking distance should be used. After the stops from the origin and destination are found, they are flagged for network analysis.

12.4.4   Data and Database Management Systems

Four data files were used in developing the application: bus route network, street network, bus stops, and time points with schedule data. These data files are stored in a relational database system using Access by Microsoft. Real-time GPS location data were not available at the time of project development, but the application was developed as an open system to incorporate real-time information when it becomes available in the near future. The bus route file is derived from the street centerline file with the addition of some attribute data, such as street length, speed limit, and travel time. A line feature street map and a point feature bus stop map were used as background layers. The street map was also used as a base map for address matching of trip origins and destinations. The bus stop map was used for defining start and end stops of a trip. The bus schedule database is separately stored from the spatial data for easy update and management.

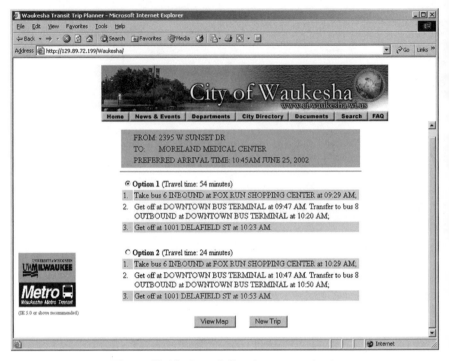

Figure 12.14 Travel direction output (text).

12.4.5 Network Analysis Component

A network analysis model is the key component to provide trip itinerary planning. However, most path-finding algorithms and programs are designed for highway usage (Dial, 1971; Ikeda et al., 1994; Martin, 1963; Moor, 1959; Zhan and Noon, 1998). Although existing network analysis and path-finding algorithms serve well for highway routing and traffic assignment, problems arise when they are applied to transit, because transit networks have significantly different characteristics from highway networks (Peng, 1997; Spear, 1994). Many researchers have pointed out the inadequacy of applying the path-finding algorithms of highway networks to solve the minimal path-finding problems for transit networks (Chriqui and Robillard, 1975; De Cea and Fernandez, 1989; Last and Leak, 1976; Le Clercq, 1972; Spiess and Florian, 1989; Tong and Richardson, 1984; Wong and Tong, 1998). First, because transit service is time dependent, different times of the day or different days of the week have different levels of transit service. Some services are available only at the peak time period. Second, one street segment may serve different bus routes and many routes may stop at the same bus stop. This is the so-called common-bus-lines problem (Chriqui and Robillard, 1975). Third, unlike the highway routing problem, where the computation of

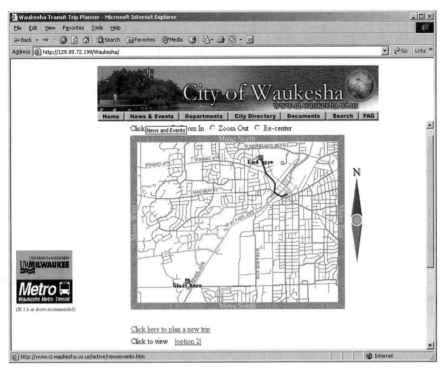

Figure 12.15 Travel direction output (map and text).

shortest path is symmetric with respect to an origin–destination pair, the routing on transit networks from origins to destinations is not symmetric with that from destinations to origins. Fourth, transit transfers depend on the arrival time of another bus. Hence, the best path between an origin and destination can change depending upon the timing of services available. Furthermore, many routes have loop routes and layover time. These unique characteristics make the minimal path-finding application for transit networks much more challenging.

In the case of highway network analysis, one street segment and one intersection have one unique value of travel time and turn weight (or turn penalty). However, in a transit system, one street segment may have several bus routes; each has its own headway. Some are regular buses and some are express buses. It is even more difficult to determine the turn weight (wait time) at each intersection in a transit system. At the same intersection, different buses may have different turn weights. Take intersection A in Figure 12.16 as an example. For riders taking bus B-3 north, the left-turn-weight time is very small because the rider does not need to transfer. But for the same turn, if the rider has to transfer from B-3 to B-2 west, the turn weight could be very large because the rider has to transfer. However, conventional

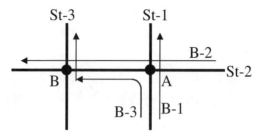

Figure 12.16 Illustration of turn weight determination. [Reprinted from Peng and Huang (2000, Fig. 4) with permission from Elsevier Science.]

path-finding programs based on the street network require a single turn weight for each turn at every intersection. This makes the actual turn weight extremely difficult to determine.

Adding more complexity to the situation, the turn weight changes over time because the bus headway changes; some buses even stop serving at certain times of the day. This problem is similar to constructing a shortest path based on the congestion level on roadways, sometimes referred to as Time-Dependent Constrained Shortest Path (TCSP) (Frank et al., 2000). Even when the turn weight is calculated for every second, it is still difficult to determine which time period to use. If the time of trip origin is used, when the bus gets to that transfer point, the transfer route may not be in service. When the expected arrival time to that intersection is used, how do you know the expected arrival time if you have not determined the path? Because of the complexity of the transit route network, conventional highway network topology and analysis methods are difficult to apply in transit networks.

Therefore, Huang and Peng (2002b) developed their own path-finding algorithms based on the bus service schedules and an object-oriented transit network. The following two sections describe the path-finding algorithms and the object-oriented dynamic transit networks.

12.5 DEVELOPMENT OF TRANSIT PATH-FINDING ALGORITHMS

Key Concepts

Headway-based path-finding algorithms *are based on the shortest-path algorithm which minimizes total travel time by taking into account the in-vehicle time and transfer time. Constant (average) headway during a time period on a segment is usually assumed, and passengers are assigned to the first arriving vehicle based on the combined frequencies of common bus lines.* **Schedule-based algorithms** *determine the time-dependent least-cost paths between origins and destinations in the tran-*

sit network. They identify the unique optimal path given a predefined schedule and trip date and time and are termed as deterministic algorithms by some authors.

Several transit network models were proposed in the literature. Most of these prior models for the shortest-path finding in transit networks can be categorized into two groups: headway based and schedule based. The headway-based path-finding algorithms assign passengers to the first arriving vehicle based on the combined frequencies of common bus lines (Spiess and Florian, 1989). Constant (average) headway during a time period on a segment is usually assumed. The shortest path-finding algorithms are usually variants of traffic assignment procedures used for highway networks that are modified to reflect the waiting time inherent to transit networks (Chriqui and Robillard, 1975; Dial, 1967; Last and Leak, 1976). Some schedule-based transit network models used a branch-and-bound algorithm to determine the time-dependent least-cost paths between all origin and destination pairs in the transit network. Because the schedule-based algorithms identify the unique optimal path given a predefined schedule and trip date and time, they are termed as deterministic algorithms by some authors (Tong and Richardson, 1984; Wong and Tong, 1998). On the other hand, the path assignment for the headway-based approach is stochastic and heuristic in nature. It does not identify transit routes for given trips. Instead, headway-based algorithms identify spatial patterns of the optimal paths. Transit routes can be assigned based on schedule and the trip data/time.

12.5.1 Schedule-Based Path-Finding Algorithm for Transit Networks

The major problem of the headway-based algorithm is that schedule coordination is not fully supported because the path is derived from a probability method. Schedule-based path-finding algorithms, on the other hand, are strictly dependent on service schedules. Tong developed a schedule-based transit model for path finding in 1986 (Wong and Tong, 1998) by using the branch-and-bound techniques. Friedrich et al. (2001) developed a timetable-based algorithm for the many-to-many transit assignment problem. Neither of these algorithms were used in the Internet transit trip-planning systems due to concerns about performance. So Huang and Peng (2002a,b) developed their own path-finding alogorithm based on service schedules and object-oriented transit networks data model.

Huang and Peng (2002b) present two newly developed schedule-based quickest-path-finding algorithms and one non-schedule-based minimal transfer path algorithm that are specific for on-line trip-planning applications. Facilitated by a particularly designed network structure and an object-oriented data model based on GIS technology (Huang and Peng, 2002a), the algorithms have been shown to be very efficient in path finding when they are implemented in an Internet transit trip-planning system.

Internet GIS Showcase: Transportation Improvement Program Information System

This Internet GIS site, developed by North Central Texas Council of Governments, is designed to provide information about Transportation Improvement Program (TIP) projects in the Dallas–Fort Worth metropolitan planning area. Currently, the system includes projects selected or programmed by the Regional Transportation Council. Information on Texas DOT-selected projects will be available at a later date.

The Web site provides two client interfaces: One is a text query using parameters that the client inputs such as city and project type, and the other uses an interactive map to identify projects in the area of interest.

This site is useful to residents, travelers, planners, and anyone else interested in planned improvements to the transportation infrastructure. Users can search for projects by location, type, or agency and can determine the cost and scheduled year of implementation for each project. The map interface, though currently somewhat rudimentary, is an adequate indicator of the types of projects planned in a given area. This application uses MapObjects IMS of ESRI.

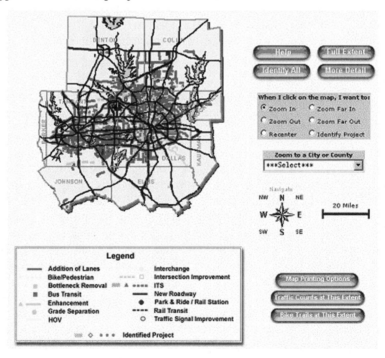

Image reprinted with permission from North Central Texas Council of Government.

Source: http://www.dfwinfo.com/trans/tipins/index.html.

Assuming a transit network G is composed of transfer nodes N and bus routes (traversals) R, we have

$$G = (N, R) \quad \text{where} \quad N = \{N_1, N_2, \ldots, N_i, \ldots, N_m\}$$

We define $N_i = (t_i, \text{wt}_i, s_i, ra_i, na_i)$, where

t_i is the arrival time, a date type variable;

wt_i is the waiting time, length of time (in minutes);

s_i is search status ($0 =$ unsearched, $1 =$ root, $2 =$ branch, $3 =$ searched node);

ra_i is arrival traversal; and

na_i is the arrival node.

For all traversal segments traveling from node i to node j, we define $R_{ij} = \{r_{ij}\}$ and $R_{ij} \in R$, where r_{ij} is a traversal segment from i to j. Forward search starts from the origin node toward the destination node with a planned departure time. We further assume that the planned departure time at the origin is t_0. All network nodes are initialized before path search:

Origin node $\qquad N_o := (t_o, 0, 1, -1, -1)$

Any node i $\qquad N_i := (\infty, \text{wt}_i, -1, -1, -1)$

Initial root nodes $\quad N_{\text{rt}} := \{N_o\}$

Once nodes are initialized, network search can be started:

```
Loop if N_rt ≠ ∅
For each root node N_i ∈ N_rt
 Find branch nodes N_br := {N_j} such that R_ij ∈ R
 Set                 N_br := N_br\{N_i}
 For each branch node N_j ∈ N_br
  t*_j :=   min   {t_j,   min   {t_j^{r_ij}}}
          j:N_j∈N_br      r_ij:R_ij∈R

     with t_j^{r_ij} > t_i + wt_i in which wt_i = 0 if ra_j = ra_i, wt_i >
     0 otherwise
     Update s_j := 2, na_j := i, ra_j := r_ij, r_ij ∈ R_ij if t_j*
     < t_j
 End for each branch node
 Set N_rt := {N_j} for all s_j = 2
 Set s_j := 1 for N_j ∈ N_rt
End for each root node
End loop
```

As an example, suppose we have a transit network that is made up of four bus routes (traversals) and four nodes (Figure 12.17). Table 12.2 is the schedule of this transit system. We further assume that if a transfer is needed at any node, a 2-minute walk time (or cushion time) must be taken into account. Now, a passenger plans to travel from node 1 to node 4 with a planned departure time of 8:10 at node 1. A forward-search algorithm initializes nodes as $N_1 = (8:10, 0, 1, -1, -1)$, $N_2 = N_3 = N_4 = (\infty, 2, -1, -1, -1)$. We set $N_{rt} = N_1 = \{1\}$. Since $N_{rt} \neq \emptyset$, for each root node (only node 1 now) in N_{rt}, find its branch nodes $N_{br} = \{2, 3\}$. For node 2, the earliest possible arrival time is the minimum of existing arrival time (∞) and all possible arrival times from node 1, which are subject to the constraint $t_j^{r_{ij}} > t_i + wt_i$, which means that they should be no earlier than the arrival time at node 1 (8:10) plus a walk time if applicable (in this case, no transfer is needed; walk time is 0). Therefore arrival time at node 2 is 8:25. Node 2 is then updated as $N_2 = (8:25, 2, 2, 3, 1)$. Similarly, branch node 3 is updated as $N_3 = (8:31, 2, 2, 1, 1)$. Since all branch nodes are searched, they are turned to root nodes. Therefore, $N_2 = (8:25, 2, 1, 3, 1)$, $N_3 = (8:31, 2, 1, 1, 1)$, and $N_{rt} = \{2, 3\}$.

The next iteration starts from root node 2. Its branch node 4 is updated as $N_4 = (8:43, 2, 2, 4, 2)$. Note that there is a transfer at node 2; therefore a 2-minute walk time must be considered at this root node. The minimum among all arrivals from node 2 that satisfies the constraint $t_j^{r_{ij}} > t_i + wt_i$ is 8:43 (8:43 >8:25+0:02). After all branch nodes of root node 2 are evaluated, root node 3 is to be searched. Node 4 is the only branch node of root node 3, and it already has an arrival time 8:43. The existing arrival values survived in the minimization because the earliest possible arrival time from node 3 via route 2 is 8:45. Therefore, values of node 4 remain unchanged, that is, $N_4 = (8:43, 2, 2, 4, 2)$. Since all branch nodes (only node 4 this time) in this iteration are searched, we turn them (it) into root node(s); therefore, $N_4 = (8:43, 2, 1, 4, 2)$.

The third iteration of search starts from root node 4 and results in no branch nodes. Therefore, no more root nodes ($N_{rt} = \emptyset$) are identified in the following search process, and the search is completed. The optimal path can be easily retrieved from a back tracking to the search tree.

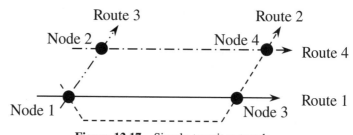

Figure 12.17   Simple transit network.

TABLE 12.2 Schedule of a Simple Transit System

Route 1		Route 2			Route 3		Route 4	
Node 1	Node 3	Node 1	Node 3	Node 4	Node 1	Node 2	Node 2	Node 4
8:01	8:21	8:00	8:23	8:30	8:05	8:15	8:00	8:15
8:11	8:31	8:15	8:38	8:45	8:15	8:25	8:14	8:29
8:21	8:41	8:30	8:53	9:00	8:25	8:35	8:28	8:43
8:31	8:51	8:45	9:08	9:15	8:35	8:45	8:42	8:57
8:41	9:01	9:00	9:23	9:30	8:45	8:55	8:56	9:11

The backward-search algorithm is very similar to the forward-search algorithm except that the search direction is from destination to origin and an expected arrival time is specified at the destination. Besides, at the initialization stage:

$$\text{Destination node:} \quad N_d := (t_d, 0, 1, -1, -1)$$
$$\text{Node } i \quad\quad\quad\quad N_i := (-\infty, \text{wt}_i, -1, -1, -1)$$
$$\text{Set root nodes} \quad N_{rt} := \{N_d\}$$

While evaluating the branch nodes in the search process, expected arrival time at an intermediate node is maximized, that is,

$$t_j^* := \max_{j:N_j \in N_{br}} \{t_j, \max_{r_{ji}:R_{ji} \in R} \{t_j^{r_{ji}}\}\}$$

with $t_j^{r_{ji}} + \text{wt}_i < t_i$ in which $\text{wt}_i = 0$ if $ra_j = ra_i$, $\text{wt}_j > 0$ otherwise. Update $s_j := 2$, $na_j := i$, $ra_j := r_{ji}$, $r_{ji} \in R_{ji}$ if $t_j^* > t_j$.

12.6 OBJECT-ORIENTED DYNAMIC TRANSIT NETWORK MODEL

Key Concepts

In the view of object-oriented data modeling, a **transit network (NET)** *is an object which consists of a set of subclass objects: a* **transfer node** *is a point object where transfer between transit routes is allowed; a* **traversal** *is a unique spatial pattern of a transit route, composed of a string of transfer nodes; and a* **time map** *controls the creation and destruction of space–time objects in the dynamic network topology.*

The transit network is a highly complex and dynamic system—to ensure a high performance in a trip-planning service on the Internet, the above algorithms depend on an efficient GIS data model. Figure 12.18 is a general

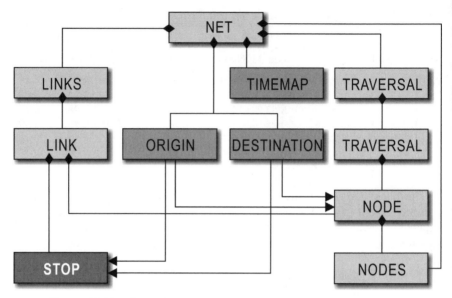

Figure 12.18   General object-oriented data model for transit network.

view of an object-oriented data model for the transit network, in which the NET object represents the entire transit network. The NET object consists of an ORIGIN and a DESTINATION object as well as a TIMEMAP object. It also includes a set of vector objects, including LINKS, TRAVERSALS and NODES, each of which is a vector that contains its corresponding objects. A NET object is defined by the net class CNET ("C" indicates class) which has a set of public properties and methods (interface) (Figure 12.19a): The constructor Net() initializes a transit network object; the SetOrigin() and

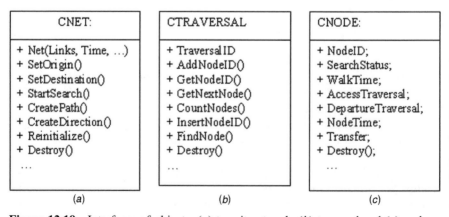

Figure 12.19   Interfaces of objects: (a) transit network, (b) traversal and (c) node.

SetDestination() methods are used to set origin and destination nodes; the StartSearch() method starts a network search session and the Destroy() method destroys the object when the task is finished.

The LINKS vector contains structural information about the network. A link object in the LINKS vector is composed of a string of bus stops and a series of constant transfer nodes. The LINKS vector is created when the application is started, and it persists until the application ends. It is passed to the NET object as a parameter when a new NET is being constructed.

A traversal object in the TRAVERSALS vector is composed of a string of transfer nodes (denoted as NODE in the diagram). It provides a set of properties and methods that can be used to set and query values in the dynamic topology construction and route search process (Figure 12.19b). For example, method AddNodeID() is used to add nodes to the traversal sequentially in creating a new traversal. Method InsertNodeID() is used to insert a node at a specific position in a traversal. It is particularly useful in creating origin and destination nodes dynamically. Moreover, method GetNodeID() gets the ID of a node at a specific position in the traversal; method GetNextNodeID() gets the next node of a specific node; method FindNode() gets the position of a specific node in the traversal, and method CountNodes() gets the number of nodes in the traversal.

As we have discussed, not all traversals are in service all the time. Active traversal objects are created in the construction of the NET object based on time of the trip. The TIMEMAP object in the NET is responsible for controlling the creation of active traversals. The TIMEMAP object looks at the date and time of the trip specified by the customer, checks against the calendar and schedule, and finally decides which traversals should be built. For example, the Waukesha Metro Transit system has 202 traversals; however, the time finds that only 32 of them should be built in a dynamic network topology at the peak hour (3:43 p.m.) of a weekday.

Similarly, a NODE object in the NODES vector is dynamically created by the TIMEMAP object. A NODE object maintains network search information; therefore, it has more public properties and methods than the NET object. Figure 12.19c illustrates some of the public properties and methods of the node class CNODE ("C" indicates class). Property SearchStatus of CNODE maintains the status of a node in the search process ("not searched," "is root node," "is branch node," "searched branch node," and "closed node"). WalkTime property sets and gets the walk time for transfers between routes at the node. AccessTraversal property records the accessing traversal and DepartureTraversal property records the departing traversal in the search process. NodeTime property maintains arrival or departure time depending on the search method. Twenty public properties and method have been designed for the CNODE class.

The object-oriented data model and schedule-based path-finding algorithms have been successfully implemented to the Internet trip-planning system for Waukesha Metro Transit, Wisconsin (http://129.89.72.199/Waukesha). The system performance of the whole trip planner system is quite good, even using a simple PC with a Pentium III 800-MHz CPU, 256 MB RAM, and a 20-GB IDE hard drive. The average computation time for an individual forward or backward search is 1.34 seconds if transfers are involved in the trip, or 0.60–0.65 second otherwise. Average computation time for a minimal transfer path search is 0.62–0.74 second. The above computation times include a 0.47-second network initialization time. Typically, if two trip options are generated and evaluated in an Internet trip plan, the average computation time is 2.21 seconds. The system performance on the Internet was tested with six different origin–destination pairs and travel time. Six requests are issued simultaneously in the test. The average response time for each request is 2.75 seconds. With a faster server, the response time could be substantially reduced.

The system design allows for displaying real-time bus location once the real-time AVL data are available. The bus locations on the map can be updated in a predefined time interval such as every 30 seconds or every minute. An example is shown in Figure 12.20 to display real-time bus locations and bus movement animation using GPS data. One important improvement to be made in the future would be using real-time bus location based on the AVL data in

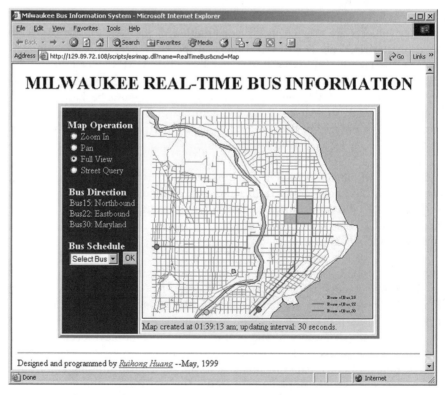

Figure 12.20   Example of displaying real-time bus locations.

the path-finding process, raising the level of service to D2 and D3. Since AVLs have been installed on buses in many transit systems, serving real-time data is a matter of technical and institutional coordination. The current on-line transit information system is designed flexibly enough to accommodate future real-time GPS data. Another improvement is to disseminate information through wireless devices using Mobile GIS.

12.7 CONCLUSIONS

Some transportation-related Internet GIS applications presented in this chapter demonstrated that Internet GIS and mobile GIS could play an important role in the development of intelligent transportation systems, particularly in displaying and disseminating transportation-related information. Transportation information is inherently spatial and time dependent. Internet GIS has a unique feature to handle spatial data and to process real-time information. It can also disseminate geospatial information in a very visual way to users. Furthermore, with the development of mobile GIS, distributing transportation-related information to the mobile travelers has a great potential to become the killer application of Internet GIS. Many automobile companies and the telecommunication industry already focus on the deployment of informatics and the intelligent use of real-time digital information inside the vehicles. In the near future, intelligent transportation systems will be packaged inside mobile GIS devices via wireless communication and provide integrated, comprehensive location-based services for the public and traditional GIS users.

WEB RESOURCES

Description	URI
U.K. Public Transport Information Web site	http://www.ul.ie/~infopolis/

REFERENCES

Batty, M. (1999). New Technology and GIS. In P. A. Longley M. F. Goodchild D. J. Maguire, and D. W. Rhind (Eds.), *Geographic Information Systems,* Chichester, U.K.: Wiley, pp. 309–316.

Casey, R. F., Labell, L. N., Prensky, S. P., and Schweiger, C. L., (1998). *Advanced Public Transportation Systems: The State of the Art Update '98.* Washington, DC: Federal Transit Administration.

Chriqui, C., and Robillard, P. (1975). Common Bus Lines. *Transportation Science, 9,* pp. 115–121.

Coleman, D. J. (1999). Geographic Information Systems in Networked Environments. In P. A. Longley, M. F. Goodchild, D. J. Maguire, and D. W. Rhind (Eds.), *Information Systems*. Chichester, U.K.: Wiley, pp. 317–329.

Coleman, D. J., and McLaughlin, J. D., (1997). Information Access and Usage in a Spatial Information Marketplace. *Journal of Urban and Regional Information Systems*, 9(1), pp. 8–19.

Conquest, J., and Speer, E. (1996). Disseminating ARC/INFO Dataset Documentation in a Distributed Computing Environment. In *Proceedings 1996 ESRI User Conference*. Redlands, California: ESRI. URI: http://www.esri.com/resources/userconf/proc96/TO200/PAP166/P165.m.

De Cea, J., Bunster, J. P., Zubieta, L., and Florian, M. (1988). Optimal Strategies and Optimal Routes in Public Transit Assignment Models: An Empirical Comparison. *Traffic Engineering and Control*, 29, pp. 520–526.

De Cea, J., and Fernandez, J. E. (1989). Transit Assignment to Minimal Routes: An Efficient New Algorithm. *Traffic Engineering and Control*, 30, pp. 491–494.

Dial, R. B. (1967). Transit Pathfinding Algorithm. *Highway Research Records*, 205, pp. 67–85.

Dial, R. B. (1971). A Probabilistic Multipath Assignment Model Which Obviates Path Enumeration. *Transportation Research*, 5, pp. 83–111.

Dijkstra, E. W. (1959). A Note on Two Problems in Connection with Graphs. *Numerische Mathematik*, 1, pp. 269–271.

Doyle, S., Dodge, M., and Smith, A. (1998). The Potential of Web-Based Mapping and Virtual Reality Technologies for Modeling Urban Environments. *Computers, Environment and Urban Systems*, 22(2), pp. 137–155.

Environmental Systems Research Institute. (1998). *NetEngine: A Programmer's Library for Network Analysis*. Redland, California: Environmental Systems Research Institute.

Frank, W. C., Thill, J.-C., and Batta, R. (2000). Spatial Decision Support System for Hazardous Material Truck Routing. *Transportation Research C: Emerging Technology*, 8, 337–359.

Friedrich, M., Hofsäss, I., and Wekeck, S. (2001). Timetable-Based Transit Assignment Using Branch & Bound. In *Proceedings of the 80th Annual Meeting of Transportation Research Board* (CD-ROM), Washington, DC.

Handler, G. Y., and Zang, I. (1980). A Dual Algorithm for the Constrained Shortest Path Problem. *Networks*, 10, pp. 293–310.

Howard, D., and MacEachren, A. M. (1996). Interface Design for Geographic Visualization: Tools for Representing Reliability. *Cartography and Geographic Information Systems*, 23(2), pp. 59–77.

Huang, R., and Peng, Z. R. (2002a). An Object-Oriented GIS Data Model for Transit Trip Planning Systems. *Transportation Research Record*, 1804.

Huang, R., and Peng, Z. R. (2002b), Schedule-Based Path Finding Algorithms for Transit Trip Planning Systems. *Transportation Research Record*, 1783, 142–148.

Ikeda, T., Hsu, M. Y., Imai, H., Nishimura, S., Shimoura, H., Hashimoto, T., Temmoku, K., and Mitoh, K. (1994). A Fast Algorithm for Finding Better Routes by AI Search Techniques. In *IEEE Vehicle Navigation & Information Systems Conference Proceedings B3-6*, pp. 291–296.

Kirby, R., and Potts, R. B. (1969). The Minimal Route Problem for Networks with Turn Penalties and Prohibition. *Transportation Research* 3(3), pp. 397–408.

Last, A., and Leak, S. E. (1976). Transept: A Bus Model. *Traffic Engineering and Control,* 17, pp. 14–20.

Le Clercq, F. (1972). A Public Transport Assignment Method. *Traffic Engineering and Control,* 13, pp. 91–96.

Moor, E. F. (1959). The Shortest Path Through a Maze. In *Proceedings of the International Symposium on the Theory of Switching,* Harvard University, Cambridge, Massachusetts, pp. 285–292.

Martin, B. V. (1963). Minimum Path Algorithms for Transportation Planning, Research Report R63-52. Department of Civil Engineering, Massachusetts Institute of Technology, Cambridge, Massachusetts.

Muro-Medrano, P. R., Infante, D., Guillo, J., Zarazaga, J., and Banares, J. A. (1999). A CORBA Infrastructure to Provide Distributed GPS Data in Real Time to GIS Applications. *Computers, Environment and Urban Systems,* 23(4), pp. 271–285.

Peng, Z. R. (1997). A Methodology for Design of GIS-based Automatic Transit Traveler Information Systems. *Computers, Environment and Urban Systems,* 21(5), pp. 359–372.

Peng, Z. R. (1999). An Assessment Framework of the Development Strategies of Internet GIS. *Environment and Planning B: Planning and Design,* 26(1), pp. 117–132.

Peng, Z. R., and Beimborn, E. (1998). Internet GIS: Applications in Transportation. *TR News,* March/April, N. 195, pp. 22–26.

Peng, Z. R., and Huang, R. (2000). Design and Development of Interactive Trip Planning for Web-Based Transit Information Systems. *Transportation Research C: Emerging Technology,* 8, pp. 409–425.

Peng Z. R., and Jan, O. (1999). An Assessment of Means of Transit Information Delivery. *Transportation Research Record,* 1666. pp. 92–100.

Peng, Z. R., and Nebert, D. (1997). An Internet-Based GIS Data Access System. *Journal of Urban and Regional Information Systems,* 9(1), pp. 20–30.

Plewe, B. (1997). *GIS Online: Information Retrieval, Mapping, and the Internet.* Santa Fe, New Mexico: OnWorld.

Ran, B., Chang, B. P., and Chen, J. (1999). Architecture Development for Web-Based GIS Applications in Transportation. Paper presented at the 78th Transportation Research Board Annual Meeting, January 10–14, 1999, Washington, DC.

Salters, T. (1996). DART on Target. *ITS World,* May/June.

Sarjakoski, T. (1998). Networked GIS for Public Participation—Emphasis on Utilizing Image Data. *Computers, Environment and Urban Systems,* 22(4), pp. 381–392.

Spear, B. D. (1994). GIS and Spatial Data Needs for Urban Transportation Applications. In D. Moyer and T. Ries (Eds.), *Proceedings of the 1994 Geographic Information Systems for Transportation (GIS-T) Symposium,* Norfolk, Virginia: American Association of State Highway and Transportation Officials, pp. 31–41.

Spiess, H., and Florian, M. (1989). Optimal Strategies: A New Assignment Model for Transit Networks. *Transportation Research B,* 23, pp. 83–102.

Tong, C. O., and Richardson, A. J., (1984). A Computer Model for Finding the Time-Dependent Minimum Path in a Transit System with Fixed Schedules. *Journal of Advanced Transportation,* 18(2), pp. 145–161.

Water, N. M. (1999). Transportation GIS: GIS-T. In D. J. Maguire, M. F. Goodchild, and D. W. Rhind. (Eds.), *Geographical Information Systems,* 2nd ed., New York: Wiley, Chapter 59.

Wong, S. C., Tong, and C. O. (1998). Estimation of Time-Dependent Origin-Destination Matrices for Transit Networks. *Transportation Research B,* 32(1), pp. 35–48.

Zhan, F. B., and Noon, C. E. (1998). Shortest Path algorithms: An Evaluation Using Real Road Networks. *Transportation Science,* 32(1), pp, 65–73.

CHAPTER 13

INTERNET GIS APPLICATIONS IN PLANNING AND RESOURCE MANAGEMENT

The evidence is there to be read. The record of cause and effect constitutes the common knowledge of natural scientists. But the status quo ante is being reconstituted without direction or constraint.
—Ian L. McHarg (1969, p. 17)

13.1 INTRODUCTION

The focus of this chapter is on the use of Internet GIS in planning and resource management activities. Because local governments tend to take part in *comprehensive* planning activities, they become involved in environmental planning, transportation planning, economic planning, and a host of other planning activities important to the continued well-being and future development of their respective communities.

Government agencies have historically been the predominant users of GIS. But as private organizations expanded, they also began using GIS to assist in their planning and management activities. Internet GIS has turned into a boon for organizations because it offers faster and more efficient methods of communication and collaboration between departments and facilities of the same organization and between different organizations, not to mention cost efficiency. Because of the opportunity for community outreach, municipalities are increasingly developing Internet GIS sites for public access. For certain functions, though, some public and private organizations utilize their Internet GIS applications within their organizations through intranets that are generally inaccessible to the general public.

As a means of illustrating the utility, versatility, and practicality of Internet GIS in planning and resource management activities, this chapter presents a number of case studies describing current uses of Internet GIS by local governments and professional organizations.

Key Concepts

Government agencies have historically been the predominant users of GIS. Because of the opportunity for community outreach, municipalities are increasingly developing Internet GIS sites for public access.

Because some applications run on private intranets, case studies discussing such applications will not present examples of the applications as snapshots from the organizations' Web sites as do those case studies whose applications are running on the WWW. For clarity under the Web Application heading, each case study will be identified as either an *intranet* and/or an *Internet* application. But note that the term "Internet GIS" can refer to either an intranet or an Internet GIS. For better understanding of current applications of Internet GIS, the case studies are organized by purpose or activity: trip planning, infrastructure planning and management, emergency planning, and community development.

13.2   CASE STUDIES IN INFRASTRUCTURE PLANNING AND MANAGEMENT

13.2.1   Case Study 1: Facilities Planning for New York State Office of Mental Health (OMH)*

BACKGROUND   The OMH manages over 1300 psychiatric facilities and over 2500 inpatient and outpatient programs across the state of New York. Various emergency, community support, residential, and family care programs overseen by OMH are available through both local governments and nonprofit agencies.

PURPOSE OF INTERNET GIS   OMH wanted to implement an interfacility network that would promote better communication and collaboration between the administrators in the main office in Albany and the facilities planners scattered in remote offices across the state. By promoting on-line collaboration and exchange of information, design plans and changes, maintenance requests, and cost estimates could be processed faster and more efficiently, thereby facilitating the best allocation of resources.

Source: http://www.mapinfo.com/community/free/library/nys_health_casestudy.pdf.

DATA SOURCES Because OMH's facilities planning incorporated architectural, site, and construction plans, as well as the associated costs, into its design plans, the proposed facilities planning system needed to be able to access photographs of buildings, floor plans, plot plans, utility costs, and other relevant information. Digital maps of OMH's facilities and their associated utilities were integrated with OMH's master plan for facilities development and other design plans that described the details of the buildings, including how the spaces within the buildings were being used and the standards that regulated their use. The integration of the plans allowed for better visualization of the development process from start to finish.

SYSTEM OPERATIONS The Facilities Information System (FIS) created for OMH's facilities planning and management activities combined Internet GIS technology with the functionality of its Oracle data management, analysis, and retrieval system. By allowing administrators and field personnel to collaborate together on-line, facilities plans could be developed more efficiently and cost effectively. FIS allowed planners to query the databases for applicable data as each project progresses. Then the queries were automatically generated as maps for visualization of the development process. For planners and administrators alike, the ability to produce maps of time-line scenarios helps in making short- and long-range plans for their facilities as well as providing a tool for the management of those facilities.

WEB APPLICATION OMH's facilities planning application runs on a private intranet networked for its facilities in the state of New York. Examples of the application are not available.

13.2.2 Case Study 2: Asset Management and Transportation Network Planning with Road Management Information System (ARMIS) GIS, Queensland, Australia*

BACKGROUND Headquartered in Brisbane, the Main Roads Department (MRD) of Queensland, Australia, manages 21,000 miles of roads from 14 district offices. MRD is responsible for overseeing all transportation, construction, and environmental projects associated with the road network.

PURPOSE OF INTERNET GIS MRD needed an Internet GIS for asset management of its corridors, properties, and projects in a multiuser environment. ARMIS provided pertinent information regarding a road's physical attributes, including the number of lanes, the width of lanes, whether the road was divided or undivided, the width of the shoulder, and the type of surface on

---

Source: Slater, S. (2002). Queensland Creates a Princely GIS. *GeoWorld*, May, pp. 42–45.

the road (i.e., concrete, asphalt, gravel), but it needed to be updated and upgraded to allow for accurate spatial analysis.

DATA SOURCES   The existing ARMIS provided 10 years of information concerning construction projects, road surface types and conditions, speed limits, and maintenance history as well as information concerning the environment, traffic counts, and traffic accidents. New location-based data had to be collected for areas mapped before 1998 because the source data came from cadastral maps that were less than accurate. The new data collection would be done using GPS, ensuring accuracy and compatibility with the developing Internet GIS. MRD also wanted aerial photographs, scanned documents, CAD files, and road videos integrated into the system.

SYSTEM OPERATIONS   The final ARMIS GIS and ARMIS Online products were a combination of ESRI, MapInfo, and Intergraph (GeoMedia) technologies where multiple users could perform "dynamic segmentation analysis" via an intranet. MRD merged aspects of the three technologies because any one technology could not meet all of the department's needs. One need MRD felt was an important tool for users was the ability to visualize ARMIS data. This required the integration of video into the system. GeoMedia provided the technology for real-time video "filmed" from a vehicle traveling down a road to be viewed by any user in the network. As a vehicle moves along a stretch of road, dynamic segmentation software allowed the user to retrieve attribute data associated with any given point along the driven segment.

FUTURE PLANS   MRD eventually wants to be able to share data with other government transportation departments in Queensland.

WEB APPLICATION   ARMIS GIS is accessible via MRD's private intranet, ARMIS Online.

EXAMPLES FROM INTRANET WEB SITE*   The snapshots of some of the ARMIS GIS features that follow were obtained from a paper posted on the Internet at http://www.intergraph.com/gis/community/geospatialworld/proceedings/doc/o56.doc.

Figure 13.1 illustrates the ARMIS GIS feature for querying attributes of roads, including road, traffic, and crash attributes.

Figure 13.2 illustrates how the existing video component of ARMIS is integrated in GeoMedia Professional, thereby allowing users to retrieve attribute data from any point location along a driven route. The user does not receive a stream of video; instead, the user receives the image at the exact time the data were called up. Dynamic segmentation software constantly up-

*Source: McGreevy (2001).

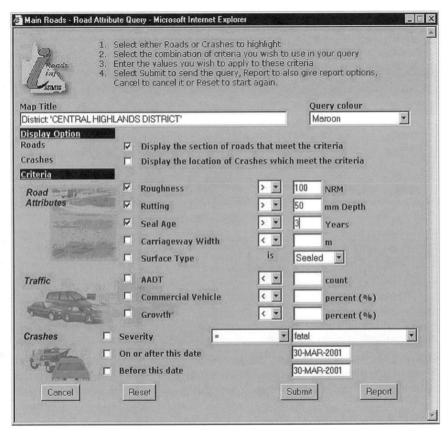

Figure 13.1 ARMIS road attribute query. (*Source:* McGreevy, 2001. Figure reprinted with permission of Queensland Department of Main Roads, Australia.)

dates the application with a point (and its associated attributes) corresponding to the location of the vehicle.

Figure 13.3 illustrates MRD's intranet, ARMIS Online, with an example of the thematic mapping capabilities of ARMIS GIS. The on-line application allows users to query map layers and create new maps of selected features.

13.2.3 Case Study 3: Transportation, Utility, and Strategic Planning with Property Land Use System (PLUS), Queensland Rail, Queensland, Australia*

BACKGROUND   Queensland Rail is the largest rail and freight operator in Australia with over 6600 miles of rail and rail corridors and over 1400 railway

Source: http://www.mapinfo.com/community/free/library/queensland_casestudy.pdf.

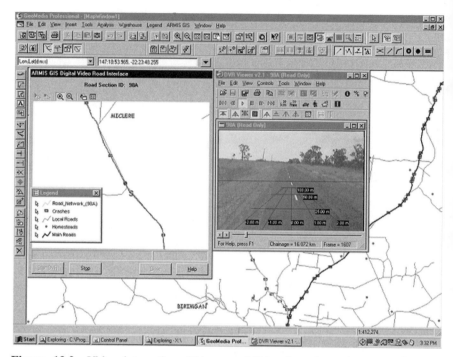

Figure 13.2   Video integration. Video capabilities have been integrated into the ARMIS GIS, allowing users to see a given location (and its attributes) along a driven route in "real time" as the video is updated by dynamic segmentation software. (*Source:* McGreevy, 2001. Figure reprinted with permission of Queensland Department of Main Roads, Australia.)

stations. It also owns and manages associated infrastructure such as utility lines (water, gas, and electric), bridges, and tunnels.

PURPOSE OF INTERNET GIS   Queensland Rail was in need of an interdivisional system that would efficiently manage its numerous land holdings and facilities infrastructure. Its Property Division was assigned the task of creating an Internet GIS that would manage the resources for each of the company's seven divisions (corporate services, technical services, infrastructure services, network access, passenger services, coal and freight services, and workshops).

DATA SOURCES   The database containing the company's rail, property, and corridor holdings was already set in place. It just needed to be integrated into a Web environment.

SYSTEM OPERATIONS   The original PLUS only let users produce detailed maps of property holdings color coded by the function each performed. As Queensland Rail's needs grew, PLUS grew into PLUSnet, which provided for

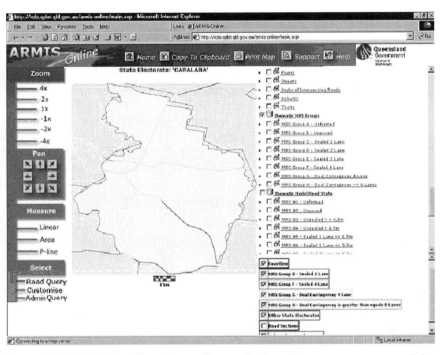

Figure 13.3 ARMIS on-line. The on-line application allows users to query and thematically map selected features. (*Source:* McGreevy, 2001. Figure reprinted with permission of Queensland Department of Main Roads, Australia.)

data management and strategic planning. PLUSnet utilized MapInfo Professional and MapInfo EasiMaps for map creation and MapInfo MapX for viewing the maps in a Web environment. Two of the main functions of PLUSnet are mapping the fastest routes by which to move freight and, through Queensland Rail's Disaster Recovery System, identifying the best and fastest routes by which to respond to accidents.

WEB APPLICATION   Queensland Rail's PLUSnet system operates on the company's private intranet.

13.2.4 Case Study 4: Utilities Planning with Pennsylvania One Call System (POCS)*

BACKGROUND   In order to ensure the safety of the public and the workers at a construction site, most states require that some type of notification system be implemented (e.g., Wisconsin's Digger's Hotline) to inform utilities when

---

Source: http://www.mapinfo.com/community/free/library/pocs_casestudy.pdf.

a dig is about to take place. The notifying contractor would then have to wait three days as a safety precaution. Because Pennsylvania's statewide Automated Notification System (ANS) was slow and inefficient, a number of resident utility companies joined together to form POCS, a nonprofit service company that created a direct communications network between contractors and utility companies.

PURPOSE OF INTERNET GIS   POCS wanted to offer an optional, subscriber-only service that would save its member utility companies time and money by removing the unnecessary processing of dig notices by companies not affected by the dig. An Internet GIS would offer a means of quickly analyzing and communicating an excavation query via the Internet.

DATA SOURCES   The main sources of data came from the mapped boundaries of the underground facilities of each member utility company and from street maps that contained address-level information. Address information was necessary to map the boundaries of the facilities and to geocode the coordinates of the dig sites.

SYSTEM OPERATIONS   POCS created a subscriber service called Saf-Call that utilized the ANS. Essentially, a contractor would call the 1-800 number to notify POCS of a proposed dig, the site data for the dig was entered into the ANS, and then the ANS would generate a list of all of the utility companies within the municipality where the dig was proposed to take place. For non-subscribing utility companies, the ANS would automatically send a notice to each, regardless of who was affected by the dig. But for subscribers, the Saf-Call department at POCS would receive the list, screen out any dig notices from excavators whose activities did not impact the subscribing utility, and then notify only those utility companies affected by the dig(s). The Saf-Call service proved to save subscribing utilities countless hours of processing needless dig notices.

POCS uses MapInfo Professional, Enterprise, and StreetPro Enhanced Address Layer (EAL) to generate maps, query data, and Web enable the Saf-Call service. StreetPro EAL's geocoding capabilities allow POCS to provide its clients with up-to-date street data.

WEB APPLICATION   The Web application is accessible over the Internet, but only to members of POCS's subscriber service, Saf-Call.

13.3 CASE STUDIES IN EMERGENCY PLANNING

13.3.1 Case Study 5: Emergency Planning with Military Software, Utah Olympic Public Safety Command, Salt Lake City*

BACKGROUND   Salt Lake City was honored with playing host to the Winter Olympics in 2002. The event brought 60 law enforcement agencies together

---

Source: Barnes (2002).

to ensure the safety of athletes and spectators alike. The agencies used an emergency evacuation system whose development had begun two-and-one-half years prior to the games.

PURPOSE OF INTERNET GIS   Because of the thousands of people that would be converging on Salt Lake City, Utah, for the Olympic Games, the Salt Lake City wanted to be prepared for any type of emergency—be it natural or man made—that might befall the city. The city commissioned Science Applications International Corp. (SAIC) to create an emergency evacuation plan in the form of an Internet GIS to manage the task. One of SAIC's responsibilities was to track the shuttles that transferred Olympic athletes from one location to another.

DATA SOURCES   The data required for the Internet GIS system included aerial photographs of Salt Lake City, footprints of buildings, elevations of the area, streets, and other visual components that aided security personnel in identifying the locations of the shuttles.

SYSTEM OPERATIONS   In order for the system to respond to emergency situations in the manner required, SAIC integrated two military software applications—Consequence Assessment Toolkit (CATS) and E-Team—with ESRI's ArcIMS.

CATS provided the ability to simulate natural and man-made disasters as well as the ability to generate casualty and damage predictions. Designed for the military to model explosions and natural disasters, CATS was implemented in Salt Lake City to model response simulations of crowds and parade routes. The E-team provided the ability for users to obtain real-time incidence reports identifying the type of incident, the location of the incident, and who was involved.

SUPPORTIVE SYSTEMS   In support of the security efforts within Salt Lake City, the Utah Highway Patrol and Valley Emergency Communications Center together purchased an AVL system from CompassCom. The AVL along with GPS receivers and wireless modems tracked the shuttles that transported the athletes to Olympic-event locations. The shuttles—each equipped with a GPS receiver and a wireless modem—were assigned specific routes along which to travel. These routes were then buffered and programmed into the AVL system so that the shuttles could be monitored for any deviations from their routes. If a shuttle moved out of its buffer zone, an automatic alarm was set off at the command center. The driver, also, could manually notify the command center in the event of an emergency.

PUBLIC SAFETY AFTER THE GAMES   The AVL monitoring system proved to be such a success that the city chose to keep it in place for managing public safety services. Hospitals, police departments, and fire departments set up their own command centers for monitoring and dispatching their new GPS and wireless modem-equipped emergency vehicles. The system now allows public safety agencies to track the current locations of emergency vehicles and to assess which vehicles would best be dispatched to an emergency call.

WEB APPLICATION   The Olympic Public Safety Command operated the system over an intranet for the duration of the Olympic Games. Currently, Salt Lake City operates the system in the same way. The wireless application of the AVL monitoring system also continues to operate via an interorganizational intranet.

13.3.2   Case Study 6: Evacuation Planning with Hurricane Evacuation Decision Support Solution, South Carolina Department of Transportation (SCDOT) and Emergency Preparedness Division (EPD)*

BACKGROUND   To prepare for Hurricane Floyd, the SCDOT and the Emergency Preparedness Division (EPD) of the South Carolina Emergency Management Division (SCEMD) joined forces to develop an evacuation plan for South Carolina residents. Responsible for the development, coordination, and execution of evacuation plans, the EPD together with the SCDOT created an Internet GIS that provided real-time weather and evacuation information over the Internet.

PURPOSE OF INTERNET GIS   Because of the destruction caused by Hurricane Hugo back in 1989, the SCDOT and the EPD wanted to create a system by which to monitor storm systems and initiate evacuation procedures should an emergency situation arise. To accomplish their goals, they looked to an Internet mapping system used by the Pentagon for assessing road, weather, and route data for emergency transport of military personnel and equipment.

DATA SOURCES   The data that the SCDOT initially incorporated into the system included traffic counts taken by 48 permanent automatic counters placed along major highways in the state, an existing Emergency Evacuation Route Map currently part of the SCDOT's existing GIS, maps of road closures and detours from other state agencies, and up-to-the-minute weather data from the NOAA. During an emergency, portable counters would be set up along minor roads from which SCDOT personnel would physically retrieve the counts and call them in by cell phone to their headquarters in Columbia, South Carolina.

SYSTEM OPERATIONS   The automatic traffic counters would call their counts in every hour to a computer that would download the data into an Excel

*Source: http://intergraph.com/gis/customers/articles/scdot.pdf.

program that generated charts and graphs of traffic counts. The system, which was programmed to constantly monitor the NOAA's Web site for real-time weather data, such as storm-tracking maps, satellite photographs, and other images that could be overlaid on the SCDOT's base map, linked the weather data with the traffic counts by time. This allowed the SCDOT to analyze the relationship between weather conditions and traffic. The weather and traffic data, along with maps of evacuation routes, road closures, and detours, provided the basis for the evacuation procedures during Hurricane Floyd. All relevant data were subsequently posted on the Internet using GeoMedia's WebMap software.

LESSONS LEARNED   The SCDOT and the EPD were not prepared for the gridlocks that occurred during the evacuations. While looking back at the incidents during the evacuation procedures, they realized that many of the vehicles on the traffic-jammed freeways were entering South Carolina from Georgia and Florida. They concluded that the lack of coordination of evacuation procedures and the lack of communication between South Carolina and neighboring states contributed to the congestion problems.

FUTURE PLANS   The SCDOT plans to incorporate a strategy to accommodate traffic flow from neighboring states. The strategy will incorporate a means of phasing the evacuations of South Carolina residents with out-of-state residents in order to stay within highway capacity thresholds, thus avoiding future incidences of gridlock.

WEB APPLICATION   The SCDOT and the EPD temporarily operated an Internet GIS on the WWW for the purpose of informing South Carolina residents of evacuation routes and procedures as well as road closures and alternative routes and providing up-to-date weather reports. The system is not accessible to the public all year round.

13.4 CASE STUDIES IN COMMUNITY PLANNING

13.4.1   Case Study 7: Economic Development with Vallejo Economic Development Information System (VEDIS), Vallejo, California*

BACKGROUND   After the loss of 15,000 jobs and $668 million in production when the Mare Island Naval Shipyard shut down in 1996, the city of Vallejo took action to mitigate the effects of the economic loss. By investing its resources in the economic redevelopment of the area, the city chose to be proactive rather than reactive in its response to the change in the economic climate.

Source: http://www.osec.doc.gov/eda/pdf/InnovLDEP.pdf.

PURPOSE OF INTERNET GIS   VEDIS was created as a means of promoting and attracting new businesses. By providing real estate agents, developers, businesses, and residents Internet access to property and other relevant information, development activities could be initiated in a more economic and timely manner. The city hopes to attract a variety of businesses and industries through the use of the Internet-based VEDIS to fill vacant properties, thereby expanding the local tax base and increasing employment.

DATA SOURCES   Data available to users of the Web site include building attributes, demographic data, traffic counts, and business lists. The specifics regarding a redevelopment project may also be acquired through the Web site. The property database is updated directly through the Web site by real estate agents, brokers, and leasing agents as soon as properties enter and exit the market. The real estate professionals add such information as the lease or purchase price of the building, how large the building is in square feet, and how close the building is to major highways. A virtual tour of the immediate area for each property is also available to supplement the written descriptions.

SYSTEM OPERATIONS   The power in the VEDIS application comes with its ability to perform demographic and market analyses. The user can analyze the area around a property and then generate a report of the results. The system uses economic indicators such as household income and consumer expenditures to determine the economic feasibility of establishing a particular type of business at a proposed location.

FUTURE PLANS   The city of Vallejo has plans to improve VEDIS by adding data to its existing databases and by adding a thematic component to the mapping application. The city also plans to obtain feedback concerning the current operations of and potential future improvements to the VEDIS system by surveying real estate professionals on an annual basis.

WEB APPLICATION   Vallejo's VEDIS application was specifically designed to stimulate economic development by providing a resource by which real estate professionals, developers, and the general public could research properties and market conditions. For this reason, the application can be found on the WWW at http://www.ci.vallejo.ca.us/econdev/econdev.htm.

EXAMPLES FROM INTERNET WEB SITE   Figures 13.4, Figure 13.5, and Figure 13.6 represent different components of the VEDIS application. The initial step in the VEDIS process is to select a category of sites (i.e., vacant, industrial) and then identify one property of interest. Figure 13.4 illustrates the map of, in this example, the vacant site and its complete property description. In order to research the feasibility of putting a particular type of business on that site, the user may want to make use of the demographics component to see how people are spending their money in that area (Figure 13.5). The user

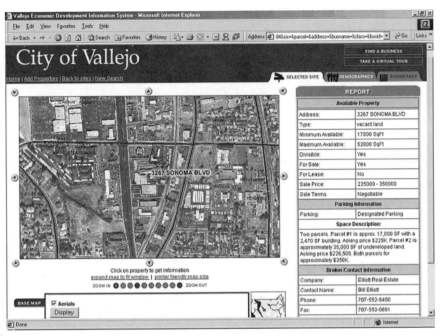

Figure 13.4 Site selection. In this example, a vacant property at 3267 Sonoma Boulevard was selected for detailed analysis. (*Source:* http://www.ci.vallejo.ca.us./ed.asp.)

can then check out the competition in the area with the business component of the application by selecting a distance from the site. A business report as seen in Figure 13.6 will be generated identifying how many establishments under each category of business are within the defined distance from the site. By knowing how people are spending their money and by seeing how many establishments of the proposed business already exist in the area, the user can determine the feasibility of establishing a particular type of business at a particular location.

13.4.2 Case Study 8: Community Development with Enterprise Geographic Information System (EGIS), U.S. Department of Housing and Urban Development (HUD)*

BACKGROUND   HUD was created as part of the Department of Housing and Urban Development Act of 1965, which promotes the philosophy that all Americans deserve a decent home. By assisting with home ownership, providing housing assistance, offering fair and affordable housing, and offering

---

Source: http://hud.esri.com/egis.

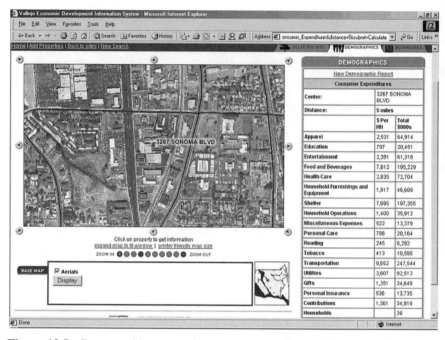

Figure 13.5 Demographic report of consumer expenditures. The report reflects how people within a five-mile radius of the site spend their money. (*Source:* http://www.ci.vallejo.ca.us/ed.asp.)

economic development assistance, HUD serves to improve the environment of disadvantaged persons, neighborhoods, and communities.

PURPOSE OF INTERNET GIS   HUD and ESRI, created EGIS for the purpose of providing governments and the general public with easy Internet access to HUD data and mapping capabilities. EGIS is meant to assist neighborhoods, cities, counties, and states in their housing and economic development programs.

DATA SOURCES   HUD's housing data include information regarding enterprise zones, community projects such as public facilities projects and anti-crime projects, and community programs such as senior and youth programs. Housing data are also represented by type (i.e., Section 8 properties). Other data sources include the EPA, the Federal Emergency Management Agency (FEMA), and the U.S. Census Bureau.

SYSTEM OPERATIONS   GIS was created using ESRI products and integrated into a Web environment with ArcIMS technology.

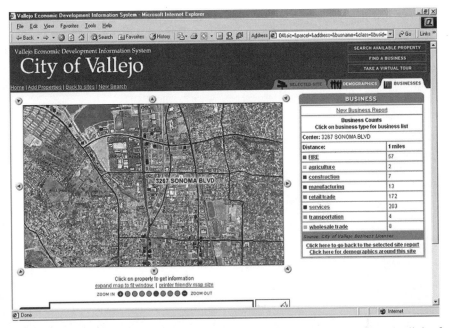

Figure 13.6 Business report of business counts by category. Clicking on the link of each category will bring up a list of all the businesses in the pre-defined range that fall under that category. (*Source:* http://www.ci.vallejo.ca.us/ed.asp.)

WEB APPLICATION   Because EGIS was specifically designed for the WWW, it can be accessed at http://hud.esri.com/egis.

EXAMPLES FROM INTERNET WEB SITE   EGIS is a useful application for determining housing and community development opportunities in a region. Figure 13.7 illustrates the types of housing units and projects within a portion of Milwaukee, Wisconsin, that covers portions of zip codes 53211 (Shorewood/Milwaukee), and 53212 (Milwaukee), and 53217 (Glendale/Whitefish Bay).

Figure 13.8 illustrates the thematic mapping process for a small section of Milwaukee just northwest of the downtown. The map shows the density of Section 8 homes by unit count in green tones with Census tract 004400 showing up as brown. This Census tract represented the tract with the highest number of Section 8 properties that was touched by the buffer zone (the circle in the east and southeast corner of the map).

Figure 13.9 illustrates an example from the Revitalization Area Locator and represents one of two revitalization areas in Wisconsin—zip code 53204 of Milwaukee. For anyone who is familiar with the Milwaukee area, Selected Revitalization Area is located on the near-south of the downtown, which includes the southeastern third of the Menomonee Valley.

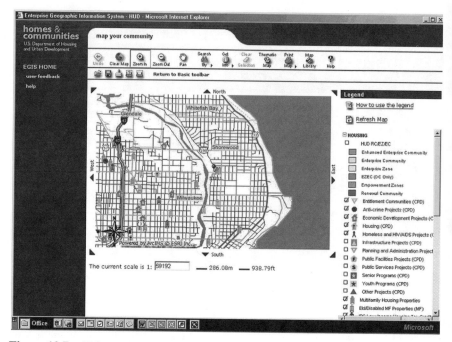

Figure 13.7 "Map your community." The area search is centered on area code 53211 in Milwaukee, WI, as illustrated by the pin on the map, but the area represented shows portions of zip codes 53211, 53212, and 53217. (*Source:* http://hud.esri.com/egis.)

Another example of HUD's location services is the RC/EZ/EC Locator. The Milwaukee, Wisconsin, Renewal Community is illustrated in Figure 13.10 and is comprised of a good part of the central city. If Figure 13.9 and Figure 13.10 are compared, one can see that many near-south side properties fall, expectedly, within both the revitalization and renewal community areas.

13.4.3   Case Study 9: Land Use Planning and Environmental Impact Assessment with Historical Maps, David Rumsey Collection, San Francisco, California*

BACKGROUND   David Rumsey (2002) of the David Rumsey Collection in San Francisco, California, owns the largest collection of historical maps in the United States. His paper collection exceeds 150,000 maps, with the majority representing North and South America from the eighteenth and nineteenth centuries. His collection also consists of rare globes, atlases, and puzzles. The value of the service from a planning perspective comes from the ability to

---

Source: Rumsey (2002).

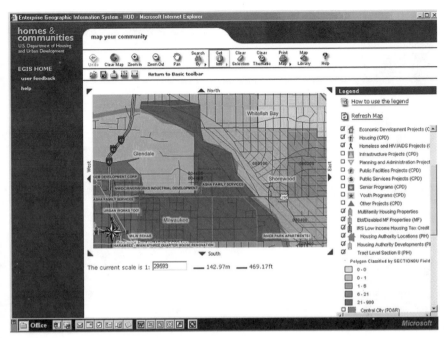

Figure 13.8 Thematic mapping abilities of EGIS. The census tract highlighted in brown represents the tract with the highest number of Section 8 properties that was touched by the circular buffer zone. (*Source:* http://hud.esri.com/egis.)

compare and overlay historical maps with contemporary orthographic images. Planners can reconstruct the past by overlaying maps from different years and taking note of the changes in the landscape. Putting an area in historical context can help identify environmental impacts over time and provide contemporary planners with a medium from which to "learn from the past."

PURPOSE OF INTERNET GIS   Because David Rumsey wanted to share his collection with the public, he chose to put his collection on the Internet where it could reach the largest audience. The Internet site was very successful as a regular site for accessing information, but David Rumsey wanted to bring more functionality to his collection through overlay and interactive functions.

DATA SOURCES   Currently, the David Rumsey Collection has 6500 historical maps for viewing and download, but only the historical maps for San Francisco, California (11 maps falling between the years 1851 and 1926), and Boston, Massachusetts (18 maps falling between the years 1776 and 1897), have been fully integrated into the Internet GIS. These maps can be overlaid on top of city data (e.g., streets), orthophotos, topographic maps, Landsat images, or three-dimensional elevation models (DEMS) from the USGS.

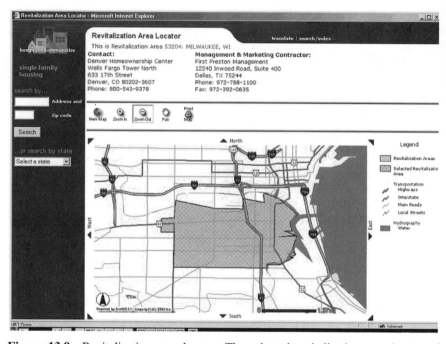

Figure 13.9 Revitalization area locator. The selected revitalization area is part of area code 53204 of Milwaukee, WI. (*Source:* http://hud.esri.com/egis.)

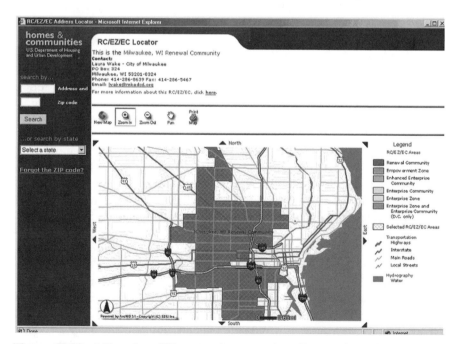

Figure 13.10 Milwaukee, WI, renewal community. (*Source:* http://hud.esri.com/egis.)

DATA CONVERSION   In order to view the historical maps over the Internet, the maps had to be scanned into digital format and then compressed for transmission. Although the process of getting just the right lighting and just the right angle to produce a scanned image with the least amount of distortion was difficult, the process of registering the less-than-accurate historical maps to modern map projections was even more so. Depending on the accuracy of the original map, the number of control points needed for registration ranged anywhere from 20 to 100.

SYSTEM OPERATIONS   The Web site offers two versions of GIS functionality depending on the user's level of expertise: GIS Basic Browser and GIS Professional Browser. The professional version just offers more functions for viewing and manipulating map data. The same data sets are available to both browsers.

One feature that is common to both browsers but is not normally found in GIS operations is the ability to perform photo image enhancements (swipe, blend, change bands, morph horizontal, and morph vertical), such as those in a paint program. Another feature not normally found is the quad-viewer. For comparative purposes, users may choose to view more than one map at a time with the quad-viewer, which, as the name implies, allows viewing of up to four maps at one time.

Other special features of the site include the new addition of a "fly-through" where a historic map can be referenced to a USGS DEM for the chosen area, creating a three-dimensional representation of the area back in, for example, 1812. For those who want to download data for their own personal uses, the map data are accompanied by georeferencing information for easy integration and registration.

FUTURE PLANS   David Rumsey plans to make more geographic areas available for viewing in the GIS browsers. The next series of maps will cover Washington DC; New York City; Los Angeles; Chicago; Denver; and Seattle. He also plans to add historical detail in the form of fire and insurance maps and hand-drawn maps of land parcels and ownership.

WEB APPLICATION   David Rumsey wanted to share his comprehensive collection of historical maps with as many people as he could, so he had his collection integrated into an Internet GIS on the WWW at http://www.davidrumsey.com.

EXAMPLES FROM INTERNET WEB SITE   The versatility in this application can be seen in Figures 13.11–13.13. Figure 13.11 illustrates Boston, Massachusetts, in 1835 overlaid by current road, water, and park features.

Figure 13.12 illustrates Boston as it looks today. The quad-viewer illustrated in Figure 13.13 shows simultaneous viewing of contemporary ortho-

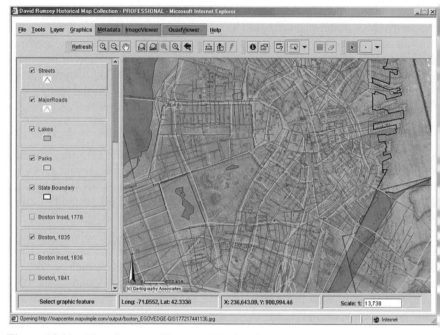

Figure 13.11   Historic map of Boston, MA, in 1835 overlaid by modern roads, water, and parks. (*Source:* http://www.davidrumsey.com.)

photo and hillshade digital elevation maps of Boston and historic maps of Boston in 1776 and in 1874.

Figures 13.14 and Figure 13.15 represent the query feature and the identify feature, respectively, as one might find in any of a number of professional desktop GIS applications.

13.4.4   Case Study 10: Community Development Planning with Smart Permits Using Sunnyvale Geographic Information System (SunGIS™) in Sunnyvale, California*

BACKGROUND   Joint Venture: Silicon Valley (JVSV) manages an interjuris-dictional permitting process over the Internet called Smart Permits. Smart Permits is a regional application networked through three counties for the purpose of simplifying and speeding up the process of applying and getting approval for building and other permits. Implementation of Smart Permits is up to the discretion of the local municipality, which is why communities in the Silicon Valley, California, network are in various stages of implementa-

Source: http://www.gisplanning.com.

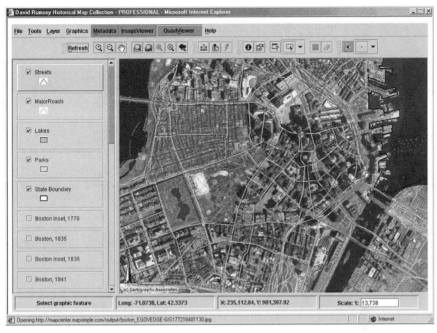

Figure 13.12   Contemporary orthophotograph of Boston, MA. (*Source:* http://www .davidrumsey.com.)

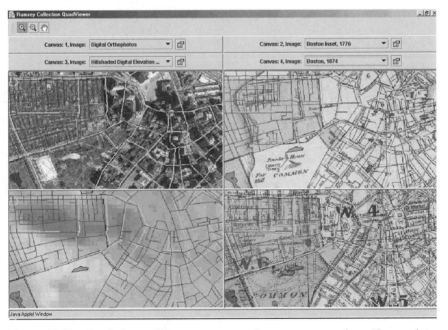

Figure 13.13   Quad viewer. Users can compare four maps at one time. (*Source:* http: //www.davidrumsey.com.)

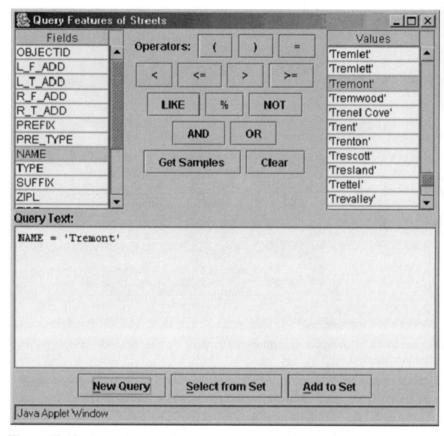

Figure 13.14 Query feature dialog box. This feature closely resembles the query features of professional desktop versions of GIS. (*Source:* http://www .davidrumsey.com.)

tion. Communities currently participating include Santa Clara, Palo Alto, Mountain View, Milipitas, and Sunnyvale. Of the participating communities, Sunnyvale is at the highest level of implementation with its fully functional Internet GIS, SunGIS.

PURPOSE FOR INTERNET GIS   Government leaders within the three-county region have the firm belief that a slow permitting process hinders community growth. Even though communities must implement their own systems, the pilot communities listed above and others have enthusiastically embraced the Smart Permits system because they realize the potential for economic growth.

The regional component of the Smart Permits system allows cities to track what is going on in neighboring cities, thus facilitating regional planning. The Internet GIS that Sunnyvale implemented for participating in the Smart Permits system serves to facilitate the city's community development efforts. The

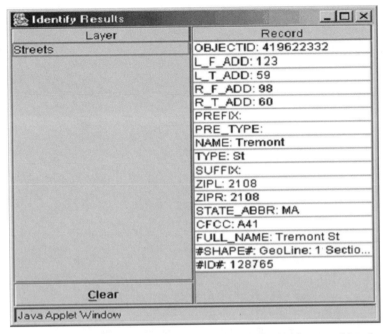

Figure 13.15 Identify feature dialog box. This feature closely resembles the identify features of professional desktop versions of GIS. (*Source:* http://www.davidrumsey.com.)

main reasons for Sunnyvale's SunGIS lie with the city's desire to share information with other municipalities and to speed up and simplify the permitting process for those with building projects.

SYSTEM OPERATIONS   In order to make Smart Permits work, JVSV had to standardize the building codes and permit applications for the region. Twenty-nine municipalities within the three counties adopted the new codes, allowing the Smart Permits system to operate consistently across jurisdictions.

FUTURE PLANS   JVSV wants to improve the system by allowing the entire permitting process to be accomplished on-line. The on-line process would not only allow builders to apply for a permit, it would also allow them to pay fees and to submit drawings to reviewers. The reviewed drawings—complete with annotations—would then be sent back to the builders electronically. JVSV also proposes to provide on-line scheduling and tracking of inspections throughout the course of a project, thereby speeding up the building process.

WEB APPLICATION   SunGIS is the most advanced Internet GIS available among the JVSV communities. It can be accessed on the WWW at http://www.ci.sunnyvale.ca.us/sungis. Other pilot communities whose applications

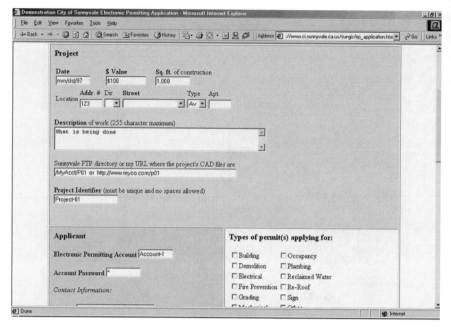

Figure 13.16   Sunnyvale's electronic permitting application. The form requests information regarding the project, the applicant, the contractor, the type of permits applied for, and pollution sources. (*Source:* http://www.ci.sunnyvale.ca.us/sungis.)

are at various stages of development can be found through the Joint Venture Web site at http://www.jointventure.org/initiatives/smartpermit/pilot.html.

EXAMPLES FROM INTERNET WEB SITE   Figure 13.16 is an example of the standardized permit application used in the three-county area utilizing Smart Permits. Figure 13.17 represents the on-line customer account listing that lets the applicant know the status of his or her permit(s).

13.4.5   Case Study 11: Property Assessment with MapMilwaukee in Milwaukee, Wisconsin*

BACKGROUND   In 1975, the Milwaukee Property File (MPROP) was created to store attribute data describing each of nearly 160,000 properties in the city. (It should be noted that the property file does not include properties of adjacent cities and villages; it only includes those properties that fall within the jurisdictional boundaries of the city of Milwaukee.) The MPROP is *the* main

---

Source: http://www.gis.ci.mil.us/isa/Map_Milwaukee.

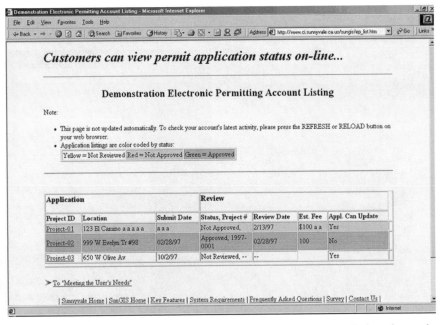

Figure 13.17 Electronic permitting account listing. The account listing shows the status of each permit application. (*Source:* http://www.ci.sunnyvale.ca.us/sungis.)

source for property information in Milwaukee and is accessed by nearly all departments in the city.

PURPOSE FOR INTERNET GIS   MapMilwaukee was created as an informational tool to provide citizens with the opportunity to learn more about the properties in their city.

DATA SOURCES   MPROP is the main source of data for MapMilwaukee, but the application also provides data regarding streets, demographics, and political boundaries.

SYSTEM OPERATIONS   The application allows the user to search by taxkey number, address, or intersection. In property queries, the taxkey number will provide a link to assessment information in tabular form and, from there, to the assessor's Web page, where other data may be downloaded for personal use. The application also possesses a thematic component in its dot map data tool so users can compare areas by the density of an attribute.

WEB APPLICATION   Because this application was designed with the public in mind, it can be accessed on the WWW at http://www.mapmilwaukee.com.

Figure 13.18   Property search using MapMilwaukee. Clicking on the taxkey link will take the user to "Property Assessments Results." (*Source:* http://www.gis.ci.mil.us/isa/ Map_Milwaukee.)

EXAMPLES FROM INTERNET WEB SITE   Figure 13.18 illustrates an example of a property search. The bottom right frame displays the parcel information with the taxkey number at the top. This link will take the user to the Property Assessment Results page illustrated in Figure 13.19. From this page, the user may link directly to the assessor's office.

13.4.6   Case Study 12: Work-in-Progress for Neighborhood Planning for Tampa, Florida*

BACKGROUND   The city of Tampa has coordinated a number of planning programs that provide guidelines for neighborhood development, urban design, and strategic planning. One of the city's most ambitious undertakings is the creation of an Internet GIS. Through the integration of the city's GIS called INDEX into a Web environment, citizens will have better opportunity to participate in neighborhood planning activities.

Source: http://www.tampagov.net/dept_planning_section/Demonstration_Project_INDEX/index.asp.

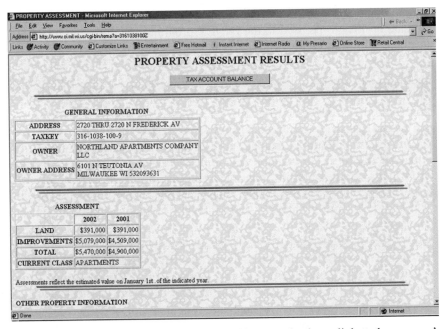

Figure 13.19 Property assessment results. This page also has a link to the assessor's home page. (*Source:* http://www.gis.ci.mil.us/isa/Map_Milwaukee.)

PURPOSE OF INTERNET GIS   In order to assist local neighborhoods in their planning efforts, Tampa is implementing an Internet GIS. The application will provide citizens with access to relevant community data and the tools for effectively assessing and analyzing that data for use in neighborhood plans.

DATA SOURCES   The data required to provide the services Tampa plans to offer in its INDEX application has entailed the recoding, updating, correction, and rectification of data from multiple sources. The majority of the data originates from the Planning Commission; the Tampa Planning and Management, Public Works, Transportation, and Management Information Systems departments; the Florida Department of Transportation; the Hillsborough Area Regional Transit Authority; and the Hillsborough County Property Appraiser. The categories of data that may be queried once INDEX is fully up and running will include demographic, land use, housing, employment, recreation, and environmental data.

PRELIMINARY SYSTEM DEVELOPMENT   In order to create a model that would assess neighborhood conditions, the city first selected six diverse neighborhood types as study areas. The neighborhood types—"mature urban village," established central city with some redevelopment, mixed use of residential and entertainment, transitional, "upswing," and new urban fringe—were used to create profiles to be used in the INDEX program. INDEX is a GIS created specifically for Tampa by Criterion Planners and Engineers.

CURRENT LIMITATIONS Two potential limitations exist with the INDEX software: (1) it poses a constraint on database development because it prefers geographic areas between 1500 and 2500 acres and (2) it generalizes land uses because of the required collapsing of original land use codes.

Source data limitations include inaccuracies in the data and missing data in the original files. For example, random field checks of properties listed in the Property Appraiser's file found that land use designations for properties in the file were 8–15% inaccurate. The main reason for the inaccuracies in the file is the age of the data. Missing data relate to the files not being updated. For example, the sign and signal database only contains records of signals installed or replaced after 1990. Those installed previously are not in the database. To rectify the data limitations, the city is systematically updating its databases.

FUTURE PLANS As of this writing, the system is not yet fully developed. Final plans are to integrate the INDEX GIS into an Internet environment, but exactly when has not yet been determined. Updates pertaining to the INDEX project can be found at: http://www.tampagov.net/dept_planning/planning_section/Demonstration_Project_INDEX/index.asp.

EXAMPLE FROM INDEX DEVELOPMENT PROCESS The INDEX GIS has the ability to generate 46 indicators within 6 different categories. Figure 13.20 illustrates the six economic indicators that can assist communities assess their employment strengths and weaknesses. Other indicators fall under the categories: demographic, land use, housing, recreation, and environmental.

13.5 CONCLUSION

The purpose of this chapter was to illustrate some of the activities within which Internet GIS plays a part. The case studies illustrated a few planning and management activities as they relate to mass transit users, infrastructure (e.g., buildings, rail systems), transportation systems, emergency procedures, and community development. A host of other activities exist that use Internet GIS ranging from protecting natural habitats to monitoring water systems to creating virtual cities. Any activity that can utilize spatial data can benefit from a GIS, and any activity that needs to communicate and collaborate with many users at different locations can benefit from a Web environment. Meld the two technologies together and you have a very powerful tool for planning and managing resources.

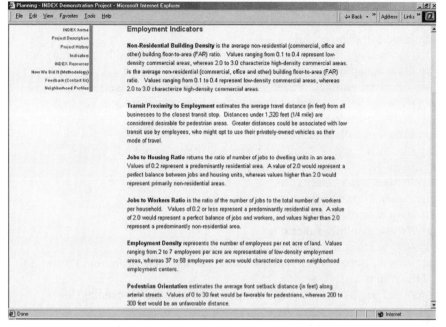

Figure 13.20 Employment indicators. Of a total of 46 indicators that INDEX can generate, 6 are employment indicators. (*Source:* http://www.tampagov.net/dept_planning/planning_section/Demonstration_Project_INDEX/Employment_Indicators.asp.)

These cases demonstrated that one of the benefits of Internet GIS is the ability to portray spatial relationships to a large audience with a picture that is easy to interpret. It has the potential to save time and provide better customer services to governmental agencies and private organizations. The limitation of Internet GIS at this stage is that some of the people may not be able to access computers and the Internet. The other limitation is the inability of current Internet GIS programs to perform many of the analytical functions of a traditional GIS. But as Internet software continues to improve and Internet GIS software gains greater functionality in a Web environment, the possibilities for planning and resource management activities become boundless.

A technology that has been around for over a decade but is only now coming into its own is wireless technology. Some organizations have already stepped into the wireless arena with GPS and wireless modems, as did the Utah Highway Patrol and the Valley Emergency Communications Center during the Olympic Games. Wireless technology offers a medium by which organizations can communicate with personnel in the field in real time. As the need for real-time information grows, wireless technology will take on an even more dominant role in planning and resource management activities.

WEB RESOURCES

http://hud.esri.com/egis
http://www.mapinfo.com/community/free/library/nys_health_casestudy.pdf
http://www.mapinfo.com/community/free/library/queensland_casestudy.pdf
http://www.mapinfo.com/community/free/library/pocs_casestudy.pdf
http://www.mapinfo.com/community/free/library/howardco_casestudy.pdf
http://www.osec.doc.gov/eda/pdf/InnovLDEP.pdf
http://www.ci.vallejo.ca.us/ed.asp
http://www.intergraph.com/gis/customers/articles/scdot.pdf
http://ims.gisresources.com/covington/covington.htm
http://www.davidrumsey.com
http://www.ci.sunnyvale.ca.us/sungis
http://www.gisplanning.com
http://www.mapmilwaukee.com
http://www.gis.ci.mil.us/isa/Map_Milwaukee
http://www.tampagov.net/dept_planning_section/Demonstration_Project_IN-
    DEX/index.asp
http://intergraph.com/gis/community/geospatialworld/proceedings/doc/
    056.doc
http://www.ul.ie/~infopolis

REFERENCES

Barnes, S. (2002). Salt Lake Hosts Spatial Olympics. *Geospatial Solutions,* April, pp. 26–33.

McGreevy, W. (2001). How Many Roads Must a Man Walk Down: Remote Investigations from the Desktop. *GeoSpatial World 2001 Conference,* Atlanta, Georgia, June.

McHarg, I. L. (1969). *Design with Nature.* New York: Wiley.

Rumsey, D. (2002). From Parchment to Ether: Fusing Historical Maps with Web GIS. *Geospatial Solutions,* April, pp. 34–39.

Slater, S. (2002). Queensland Creates a Princely GIS. *GeoWorld,* May, pp. 42–45.

CHAPTER 14

CONCLUSIONS AND EPILOGUE

Access to information via the media, for example, and the qualities and controls on information flow play an important role in how we can hope to understand and change the world.
—David Harvey (2000, p. 236)

14.1 OVERVIEW

Thirty-six years ago, when the first GIS project, Canada Geographic Information System, was established by Tomlinson, the computer system was an IBM 360, the mainframe computer, which required several rooms to place its hardware components, including card reader/punch, CPU, display terminal, console station, etc. At that time, no one could imagine the dramatic development of GIS in subsequent decades. Today, people can hold a palm-size PC in their hands and search for the location of a nearby coffee shop or use a Web map application to plan their vacations in California. Dynamic and interactive GIS applications have permanently changed the way people live and work today.

Internet geographic information services have become more and more important to the public and to researchers. Currently, thousands of Internet map servers provide on-line mapping functions, including urban plans, natural resource management, and census data (Coleman, 1999; Limp, 1997). The need for global access and decentralized management of geographic information is pushing the GIS community to distribute GIS Web services on the Internet.

This book has provided an overview of current developments in distributed computing and Internet GIServices. The discussions in this book are intended to help the GIS community adopt a sustainable, integrated strategy in developing open and distributed GIServices. By combining GIS components and data objects dynamically across a network, computing resources may be utilized more effectively for GIS applications.

The real value of distributed GIServices lies not in the technologies but in the actual content underlying the technologies. The members of the GIS community, including geographers, GIS professionals, and GIS users, are the people who actually create the valuable content in distributed GIServices. Well-designed applications will energize the growth of GIServices networks. Content creators are the key to the success of distributed GIServices. This final chapter will argue that as more people participate in distributed GIServices, the GIServices networks will become more valuable. Hopefully, the increasing value of GIServices will attract more people to participate and ensure the sustainable development of GIServices networks in the long run.

A central concept in this book involves the design of a sustainable framework that can cope with rapidly changing needs and technologies in the twenty-first century. In fact, our world is changing so fast, new technologies are invented every month, and new milestones are reached in scientific research every year. Along with the rapid change of information technology, traditional computing systems can no longer survive in the jungle of ITs. A flexible and upgradeable architecture for distributed GIServices is essential to ensure sustainable development of GIS in the next few decades.

Another underlying reason for the adoption of Internet-based GIServices is the need for communication in the GIS community. The reason for distributing GIServices on the Internet is not that the Internet technology is fancy and popular, but that the Internet can help people communicate more efficiently and effectively. For example, local and federal governments can use the Internet to disseminate maps and geospatial data to citizens. GIS software vendors can use the Internet to deliver upgraded software products or get feedback from users. Geographers can use the Internet to exchange hydrological models and spatial analysis procedures. The Internet-based communication of GIServices will facilitate the synergy of the GIS communities by sharing geospatial data, programs, models, and knowledge.

14.2 FUTURE IMPACT

Along with the popular use of the Internet and telecommunications technologies, on-line GIServices are essential in providing us with an informative and convenient living environment. This section will discuss the major impact on three groups in the GIS community: the GIS industry, geographers, and the public.

14.2.1 Future Impact on the GIS Industry

The development of distributed GIServices has three major impacts on the GIS industry. By adopting these dynamic, modularized frameworks, distributed GIServices will exploit the reusability and compatibility of GIS software and data objects. Currently, the software design in traditional GISystems rarely emphasizes reusability due to a framework that is closed, isolated, and vendor proprietary. Heterogeneous software and database engines have caused serious problems in data sharing and processing incompatibility. With the help of modern software engineering tools, software reusability will generate higher productivity, increase the efficiency of software programming, and provide a mechanism for prototyping when developing a new system or adopting a new technology (Yourdon, 1993). Reusable GIS software can reduce the programming workload significantly. GIS users will have more choices in developing their own GIS applications without the constraints of GIS software vendors.

The second impact on the GIS industry is that the design of distributed GIServices can help the GIS industry migrate gradually from legacy systems and adopt new technologies. It is expensive for traditional GISystems to adopt new technologies due to the ad hoc design and the lack of modularized frameworks. The LEGO-like GIS components and operational metadata scheme proposed in this book can facilitate the GIS industry to migrate from the legacy GISystems to a new framework of distributed GIServices.

> Future IS [Information System] technology should support continuous, incremental evolution. IS migration is not a process with a beginning, middle, and end. Continuous change is inevitable. Current requirements rapidly become outdated, and future requirements cannot be anticipated. The primary challenge facing IS and related support technology is the ability to accommodate change (e.g. in requirements, operation, content, function, interfaces, and engineering) (Brodie and Stonebraker, 1995, p. xvii).

The design of a dynamic GIServices framework proposed in this book will provide a possible solution for the migration challenge and facilitate a long-term, sustainable development of GIServices. By modularizing the dynamic framework of distributed GIServices into three containers (agent containers, component containers, and data containers) within GIS nodes, the GIS industry can gradually and easily migrate these components from the legacy GISystems into a new framework. For example, the migration process can start from the user interface components, then the upgrade of GIS database engines, and finally, the replacement of the core GIS programs. Brodie and Stonebraker (1995) call this type of migration the incremental approach, which can be carried out by small incremental steps until the desired long-term objective is reached. Each step requires a relatively small resource allocation. Another type of migration mentioned in their book is the Cold Turkey approach, which attempts to rewrite the legacy information system from scratch to produce the target information system using modern software techniques and the hardware of the target environment. Their conclusion is that the incremental approach is much better than the Cold Turkey approach, as the latter usually fails in real-world cases. In short, the modularized framework of distributed GIServices will facilitate the GIS industry to adopt the incremental approach for the migration from the legacy GISystems and the adoption of new technologies.

The third impact on the GIS industry is that distributed GIServices will change the development strategies of GIS software vendors and will transform the current monopolized GIS market into an open, competitive environment. Traditional marketing for GISystems is usually targeted to the GIS professionals, the GIS consulting companies, government agencies, and high education institutions. Along with the development of distributed GIServices, the major markets of GIS software will shift toward the public and end users. Small, modularized, Web-based GIS components will replace huge, workstation-type, and all-in-one GIS software packages. The design of GIS components will emphasize extensibie GIS operations and customizable tools for different applications. The pricing for GIS software will also change from year-based site licensing to the usage-based, individual charges. For example, a user may be charged for a 10-time use or a three-day use of a buffering module. The different pricing schemes of GIS software will encourage more people to use GIServices in their daily lives and short-term activities.

Moreover, distributed GIServices will release the power of control in GIS software from the major GIS vendors to individual GIS software programmers and small software companies. Since the future development framework of GIS software will provide APIs in a distributed-component framework, such as the Microsoft .NET-based applications and the Java platform, individual programmers can access or invoke GIS programs without knowing the original source code. Thus, GIS software programmers can easily extend the functionality of GIS packages by adding new GIS components and developing their own specialized products. In the future, the traditional GIS software vendors who currently control the major GIS markets will shift their focus of software development from the all-in-one types of GIS products to the main GIServices engines and frameworks. The GIServices framework will include the core programs for major GIS operations, the Internet map server for disseminating GIServices, and the spatial database engines for efficient data storage and management. These products developed by the major vendors will allow third-party companies to develop extensions and additional functions for these core products.

Distributed GIServices will change the market focus in the GIS industry and reassign the new responsibilities for its players, including the consulting companies, software vendors, and application users. The dynamic architecture and open APIs will facilitate more effective and collaborative software development among different GIS companies and programmers. GIS users will have more choices of GIS software in the future due to the open and free competition GIS market. The distributed, modularized architecture can provide a long-term, sustainable development strategy for the GIS industry and adopt new technologies over time. Since "today's new system will be tomorrow's maintenance problem" (Yourdon, 1993, p. 276), methods to up-grade legacy systems and take advantage of new technologies will be the major challenge for the GIS industry in the twenty-first century. The design of distributed GIServices may provide a possible solution to the challenge and help the GIS industry adopt new technologies in a smooth and efficient fashion.

Key Concepts

By adopting dynamic, modularized frameworks, distributed GIServices will facilitate the reusability of GIS software and data objects. **Software reusability** *will generate higher productivity, increase the efficiency of software programming, and provide a mechanism for prototyping when developing a new system or adopting a new technology*

14.2.2 Future Impact on Geographers

Contemporary GIS promises geographers comprehensive spatial analysis functions and modeling techniques as a powerful tool for synthesis (Abler, 1987). However, spatial analysis and modeling functions are in fact weak and premature in traditional GISystems. Even if some GISystems do provide limited spatial analysis functions, most users rarely use the analysis or modeling functions in their GIS application, which causes the problem of system underuse (Davies and Medyckyj-Scott, 1996). The possible explanation would be that "developments in the GIS industry largely reflect the demands of the GIS marketplace. This has been dominated for the past decade by applications in resource management, infrastructure and facilities management, and land information. In these areas, GIS tends to be used more for simple record-keeping and query than for analysis" (Goodchild et al., 1992, p. 410). The market-oriented development of GISystems is inappropriate for individual users, such as geographers and spatial scientists. What geographers really need is the spatial-analysis-oriented and question-driven GIServices. By adopting the dynamic framework of distributed GIServices, geographers may be able to develop the specialized programs and models that can really focus on the spatial analysis theories and geographic problems. The following paragraphs will describe three major impacts on geographers and spatial scientists in adopting the dynamic framework of distributed GIServices in the future.

First, geographers and spatial scientists can build more realistic models to solve their research problems by combining the LEGO-like GIS components and models in distributed GIServices. Currently, geographic research and scientific problems focus on large-scope, multidisciplinary issues—such as global change, sustainable development, and urban growth—that involve many experts and specialists from different disciplines. However, traditional

GISystems cannot easily integrate their specialized knowledge and models into an integrated framework. For example, the urban growth research may require the spatial statistic analysis functions, the graphic display of population profiles, network analysis for transportation, digital elevation model, and hydrological models for underground water conditions. In the past, these specialized models and tools were developed separately in different software and platforms and were not compatible. By using the distributed GIServices and LEGO-like component framework, different types of GIS models can be easily integrated and provide a more realistic modeling environment for scientists. Geographers can combine different GIS components in order to provide better explanations for geographic problems. Moreover, geographers can share their expertise and models with other scientific communities by distributing their models and programs over the Internet. In short, the LEGO-like frameworks of distributed GIServices will facilitate GIS modeling in a question-driven and exploratory method (Fischer et al., 1996). Geographers and spatial scientists can then easily share or exchange their GIS models and analysis methods under the dynamic architecture of distributed GIServices and facilitate the continuous progress of geographic information science.

Second, distributed GIServices will help scientists and geographers focus on the domain of problems rather than the mechanism of system implementations. Traditional GISystems with unfriendly user interface and complicated programming tools prevent many geographers from adopting GIS modeling tools to solve their research problems. In the past, the primary obstacle of using GIS tools was the mechanism of GIS model implementation. To construct a GIS model, geographers had to understand the details of implementation mechanisms, including macrolanguage programming, database management, data conversion, and hardware device configuration. These tasks require comprehensive programming experience and computer literacy, which may go beyond the regular training geographers receive. By adopting distributed GIServices, the distributed GIS components can help geographers construct their GIS models more easily and relieve most of the implementation tasks from geographers' shoulders. The collaborations among distributed components will take care of system-level problems and the details of model implementation, such as database management, data conversion, hardware configuration, and software compatibility. Geographers will have more energy to concentrate on spatial analysis issues and the examination of geographic problems.

Finally, the flexible data access approach and the operational metadata scheme in distributed GIServices will help geographers utilize on-line information more efficiently and facilitate the reusability of geospatial data for geographic research. In a traditional GISystems environment, the most expensive cost of GIS implementation is in data input and data conversion (Korte, 1994). Most geographic research requires multiple geographic data sets, which need to be generated by costly procedures, including map digitizing, image scanning, or the classification of remote-sensing images. These procedures cost both hardware investment and labor. For an individual geographer, the high cost of digital data hinders the construction of GIS models and spatial analysis procedures. With the help of distributed GIServices, geographers will be able to exchange and share spatial data sets and be free from the constraints of heterogeneous data models and isolated system environments. Moreover, the operational metadata described inside geodata objects will improve the efficiency of data modeling and help geographers get the appropriate data sets for their research projects. Such a dynamic, distributed GIServices architecture will provide a cost-effective way for geographers to download, exchange, or share spatial data sets in distributed-network environments. Geographers will get better data for better use with the help of distributed databases and operational metadata.

To summarize, distributed GIServices will help geographers in the establishment of GIS models, the design of spatial analysis tools, and the better use of geospatial data sets. Moreover, the adoption of distributed components and operational metadata will encourage geographers to formalize the geographic knowledge for the training of software agents and the definition of object behaviors for geodata objects and components. The construction of a geographic knowledge base will facilitate the future process of geographic information science. Geographers will have more energy to dedicate their effort to geographic research and new scientific contributions.

14.2.3 Future Impact on the Public

One of the major differences between traditional GISystems and distributed GIServices is the extension of target users. Traditional GISystems are designed for the GIS professionals, geographers, or spatial analysts associated with consultant companies, universities, or government organizations. Distributed GIServices are designed for the public and provide useful information in our daily lives. The public will be the major group with the most to gain if the whole society adopts the dynamic framework of distributed GIServices. However, there are also some negative impacts on the public that come with the development of distributed GIServices. The following sections will identify both positive and negative aspects of GIServices from a public service perspective.

14.2.3.1 Positive Aspects There are two major benefits of distributed GIServices for the public services. The first benefit is to provide transparent, ubiquitous GIServices in daily life. Distributed GIServices will be intricately connected into the web of everyday life, along with ubiquitous computing (Armstrong, 1997; Weiser, 1993). With the popular use of cellular phones, PDA, auto-PCs, and GPS receivers, daily use of distributed GIServices will be essential in a transparent way. For example, many travel-based Web sites, such as Microsoft Expedia and MapQuest, can provide integrated services for travel plans, including the purchase of airplane tickets, hotel reservations, and car rentals. Another example is automobile navigation systems, which have

become standard equipment in taxis, rental cars, police cars, and luxury se-
dans. In the future, more and more GIServices will be available in a more
transparent way. For example, each bus stop may set up a digital display
board, which can indicate when the next bus arrives, whether the bus will be
delayed, and the real-time location of the bus. The information can help the
passengers know how long they will wait before their bus comes and improve
the efficiency of public transportation. Another example is the automatic parking arrangement in shopping malls. When people drive to the shopping mall,
each car with its auto-PC may build network connections with the parking
lot server when the car enters the parking lot of the mall. The parking lot
server will tell the car whether there is parking space available and indicate
where the nearest parking lot is from a favorite shopping store. Such kinds
of GIServices will be very popular and feasible in the future with the help
of the network communication and intelligent framework of distributed
GIServices. The public may not even recognize that these services are GIS
applications. With the help of distributed GIServices, such public services
will become more friendly and useful for the public.

The second benefit of distributed GIS is to deliver real-time, integrated
services for emergency events. For example, the dispatch of emergency vehicles (EMT, police, and fire) with real-time road condition reports can avoid
unnecessary delays and ensure the rescue team arrives promptly at the scene.
Another example of real-time services is natural hazard reports and
evacuation/rescue plans. Distributed GIServices can facilitate efficient hazard
management and rescue/relief plans for floods, tornadoes, earthquakes, or
typhoons. The intelligent network of hazard management services with real-
time data-gathering devices, such as video cameras, GPS receivers, and wired
weather stations, will provide essential information for the local and federal
governments in building effective warning systems and quick-response rescue/
relief plans. With the help of telecommunications networks, the real-time data
reports and monitoring for natural hazard damages can be distributed to related departments immediately. Different departments, such as the police department, transportation department, and fire department, can work together
based on real-time geographic information and provide the necessary services
to the damaged areas immediately. With the help of real-time distributed
GIServices, federal and local governments can provide a more secure and
safe environment for their citizens. In general, distributed GIServices can
provide the public a safe civil life and a convenient way to live in the twenty-
first century.

14.2.3.2 *Negative Impact*   Although distributed GIServices can provide
useful and essential information for the public, they also have some negative
aspects. The major problem in adopting distributed GIServices in public services is the contribution to the "information ghetto" (Graham and Marvin,
1996) and the Digital Divide [National Telecommunications and Information
Administration (NTIA), (1999)]. Even though society uses computer tech-

nology and network facilities as the major tools to provide public information services for its citizens, some people will be deprived of their civil rights and miss the opportunities for success due to the lack of tools for accessing public information services.

While affluent and elite groups are beginning to orient themselves to the Internet and home informatics and telematics systems, other groups are excluded by price or lack of skills or are exploited by such new technologies. Advanced telecommunications and transport networks open up the world to be experienced as a single global system for some. But others remain physically trapped in information ghettos where even the basic telephone connection is far from a universal luxury (Graham and Marvin, 1996, p. 37).

A similar problem was also identified in research on the "digital divide" carried out by the NTIA, U.S. Department of Commerce. NTIA initiated its first survey in 1994, then continued it in 1997 and 1998. These surveys discovered that "the situation of Digital Divide—the divide between those with access to new technologies and those without—is now one of America's leading economic and civil right issues" (NTIA, 1999, p. xiii). The survey identified user profiles of telephone services and Internet access and cross-tabulated the information according to several variables (such as income, race, age, and education) in three geographic categories—rural, urban, and central city. The first report identified the problem of a disproportionate lack of access to the Internet in rural areas and central cities: "Black households in central cities and particular rural areas have the lowest percentages of PCs, with central city Hispanics also ranked low" (NTIA, 1995, p. 3). Two years later, the second survey indicated the digital divide gap widened with the popular growth of the Internet and PCs. The results also indicated that female-headed households lag significantly behind the national average of Internet access and PC usage (NTIA, 1998). The third report released in 1999 found that "a Digital Divide still exists, and, in many cases, is actually widening over time. Minorities, low-income persons, the less educated, and children of single-parent households, particularly when they reside in rural areas or central cities, are among the groups that lack access to information resources" (NTIA, 1999, p. xiii). For example, "urban households with incomes of $75,000 and higher are more than twenty times more likely to have access to the Internet than rural households at the lowest income levels, and more than nine times as likely to have a computer at home" (NTIA, 1999, p. xv).

Key Concepts

The major problem in adopting distributed GIServices in public services is the contribution to the **information ghetto** *(Graham and Marvin, 1996) and the* **digital divide** *(NTIA, 1999). The situation of digital divide—the divide between those with access to new technologies and those without—is now one of America's leading economic and civil right issues.*

In general, the digital divide reports indicate that provision of public services requires in-depth considerations beyond the deployment of technology. The social aspect of public access needs to be considered while local or federal governments deploy the framework of distributed services. Internet technology and distributed GIServices should help the public, especially those who need additional support from the governments and social welfare. NTIA's research also reveals that "many of the groups that are most disadvantaged in terms of absolute computer and modem penetration are the most enthusiastic users of on-line services that facilitate economic uplift and empowerment" (NITA, 1995, p. 4). Besides the adoption of network technology and real-time information services, governments may need to provide public facilities for the public to access this information, as by offering Internet access in local public libraries, schools, and community centers. Also, some essential geographic information may use alternative media for the public, such as paper maps, telephone services, or public broadcast systems. In general, distributed GIServices should allow the public to access essential information from both public and private places and should not be limited to certain societal groups or classes.

14.3 VISION OF FUTURE DISTRIBUTED GIS

It seems that, without doubt, the future of distributed GIS is in component-based applications (OGC, 2001; Vckovski, 1998). The idea is that the user can discover, access, and retrieve geospatial data and geoprocessing tools from anywhere and from any Web site. Components from different sources are interoperable and can communicate with each other. For example, the user can discover, retrieve, and use the buffer component from company A, spatial overlay component from company B, and street network data from the Department of Transportation with one projection and land use data from the planning department with another projection. These components and different data sets are interoperable and can be assembled on demand at the user's client site.

However, the data and application on the Web are so diverse that it is a great challenge to locate and access these diverse resources, manage them, and deliver them through different channels. Clearly, there needs to be some sort of standard so that these components and databases are interoperable and reusable. These standards would require (Kirtland, 2001)

- a standard way to represent data,
- a standard way to access data and other components,
- a common way to describe available components and databases,

- a way to discover components and databases located on different Web sites, and
- a way for service providers to announce or advertise the availability of their services.

There are currently two ways to implement distributed components and data over the Web. One option is to use the component-based technology to construct distributed GIS, as discussed in Chapter 5. The other option is to use Web services to construct GIS Web services. Distributed GIS is based on either the CORBA/Java model or Microsoft's COM+/architecture. A distributed GIS in the CORBA/Java world is one in which the Java applets on the Web client communicate with objects on the GIS server via object-to-object interaction methods. Distributed GIS based on the COM+/ActiveX model is one in which the ActiveX controls on Web browsers communicate directly with COM+ objects on the server.

However, there are some problems with these two distributed-component approaches. First, the CORBA/Java model is not interoperable with the COM+/ActiveX model. For instance, components created using Java beans cannot communicate with components created using ActiveX controls. Second, there is no mechanism set up to discover distributed components over the Web and for service providers to register and advertise the availability of their resources.

The other approach, Web service, according to IBM's Web services tutorial, is "a new breed of Web application. They are self-contained, self-describing, modular applications that can be published, located, and invoked across the Web. Web services perform functions, which can be anything from simple requests to complicated business processes. . . . Once a Web service is deployed, other applications (and other Web services) can discover and invoke the deployed service."

These two options are very similar. Both adopt the concept of reusable module applications; that is, functions of module applications can be used by other modules without requiring an understanding of the details of the implementation inside the components. However, there are differences between the component-based technology and the Web services. The component-based technology uses the object-model-specific protocols to access remote components, such as DCOM, RMI, or IIOP. The Web services are accessed via Web protocols and interoperable data formats, specifically, HTTP and XML. Furthermore, the GIS Web services approach also has service registry and discovery, which makes it easier to locate services and objects across the Web.

The Web service is emerging and has a potential to become the next big application on the Web because it takes advantage of the strengths of both the distributed-object technology and the Web. The simplicity of access and ubiquity of the Web as well as the omnipresent HTTP make the Web an important information distributor, component container, and user interface. The traditional middleware platforms in the distributed-object technology provide great implementation vehicles for Web services. In other words, the Web provides a uniform and widely accessible interface and access glue over services that are more efficiently implemented in a traditional middleware platform.

In its request for technology, the OGC's Web Services Initiative (March 2001) adopted the Web services framework. We have discussed distributed GIS in Chapter 5; in this chapter we will focus on the GIS Web services.

14.3.1 What Are GIS Web Services?

Web services are interoperable, self-contained, self-describing, module components that can communicate with each other over the Web services platform. OGC envisions that GIS "Web Services will allow future applications to be assembled from multiple, network-enabled geoprocessing and location services." The GIS Web service is "a vendor-neutral interoperable framework for Web-based discovery, access, integration, analysis, exploitation and visualization of multiple online geodata sources, sensor-derived information, and geoprocessing capabilities" (OGC, 2001).

Key Concepts

Web services *are interoperable, self-contained, self-describing, module components that can communicate with each other over the Web services platform.* **GIS Web services** *will allow future applications to be assembled from multiple, network-enabled geoprocessing and location services.*

The basic platform for Web services is XML plus HTTP (Vasudevan, 2001). HTTP is a ubiquitous Internet protocol that is supported by almost all Internet applications. XML provides a standard way to represent data, and its XML schemas provide a way to describe and validate data types. The combination of HTTP and XML is a great way to achieve interoperability over the Web. But HTTP and XML by themselves are not sufficient for fully functional GIS Web services. We also need other standards to address common means to access XML data and to describe, discover, and advertise services. The IT industry is developing such a standard to address these issues. Some important standards are emerging, particularly SOAP, UDDI, and WSDL.

SOAP defines a protocol for information exchanges among programs in the heterogeneous Internet environment by using HTTP and XML. Specifically, SOAP defines a uniform way of encoding an HTTP header and passing XML-encoded data. It also defines a way to perform RPCs so that a program in one computer can call a program in another computer and exchange information.

In order for SOAP to access Web services, we first need to find them. This is the function of UDDI, an XML-based registry that allows service providers or businesses to advertise and list their Web services on the Internet. It has three functions—publish or describe, discover, and bind or integration (Vasudevan, 2001):

- *Publish* UDDI specifies a mechanism for Web service providers to advertise and describe the existence of their Web services. This is like the function of the White Pages in the traditional telephone book. It allows the provider of Web services to register itself.
- *Discover* UDDI provides a way for Web service consumers (clients) to dynamically find other Web services. This is like the function of the Yellow Pages. It allows the service consumers (applications) to find a particular Web service across the Internet.
- *Bind* Once the Web service has been found, UDDI specifies ways for application to connect to and interact with the Web service of interest. This is commonly referred to as the function of the Green Pages.

The ultimate goal of UDDI in business and e-commerce is to streamline on-line transactions by enabling companies to find one another on the Web and make their systems interoperable for on-line transaction.

While UDDI is an XML-based registry for providers of Web services to list their services on the Internet, WSDL is the language used to do this. WSDL is an XML-based contract language used to describe the Web services, including what a Web service can do, where it resides, how to invoke it, and what messages the Web service accepts and generates. Essentially, WSDL is a template for how Web services should be described and bound by clients (Vasudevan, 2001).

OGC envisioned the future distributed GIS as distributed geoprocessing systems that can be seamlessly integrated and be able to communicate with each other using technologies such as XML and HTTP. OGC adopted the concept of Web services in its OGC Web services Initiative and encouraged leveraging IT industry efforts such as SOAP, UDDI, and WSDL in its Request for Technology in Support of an OGC Web Services Initiative (2001). It is developing standards to describe "operation" service components and "data" components. The *operation* components provide operations such as spatial search, query, and overlay and include four groups—viewers and editors, catalogs, repositories, and operators:

- *Viewer and editors* are user interface components that allow the user to view, edit, and conduct other operations on the underlying data. The difference between a viewer and an editor is that the editor enables users to edit the data and save changes, while the viewer can only display the data. The viewer can also become a component container that users can use to locate, retrieve, and assemble other operator components.

- *Catalogs* are collections of metadata, or information repositories, about other data objects and operators. The user can use catalog components to search for metadata or the names of the objects. Catalogs provide a mechanism for data and service providers to list and advertise the availability and types of data and services. The user searches for catalogs first before getting to the data repositories. Catalogs can be implemented by using WSDL.

- *Repositories* are collections of data. These are places where data and other resources reside. To help speed up the process of finding items, each item is given a name within the repository and each repository maintains an index. Therefore, given a name, a repository can find the resource. Repositories can be implemented using UDDI.

- *Operators* are operation or processing components that offer functions of geoprocessing and analysis tools, for example, buffer analysis, spatial overlay, and projection conversion. They can transform, combine, or create data and generate outputs. Operators can be implemented by using SOAP.

The *data* components provide contents for operations and include data, metadata, names, and relationships (OGC, 2001):

- *Data* are raw information about an object.
- *Metadata* are descriptions about the data.
- *Names* are used to register and find data. A name makes sense only within a content or namespace. In a repository, a data name together with the repository name can be used to identify and locate data items. Names can refer to data or to operators.
- *Relationships* are links between named items.

Names and relationships can be implemented by using URI, XLink, and XPointer.

14.3.2 Putting Web Services Together: Web Services Architecture

Now that we have all these Web service components, how do we put them together? From an *n*-tier application architecture perspective, the concept of the Web service is simply a framework to access different service components that are implemented by other kinds of middleware (Doyle et al., 2001). We can break down this framework into three tiers: presentation, business logic, and database, as shown in Figure 14.1.

It can be seen from Figure 14.1 that the computing devices are very diverse, from regular desktop and laptop computers that use Web browsers in HTML format to PDAs and mobile cell phones that may be in the WML format. To provide GIS Web services to these diverse access devices requires different

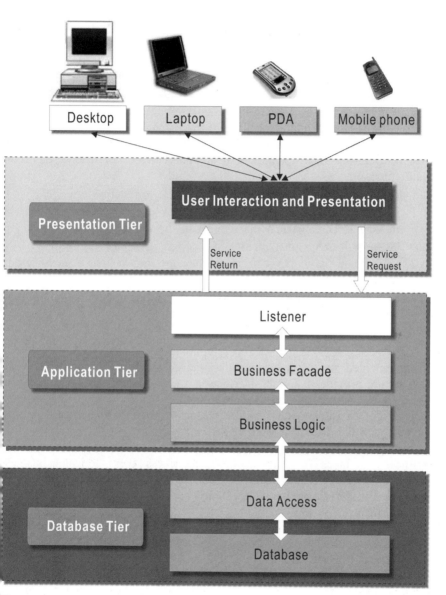

Figure 14.1 Generic architecture for GIS Web service. (*Source:* Adapted from Kirt-
and, 2001, reprinted with permission.)

information rendering or presentation methods with different access platforms.

Users from a variety of computing devices or access devices make a service request through the user interface with the presentation tier. That request is usually encoded in XML and sent to the service-neutral request handler or a listener. The listener receives and parses the user requests and then forwards them to a business facade that exposes the operations supported by the business logic. Depending on the nature and complexity of the user request, the request is processed either by the business facade or business logic. The logic itself is implemented by a traditional middleware platform (Vasudevan, 2001).

The listener is an important element in the application tier that links the user request to the Web services. In fact, the listener is the only part of the Web service that knows it is part of a Web service. The listener has three major functions:

1. to receive incoming messages containing requests for service, parse the messages, and dispatch the request to the appropriate method on the business facade;
2. to package the response from the business facade into a message and send that back to the client; and
3. to handle requests for contracts and other documents about the Web service.

The business logic (e.g., geoprocessing, operations) of the GIS Web service is implemented into two parts: business facade and business logic. The business facade provides a simple interface that maps directly to operations provided by the business logic layer. In a simple Web service, all the business logic might be implemented by the business facade, which would interact directly with the data access layer (Kirtland, 2001).

The data access layer presents a logic view of the physical data to the business logic layer. The purpose of the data access layer is to isolate business logic from changes to the underlying data stores and ensure the integrity of the data. The data layer stores information required by the Web service (Kirtland, 2001).

Now how do those operation and data components fit into this Web service architecture? Viewers and editors are usually located in the presentation tier that interacts directly with users. Catalogs and repositories of operators can be located in the business facade to indicate the availability of the operator services. Operators are located in the business logic layer. For data components, data are located in the database layer. Catalogs and repositories of data as well as metadata, names, and relationships components are located in the data access layer.

Figure 14.2 is OGC's version of OGC Web services (Doyle et al., 2001) "OGC Web Services provide a vendor-neutral, interoperable framework for web-based discovery, access, integration, analysis, exploitation and visuali

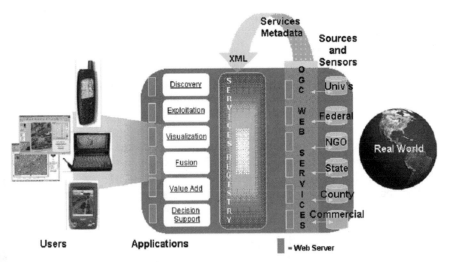

Figure 14.2 OGC Web services. (*Source:* Doyle et al., 2001, with permission from OGC.)

zation of multiple online geodata sources, sensor-derived information, and geoprocessing capabilities" (Doyle et al., 2001). For Web service providers, OGC Web services provide a framework for them to develop their own service components, such as a network analysis. They can then be published, located (via metadata), and invoked across the Web by other users. Since these service components are self-contained, self-describing, modular applications, they can be treated as "blackboxes" or operations that can perform certain tasks. The developer can choose any development tools to build the service components, and the user does not care much how they were built and simply uses them as needed.

How can the user use these GIS Web services? We will consider an example. Suppose a worker commutes to work by driving to a park-and-ride lot and then transfers to an express bus to get to the workplace. He or she has two choices in selecting the park-and-ride lot based on the availability of parking space and the arriving time of the express bus. To determine which park-and-ride lot to use, the worker makes a request from a cell-phone, that request is then sent to the listener in a server, and the server searches data simultaneously from a Web site that provides real-time transit information and from another Web site that offers parking information. Once the real-time data have been found, the listener then sends the data sources URL and the user request to a decision support system, a business logic operator on another Web site, which calculates the availability and the total travel time and offers the user a couple of options. The user receives the recommendations from the cell phone and makes the final decision.

The same information is also very useful for transit dispatches. For example, the bus dispatcher in the control center can use real-time information about the number of persons waiting at the park-and-ride lots from one Web site and real-time traffic information from the other to determine whether to dispatch a new bus service. If he or she does dispatch a new bus, the quickest route for the bus to get there can be found using a decision support service on yet another Web site. The decision support service incorporates real-time traffic information and bus load and delay information into its decision-making model to make a recommendation for the bus dispatcher.

As bandwidth and storage continue to get cheaper and Web-based technologies such as GML, SOAP, UDDI, and WSDL mature, GIS Web services will have more widespread applications. A true component-based, interoperable distributed GIS will emerge into the mainstream of IT.

Web technology will continue to evolve. Regardless of the ever-changing nature of the Web, one thing is almost certain—diverse data and business logic tools will proliferate and will be accessed from an equally diverse array of devices. Distributed GIS that focuses on distributing data and geoprocessing tools in a standard and open manner to different clients using a variety of devices will be the future of geospatial technology. Furthermore, by adopting the standard IT, distributed GIS will blur into other ITs.

14.4   ALTERNATIVE FUTURES

The previous discussions in this chapter highlight the promising future of distributed GIServices and the possible on-line GIS applications and research directions for the GIS community. However, it does not guarantee that comprehensive, distributed GIServices will be automatically available in the future. If the GIS industry does not apply distributed-component technologies in their products or if the leading GIS vendors and their Internet map server packages do not adopt the OGC specifications and standards, the comprehensive GIServices may not be available. The following paragraphs will provide such a different perspective and discuss the alternative prospect for the development of GIServices.

There are three possible paths for the GIS community. The first path is that the GIS community follows the traditional paradigm of GISystems, which focuses on stand-alone workstations with limited network capability, instead of distributed GIServices. The second path is that the GIS community will develop distributed GIServices but the software frameworks will be developed by private GIS vendors without providing an open environment for the integration of different GIServices. The third path is that the GIS community will be able to develop distributed GIServices in an open and distributed software framework, as was proposed in the previous chapters. The following discussion will illustrate the possible futures suggested by the first two paths by focusing on the impact on three GIS user groups: the public sector, the private sector, and the scientific community.

14.4.1 First Path: Centralized GISystems

Many people in the GIS community argue that centralized GISystems are better than distributed GIServices because stand-alone GISystems are stable, robust, powerful, and worry free. Network-based, distributed GIServices are slow to respond, complicated, insecure, unreliable, and troublesome. Some GIS professionals and research scientists may prefer to continue the traditional development of GISystems rather than explore the possible applications of distributed GIServices. What will happen if the GIS community chooses to retain the paradigm of GISystems? The following discussion will analyze the possible impact on three user groups in the GIS community and the advantages and disadvantages of this alternative future.

For the public sector, one advantage for adopting traditional GISystems is that federal agencies and local governments can continue to use existing GISystems and computer facilities without worrying about retraining their IT experts, purchasing new hardware and software, or changing system administrative tasks such as setting up new Web servers and reorganizing local network connections. Also, federal agencies and local governments can keep their data production and process procedures in a traditional GISystems approach, which makes it easier to set up their annual budget plans. If federal agencies or governments decide to provide distributed GIServices for their users, the initial cost of the hardware/software will be very expensive, including purchasing enterprise Web servers, testing and installing map server packages, training programs for adopting new technology, and the maintenance of Web servers and GIServices.

There are three major disadvantages for the public sector. First, the dissemination of geographic information using traditional media such as paper maps and CD-ROMs will continue to be expensive and difficult to update. Second, the centralized GIS database management may cause a management problem along with the increase in the size of GIS databases. For example, with the modern remote-sensing technology and GPS-based data collections, the size of a working GIS database may increase with new information being added every day. Without a distributed GIServices architecture, the maintenance of traditional GIS databases will become more difficult. Third, without providing catalogue services or metadata along with distributed GIServices, many GIS projects may duplicate GIS data sets, which may have already existed in other agency's databases. For example, a transportation department in a local county government may want to create city road data sets by digitizing a paper map generated from the USGS. However, the USGS may already have the city roads in a digital format. Under the current GISystems framework, it is really difficult to facilitate the sharing and exchanging of geospatial information among different agencies and governments.

For the private sector, one advantage for keeping the traditional GISystems paradigm is that the pricing policy for current GIS software is easier to handle. In general, software companies can charge users for a single GIS package or an annual licensing fee even though the users may only use a fraction of the total software functions or just one month of GIS projects during the whole year. Also, with stand-alone GIS packages it is easier to keep track of the licensing policy and charges for each user and machine. The pricing policy for distributed GIServices is more difficult to make, and the usage of GIServices is harder to track. Also, the data protection, system security, and copyright issues are simpler in the GISystems than for distributed GIServices.

There are three disadvantages for the private sector. First, the cost of software development and prototyping is higher and the cycle of new products is longer. Distributed GIServices with distributed-component technologies and modularized software frameworks such as COM, Java, or CORBA can improve the processes of software development for GIS vendors and software companies. Second, the number of users in traditional GISystems is small compared to other types of IT applications. By adopting distributed GIServices, the number of potential GIS users will increase dramatically. These GIServices include, for example, wireless car navigation systems, mobile mapping devices, and virtual shopping guides. If the private sector does not provide new types of distributed GIServices, the growth of GIS users will be very limited in the future. Finally, it is difficult to integrate GISystems with other types of computer applications, such as visualization software and statistic packages. The problems of software integration may prevent the further adoption of GIS in many software applications.

For the scientific community, the first advantage for GISystems is that traditional GISystems are easier to handle and learn than distributed GIServices. Scientists can quickly apply GISystems to their research projects. For example, oceanographers can use a basic GIS package (such as ArcView or MapInfo) to display their research findings. However, it is more difficult to ask a scientist to set up a Web site or Internet map server.

However, without adopting distributed GIServices in the future, the scientific community may need to pay a high price for GIS software packages and yet may only use 10% of its GIS functions. Also, the nature of traditional GISystems (stand alone) may prevent or discourage scientists from sharing their data results with other disciplines because of the lack of collaboration between heterogeneous GIS databases.

14.4.2   Second Path: Private, Vendor-Specialized GIServices

Another possible path for the GIS community is to develop distributed GIServices in a private, vendor-specialized framework instead of an open standardized architecture. People, especially from the private sector, may argue that the vendor-based framework is better because GIS vendors can adopt the most advanced IT and respond to their customers faster, which means better IT support and services. Also, vendor-specialized products can provide customized services on different domains, such as utilities management, traffic controls, or emergency dispatch. The following discussion will illustrate the advantages and disadvantages in the same three groups: the public sector, the private sector, and the scientific community.

For the public sector, the major advantage is that the vendor-based software model may be able to provide better IT support and customer services. Since different software companies have their own software coding and specialized data structure, private GIServices frameworks may provide a better data security model and content protection for sensitive geospatial data and remote sensing images.

There are three disadvantages for the public sector. First, if different federal agencies or local governments adopt different vendor-based frameworks, it will be very difficult to exchange or integrate GIServices from two different software frameworks. Second, with the development of NSDI, it is essential to have both the vertical integration (different locations with the same spatial themes) and horizontal integration (different data layers in a single area) of geodata clearinghouses. With the private, vendor-based GIServices framework, it is very unlikely to ask all government agencies and data clearinghouses to adopt the same GIS vendor's solution. Third, in the long run, it is extremely difficult to migrate GIServices from one vendor-based platform to another. Without open and standardized GIServices framework, the migration from one vendor package to another will need to replace many software components, including GIS databases, map server engines, client-side viewers, middlewares, and analysis procedures.

For the private sector, there are many advantages for adopting vendor-based GIServices. First, the leading GIS vendors who control the majority of the GIS software market will get the most benefits. The private GIServices framework can allow these vendors to integrate and monopolize the whole spectrum of GIServices from data production and data analysis to data distribution within a single vendor-based architecture. The second advantage is the ability to keep customers and users forever, since the migration to another system architecture is extremely difficult. The private, vendor-based GIServices can secure the profit growth of major GIS software vendors and their market value in the long run.

The disadvantage of vendor-based GIService framework for the private sector can only be experienced by small GIS companies or individual GIS consultants. Because it is difficult to develop a customized GIS component or extension under a vendor-specialized development framework, the small software vendors and GIS programmers will not be able to develop specialized GIS functions or components for their customers. The whole market of distributed GIServices will be monopolized by a few leading GIS vendors.

For the scientific community, the advantages of the private, vendor-based framework are similar to those of the public sector. The vendor-based GIServices may provide better IT support and services. Also, it is easier for scientific researchers to rely on the commercial products to set up the whole GIServices framework from a single package rather than to build complicated, modularized GIServices components step by step.

However, first, under the vendor-based GIServices framework, scientists with programming capabilities will not be able to customize GIServices functions by themselves unless the private vendors provide their source codes. Second, the functionality of the commercial GIServices packages will not focus on scientific research tools but on the basic map display and query functions which can attract bigger user groups, such as the AM/FM industry, local governments, and federal agencies. Third, different GIS projects developed in the scientific community may not be able to share with each other if different projects adopt different frameworks. For example, an Arctic research project may involve multiple nations and different governments. Without an open, standardized GIServices framework, it is very difficult to gather and analyze GIS data from different sources.

To sum up, the first alternative path, GISystems, is very unlikely to be the future trend because the majority of the GIS community already realizes the importance of distributed GIServices. Currently, many federal governments, GIS companies, and scientists are already providing distributed GIServices via the WWW. The second path, vendor-based GIServices, is likely to be the trend as many private software companies are pushing the GIS community in this direction. The public sector and the scientific community may not realize the serious problems that may occur if they adopt the private, vendor-based GIServices solution. The only way to prevent a privatized GIServices framework is to educate people about these negative impacts on the public and scientific community and push the private sector to adopt open and standardized GIServices frameworks in the future. Hopefully, by doing so, the GIS community will choose the third path, wherein distributed GIServices will be developed in an open and distributed software framework.

14.5 CONCLUSION

The adoption of distributed GIServices not only is a technology migration or a new framework of GIS but also has significant impact on society, economics, and daily life. Distributing GIServices on the Internet will facilitate the development of reusable, compatible, and upgradeable GIS software and databases. By adopting modularized, real-time Web-based GIServices, geographers and spatial scientists can build more realistic models to solve research problems and focus on the domain of geographic problems rather than on the mechanisms of system implementation. Geographers can utilize geographic information and share research results and models more efficiently.

In general, the deployment of GIServices will facilitate scientific data management in the geographic domain. The scientific community, through government science agencies, professional societies, and the actions of individual scientists, should improve technical organization and management of scientific data in the following ways (National Research Council, 1998, p. 3):

(a) Work with the information and computer science communities to increase their involvement in scientific information management.

(b) Support computer science research in database technology, particularly to strengthen standards for self-describing data representations, efficient storage of large data sets, and integration of standards for configuration management.

(c) Improve science education and reward systems in the area of scientific data management.

(d) Encourage the funding of data compilation and evaluation projects and of data rescue efforts for important data sets in transient or obsolete forms, especially by scientists in developing countries.

Hopefully, the network-based GIServices framework will be able to provide an effective approach for scientific data management and facilitate the creation of self-describable geospatial data sets.

One of the main goals for distributing GIServices on the Internet is to encourage people and organizations to use geographic information and to make more informed decisions. The comprehensive framework of distributed GIServices introduced in this book will ensure high-quality GIServices provided from both the public and private sectors. The power of GIServices, as decision-making tools and query engines, will be released from the GIS professionals to the public. In addition to technological implementation, challenges remain for the GIS community to address. One challenge is to make sure that everyone is connected and has access to geographic information services: "Traditionally, our notion of being connected to the nation's communications networks has meant having a telephone. . . . To be connected today increasingly means to have access to telephones, computers, and the Internet. While these items may not be necessary for survival, arguably in today's emerging digital economy they are necessary for success" (NTIA, 1999, p. 77). How can we ensure that the on-line resources are available to the public? How can we provide more friendly interfaces for information search and query? How can we encourage the GIS industry to collaborate with local communities? These questions will need to be answered in the near future by the members of the GIS community.

The above discussions have illustrated some research directions and problems for distributed GIServices in the future. All the suggestions, frameworks, and tools designed and specified in this book are aimed to make distributed GIServices more flexible, intelligent, and feasible. In this new millennium, there is a fundamental research topic implied in the design of distributed GIServices, which is the emergence of GIServices networks. In the future, the network of distributed GIServices will connect hundreds of thousands of GIS nodes and workstations together, with collaborations among millions of agents, data objects, and GIS components. Will the collaborations and interactions inside the whole network of GIServices create the emergence of geographic information science? John Holland (1998, p. 225) said that "emergence occurs in systems that are generated. The systems are composed of copies of a relatively small number of components that obey simple laws. . . . The interactions between the parts are non-linear, so the overall behavior cannot be obtained by summing the behaviors of the isolated components." Thus, when sophisticated networks are created for distributed GIServices, when thousands of GIS workstations are connected together, when millions of GIS software agents are roaming on/across the GIServices network and accessing on-line data, it is important for the GIS community to watch very closely the evolution of GIServices networks.

The next consideration for distributed GIServices is the development of standards and specifications from the GIS industry and government agencies. This book introduces the major standards of distributed GIServices developed by OGC and ISO/TC 211. These standards and specifications will have significant impacts on the future development of GIServices. However, the GIS community should be aware that these standards may not necessarily be adopted widely in the future. If these standards are not feasible for the actual GIS tasks and implementations, they may be replaced by other standards. One of the most famous standards examples is the design of IP between TCP/IP and OSI.

This drawback became very clear in the United Kingdom in the 1980s when the governments of the world were saying that OSI is the standard for networks; many companies fell into the trap of believing this and spent huge sums of money to support OSI. Of course, history shows that TCP/IP became the de facto world standard for a number of reasons, mostly pragmatic and concerning practical working and speed (OSI was so overdeveloped, it was too slow) (Murch and Johnson, 1999, p. 66).

The lessons from the history of TCP/IP indicate that no matter how carefully articulated, standards must be adopted by a wide enough range of suppliers and corporate bodies to be meaningful (Murch and Johnson, 1999). Currently, the OGC specifications and the ISO standards are under development by hundreds of experts and specialists. However, these standards and specifications have not been widely examined in actual implementation. From the authors' personal perspective, these standards will need to be revised based on real-life applications, including practical working and performance tests of actual GIS tasks to ensure feasibility. Without considering actual use and practical tasks, standards developed by OGC and ISO/TC 211 may not be adopted widely. It is important for the GIS community to have feasible standards for distributed GIServices and, at the same time, prevent the standards and specification from overdeveloping.

In the twentieth century, the visible world was reshaped by the development of engineering technology. People used technology to create skyscrapers, airplanes, space shuttles, water dams, and so on. In the twenty-first century, the invisible world is going to change by the progress of IT. The developments of the Internet, wireless phones, biochemistry engineering, DNA research, e-commerce, virtual reality, and the Web have already made our lives very different and will continue to do so. However, when technology is changing the world inside out, one needs to consider whether to use the technology: "The more interconnected a technology is, the more opportunities it spawns for both use and misuse" (Kelly, 1998, p. 45.). While enjoying the power of network-based GIServices in the twenty-first century, one may want to ask if technology is being used or misused. The GIS community will have to ensure that these new technologies will be used appropriately to facilitate research in geospatial science, the growth of the GIS industry, and better quality of life for the public.

WEB RESOURCES

Descriptions	URI
Web services	http://msdn.microsoft.com/library/techart/websvcs_paltform.htm http://www.xml.com/pub/a/2001/04/04/webservices/index.html
Digital Divide	http://www.ntia.doc.gov/ntiahome/digitaldivide
OGC Web services	http://ip.opengis.org/ows

REFERENCES

Abler, R. F. (1987). The National Science Foundation National Center for Geographic Information and Analysis. *International Journal of Geographical Information Systems*, 1(4), pp. 303–326.

Armstrong, M. P. (1997). Emerging Technologies and the Changing Nature of Work in GIS. In *Proceedings of GIS/LIS'97*, Cincinnati, Ohio, American Society for Photogrammetry and Remote Sensing, pp. 800–805.

Brodie, M. L., and Stonebraker, M. (1995). *Migrating Legacy Systems*. San Francisco: California: Morgan Kaufmann.

Coleman, D. J. (1999). Geographical Information Systems in Networked Environments. In P. A. Longley, M. F. Goodchild, and D. J. Maguire (Eds.), *Geographical Information Systems: Principles, Techniques, Applications and Management*, 2nd ed. New York: Wiley, Chapter 22, pp. 317–329.

Davies, C., and Medyckyj-Scott, D. (1996). GIS Users Observed. *International Journal of Geographical Information Systems*, 10(4), pp. 363–384.

Doyle, A., Reed, C., Harrison, J., and Reichardt, M. (2001). *Introduction to OGC Web Services: OGC Interoperability Program White Paper.* URI: http://ip.opengis.org/ows/010526_OWSWhitepaper.doc.

Fischer, M. M., Scholten, H. J., and Unwin, D. J. (Eds.). (1996). *Spatial Analytical Perspectives on GIS.* GISDATA No. IV. London: Taylor & Francis.

Foresman, T. W. (Ed.). (1998). *The History of Geographic Information Systems: Perspectives from the Pioneers.* Upper Saddle River, New Jersey: Prentice-Hall.

Goodchild, M. F., Haining, R., and Wise, S. (1992). Integrating GIS and Spatial Data Analysis: Problems and Possibilities. *International Journal of Geographical Information Systems,* 6(5), pp. 407–423.

Graham, S., and Marvin, S. (1996). *Telecommunications and the City: Electronic Spaces, Urban Places.* London: Routledge.

Harvey, D. (2000). *Spaces of Hope.* Berkeley, California: University of California Press.

Holland, J. H. (1998). *Emergence: From Chaos to Order.* Reading, Massachusetts: Addison-Wesley.

Kelly, K. (1998). *New Rules for the New Economy: 10 Radical Strategies for a Connected World.* New York: Penguin Books.

Kirtland, M. (2001). *A Platform for Web Services.* URI: http://msdn.microsoft.com/library/default.asp?url=/library/en-us/dnwebsrv/html/websvcs_platform.asp, January.

Korte, G. B. (1994). *The GIS Book,* 3rd ed. Santa Fe, New Mexico: OnWord.

Limp, W. F. (1997). Weave Maps across the Web. *GIS World,* September, 10(9), pp. 46–55.

Murch, R., and Johnson, T. (1999). *Intelligent Software Agents.* Upper Saddle River, New Jersey: Prentice-Hall.

National Research Council. (1998). *Bits of Power: Issues in Global Access to Scientific Data.* Washington, DC: National Academy Press.

National Telecommunications and Information Administration (NTIA). (1995). *Falling through the Net: A Survey of "Have nots" in Rural and Urban America.* Washington, DC: NTIA, U. S. Department of Commerce.

National Telecommunications and Information Administration (NTIA). (1998). *Falling through the Net II: New Data on the Digital Divide.* Washington, DC: NTIA, U. S. Department of Commerce.

National Telecommunications and Information Administration (NTIA). (1999). *Falling through the Net: Defining the Digital Divide: A Report on the Telecommunications and Information Technology Gap in America.* Washington, DC: NTIA, U. S. Department of Commerce. URI: http://www.ntia.doc.gov/ntiahome/digitaldivide, May 11, 2000.

Open GIS Consortium (OGC). (2001). *OGC Web Services Initiative Request for Quotation.* URI: http://ip.opengis.org/ows.

Vasudevan, V. (2001). *A Web Services Primer.* URI: http://www.xml.com/pub/a/2001/04/04/webservices/index.html, April 4.

Vckovski, A. (1998). *Interoperable and Distributed Processing in GIS*. London: Taylor & Francis.

Weiser, M. (1993). Hot Topics: Ubiquitous Computing. *IEEE Computer,* October, p. 71–72.

Yourdon, E. (1993). *Decline and Fall of the American Programmer.* Englewood Cliff, New Jersey: Prentice-Hall.

ACRONYMS

1G, . . ., 4G   First- to Fourth-Generation cellular networks

ACE	Annotated Centerline Engine
ACL	Agent Communication Language
ACP	Agent Communication Protocol
ADA	Americans with Disabilities Act
ADEPT	Alexandria Digital Earth Prototype
ADL	Alexandria Digital Library
ADSL	Asymmetric Digital Subscriber Line
ADO	ActiveX Data Object
ADRG	Arc Digitized Raster Graphics
AE	Application Entity
AI	Artificial Intelligence
ALE	Application Launching and Embedding
ALRIS	Arizona Land Resource Information System
AMPS	Advanced, or American, Mobile Phone System
ANSI	American National Standards Institute
AOL	American Online
API	Application Programming Interface
ASCII	American Standard Code for Information Interchange
ASCII CSV	ASCII Comma Separated Variable
ASP	Active Server Page
ASRF	ARC Standardized Raster Product
ASTER	Advanced Spaceborne Thermal Emission and Reflection Radiometer
Atlas BNA	Atlas Boundary File

ATM	Asynchronous Transfer Mode
AVHRR	Advanced Very High Resolution Radiometer
AVL	Automatic Vehicle Location
AVM	Automated Vehicle Monitoring
AWT	Abstract Window Tool Kit
B2B	Business to Business
BBS	Bulletin Board System
BCL	Base Class Library
BTS	Bureau of Transportation Statistics
CA	Certificate Authority
CAC	Channel Access Control
CAD	Computer-Aided Design
CADRG	Compressed Arc Digitized Raster Graphic
CCOGIF	Canadian Council of Geomatics Interchange Format
CDC	Connected Device Configuration
CDK	Conduit Development Kit
CDMA	Code Division Multiple Access
CDPD	Cellular Digital Packet Data
CERES	Clouds and Earth's Radiant Energy System
CFML	ColdFusion Markup Language
CGI	Common Gateway Interface
CGM	Computer Graphics Metafile
C-HTML	Compact HTML
CLDC	Connected Limited Devices Configuration
CLI	Call-Level Interface
CLIE	Communication, Link, Information, and Entertainment
CLR	Common Language Runtime
CLS	Common Language Specification
CNIDR	Clearinghouse for Networked Information Discovery and Retrieval
CODOT	Colorado Department of Transportation
COM	Component Object Model
CORBA	Common Object Request Broker Architecture
COTS	Commercial-off-the-Shelf
CPU	Central Processing Unit
CRC	Cyclical Redundancy Check
CRM	Customer Relationship Management
CRT	Cathode-Ray Tube
CS	Communication Service
CSDGM	Content Standard for Digital Geospatial Metadata
CSDM	Collaborative Spatial Decision Making
CSI	Communication Services Interface
CSLIP	Compressed Serial Line Internet Protocol
CSMA/CD	Carrier Sense with Multiple Access and Collision Detection

CSS	Cascading Style Sheet
CSU	Channel Service Unit
CTS	Common Types System
CVM	CVirtual Machine
DAAC	Distributed Active Archive Center
D-AMPS	Digital Advanced Mobile Phone System
DAR	Directed Asymmetric RowSet
DBMS	Database Management System
DCE	Distributed-Computing Environment
DCOM	Distributed-Component Object Model
DCP	Distributed-Computing Platform
DD	Decimal Degree
DDCF	Distributed-Document Component Facility
DDE	Dynamic Data Exchange
DDS	Digital Data Service
DEG	Display Element Generation
DEM	Digital Elevation Model
DES	Data Encryption Standard
DFS	Distributed File Service
DGI	Distributed Geographic Information
DGIWG	Digital Geographic Information Working Group
DGN	Director General de Normas
DGPS	Differential GPS
DHTML	Dynamic HTML
DLG	Digital Line Graph
DLI	Digital Library Initiative
DLL	Dynamic Link Library
DMS	Degrees, Minutes, Seconds
DNS	Domain Name System
DOD	Department of Defense
DOI	Digital Orthorectified Imagery
DOM	Document Object Model
DOQ	Digital Orthophoto Quadrangle
DOS	Denial of Service
DRDA	Distributed Relational Database Architecture
DRG	Digital Raster Graphic
DSL	Digital Subscriber Line
DSSS	Direct-Sequence Spread Spectrum
DSU	Data Service Unit
DTD	Document Type Definition
DTED	Digital Terrain Elevation Data
DWF	Drawing Web Format
DWG	AutoCAD Drawing Format
DXF	Drawing Interchange Format

EAL	Enhanced Address Layer
EBCDIC	Extended Binary Coded Decimal Interchange Code
ECW	Enhanced Compressed Wavelet
ECWP	ECW Protocol
EDCEROS	Data Center
EDGEOS	Data Gateway
EDGE	Enhanced Data Rates for GSM Evolution
EJB	Enterprise Java Bean
EM	Enterprise Manager
EOS	Earth Observing System
EOSDIS	Earth Observing System Data and Information System
EPA	U.S. Environmental Protection Agency
EPSG	European Petroleum Survey Group
ERDAS	Earth Resource Data Analysis System
EROS	Earth Resources Observation Systems
ESRI	Environmental Systems Research Institute
ERP	Enterprise Resource Planning
ETSI	European Telecommunications Standards Institute
EWP	Emergency Watershed Protection
FCC	Federal Communications Commission
FDD	Frequency Division Duplex
FDDI	Fiber-Distributed Data Interface
FDMA	Frequency Division Multiple Access
FGDC	Federal Geographic Data Committee
FHMA	Frequency-Hopped Multiple Access
FHSS	Frequency-Hopping Spread Spectrum
FTP	File Transfer Protocol
Gbps	Gigabits per second
GCMD	NASA's Global Change Master Directory
GDCA	Geospatial Data Clearinghouse Activity
GDO	Geographic Data Objects
GES	Goddard Earth Sciences
GeoTIFF	Geographic Tagged Image File Format
GID	Geographic Identifier
GIF	Graphics Interchange Format
GIS	Geographic Information Systems
GIS-T	GIS in Transportation
GLIS	Global Land Information System
GML	Geography Markup Language
GPRS	General Packet Radio Service
GPS	Global Positioning System
GSDI	Global Spatial Data Infrastructure

GSM	Global System for Mobile Communications
GUI	Graphic User Interface
HDML	Handheld Device Market Language
HPERLAN	High-Performance European Radio LAN
HS	Human Interaction Services
HSCSD	High-Speed Circuit Switched Data
HTI	Human Technology Interface
HTML	HyperText Markup Language
HTTP	HyperText Transfer Protocol
HTTPD	HTTP Demon
HUD	Housing and Urban Development
IAS	Internet Application Server
iDEN	Integrated Digital Enhanced Network
IDL	Interface Definition Language
IEC	International Electrotechnical Comission
IEEE	Institute of Electrical and Electronics Engineers
IHO	International Hydrographic Organization
IID	Interface ID
IIOP	Internet Inter-ORB Protocol
IIS	Internet Information Server
IMS	Internet Map Server
I/O	Input/Output
IOR	Interoperable Object Reference
IP	Internet Protocol
IPMA	Internet Performance Measurement and Analysis
IPX	Internetwork Packet Exchange
IrDA	Infra-red Data Association
IRS	Indian Remote Sensing
ISAPI	Internet Server Application Program Interface
ISDN	Integrated Services Digital Network
ISI	Information Services Interface
ISO	International Standards Organization
ISP	Internet Service Provider
IT	Information Technology
ITS	Intelligent Transportation System
ITU	International Telecommunications Union
J2EE	Java 2 Enterprise Edition
J2ME	Java 2 Micro Edition
J2SE	Java 2 Standard Edition
JAR	Java Archive
JDBC	Java Database Connectivity

JDC	Japanese Digital Cellular
JDK	Java Development Kit
JIT	Just-in-Time
JPEG	Joint Photographic Experts Group
JSP	Java Server Page
KIF	Key Index Files
KQML	Knowledge Query and Manipulation Language
KVMK	Virtual Machine
LAN	Local-Area Network
LBS	Location-Based Service
LCP	Link Control Protocol
LIHTC	Low-Income Housing Tax Credit
LLC	Logical Link Control
LSN	Large-Scale Networking
MAC	Media Access Control
MACDIF	Map And Chart Data Interchange Format
MAN	Metropolitan-Area Network
MCDA	Micro Channel Developers Association
MCDN	Microcellular Data Network
MC&G	Mapping, Charting, and Geodesy
MDF	Map Definition File
MExE	Mobile Execution Environment
MID/MIF	MapInfo Data/Map Info Interchange File
MIDP	Mobile Information Device Profile
MIM	Management Information System
MIME	Multipurpose Internet Mail Extensions
MIS	Management Information System
MISR	Multi-angle Imaging Spectroradiometer
MME	Microsoft Mobile Explorer
MMIT	Microsoft Mobile Internet Took Kit
MODIS	Moderate-Resolution Imaging Spectroradiometer
MOM	Message-Oriented Middleware
MOPITT	Measurements of Pollution in the Troposphere
MPC	Mobile Positioning Center
MPO	Metropolitan Planning Organization
MrSID	Multi-Resolution Seamless Image Database
MS	Model/Information Management Services
MSAU	Multistation Access Unit
MSC	Mobile Switching Center
MSN	Microsoft Network
MTBF	Mean Time Between Failure
MTS	Microsoft Transaction Server

MTTR	Mean Time to Restore Services
MTU	Maximum Transmission Unit
MWF	Map Window File
MWX	Map Window XML
NAP	Network Access Point
NASA	U.S. National Aeronautics and Space Administration
NATO	North Atlantic Treaty Organization
NCP	Network Control Protocol
NDIS	Network Driver Interface Specification
NetBEUI	NetBIOS Extended User Interface
NGDC	U.S. National Geospatial Data Clearinghouse
NGI	Next Generation of the Internet
NIC	Network Interface Cards
NII	National Information Infrastructure
NIMA	National Imagery and Mapping Agency
NISE	Near-Real-Time Ice and Snow Extent
NIST	National Institute of Standards and Technology
NNI	Network-to-Network Interface
NOAA	National Oceanic Atmospheric Administration
NOS	Network Operating System
NRCS	Natural Resources Conservation Service
NSAPI	Netscape Server Application Program Interface
NSDI	National Spatial Data Infrastructure
NSF	National Science Foundation
NTIA	National Telecommunications and Information Administration
OC	Optical Carrier
ODBC	Open Database Connectivity
ODL	Object Definition Language
ODP	Open Distributed Processing
OGC	OpenGIS Consortium
OGF	Open GRASS Foundation
OGIS	Open Geodata Interoperability Specification
OGM	Open Geodata Model
OLE	Object Linking and Embedding
OLE DB	Object Linking and Embedding Database
OLTP	On-Line Transaction Processing
OMA	Object Management Architecture
OMB	Office of Management and Budget
OMG	Object Management Group
OO	Object Oriented
OOM	Object-Oriented Modeling
OOUI	Object-Oriented User Interface
ORB	Object Request Broker

OSD	OnSite Drawing
OSE	Open Systems Environment
OSI	Open Systems Interconnection
OSM	OnSite Markup
OTM	Object Transaction Monitor
P2P	Peer to Peer
PCMCIA	Personal Computer Memory Card International Association
PCS	Personal Communication Service
PDA	Personal Digital Assistant
PDC	Personal Digital Cellular
PDE	Position-Determining Equipment
PDF	Portable Document Format
PDM	Product Data Management
PDN	Public Data Network
PERL	Practical Extraction and Reporting Language
PIT	Public Interface Terminal
PKI	Public Key Infrastructure
PMR	Private Mobile Radio
PNG	Portable Network Graphics
POMS	Property Owners and Managers Survey
POTS	Plain Old Telephone Service
PPGIS	Public Participation GIS
PPP	Point-to-Point Protocol
PQA	Palm Query Application
PS	Processing Services
PSTN	Public Switched Telephone Network
PUSH	Publish-and-Subscribe Glue
PVA	Property Valuation Administration
PVC	Permanent Virtual Circuit
PVT	Position/Velocity/Time
QOS	Quality of Service
RAD	Rapid Application Development
RAM	Random-Access Memory
RAS	Remote Access Service
RD-LAP	Radio Data Link Access Protocol
RDT	Regional Transportation District
REGIS	Research Program in Environmental Planning and GIS
RF	Radio Frequency
RFI	Request for Information
RFP	Requests for Proposal
RFQ	Request for Quotation
RIC	Raster Image Catalog
RMI	Remote Method Invocation

RML	Redline Markup Language
RM-ODP	Reference Model for ODP
ROM	Read-Only Memory
ROMANSE	Road Management System for Europe
RPC	Remote Procedure Call
RSA	Rivest, Shamir, and Adleman Key System
RTA	Regional Transit Authority
RTF	Rich Text Format
SAIF	Safe Access Inspection Fitting
SAX	Simple API for XML
SCOTS	Standards-Based Commercial-off-the-Shelf
SDE	Spatial Data Engine
SDF	Spatial Data File
SDH	Synchronous Digital Hierarchy
SDK	System Development Toolkits
SDP	Spatial Data Provider
SDSU	San Diego State University
SDTS	Spatial Data Transfer Standard
SGML	Standard Generalized Markup Language
SHP	Shape File
SIAS	Smallworld Internet Application Server
SIF	Spatial Index File
SIG	Special Interest Groups
SLIP	Serial Line Internet Protocol
SMDS	Switched Multimegabit Data Service
SMP	Symmetric Multiprocessing
SMS	Short Message Service
SMTP	Simple Mail Transfer Protocol
SNA	Systems Network Architecture
SNMP	Simple Network Management Protocol
SOAP	Simple Object Access Protocol
SOCDS	State of the Cities Data Systems
SONET	Synchronous Optical Network
SQL	Structured Query Language
SQL/CLI	SQL Call-Level Interface
SRI-NIC	Stanford Research Institute's Network Information Center
SRS	Spatial Reference System
SS	System Management Service
SST	Sea Surface Temperature
STS	Synchronous Transfer Signal
STS-1	Synchronous Transport Signal level-1
SVG	Scalable Vector Graphic
TC	Technical Committee
TCP	Transmission Control Protocol

TCP/IP	Transmission Control Protocol/Internet Protocol
TCSP	Time-Dependent Constrained Shortest Path
TDD	Time Division Duplex
TDMA	Time Division Multiple Access
TETRA	Trans-European Trunked Radio
TIFF	Tagged Image File Format
TIGER	Topologically Integrated Geographic Encoding and Referencing
TIN	Triangular Irregular Network
TP	Transaction Processing
UARS	Upper Atmosphere Research Satellite
UCAID	University Corporation for Advanced Internet Development
UDA	Universal Data Access
UDDI	Universal Discovery, Description, and Integration
UML	Unified Modeling Language
UMTS	Universal Mobile Telecommunications System
UPS	Uninterrupted Power Source
URI	Uniform Resource Identifier
URL	Uniform Resource Locator
USB	Universal Serial Bus
USDC	U.S. Digital Cellular
USGS	U.S. Geological Survey
USRP	UPS Standardized Raster Product
UTP	Unshielded Twisted Pair
UWC	Universal Wireless Communications
vBNS	Very High Performance Backbone Network Service
VM	Virtual Machine
VMDS	Smallworld Version Managed Data Store
VML	Vector Markup Language
VPF	Vector Product Format
VPN	Virtual Private Network
W3C	World Wide Web Consortium
WAE	Wireless Application Environment
WAIS	Wide-Area Information Server
WAN	Wide-Area Network
WAP	Wireless Application Protocol
W-CDMA	Wideband CDMA
WG	Working Group
WKTs	Well-Known geospatial data Types
WLAN	Wireless LAN
WML	Wireless Markup Language
WMS	Web Map Server

WMT	Web Mapping Testbed
WPPS	Web-Based Public Participation System
WS	Workflow/Task Services
WSDL	Web Service Definition Language
WTA	Wireless Telephony Application
WWW	World Wide Web

XHTML	eXtensible HTML
XHTMLMP	XHTML Mobile Profile
XLink	XML Linking Language
XML	eXtensible Markup Language
XPointer	XML Pointer Language
XQL	XML Querry Language
XSD	XML Schema Definition
XSL	Extensible Stylesheet Language
XSLT	XML Transformation Language

INDEX